T0175902

Climate Change and Displacement Reader

Estimates put the future scale of human displacement as a result of climate change at anywhere from 150 million to one billion persons. Despite this crisis in the making, few countries or international institutions are adequately prepared to address and resolve this emerging human rights crisis.

This compilation brings together 50 of the leading texts on climate change and displacement. It provides a consolidated source and substantive overview of the key issues relating to climate change and displacement, including:

- the reality of climate displacement
- the shape of current and proposed international law on this matter
- the institutional and governance framework that will address and respond to this crisis
- an analysis of what a cross-section of governments and civil society organizations are already doing to prepare for and act against climate displacement.

This volume is an indispensable resource for academics researching this issue, community and international advocates, officials of the United Nations and international human rights and humanitarian organizations, national and municipal governments as well as all people concerned about the human rights of climate displaced persons throughout the world.

Scott Leckie is the Director and Founder of Displacement Solutions (www. displacementsolutions.org). He frequently lectures and teaches several human rights courses.

Ezekiel Simperingham is an international human rights lawyer.

Jordan Bakker is currently a student at Monash University Law School, and works as a research associate with Displacement Solutions.

Climate Change and Displacement Reader

Edited by
Scott Leckie, Ezekiel Simperingham and Jordan Bakker

publishing for a sustainable future

First published 2012
by Earthscan
2 Park Square, Milton Park, Abingdon, Oxon OX14 4RN

Simultaneously published in the USA and Canada
by Earthscan
711 Third Avenue, New York, NY 10017

Earthscan is an imprint of the Taylor & Francis Group, an informa business

© 2012 selection and editorial material, Scott Leckie, Ezekiel Simperingham and Jordan Bakker; individual chapters, the contributors

The right of the editors to be identified as the authors of the editorial material, and of the authors for their individual chapters, has been asserted in accordance with sections 77 and 78 of the Copyright, Designs and Patents Act 1988.

All rights reserved. No part of this book may be reprinted or reproduced or utilised in any form or by any electronic, mechanical, or other means, now known or hereafter invented, including photocopying and recording, or in any information storage or retrieval system, without permission in writing from the publishers.

Trademark notice: Product or corporate names may be trademarks or registered trademarks, and are used only for identification and explanation without intent to infringe.

British Library Cataloguing in Publication Data
A catalogue record for this book is available from the British Library

Library of Congress Cataloging in Publication Data
Leckie, Scott.
  Climate change and displacement reader / Scott Leckie, Ezekiel Simperingham, and Jordan Bakker.
      p. cm.
  Includes bibliographical references and index.
  1. Environmental refugees. 2. Forced migration. 3. Climatic changes--Social aspects.
  I. Simperingham, Ezekiel. II. Bakker, Jordan. III. Title.
  HV640.L36 2012
  362.87–dc23   2011024023

ISBN: 978-0-415-69133-8 (hbk)
ISBN: 978-0-415-69134-5 (pbk)
ISBN: 978-0-203-15228-7 (ebk)

Typeset in Perpetua
by HWA Text and Data Management, London

For climate-displaced persons everywhere
Past ... current ... future

Contents

Foreword

Climate change presents the world with daunting challenges. However, perhaps uniquely among global threats, it also presents opportunities to act preemptively to reduce its impact.

In terms of challenges, UNHCR, and the international community more broadly, are deeply concerned by the human displacement that is projected to occur as a consequence of climate change. Even the most conservative predictions would expand the world's displaced population in the coming decades to a level at least three times the current number. Estimates on the upper end would put the number of displaced twenty times higher than the world's refugee and IDP population in 2010.

Clearly, much remains to be done to ensure that the world – particularly the most vulnerable nations and citizens – is adequately prepared.

This timely and useful Reader, compiled by Scott Leckie, Ezekiel Simperingham and Jordan Bakker of Displacement Solutions, provides an overview of many of the key issues arising in connection with climate-induced displacement, including the legal and institutional dimensions.

While some of the available literature documents the fact that migration and displacement are forms of adaptation to climate change, a large amount of work remains to be done to understand the dynamics of such movements and their implications for the humanitarian and development communities.

Given the global nature of the threat, we – as an international community – must come together to address climate change. In so doing, a key concern must be to ensure that the rights of all of those affected are respected and protected in full.

Volker Türk
Director, Division of International Protection (DIP)
United Nations High Commissioner For Refugees
7 May 2011, Geneva

Acknowledgements

The editors are grateful to many persons, each of whom played a part in the completion of this Reader. First, we would like to thank Volker Türk of UNHCR for his excellent foreword and insightful views on the role of the international community in addressing climate displacement. We would also like to express our gratitude to each of the authors and publishers who have provided their approval for re-printing their works in this volume. Special thanks go both to Manuel Jacob of the Federal Department of Foreign Affairs of the Government of Switzerland for financially supporting Displacement Solutions to prepare this book.

Scott Leckie, Ezekiel Simperingham and Jordan Bakker
May 2011

Introduction

Estimates routinely put the future scale of human displacement as a result of the effects of climate change at anywhere from 150 million to one billion persons. However, despite this impending human rights crisis, few people, countries or institutions are adequately equipped to prepare for and respond to this mass displacement.

In part this is a result of the immense controversy surrounding this issue. Indeed, there is not only dispute about the scope of this future crisis as well as the appropriate legal, institutional and humanitarian preparation and response, but also the very existence of climate change displacement.

To date, while a number of publications have alerted the world to the challenge of climate change displacement and the global nature of this crisis, few of these publications have focused on solutions to the human rights dimensions of this global conundrum.

Human displacement caused by the consequences of climate change – rising sea levels, increasing storm frequency and intensity, flooding, drought, famine and other environmental degradation – will cause, and has already begun to cause, forced displacement at a truly global level.

All of the estimated 150 million to one billion persons who will be displaced will require respect for, protection of and fulfilment of the full range of their human rights. Climate displaced persons will especially require housing, land, property and livelihood solutions to their temporary or potentially permanent displacement.

In preparing this compilation, the editors have selected the full texts or extracts of 50 publications that address various key aspects of climate change and displacement. We have endeavoured to provide a comprehensive overview to the key issues; however, our particular focus has been on those authors and institutions who have proposed constructive and practical solutions to address the human rights dimensions of this crisis, wherever these solutions might occur in the world and at whatever community, national, regional or institutional level that they need to be addressed.

This compilation is divided into five parts, respectively addressing: the reality of climate change and displacement (Part 1), the international legal and institutional framework (Part 2), proposed new legal standards (Part 3), affected countries (Part 4), and community and NGO responses and proposed solutions (Part 5).

In selecting these five themes we have sought to represent the key concerns, the central issues of dispute and the next steps required to deal adequately with climate change and displacement.

Part 1 includes seven contributions, each of which addresses and explores the reality of climate change and displacement. This part provides an overview and analysis of the broader dynamics of climate displacement, the likely scale and scope of this crisis, the groups who will be especially vulnerable to climate displacement from a human rights perspective, as well as the potential implications for regional and global peace and security resulting from this mass displacement of people.

Part 2 outlines the international legal and institutional framework relevant to climate displacement, particularly focusing on refugee law, human rights law and the law on statelessness. This section begins by addressing the relevant normative and institutional framework and identifying the key gaps that need to be addressed in order to provide for comprehensive and effective protection of climate displaced persons. The section then provides an overview of the approach of various institutions – including the Security Council and the Human Rights Council – and concludes with three selected climate change declarations and standards that could be used as the inspiration for effective legal and institutional solutions to climate displacement.

Part 3 includes substantive proposals for addressing the gaps in the legal framework for the protection of climate displaced persons. The two contributions address various proposals for a new or modified legal framework, including: a separate legal and political regime created under a protocol to the UNFCCC; a regime utilizing the Guiding Principles on Internal Displacement as a model for aggregating and adapting existing human rights principles to protect the rights of climate displaced persons; as well as a detailed proposal for a new and comprehensive international convention for the protection of climate displaced persons.

Part 4 explores direct national experiences in grappling with current and future displacement caused by the effects of climate change. This section includes country-specific contributions on Australia, Bangladesh, India, Indonesia, Kiribati, Maldives, New Zealand, Papua New Guinea (Carteret Islands), Tuvalu, Vanuatu and Vietnam. These individual contributions reveal the extent to which domestic measures have already been taken in many of the regions of the world to address past, current and future displacement caused by climate change. This section also highlights the ad-hoc nature of many measures taken to date and the absence of adequate preparation for this crisis by the vast majority of countries and international institutions.

Part 5 concludes with five contributions prepared by community-based organizations and non-governmental organizations that outline strategies to prepare for and address the human rights dimensions of climate displacement. Much of the most creative and solution-orientated work on climate displacement in 2011 is being undertaken at the level of communities already being directly affected by the effects of climate change. This section emphasizes the need for governments and international institutions to take immediate and effective steps to prepare for this impending human rights crisis.

* * *

Obviously, the editors have been selective in choosing the sources presented here, and we apologize from the outset to everyone who has written on the subject of climate displacement but whose excellent contributions we were not able to include in this volume.

In selecting the analyses that have been reproduced below, we have aimed to provide a broad range of views on the complex issue of people and communities being uprooted due to the consequences of climate change. We have attempted, and hopefully succeeded, in putting together a consolidated text that will provide both decision-makers and advocates with a solid cross-section of the information required to comprehensively address and solve climate displacement wherever it may occur.

In terms of the country-specific works selected, we have decided to focus on displacement caused by sea-level rises and storm inundation. We are certainly fully cognizant of the fact that extensive displacement is already occurring due to a wide multitude of other environmental causes, including climate change. Nonetheless, we have chosen to emphasise those cases of current or future displacement which may, in fact, prove the most difficult to resolve; in particular when domestic relocation or resettlement within the borders of a state is no longer possible either due to the non-existence of territory as in the case of a small island nation such as Tuvalu or Kiribati, or when land shortages within an affected country are so severe that alternative land for climate displaced persons is not possible to access, such as in Bangladesh.

If we attempt to identify what motivated us to compile this Reader, it is our strong belief that, far from being an impossible task leading inevitably to global catastrophe, resolving forced climate displacement is possible and can be achieved within a human rights-based framework. We see the global climate displacement crisis not only as an outcome of decades of political and ecological mis-management by the world's most polluting nations, but also as an opportunity for governments and civil society the world over to build the policies, laws and projects required to simultaneously mitigate climate change and to ensure that adaptation measures to climate change have at their core the resolution of any displacement that occurs.

We believe that in the vast majority of cases, as severe and daunting as these problems are, that local and national solutions to climate displacement, which are fully consistent with the human rights of those affected, can be found; that is, if the political will can be generated to do so. This does not mean that such solutions will invariably be either cheap or able to be implemented quickly. What it does imply, however, is that appropriate long-term planning, targeted resource allocations and well-organised popular movements can together determine what measures are required, where resources can be found, and ultimately, how to achieve them.

In this light, we are particularly inspired by the emergence of a growing number of community-based organizations throughout the world that are uniformly dedicated to working with communities affected by climate change to find permanent new homes and lands for this rapidly expanding group of displaced persons. The work of groups such as Tulele Peisa from the Carteret Islands in Papua New Guinea, the Association for Climate Refugee in Bangladesh and the Kiribati Climate Action Network to name a few, are extraordinary efforts of grassroots-led movements filling the gaping holes in law and policy left by governments too ill-equipped or unmoved to act to protect the rights of those forced to flee their ancestral homes.

Similarly, we are enlivened by citizen-led gestures of solidarity from around the world in support of climate-displaced persons. Despite the fact that many of these efforts have come about solely due to government intransigence and ineptitude, as well as pure self-interest, these offers of goodwill and solutions can act as powerful inspirations for more universal approaches to resolving climate displacement. When the Torres Strait Islanders in Australia offered land on certain islands to the people of Tuvalu we were greatly heartened. When the people of Nauru offered to provide massive amounts of soil to Pacific island nations to use in land reclamation efforts we were inspired. When the government of East Timor publicly indicated their willingness to provide land to climate-displaced people and when Indonesia undertook to provide uninhabited islands for lease to affected populations, we were equally encouraged. In these and many other cases, we have found reason to be as hopeful as we are alarmed at the prospects of climate displacement for they reveal – again – that as a planet we can solve this ever growing calamity.

As serious a problem as climate displacement is, citizens, governments and the international community working together in a coordinated and forthright manner, have the resources and skills necessary to secure all human rights, in particular, housing, land and property rights, to everyone affected by climate change. The parameters for such action have already begun to take shape. Countries everywhere need to begin planning today for the looming spectre of climate displacement; every government should have in place not only adaptation plans of action, but displacement plans of actions as well.

Every country, with a few notable exceptions, can begin today to set-aside land through processes of social land-banking programmes to ensure that the land required to relocate climate displaced persons is available in the short, medium and longer terms. Every country can today review its domestic law to determine if the current legal code is capable of grappling with climate displacement in a way that explicitly protects the rights of those affected. Likewise, every government can (and should) establish public agencies that will be responsible for looking after climate displaced persons and providing solutions for them. Every country should equally realize that popular grassroots movements, comprised of climate displaced persons, their advocates, NGOs and others, may hold out the best hope in the quest for housing, land and property solutions for the climate affected population. Such movements and the governments responsible for the territory where they operate can convene national climate displacement dialogues with a view to developing one, five and ten year action plans to address the precise needs of every person, family and community likely to be displaced due to climate change.

These and many other steps can be initiated immediately, and it is our fervent hope that this book will gently nudge the lethargic into action, and the already active ever further in the right direction. Although climate displacement, in so many graphic ways, is a problem the like of which humankind has rarely, if ever, faced, the ingredients required to solve it are – for the most part – already in place. Human rights laws faithfully respected and enforced, can go a long way in this regard, as can pro-active political, social and economic planning by governments the world over. Existing international organizations responsible for refugees and displaced persons, most notably UNHCR, can be mandated to expand their protective reach in support of climate displaced persons who are unable to find durable domestic solutions to their plight. Similarly, citizen groups of all types can emerge to do the job when governments and the UN are unable or otherwise unwilling to do it themselves.

The next decades will not be easy as far as climate displacement is concerned. Of the 6000 islanders in Papua New Guinea who require new lands and livelihoods on Bougainville (to where the government has agreed to resettle them), only 21 have thus far found permanent solutions. Of the millions already displaced in Bangladesh, very few have found or been offered long-term resettlement or relocation options other than finding minute amounts of space in the already overcrowded (and effectively full) slums of Dhaka, Cox's Bazaar and Chittagong. Slowly, one by one, citizens of Kiribati, Tuvalu, the Maldives and other small island nations are finding their ways to destinations overseas; some through organized labour mobility programmes and others simply on their own with family members who have already emigrated or as adherents to the world's growing global migration. No government or international agency to date has done everything it can or must do to address all of the features of climate displacement, but it is hoped that this book will act as a catalyst to those who have thus far failed to act.

PART 1

The reality of climate displacement

Chapter 1

Care, CIESIN, UNHCR, UNU-EHS and the World Bank

IN SEARCH OF SHELTER
Mapping the Effects of Climate Change on Human Migration and Displacement

[…]

Key findings

- Climate change is already contributing to displacement and migration. Although economic and political factors are the dominant drivers of displacement and migration today, climate change is already having a detectable effect.
- The breakdown of ecosystem-dependent livelihoods is likely to remain the premier driver of long-term migration during the next two to three decades. Climate change will exacerbate this situation unless vulnerable populations, especially the poorest, are assisted in building climate-resilient livelihoods.
- Disasters continue to be a major driver of shorter-term displacement and migration. As climate change increases the frequency and intensity of natural hazards such as cyclones, floods, and droughts, the number of temporarily displaced people will rise. This will be especially true in countries that fail to invest now in disaster risk reduction and where the official response to disasters is limited.
- Seasonal migration already plays an important part in many families' struggles to deal with environmental change. This is likely to become even more common, as is the practice of migrating from place to place in search of ecosystems that can still support rural livelihoods.
- Glacier melt will affect major agricultural systems in Asia. As the storage capacity of glaciers declines, short-term flood risks increase. This will be followed by decreasing water flows in the medium- and long-term. Both consequences of glacier melt would threaten food production in some of the world's most densely populated regions.
- Sea level rise will worsen saline intrusions, inundation, storm surges, erosion, and other coastal hazards. The threat is particularly grave vis-à-vis island communities. There is strong evidence that the impacts of climate change will devastate subsistence and commercial agriculture on many small islands.

- In the densely populated Ganges, Mekong, and Nile River deltas, a sea level rise of 1 meter could affect 23.5 million people and reduce the land currently under intensive agriculture by at least 1.5 million hectares. A sea level rise of 2 meters would impact an additional 10.8 million people and render at least 969 thousand more hectares of agricultural land unproductive.
- Many people won't be able to flee far enough to adequately avoid the negative impacts of climate change—unless they receive support. Migration requires resources (including financial, social, and political capital) that the most vulnerable populations frequently don't have. Case studies indicate that poorer environmental migrants can find their destinations as precarious as the places they left behind.

Policy Recommendations

New thinking and practical approaches are needed to address the threats that climate-related migration poses to human security. These include the following principles and commitments for action by stakeholders at all levels.

Avoid dangerous climate change

Reduce greenhouse gas emissions to safe levels

The international community has until December 2009, at the Conference of Parties to the United Nations Framework Convention on Climate Change (UNFCCC), to agree on a way forward. If this deadline isn't met, we will almost surely shoot past any safe emissions scenario and commit future generations to a much more dangerous world in which climate change-related migration and displacement, on a truly massive scale, is unavoidable.

Focus on human security

Protect the dignity and basic rights of persons displaced by climate change

Climate-related displacement and migration should be treated, first and foremost, as a "human security" issue. Sensationalist warnings must not be permitted to trigger reactionary policies aimed at blocking the movement of "environmental refugees" without genuine concern for their welfare.

Invest in resilience

Increase people's resilience to the impacts of climate change so that fewer are forced to migrate

The breakdown of natural-resource dependent livelihoods is likely to remain the premier driver of long-term migration during the next two to three decades. Climate change will exacerbate the situation unless vulnerable populations, especially the poorest, are assisted in building climate-resilient livelihoods. This will require substantial investment in:

- in situ adaptation measures including, for instance, water-wise irrigation systems, low/no-till agricultural practices, income diversification, and disaster risk management;
- the empowerment of women and other marginalized social groups to overcome the additional barriers they face to adaptation; and
- inclusive, transparent, and accountable adaptation planning with the effective participation of especially vulnerable populations.

Prioritize the world's most vulnerable populations

Establish mechanisms and binding commitments to ensure that adaptation funding reaches the people that need it most

Negotiations under the United Nations Framework Convention on Climate Change (UNFCCC) are currently focused on how to generate sufficient funds for adaptation in developing countries and how the funds should be managed. These are important questions. However, it is equally important to determine how funds will be channeled so that they reach the people who need them most. Objective criteria for assessing vulnerability to the negative impacts of climate change—including people's risk of displacement—should be developed to guide priority assistance.

Include migration in adaptation strategies

Recognize and facilitate the role that migration will inevitably play in individual, household and national adaptation strategies

For millennia, people have engaged in long- and short-term migration as an adaptive response to climatic stress. Millions of individuals and households are employing a variant of this strategy today. Human mobility—permanent and temporary, internal and cross border—must be incorporated into rather than excluded from international and national adaptation plans. This can be done in a variety of ways at a number of levels and may include:

- measures to facilitate and strengthen the benefits of migrant remittances;
- the rights-based resettlement of populations living in low-lying coastal areas and small island states.[1]

Environmentally, socially and economically sustainable resettlement meeting human rights standards (as reflected inter alia in the Guiding Principles on Internal Displacement) can be costly; and international agreements must address how these and related needs will be met. Existing mechanisms for adaptation funding, which rely on voluntary contributions, have failed to deliver. Therefore, future agreements under the UN Framework Convention on Climate Change must establish binding commitments for historic high emitters. These funds must be new and additional to existing commitments, such as those for Official Development Assistance.

Close the gaps in protection

Integrate climate change into existing international and national frameworks for dealing with displacement and migration

The unique challenges posed by climate change must be factored into norms and legal instruments dealing with displacement and migration. Especially important conundrums surround:

- disappearing states and non-viable homelands. Unlike some people displaced by conflict or persecution who may one day return home, those displaced by the chronic impacts of climate change (e.g. inadequate rainfall and sea level rise) will require permanent resettlement.
- irrevocably deteriorating living conditions. Climate change will result in cases that do not fit into current distinctions between voluntary and forced migration. At present, people who move due to gradually worsening living conditions may be categorized as voluntary economic migrants and denied recognition of their special protection needs.

In order to satisfactorily address such challenges, duty-bearers will need clear guidelines for protecting the rights of environmentally-induced migrants.

Strengthen the capacity of national and international institutions to protect the rights of persons displaced by climate change.

Institutions tasked with protecting the basic rights of migrants and displaced persons are already under-funded and overstretched. Climate change will add to their strain, making the practice of protection even more difficult. The international community must, therefore, begin substantial discussions about how to realize its duties to protect migrants and displaced persons under conditions of radical environmental change.

[…]

Conclusions

Climate change is happening with greater speed and intensity than initially predicted.[110,111] Safe levels of atmospheric greenhouse gases may be far lower than previously thought, and we may be closer to an irreversible tipping point than had been anticipated.[112] Meanwhile, global CO_2 emissions are rising at steeper and steeper rates.[113] Emissions reductions efforts have been too little, too late. Therefore, the challenges and complex politics of adaptation are joining those of mitigation at the centre of policy debates. One of the most important issues to address is how climate change will affect human migration and displacement—and what we will do about it.

There are many messages to be taken from the empirical evidence and maps presented in this Report. The following are especially important.

Environmental change, displacement and migration

The reasons why people migrate are complex but frequently reflect a combination of environmental, economic, social, and/or political factors. The influence of environmental change on human mobility is discernible and growing. Current and projected estimates vary widely, with figures ranging from 25 to 50 million by the year 2010 to almost 700 million by 2050. The International Organisation for Migration (IOM) takes the middle road with an estimate of 200 million environmentally induced migrants by 2050.

Livelihoods and human mobility

Environmental change is most likely to trigger long-term migration when it undermines the viability of ecosystem-dependent livelihoods (such as rainfed agriculture, herding and fishing) and there are limited local alternatives. The degradation of soil, water and forest resources, as well as the direct impacts of climate change (e.g. shifting rainfall), are playing important roles in emergent patterns of human mobility.

Differential vulnerability

People's vulnerability to environmental change reflects a combination of their exposure, sensitivity and adaptive capacity. As a result, degree of vulnerability varies widely within countries, communities and even households. For instance, poor people's exposure to the impacts of climate change is often higher than others because economic and political forces confine them to living in high-risk landscapes (e.g. steep hillsides prone to slippage). Meanwhile, one of the most important factors shaping adaptive capacity is people's access to and control over natural, human, social, physical, political and financial resources. Their striking lack of these things is a major reason why poor people – especially those in marginalised social groups – are much more vulnerable to the impacts of climate change than others.

Women contend with an especially wide array of constraints on their adaptive capacity. Gendered roles, as well as cultural prescriptions and prohibitions, make it far more difficult for most women and female-headed households to migrate in response to environmental change.

Government action and risks

Some forms of environmental change, including sea-level rise and glacier melt, may require large-scale government action. However, interventions can leave people no better off, or even worse, than before. For instance, the government of Vietnam is currently relocating some people living in areas threatened by riverbank erosion, flooding and storm surges. Though the intention is commendable, resettlement can carry high costs including cultural degradation, lost livelihoods, reduced access to social services, and the loss of employment networks. In sum, top-down responses to environmental change carry substantial risk, including the risk of "mal-adaptation."

The importance of inclusive, transparent and accountable adaptation processes

The scale of current and projected environmental changes necessitates a crucial role for central governments. Yet we have learnt from experience that benefits can be maximised and risks minimised if vulnerable populations are meaningfully involved in the planning, implementation, monitoring and evaluation of coordinated responses to environmental change.

This points towards one of the most important conclusions to draw from this report. Namely, that the scope and scale of challenges we face may be unprecedented; but we meet them already having many of the resources—including knowledge, skills and relationships— needed to protect the dignity and basic rights of persons threatened by displacement from environmental change.

[…]

Care, CIESIN, UNHCR, UNU-EHS and World Bank Social Dimensions of Climate Change, '*In Search of Shelter: Mapping the Effects of Climate Change on Human Migration and Displacement'*, May 2009, pp. 6–7 and 26–29.

Notes

1   For rights-based approach see Kolmannskog, V. 2009. Dignity in disasters and displacement— exploring law, policy and practice on relocation and return in the context of climate change. Paper prepared for the GECHS Synthesis Conference, "Human Security in an Era of Global Change," June 22–24, 2009, University of Oslo, Norway. In relation to the use of the term "resettlement" in this report, it is not restricted to the meaning it has assumed in refugee law and policy.

[…]

110   McCarthy, M. 2007. Earth's natural defenses against climate change 'beginning to fail.' The Independent, 18 May.

111   Alexander, C. 2008. World may be heating quickly: Scientist. The Sydney Morning Herald, 7 May. http://news.smh.com.au/world-may-be-heating-quickly-scientist/20080507-2bul.html.

112   McKibben. B. 2007. Remember this: 350 parts per million. Washington Post, 28 December.

113   ScienceDaily. 2008. Greenhouse gases, carbon dioxide and methane, rise sharply in 2007. 24 April. http://www.sciencedaily.com/releases/2008/04/080423181652.htm.

Karen O'Brien, Linda Sygna, Robin Leichenko, W. Neil Adger, Jon Barnett, Tom Mitchell, Lisa Schipper, Thomas Tanner, Coleen Vogel and Colette Mortreux

DISASTER RISK REDUCTION, CLIMATE CHANGE ADAPTATION AND HUMAN SECURITY
A Commissioned Report for the Norwegian Ministry of Foreign Affairs

[...]

3. Vulnerability to Climate Change and Extreme Events

> ... understanding who is vulnerable, and why, can help us to prevent our neighbour's home from washing into the sea, a family from suffering hunger, a child from being exposed to disease and the natural world around us from being impoverished. All of us are vulnerable to climate change, though to varying degrees, directly and through our connections to each other. (Leary et al. 2008)[60]

Climate change is associated with a myriad of socioeconomic and biophysical shifts, but potential and projected changes in climate variability, including increases in extreme event frequency or intensity, is well recognized as a central societal concern.[61] This has led to a growing body of research on the aggregate estimates of the economic and social costs of climate change in terms of human mortality and morbidity, GDP, infrastructure, and capital resources that may be affected by extreme events.[62] There is also a growing recognition of the need to prepare for and manage the effects of extreme weather events under climate change.[63] Although technical responses related to hazards and climate impacts have long been considered important, over the past decades attention has shifted to a focus on vulnerability, and particularly on the role that climate change adaptation and disaster risk reduction can play in reducing vulnerability to climate variability, hazards and extreme events.

It is important to note that definitions, conceptualizations and interpretations of vulnerability differ both between and within the disaster risk and climate change communities.[64] Several definitions of vulnerability are presented in Box 2. The IPCC definition focuses on vulnerability as a function of: 1) exposure to a climate risk; 2) sensitivity or susceptibility to damage; and

3) adaptive capacity, including the capacity to recover from impacts.[65] Vulnerability can also be explained by different causal factors, including biogeophysical and technological conditions, institutional failures, and social, economic and political conditions and inequalities.[66] A "physical vulnerability" approach emphasizes biogeophysical and technological interpretations that relate vulnerability to locations in high-risk areas (e.g., low-lying coastal areas), high concentrations of population and physical capital in small areas, a dependency on large-scale infrastructure projects, an increased risk of disease transmission due to crowded conditions, and location in fragile or vulnerable environments, such as deforested mountain slopes.[67] This hazard-centered or impact-oriented paradigm focuses largely on the physical processes underlying vulnerability to climate change and disasters. Consequently, vulnerability reduction strategies often seek to control outcomes through monitoring and predicting, as well as through engineering projects and technological interventions that contain or reduce their effects.[68] A "social vulnerability" approach, in contrast, focuses on vulnerability as the result of an interplay among many contextual factors, including biophysical, social, economic, political, institutional, technological and cultural conditions that generate unequal exposure to risk and create differential capacities to respond to both shocks and long-term changes.[69] This vulnerability context is described in more detail below.

[…]

Box 2. Defnitions of Vulnerability

"Vulnerability is the degree to which a system is susceptible to, and unable to cope with, adverse effects of climate change, including climate variability and extremes. Vulnerability is a function of the character, magnitude, and rate of climate change and variation to which a system is exposed, its sensitivity, and its adaptive capacity." (IPCC 2007)[70]

"Vulnerability is the state of susceptibility to harm from exposure to stresses associated with environmental and social change and from the absence of capacity to adapt." (Adger 2006)[71]

"The conditions determined by physical, social, economic, and environmental factors or processes, which increase the susceptibility of a community to the impact of hazards." (UNISDR 2007)[72]

"The characteristics of a person or group and their situation that influence their capacity to anticipate, cope with, resist and recover from the impact of a natural hazard (an extreme natural event and process)." (Wisner et al. 2004)[73]

3.1. The vulnerability context

The vulnerability literature provides important insights regarding how and why some individuals, households, social groups, and public institutions are likely to be disproportionately affected by climate change, extreme events, and disasters.[74] Numerous vulnerability frameworks emphasize specific contextual factors that influence exposure and the capacity to respond to change.[75] For example, Turner et al. developed a place-based framework that focuses on the coupled human-environment system and examines how hazards can potentially affect the system.[76] Their framework recognizes that responses and their outcomes collectively determine the resilience of the coupled system and may, in fact, transcend the system or location of analysis to affect other scalar dimensions of the problem, creating potential feed-backs to the original system. Other types of vulnerability frameworks include capabilities, assets, and livelihoods approaches that focus on the factors that constrain or enable people in pursuing outcomes that they value.[77] The DFID framework on sustainable livelihoods views people as operating in a context of vulnerability, where they have access to certain assets or poverty-reducing factors that are influenced by the prevailing social, institutional, and organizational environment.[78] The Pressure and Release (PAR) model of Wisner et al.[79] explicitly discusses how "unsafe conditions" are transformed into disasters given exposure to biophysical, social, political, and economic stressors. This model describes how vulnerability is rooted in social processes and underlying causes (called dynamic pressures and root causes), which may often be quite remote from the disaster event itself.

An individual or group's vulnerability to climate change and climate-related disasters is thus influenced by the complex array of social, economic, political and environmental factors operating at a variety of levels that in combination affect vulnerability.[80] Consequently, vulnerability is not evenly distributed across society, and some individuals, households, or groups are likely to be disproportionately affected by climate change or disasters. Box 3 discusses some of the issues surrounding vulnerability of two important demographic groups: the elderly and children. Interestingly, most vulnerable people do not perceive themselves to be vulnerable – they instead refer to vulnerability in terms of "weakness," "problems," and "constraints".[81] Although structural and systemic factors can contribute considerably to vulnerability, it is recognized that people and institutions act from diverse histories and worldviews and consequently have different interpretations and perceptions of risk and vulnerability, hence they may develop differential responses to similar conditions and processes.[82] Some individuals may consider that no risk is tolerable, and thus hold their government responsible for insulating them from all risks, whereas others may be willing or forced to live with considerable risk.[83] Nonetheless, as Hilhorst points out, "[n]arratives that people create about risk, vulnerability and disasters are not just statements about nature, but are also statements about state-society relations".[84]

In short, the possible effects of climate change extremes cannot be understood independently of larger social, economic and cultural changes. It is widely recognized within the disaster risk community that hazards themselves rarely create disasters, but instead it is the context in which the hazard occurs that contributes to disastrous outcomes.[85] This is relevant to climate change-related extreme events as well. Yet it is also important to recognize that the context in which climate extremes and hazards occur is constantly changing as the result of many factors, including rates of economic development and natural resource exploitation, urbanization, deforestation and land use changes. Among the many environmental and social processes that structure vulnerability, rising global food prices, warfare, corruption, trade dependency,

Box 3. Children and the Elderly: Extremely Vulnerable to Extremes?

The climate change and disaster risk communities are paying increasing attention to differential vulnerability among demographic groups, particularly children and the elderly. The IPCC Fourth Assessment Report from Working Group II, for example, noted that the health risks associated with changing incidence of weather extremes were most concentrated in vulnerable populations that include the elderly and young children.[87]

More elderly will be exposed to climate change in the coming decades, particularly in OECD countries. By 2050, it is estimated that 1 in 3 people will be above 60 years in OECD countries, as well as 1 in 5 at the global scale.[88] The factors that contribute to the vulnerability of people over 60 years of age to climate change are similar to factors that make them vulnerable to hazards: deterioration of health, personal lifestyles, loneliness, poverty, or inadequate health and social structures are all elements that can contribute to vulnerability.[89] The context in which people are aging will also influence future vulnerability to climate change. This context includes changing health conditions, as well as issues of social exclusion; welfare programme reforms and their impact on the elderly income; developments in the health and social care system; and finally, the evolution of family structures.

Children constitute a very large percentage of those who are most vulnerable to climate change. The effects of extreme events, especially for the youngest children, can be long term.[90] In explaining why children as a group are particularly vulnerable to challenges associated with climate change, Bartlett points out that children are in a rapid stage of development and are less equipped to deal with deprivation and stress, due to rapid metabolisms, immature organs and nervous systems, developing cognition, limited experience and various behavioral characteristics.[91] The adversity experienced by affected children tends to be intensified by poverty and the difficult choices low-income households make as they try to adapt to hardship. With climate change and the need to handle multiple stressors at various levels, children's voices and participation in policy and decision making is likely to become even more pressing and important, as their capacity to contribute to adaptation and disaster risk reduction has been largely overlooked.[92]

macroeconomic policies, and a host of large-scale processes associated with globalization shape the social and economic entitlements that influence vulnerability.[86] There are also important path dependencies related to vulnerability; past processes such as colonization and war shape present insecurities, while ongoing processes such as climate change and changes to ecosystem services shape future insecurities.

Hazards and extreme events themselves can alter the context for economic and social development, which can in turn reduce the capacity to respond to future extremes.[93] Cumulative effects of events such as hurricanes, floods, or droughts not only damage or destroy material assets and human lives, but they may also influence the capacity and resilience of individuals to recover their sense of well-being. Common emotional reactions after a disaster include shock, fear, grief, anger, guilt, shame, helplessness, numbness and sadness, which in combination with cognitive reactions such as confusion, indecisiveness, worry and difficulty concentrating, can make recovery a challenge for days, weeks, months, or years following a disaster.[94] The long-term implications of post-traumatic stress disorder (PTSD) have been witnessed in the aftermath of recent disasters such as Hurricane Mitch, Hurricane Katrina, and the Asian tsunami.[95] There is also an increasing body of research assessing the prevalence and severity of children's distress after an extreme event in the months following a disaster.[96] Kar finds that high exposure, lower educational levels and middle socioeconomic status significantly predicted the outcome of PTSD.[97] Whereas the shock just after a disaster is readily evident, children and their families report that the aftermath of traumatic events and the deprivations and humiliations associated with slow recovery process are particularly stressful.[98]

Vulnerability reduction is thus recognized as an important strategy for reducing disaster risks and minimizing the impacts of climate change. However, despite increased emphasis on the importance of social, political and economic contexts, climate change adaptation and traditional disaster risk management activities remain largely delinked from vulnerability reduction.[99] In fact, a synthesis of evaluation findings on humanitarian responses to natural disasters found relatively few examples of good practices related to vulnerability reduction.[100] There tends to be, instead, a disproportionate emphasis on relief and recovery processes that prioritize a return to 'normalcy', rather than focusing on the conditions that cause risk and vulnerability. In many cases, these 'normal' conditions are directly or indirectly contributing to risk and vulnerability.[101]

4. Disaster Risk Reduction and Climate Change Adaptation

> Holistic management of disaster risk requires action to reduce impacts of extreme events before, during and after they occur, including technical preventive measures and aspects of socio-economic development designed to reduce human vulnerability to hazards. Approaches toward the management of climate change impacts also have to consider the reduction of human vulnerability under changing levels of risk. A key challenge and opportunity therefore lies in building a bridge between current disaster risk management efforts aimed at reducing vulnerabilities to extreme events and efforts to promote climate change adaptation. (Few et al. 2006)[102]

Recognition of the linkages between climate variability, climate change, and extreme events has fostered a small but growing literature on the connections between disaster risk reduction and climate change adaptation.[103] This literature shows that there is a great potential for coordinated efforts towards addressing adaptation. The disaster risk community advocates using the tools, methods and policies of disaster risk reduction as a basis for addressing the risk aspects of climate change. Methodologies and experiences in working with vulnerable people and their needs through community-based initiatives are emerging as a cornerstone for disaster risk reduction.[104] At the same time, the climate change community offers a growing body of research and experience on adaptation as a social process, with an emphasis on strategies and measures to reduce vulnerability and enhance the capacity to adapt to shocks and stressors.[105]

This includes initiatives aimed at building resilience through community-based adaptation. Given these overlapping areas of expertise and empirical experience, there have been numerous calls for increased collaboration between the two communities.

Yet strategies for disaster risk reduction and climate change adaptation have until now evolved largely in isolation from each other through different conceptual and institutional frameworks.[106] The disaster risk management community has gone through various paradigm shifts since the early 1970s.[107]

Throughout these stages the "disaster" or humanitarian community has refined its practical and conceptual approach from managing disasters by addressing the hazards, to understanding and addressing the underlying factors and vulnerabilities that turn hazards into disasters, culminating in the disaster risk reduction framework.[108] The Hyogo Framework for Action (HFA) was adopted by 168 countries in 2005, and provides a technical and political agreement on the areas that needs to be addressed to reduce risk. The HFA presents five priorities for action: 1) ensure that disaster risk reduction is a national and a local priority with a strong institutional basis for implementation; 2) identify, assess and monitor disaster risks and enhance early warning; 3) use knowledge, innovation and education to build a culture of safety and resilience at all levels; 4) reduce the underlying risk factors; and 5) strengthen disaster preparedness for effective response at all levels.

Climate change adaptation has a somewhat shorter history, emerging in the United Nations Framework Convention on Climate Change (UNFCCC) signed in 1992. However, the UNFCCC and the Kyoto protocol predominantly addressed climate change mitigation and policies and measures to reduce the emissions of greenhouse gases. It was not until quite recently that adaptation came to the forefront as a key concern within the UNFCCC.[109] The possibilities for Least Developed Countries to develop National Adaptation Programmes of Actions (NAPAs) and the Nairobi Work Program – a 5-year (2005–2010) initiative under the UNFCCC,[110] were important first steps towards both enhancing the understanding of adaptation and catalyzing action on adaptation. The Bali Action Plan (BAP), agreed upon at the UNFCCC Conference of Parties (COP) in Bali, provides a roadmap towards a new international climate change agreement to be concluded by 2009 as successor to the Kyoto Protocol.[111] The BAP puts adaptation on an equal footing with mitigation. In the BAP, risk management and disaster risk reduction are identified as important elements of climate change adaptation. Further, the BAP emphasizes the importance of "building on synergies among activities and processes, as a means to support adaptation in a coherent and integrated manner".[112]

No comprehensive formal scientific assessment has been undertaken yet of the research findings and empirically-based activities that are emerging from the two communities. With increased attention to climate change and associated impacts within the disaster risk community, and growing recognition of the links between disaster risk reduction and adaptation within the climate change community, there is now both a need and an opportunity to learn from the experiences of both the disaster risk and climate change research and practice. Important lessons can be drawn from such an assessment, which can be used to better inform society on how to adapt to a changing climate, and to better integrate and coordinate adaptation and disaster risk reduction across different levels of governance.[113]

What might such lessons look like? Below, we extract and discuss four points that can be gleaned from a brief review of disaster risk reduction and climate change adaptation literatures. These points illustrate the potential synergies that might emerge from a more in-depth scoping or formal scientific assessment. In section 5, we then argue that such synergies are urgently needed to guide insights and actions that increase human security in the face of climate change.

4.1. Disaster risk reduction and climate change adaptation must be closely linked to development

As the uneven distribution of impacts and opportunities presented by climate change and disasters come into sharper focus, both disaster risk reduction and climate change adaptation have become core development issues. There are instrumental concerns about minimizing threats to progress on poverty reduction and the MDGs, but also justice and equity concerns because the impacts of climate change are often hardest felt by those who have contributed least to the problem.[114] For the climate change community, a collaboration with development researchers and practitioners has already contributed to a shift from a theoretical focus on adaptation based on future scenarios of climate change, towards identifying broad policy needs and a variety of practical adaptations than can reduce vulnerability.[115] For the disaster risk community, collaboration with the development community has played an important role in identifying vulnerability reduction strategies. Enhancing collaboration across the disaster risk, climate change and development communities may be the most effective means of promoting sustainable adaptation to climate change.

However, in an analysis of the links between adaptation, disaster risk management and development, Schipper and Pelling point to the difficulty of integrating the three agendas because of the distinct sets of actors and institutions involved […].[116] Rather than consulting each other on common topics, these groups often "reinvent the wheel" and come up with separate frameworks within the same meta-narratives. Yet a key contribution to all of these frameworks from development researchers and practitioners is the recognition that risk reduction and adaptation strategies must be carefully tailored to individual, household and community needs. Approaches that treat communities as homogeneous (i.e., able to adapt or reduce risks as a group) are prone to failure, as are adaptation and disaster risk reduction measures that do not explicitly and simultaneously address poverty.[117]

[…]

4.2. Disaster risk reduction and climate change adaptation must address local needs

Local-level experiences can be considered the front-line of impacts from hazards and extreme events, thus they can provide important insights on the most urgent challenges associated with extreme weather events in a changing climate. The disasters community has a long history of experience in working at the local level, and a body of work on community-based adaptation is also emerging to link climate change, disaster risk reduction and development.[118] Numerous examples of local needs and challenges have emerged from community level case studies carried out in relation to both disaster risk reduction and climate change adaptation [...].

These case studies have much in common, particularly in their emphasis on vulnerability and capacity assessments at the local level to identify existing coping capacities as the basis for meeting future hazards. Such community-based assessments draw from participatory methods to link vulnerability with entitlements and access to resources, often employing a sustainable livelihoods framework.[119]

Early lessons from community-based adaptation and disaster risk reduction suggest that there is considerable potential for reducing vulnerability at the local level.[120] These lessons have stressed the need for working with trusted local intermediaries who have a firm understanding of community circumstances and dynamics, basing new activities, technologies or practices on existing coping practices. Good local development practice is crucial to the process, allowing the introduction of knowledge around current and future climate risks based on existing activities and knowledge. Addressing deficits in current coping and risk management to climate-related hazards is crucial to this approach, particularly regarding extreme events. The social vulnerability approach in particular has played a critical role in reorienting traditionally top-down methods in both disaster risk reduction and climate change adaptation towards the community or local level.[121] Engagement at the community level is underpinned by a reframing of vulnerable people not as passive victims but as capable of preventing disasters and adapting to climate change within their own communities. Bottom-up approaches promote locally-appropriate measures, empower people to change their own lives, and encourage greater ownership of disaster risk reduction and adaptation actions. Communications have been highlighted as extremely important, which suggests an emphasis on presenting knowledge in a community's own language, through innovative media, and in understandable non-scientific terms.

[...]

4.3. Climate information must capture complexity and uncertainty to support adaptation and disaster risk reduction

Research has shown that scientists need to collaborate more closely with local knowledge networks and take into account people's risk perceptions, as well as the decision-making processes these communities use.[122] However, reducing disaster risk and vulnerability also requires close interaction between scientists who produce knowledge about changing patterns of risk, and researchers and practitioners who use such information for disaster risk reduction and climate change adaptation. Currently, the spatial resolution of many climate change projections is too coarse to enable effective disaster risk reduction at the local or regional scale. The gap between climate forecasts and projections and the needs of resource managers may pose some challenges to effective responses. Past experiences with reducing risks associated with climate variability can provide some important insights into disaster risk reduction and climate change adaptation.[123] In southern Africa, for example, research has demonstrated strong linkages between El Niño Southern Oscillation (ENSO) and rainfall patterns. In particular, drought events in parts of southern Africa in the early 1980s were closely correlated to ENSO events. However, more recent evidence (particularly from the late 1990s ENSO events) suggests that the relationship between ENSO and summer rainfall does not always hold in this region, particularly at the local scale where many important livelihood decisions are made.[124]

One lesson from this area of research is that over-reliance on only one indicator (e.g. ENSO signals) can be problematic for effective disaster risk reduction and climate change adaptation. Consequently, there is a need for a better understanding of complex and compound hazards, both from physical and social perspectives. The complexity of future extreme events, which are likely to be characterized by one or more hazard that is compounded by other factors (e.g. flooding combined with a cholera outbreak that coincides with an economic crisis), requires more robust and flexible disaster risk strategies and institutional responses than has been typically used in the past. Indeed, a recent report on disaster risk reduction in sub-Saharan Africa calls for better identification, assessment and awareness of disaster risks, which will require efforts from both the disaster risk reduction community and climate scientists.[125] Communication about climate change needs to be made accessible in order to engage vulnerable people without compromising scientific credibility.[126]

4.4. There are thresholds and limits to disaster risk reduction and adaptation

There are likely to be some thresholds and limits to the potential for disaster risk reduction and adaptation to enhance human security in the face of climate change. Schneider et al. note that "the risk-reducing potential of planned adaptation is either very limited or very costly for some key vulnerabilities, such as loss of biodiversity, melting of mountain glaciers or disintegration of major ice sheets."[127] In other words, there are absolute limits that are faced by many ecosystems and individual species in adapting to new climatic conditions, particularly given constraints of urban land use and conversion of natural habitats to agriculture; over-exploitation of resources such as fisheries; and other stresses such as pollution loading to terrestrial and marine environments.[128] Hence there are major non-linearities and uncertainties related to climate change. Schneider et al. also argue that "adaptation assessments need to consider not only the technical feasibility of certain adaptations but also the availability of required resources, the costs and side effects of adaptation, the knowledge about those adaptations, their timeliness, the incentives for the adaptation actors to actually implement them, and their compatibility with individual or cultural preferences."[129] Adger et al. elaborate on this by discussing six broad categories of limits to adaptation closely linked to the rate and magnitude of climate change, as well as associated key vulnerabilities: physical and ecological limits, technological limits, informational and cognitive limits, social and cultural limits, institutional political limits, and financial limits.[130]

Financial barriers to both adaptation and disaster risk reduction have been highlighted, but primarily in policy documents around the international climate regime, rather than in scientific and economic literatures. The SBSTA body of the UNFCCC, the Stern Review, the World Bank, OECD, Oxfam and UNDP have estimated adaptation costs for developing countries.[131] The Stern Review presents the lowest estimate of USD 4 billion per year to adapt to climate change.[132] The highest estimate is made by UNDP which estimates adaptation costs to USD 86–109 billion a year. An OECD study on the economics of adaptation demonstrates that these numbers, which have already been widely used in political statements and demand for more funds, should be handled with caution.[133] Baer, and Paavola and Adger, have discussed principles by which such estimates could be derived (compensation for damage; transfers to the most vulnerable, fair allocation and others).[134] The important point made by all of these analyses is that the costs of adaptation are significant and hence there are real financial barriers, especially in developing countries, to implementing adaptation in a sustainable manner. This area is significantly under-researched and emerging insights from public choice theory and other could be applied to enlighten the international costs of adaptation and disaster risk reduction in the context of the international strategies for mitigation.[135]

Reducing vulnerability to weather-related disasters also faces constraints associated with behavior and cognition of risk.[136] New research from social psychology, some highlighted in the IPCC Fourth Assessment Report, has shown that individuals deny risks, feel powerless to act, or have little adaptive capacity. For example, by examining elderly people's perceptions of heat wave risks, Wolf et al. show that individuals with low self-efficacy do not perceive themselves as able to act on perceived threats. Because they do not perceive their own vulnerability, they do little to adapt.[137] These studies also demonstrate that decision-making is not a uni-directional and sequential process; instead it is incremental and at times multi-directional. In other words, one step towards a decision may be contradicted by new information and experiences. This suggests that individual responses to climate change may not be as rational as many assessments of adaptive behavior assume.[138] Such findings have important consequences for both disaster risk reduction and climate change adaptation.

5. Human Security Implications of Climate Change

> ... climate shocks also erode long-term opportunities for human development, undermining productivity and eroding human capabilities. No single climate shock can be attributed to climate change. However, climate change is ratcheting up the risks and vulnerabilities facing the poor. It is placing further stress on already over-stretched coping mechanisms and trapping people in downward spirals of deprivation. (UNDP 2007/2008)[139]

There is growing recognition among scientists, practitioners, and policy-makers that climate change will increase the frequency and magnitude of extreme hydro-meteorological events with potentially devastating economic and social impacts at the local and regional levels.[140] Disasters are increasing in impacts and scope, not due to hazards alone, but because of the combined effects of large-scale environmental, economic, social, demographic, and technological changes.[141] Climate change and the potential for increased disasters related to extreme events also raise critical concerns for long-term human security.[142] Human security, broadly defined, includes the means to secure basic rights, needs, and livelihoods, and to pursue opportunities for human fulfilment and development.[143] The promotion of human security is also closely linked to a "positive vision" of society that is encapsulated in notions such as well-being, quality of life, and human flourishing.[144] This positive vision has been elaborated through the capabilities approach, which emphasizes the freedom of people to choose among different ways of living, and to pursue opportunities to achieve outcomes that they value.[145]

A number of recent studies have assessed the relationship between climate change and human security, demonstrating that the linkages are often both complex and context-dependent. For example, negative impacts of climate change on food security over the medium- and long-term are likely to create greater emergency food aid needs in the future.[146] Among the most widely-discussed humanitarian and human security issues surrounding climate change are the possibilities of mass migration and/or violent conflict as the result of biophysical or ecological disruptions associated with climate change. Below, we discuss how migration and conflict, both of which are emerging as key security concerns among national governments and international institutions, are intricately tied to the vulnerability context that disaster risk reduction and climate change adaptation are targeting.

5.1. Climate change and migration

Concerning migration, disasters linked to both extreme events and more gradual changes often lead to displaced people, refugees, relocated communities, and temporary or permanent migration. The relationship between climate risk and displacement is a complex one and there are a myriad of factors that affect displacements and migration. However, recent studies suggest that climate change and associated adverse environmental impacts have the potential to trigger displacement of an increased number of people.[147] Research further suggests that the bulk of migration will take place internally in individual countries; that the majority of migration will come as a result of gradual changes in climate and not so much from individual catastrophic events; that in most cases when hydro-climatic disasters occur in developing countries they will not lead to net out-migration because people tend to return to re-establish their lives after a disaster; and that long term environmental changes are likely to cause more permanent migration.[148]

Recent studies distinguish between migration driven by 1) the increasing frequency and intensity of slow onset disasters such as drought and desertification; 2) rapid onset disasters such as floods and cyclones and 3) incremental changes driven by sea-level rise.[149] Most studies agree that the most important climate change-related driver of migration will be sea-level rise, with the more careful assessments recognizing that the severity of migration will depend critically on the rate of localized changes in sea-level, and the degree to which adaptation takes place and is successful.[150] These studies also recognize that the rate of migration driven by sea-level rise is likely to be slow, but steady, which suggests that disaster risk reduction and adaptation strategies may help avoid humanitarian crises and political instability.

Some studies also recognize that there may be some degree of exaggeration surrounding discussions of "environmental refugees" driven by climate change, creating the danger of inappropriate policy responses that do little to ensure the rights of those most at risk from climate change.[151]

While it does seem likely that climate change will be an additional contributor to migration, many studies emphasize that it is very unclear how many migrants there may be, where they may move from and to, and over what time scale. This uncertainty suggests that some of the more alarmist predictions, including those by Myers and Christian Aid, should not be used as a basis for policy.[152]

It is also widely recognized that environmental change is never a sole cause of migration, and that there are always one or more underlying economic, political or other social factors that make environmental change a proximate trigger, rather than an underlying driver of migration.[153] Whether an individual may migrate due to climate change depends on what is understood of the risks posed by climate change, and to what extent the benefits and costs arising from migration are understood by the individual.[154] Many variables shape an individual or family's decision to migrate, including factors at the point of origin, factors at the destination, intervening obstacles such as distance and institutional constraints, and personal circumstances.[155] Many studies also show that in most cases migration in response to disasters is only possible after a certain level of wealth is reached, meaning that the larger humanitarian problems may be in places where people cannot afford to move, rather than the places to which they do move.[156] In terms of slow-onset disasters such as drought, the evidence is more mixed: repeated drought events such as occurred in the Sahel in the 1970s and 1980s did lead to large scale migration, although it is more often the case that drought was only a trigger, with the underlying drivers being changes in livelihood systems driven by dependence on exports of a few primary commodities as a result of colonization.[157] In other cases, such as drought in Bangladesh in 1994, large-scale migration was not an outcome.[158]

It is important to point out that migration as a form of adaptation is not unproblematic. For example, if recent estimates of a 140cm rise in sea-level and annual coral bleaching are correct,[159] then there is little that can be done to avoid or adapt to losses of land on low-lying atoll islands, with a worst case outcome being the collapse of the ability of island ecosystems to sustain human habitation and subsequent risks to the sovereignty of the world's five atoll-island states. The result may be increases in morbidity and mortality, as well as an increased demand for migration.[160] In the Arctic, too, there is arguably little that can be done to avoid or adapt to absolute losses of snow and ice, melting of permafrost, and resultant changes in social-ecological systems.[161] As with low-lying atoll islands, increased morbidity, mortality and migration may result. In both cases there are other significant losses as well, including of place and culture and the right to a nationality and a home.[162] In each case migration cannot be seen as an 'adaptation' but rather as a loss of culture, livelihood, place and the right to a home.

5.2. Climate change and conflict

The magnitude of environmental changes expected to result from even 2°C of warming above pre-industrial levels may cause significant negative social outcomes in certain social systems – in particular low income and resource-dependent societies. In recent years there has been considerable attention to the relationship between climate change and violent conflicts. Many studies propose that climate change heightens the risk of violent conflict between countries.[163] Others, however, are more circumspect, arguing that while there is cause for concern, there is as yet only limited research to substantiate the argument that climate change will increase violent conflict.[164] These debates notwithstanding, some recent research suggests that certain aspects of climate do influence the likelihood of violent conflict. Miguel et al. use rainfall variation as a proxy for economic growth in 41 African countries and find that decreases in rainfall strongly increase the likelihood of conflict in the following year.[165] Hendrix and Glaser, and Meier et al. also find associations between rainfall variability and violent conflict.[166] Nel and Ringharts show that rapid onset disasters related to climate and geology increase the risk of violent civil conflict, particularly in low and middle income countries.[167] All of these studies use aggregated data sets, and are not without their empirical and methodological problems as explained by Buhaug et al.[168] Yet they do indicate the possibility of a connection between climate and conflict, and justify grounds for concern about the possibility that climate change may increase the risk of violent conflict.

There is some evidence that some of the likely outcomes, such as dwindling resource stocks, a decline in livelihoods, decreasing state revenues, and increasing inequality across space and class, may create opportunities for some elites to harness resentment and mobilize people to fight, and this is more likely in states where regimes are weakened by decreasing revenues from resource-based rents or taxes.[169] If climate change causes migration, this too may be a cause of violent conflict in certain circumstances.[170]

Many studies recognize that there are multiple options for reducing the risk of conflict arising from climate change.[171] It is also important to recognize that conflicts resulting from climate change will not necessarily be violent and can instead lead to changes in the distribution of power and resources, and protection of the things that are valued. Furthermore, research on international river basins shows that issues of water access and water scarcity in many cases lead to cooperation, rather than conflict.[172] In short, the evidence about the links between environmental change and violent conflict is currently inconclusive. Neither qualitative examination of cases, nor research seeking generalizable findings based on statistical data, have produced robust findings.[173] There is, however, ample evidence that human insecurities associated with a lack of basic needs such as food, water, and shelter, limit capabilities and freedoms, and thus have negative implications for human development.[174]

6. Conclusions

> Both mitigation and adaptation should be seen as human security imperatives in a broader sense. (UNDP 2007/2008)[175]

Adaptation to climate change will be an enormous challenge for society over the next several decades. While mitigation measures are expected to reduce or slow the growth of future emissions, these efforts will not halt climatic changes that are already underway due to carbon dioxide and other greenhouse gases that are currently present in the atmosphere.[176] Key points emphasized throughout this report are that disaster risk reduction and climate change adaptation are of critical importance to the security of millions of people, and that vulnerability reduction can serve as a cornerstone for strategies to reduce the negative outcomes of climate change. There is a considerable body of knowledge on disaster risk and climate change that can be used as a basis for developing coordinated efforts for climate change adaptation. However, this literature has not yet been systematically assessed.

There are also many areas where new interdisciplinary research is needed. For example, the increasing occurrence of "complex extremes" and "complex emergencies" is likely to pose pressing challenges for the climate change adaptation and disaster risk communities and the development community at large.[177] The risk of more complex, frequent and intense extreme weather events will be exacerbated by both gradual and non-linear changes in climate and climate variability, suggesting the need for a renewed focus on the ways that disaster risk reduction and other adaptation strategies can influence the context in which climate change is experienced. Such research efforts must take into account the critical role that non-climatic factors, such as development levels, inequality, and cultural practices play in these complex extremes.[178] It is becoming clear that neither disaster risk reduction nor climate change adaptation is about addressing disasters or climate change alone, but rather about confronting the societal context in which these changes are occurring.[179] An assessment of the literature on disaster risk reduction and climate change adaptation represents an important first step towards identifying the strategies and frameworks for meeting present and future challenges related to climate change.

In considering the linkages between disaster risk reduction, climate change adaptation and human security, it is important to recognize that human security is not simply about freedom from conflict or prevention of population displacement.[180] Human security is closely linked to the development of human capabilities in the face of change and uncertainty. Individuals and communities faced with both rapid change and increasing uncertainty are challenged to respond in new ways that protect their social, environmental, and human rights. Considering human security as a rationale for disaster risk reduction and climate change adaptation in the face of climate change emphasizes both equity issues and the growing connections among people and places in coupled social-ecological systems.[181] Never in history has the management of threats to the environmental, social and human rights of individuals and communities been as important at local, regional and global scales, and never before have human security concerns been so closely interlinked across regions, groups and generations. As many references cited in this report convincingly show, it is possible to reduce risk and vulnerability to disasters of our own making.

[...]

From Karen O'Brien, Linda Sygna, Robin Leichenko, W. Neil Adger, Jon Barnett, Tom Mitchell, Lisa Schipper, Thomas Tanner, Coleen Vogel and Colette Mortreux, University of Oslo, *Disaster Risk Reduction, Climate Change Adaptation and Human Security: A Commissioned Report for the Norwegian Ministry of Foreign Affairs'*, GECHS Report 2008:3, pp. 13–26

Notes

60   N. Leary, C. Conde, J. Kulkarni, A. Nyong, and J. Pulhin (eds). Climate change and vulnerability (Earthscan, London, 2008, p. 3).

61   D. Liverman, Assessing impacts, adaptation and vulnerability: Reflections on the Working Group II Report of the Intergovernmental Panel on Climate Change (Global Environmental Change 18/1, 4–7, 2008); Stern (2006); Bouwer et al. (2007); Association of British Insurers (ABI) (2005); S. Changnon, Present and future economic impacts of climate extremes in the United States (Environmental Hazards 5, 47–50, 2003); M.M.Q. Mirza, Climate change and extreme weather events: can developing countries adapt? (Climate Policy, 3/3233–248, 2003).

62   M. Beniston, Linking extreme climate events and economic impacts: Examples from the Swiss Alps (Energy Policy 35, 5384–5392, 2007; S. Hallegatte et al. Why economic dynamics matter in assessing

climate change damages: Illustration on extreme events (Ecological Economics 62, 330–340, 2007); Stern (2006); R.A. Pielke Jr, Future economic damage from tropical cyclones: Sensitivities to societal and climate changes (Phil. Trans. R. Soc. A, 365, 2717–2729, 2007; H, Toya and M. Skidmore, Economic development and the impacts of natural disasters (Economics Letters, 20–25, 2007); W. Nordhaus, The Economics of Hurricanes in the United States (Department of Economics, Yale University. Working Paper 2006); C. Benson and E.J. Clay, Understanding the economic and financial impacts of natural disasters (Washington DC: World Bank, Disaster Risk Management Series No. 4. 134 p, 2004); see http://www.cred.be.

63   UNISDR (2008); Auld (2008); Venton and La Trobe (2008); F. Thomalla et al. Reducing hazard vulnerability: towards a common approach between disaster risk reduction and climate adaptation (Disasters 30/1, 39–48, 2006; M. Helmer and D. Hilhorst, Natural disasters and climate change (Disasters 30/1, 1–4, 2006).

64   Schipper (2008); O'Brien et al. (2007).

65   J.J. McCarthy, O.F. Canziani, N.A. Leary, D.J. Dokken, and K.S. White, eds, Climate Change 2001: Impacts, Adaptation & Vulnerability (Contribution of Working Group II to the Third Assessment Report of the Intergovernmental Panel on Climate Change (IPCC). Cambridge: Cambridge University Press, 2001).

66   T. Cannon, Vulnerability Analysis and the Explanation of "Natural" Disasters (Disasters, Development, and Environment, ed. Ann Varey, 3–30. Chichester: John Wiley, 1994; K. Hewitt, Regions of Risk. A Geographical Introduction to Disasters (Harlow: Addison Wesley Longman, 1997); Mitchell (1999b); D. Mileti, Disasters by Design: A Reassessment of Natural Hazards in the United States (Washington, D.C.: Joseph Henry Press, 1999); F. Kraas, Megacities as global risk areas (Petermanns Geographische Mitteilungen, 147, 6–15, 2003); Pelling (2003a); D. Hilhorst, Complexity and Diversity: Unlocking Social Domains of Disaster Response (In Mapping Vulnerability: Disasters, Development & People, ed. G. Bankoff, G. Frerks, and D. Hilhorst, 52–66. London: Earthscan, 2004); K.M. O'Neill, Rivers by Design: State Power and the Origins of U.S. Flood Control (Durham, NC: Duke University Press, 2006).

67   Mitchell (1999b).

68   Hilhorst (2004).

69   Cannon (1994); Bankoff et al. (2004); J.T. Roberts and B. Parks, A Climate of Injustice: Global Inequality, NorthSouth Politics, and Climate Policy (Cambridge: MIT Press, 2006).

70   IPCC (2007b, p. 27).

71   W.N. Adger, Vulnerability (Global Environmental Change 16/3, 268–281. p. 268, 2006).

72   UNISDR (2007).

73   B. Wisner et al. At Risk: Natural Hazards, People's Vulnerability and Disasters (2nd ed) (London and New York: Routledge, 2004).

74   Leichenko and O'Brien (2008); Committee on Disaster Research in the Social Sciences: Future Challenges and Opportunities, National Research Council (Facing hazards and disasters: understanding human dimensions. The National Academies Press, 2006); Schipper and Pelling (2006); G. O'Brien et al., Climate change and disaster management (Disasters, 30/1, 64–80, 2006); K.L. O'Brien et al., Mapping Vulnerability to Multiple Stressors: Climate Change and Globalisation in India (Global Environmental Change 14/4, 303–313, 2004); Tierney et al. (2001); Wisner, et al. (2004); C. Polsky et al., Building comparable global change vulnerability assessments: The vulnerability scoping diagram (Global Environmental Change 17/3–4, 472–485, 2007); W.N. Adger, Vulnerability (Global Environmental Change 16/3, 268–281, 2006); D. Schröter et al., Assessing vulnerability to the effects of global change: An eight step approach (Mitigation and Adaptation Strategies for Global Change 10, 573–595, 2004); S.L. Cutter and C. Finch, Temporal and spatial changes in social vulnerability to natural hazards (PNAS 105, 2301–2306, 2008); S. Eriksen and P.M. Kelly, Developing Credible Vulnerability Indicators for Climate Adaptation Policy Assessment (Mitigation and Adaptation Strategies for Global Change 12(4), 495–524, 2007); S.C. Moser, Impact assessments and policy responses to sea-level rise in three US states: An exploration of human-dimension uncertainties (Global Environmental Change 15, 353–369, 2006); K.M. O'Neill, Rivers by Design; State Power

and the Origins of U.S. Flood Control (Durham, NC: Duke University Press, 2006); G. Bankoff et al. (eds), Mapping Vulnerability: Disasters, Development and People (London: Earthscan. 236 pp., 2004); Pelling 2003a; Mitchell 1999a; D. Mileti, Disasters by Design: A Reassessment of Natural Hazards in the United States (Washington, D.C.: Joseph Henry, 1999); K. Hewitt, Regions of Risk, A Geographical Introduction to Disasters (Harlow, Essex: Longman, 1997); Eakin and Luers (2006).

75   B.L. Turner II, A Framework for Vulnerability Analysis in Sustainability Science (PNAS) 100/14, 8074–8079, 2003a; Wisner et al. (2004); C. Ionescu et al., Towards a Formal Framework of Vulnerability to Climate Change (Potsdam, PIK, 2005); A.L. Luers, The Surface of Vulnerability: An Analytical Framework for Examining Environmental Change (Global Environmental Change 15, 214–223, 2005); Eakin and Luers (2006); Eriksen and Kelly (2007); O'Brien et al. (2007).

76   Turner, et al. (2003).

77   C.O.N. Moser, The Asset Vulnerability Framework: Reassessing Urban Poverty Reduction Strategies (World Development 26/1, 1–19, 1998; A. Sen, Development as Freedom (New York: Anchor Books, 1999); DFID. Sustainable Livelihood Fact Sheets: Introduction.(London: U.K. Department for International Development, 1999. Accessed 18 August 2008 http://www.livelihoods.org/info/guidance_sheets_pdfs/section1.pdf); G. Wood, Staying Secure, Staying Poor: The "Faustian" Bargain (World Development 31/3, 455–471, 2003).

78   DFID (1999).

79   Wisner et al. (2004).

80   Eakin and Leurs (2006); Leichenko and O'Brien (2008).

81   Heijmans and Victoria, 2001, cited in Z. Delica-Willison and R. Willison 2004, Vulnerability Reduction: A Task for the Vulnerable People Themselves. (In G. Bankoff, G. Frerks, and D. Hilhorst, eds., Mapping Vulnerability: Disasters, Development & People. London: Earthscan, pp. 145–158, 2004); J. Wolf et al., Conceptual and practical barriers to adaptation: An interdisciplinary analysis of vulnerability and response to heat waves in the UK (In Adger, W.N., Lorenzoni I., and O'Brien K. (eds.), Adapting to Climate Change: Thresholds, Values, Governance. Cambridge: Cambridge University Press).

82   Hilhorst (2004).

83   UNISDR (2008).

84   Hilhorst (2004, p. 61).

85   Wisner et al. (2004); Pelling (2003a); Bankoff et al. (2004).

86   W.N. Adger and M. Kelly, Social vulnerability to climate change and the architecture of entitlements. (Mitigation and Adaptation Strategies for Global Change, 4, 253–266, 1999).

87   Confalonieri et al. (2007).

88   United Nations, World Population Ageing 1950–2050. Executive Summary (New York: United Nations, Department of Economic and Social Affairs, Population Division, 2002).

89   OECD, Declaration on integrating climate change adaptation into development cooperation (Paris, 2006).

90   S. Bartlett, Climate Change and Urban Children: Impacts and Implications for Adaptation in Low and Middle Income Countries (Human Settlements Discussion Paper – Climate Change 2, 2008).

91   Bartlett (2008).

92   K. Haynes et al., Children's voices for disaster risk reduction: lessons from El Salvador and the Philippines (IDS working paper, Brighton, IDS, 2008).

93   Leichenko and O'Brien (2008).

94   K. O'Brien et al., Hurricane Katrina Reveals Challenges to Human Security (GECHS AVISO no. 14, 2005); F. H. Norris, Range, Magnitude, and Duration of the Effects of Disasters on Mental Health (Review Update 20, 2005).

95   O'Brien et al. (2005).

96   N. Kar et al., Post-traumatic Stress Disorder in Children and Adolescents One Year After a Super-Cyclone in Orissa, India: Exploring Cross-Cultural Validity and Vulnerability Factors (BMC Psychiatry 14/7, 8, 2007).

97   Kar (2007).

98   Barlett (2008).
99   Schipper (2008).
100   K. Stokke, Humanitarian response to natural disasters: a synthesis of evaluation findings (Synthesis Report 1/2007. Norwegian Agency for Development Cooperation (Norad), 2007).
101   Schipper (2008).
102   R. Few et al., Linking climate change adaptation and disaster management for sustainable poverty reduction (Synthesis Report for Vulnerability and Adaptation Resource Group, 2006).
103   Schipper (2008); UNISDR (2008); Venton and La Trobe (2008); Helmer and Hilhorst (2006); O'Brien et al. (2006); Schipper and Pelling (2006); S. Hallegatte, Accounting for extreme events in the economic assessment of climate change (FEEM Working Paper No. 01.05, 2005); ProVention Consortium, Harmonization Portal: Exploring synergies between Climate Change Adaptation and Disaster Risk Reduction, 2008, www.proventioncon-sortium.org.
104   LCA, Reducing Disaster Risk while Adapting to Climate Change (LCA Discussion Background Paper); UNISDR, Links between Disaster Risk Reduction, Development and Climate Change (Report prepared for the Commission on Climate Change and Development, Sweden, 2008).
105   E. Penning-Rowsell, Signals' from pre-crisis discourse: Lessons from UK flooding for global environmental policy change? (Global Environmental Change 16, 323–339, 2006); E.L. Tompkins, Planning for climate change in small islands: Insights from national hurricane preparedness in the Cayman Islands (Global Environmental Change Part A 15/2, 139–149, 2005); L.O. Næss et al., Institutional adaptation to climate change: flood responses at the municipal level in Norway (Global Environmental Change Part A, 15/2, 125–138, 2005).
106   See F. Yamin et al., Vulnerability, Adaptation and Climate Disasters: A Conceptual Overview (In Yamin, F. and Huq. S., Vulnerability, Adaptation and Climate Disasters. Institute of Development Studies Bulletin Volume 36, Number 4, October 2005).
107   Hilhorst (2004).
108   UNDP, A Climate Risk Management Approach to Disaster Reduction and Adaptation to Climate Change (UNDP Expert Group Meeting Integrating Disaster Reduction with Adaptation to Climate Change, Havana, 2002 there are two UNDP 2002; LCA 2006.
109   UNFCCC, Climate Change: Impacts, vulnerabilities and adaptation in Developing Countries (Climate Change Secretariat, UNFCCC, Bonn, 68pp, 2007).
110   UNFCCC, The Nairobi work programme on impacts, vulnerability and adaptation to climate change (Climate Change Secretariat, UNFCCC, Bonn, 16pp, 2007).
111   See UNFCCC: http://unfccc.int/adaptation/items/4159.php.
112   UNFCCC (2007, p. 3).
113   Yamin et al. (2005); S. Huq et al., Reducing risks to cities from disasters and climate change (Environment and Urbanization, 19/1, 39–64, 2007); E. Polack and E. Choi, Building climate change resilient cities (In-Focus 2/6, Institute for Development Studies (IDS), UK, 2007).
114   E. Polack 2008, forthcoming. A right to adaptation: securing the participation of marginalized groups (In Tanner, T. and Mitchell, T. (eds), Poverty in a changing climate, IDS Bulletin, 2008); International Council on Human Rights Policy (ICHRP), 2008. Climate change and human rights: a rough guide, Versoix, Switzerland; T.M. Tanner and T. Mitchell (eds.), Poverty in a Changing Climate (IDS Bulletin 39/4, Institute of Development Studies, University of Sussex, UK, 2008b); W. N. Adger et al., Assessment of adaptation practices, options, constraints and capacity (In Parry, M. L., Canziani, O. F., Palutikof, J. P., Hanson, C. E., and van der Linden, P. J. (eds), Climate Change 2007: Impacts, Adaptation and Vulnerability. Contribution of Working Group II to the Fourth Assessment Report of the Intergovernmental Panel on Climate Change. Cambridge University Press: Cambridge, 719–743, 2008).
115   K. Ulsrud et al., More than rain: Identifying sustainable pathways for climate adaptation and poverty reduction (Report prepared by GECHS for the Development Fund, 2008).
116   Schipper and Pelling (2006).
117   Tanner and Mitchell (2008a); Tanner and Mitchell (2008b).
118   ProVention Consortium, Tools for Mainstreaming Disaster Risk Reduction (Guidance Notes for

Development Organisations 2007) http://www.proventionconsortium.org/?pageid=37&publ icationid=132#132; S. Huq and H. Reid, Community-Based Adaptation. A Vital Approach to the Threat Climate Change Poses to the Poor (IIED Briefing Papers, IIED, London, 2007); see www.cba-exchange.org.

119   M.K. van Aalst et al., Community level adaptation to climate change: The potential role of participatory community risk assessment (Global Environmental Change, 18/1, 165–179, 2008); A. Sen, Poverty and Famines: An Essay on Entitlement and Deprivation (Oxford: Oxford University Press, 1981); Wisner et al. (2004); D. Carney (ed.), Sustainable Rural Livelihoods: What Contribution Can We Make? (DFID, London, UK, 1998).

120   See www.cba-exchange.org; Huq and Reid, (2007).

121   Adger (2006); Wisner et al. (2004).

122   F. Sperling et al., Transitioning to Climate Resilient Development: Perspectives from Communities in Peru (World Bank Environment Department papers, 115, 2008).

123   K.L.O'Brien, K.L. and H.C. Vogel, Coping with climate variability: The use of seasonal climate forecasts in Southern Africa (Aldershot: Ashgate Publishing, 2003).

124   W.A Landman and S.J. Mason, Change in the association between Indian Ocean sea-surface temperatures and rainfall over South Africa and Namibia (International Journal Climatology, 19, 1477–1492, 1999); J.C. Reason, Seasonal to decadal prediction of southern African climate and its links with variability in the Atlantic Ocean (Bulletin of the American Meteorological Society, 87, 7, 941–955, 2006).

125   African Union et al., Report on the Status of Disaster Risk Reduction in the Sub-Saharan Africa Region (Report 2008).

126   A. Patt, Effects of seasonal climate forecasts and participatory workshops among subsistence farmers in Zimbabwe (Proceedings of the National Academy of Sciences, 102, 12623–12628, 2005).

127   Schneider et al. (2007).

128   L.J. Gordon et al., Agricultural modifications of hydrological flows create ecological surprises (Trends in Ecology & Evolution, 23, 211–219, 2008); T.E. Lovejoy and L.J. Hannah (eds), Climate Change and Biodiversity (Yale University Press: New Haven, 2005).

129   Schneider et al. (2007).

130   Adger et al. (2007).

131   UNFCCC (2007); Stern Review (2006); World Bank. An Investment Framework for Clean Energy and Development: A Progress Report (DC2006–0012. Washington, DC: The World Bank, 2006); Oxfam, Adapting to Climate Change: What's Needed in Poor Countries and Who Should Pay. (Oxfam Briefing Paper 104, 2007). OECD, Economic aspects of adaptation to climate change: an assessment of costs, benefits, and policy instruments (Working Party on Global and Structural Policies. ENV/EPOC/GSP(2008)7. Paris: OECD, 2008); UNDP (2007); Fighting climate change: Human solidarity in a divided world. Human Development Report 2007/2008, New York, UNDP, 2007.

132   Stern Review (2006).

133   OECD (2008).

134   P. Baer (2006) Adaptation: Who Pays Whom (In Fairness in Adaptation to Climate Change, (eds), W.N. Adger, J. Paavola, S. Huq, and M.J. Mace, Cambridge Mass.: MIT Press); J. Paavola and W.N. Adger, Fair adaptation to climate change (Ecological Economics 56/4, 594–609, 2006).

135   See A. Ulph and D. Ulph, Climate change–environmental and technology policies in a strategic context (Environmental and Resource Economics, 37, 159–180, 2007).

136   T. Grothmann and A. Patt, Adaptive capacity and human cognition: the process of individual adaptation to climate change (Global Environmental Change Part A 15/3, 199–213, 2005); S. Marx et al., Communication and mental processes: experiential and analytic processing of uncertain climate information (Global Environmental Change, 17, 47–58, 2006).

137   Wolf et al. (2009).

138   Grothmann and Patt (2005).

139   UNDP (2007/2008, p. 8).

140   Stern (2006); Bouwer et al. (2007); ABI (2005).

141   Leichenko and O'Brien (2008); Pielke Jr (2007); A. Oliver-Smith, Disasters and forced migration in the 21st Century (In Social Science Research Council, Understanding Katrina: Perspectives from the Social Sciences, 2006). Accessed 15 June 2007 http://understandingkatrina.ssrc.org/Oliver-Smith/; Kraas (2003).

142   J. Barnett and W.N. Adger, Climate change, human security and violent conflict (Political Geography 26/6), 639–655, 2007); A. Carius et al., Climate change and security: challenges for German Development Cooperation (Deutsche Gesellschaft fur Technische Zusammenarbeit (GTZ) GmbH: Eschborn, 2008); J. Schreffran, Climate Change and Security (Bulletin of the Atomic Scientists 64/2, 19–25, 2008).

143   K.L. O'Brien and R.M. Leichenko, Human Security, vulnerability, and sustainable adaptation, (Background Paper commissioned for the Human Development Report 2007/2008: Fighting Climate Change: Human Solidarity in a Divided World. New York: UNDP, 2007); S. Khagram, W. Clark, and D. Riras Raad, From the environment and human security to sustainable security and development. (Journal of Human Development, 4/2, 289–313, 2003).

144   R. Lister, Poverty (Cornwall, UK: Polity Press, 2004).

145   Sen (1999, p. 291).

146   M.J. Cohen, Food Security: Vulnerability Despite Abundance (Coping with Crisis Working Paper Series, International Peace Academy, July 2007).

147   V. Kolmannskog, Future floods of refugees: A comment on climate change, conflict and forced migration (Report by the Norwegian Refugee Council, Oslo, Norway, 2008); E. Ferris, Making sense of climate change, natural disasters and displacement: a work in progress (Lecture at the Calcutta Research Group Winter Course, 14 December 2007. Washington DC: Brookings Institution. 13 pp, 2007).

148   E. Piguet, Climate change and forced migration (Research Paper No 153. Evaluation and Policy Analysis Unit, UNHCR. 15 pp, 2008).

149   Piguet (2008).

150   O. Brown, Migration and climate change (IOM Migration Research Series No. 31. Geneva: International Organisation for Migration, 2008); Piguet (2008); N. Gleditsch, Climate Change and Conflict: The Migration Link (Coping with Crisis Working Paper: International Peace Academy: New York, 2007); E. Meze-Hausken, Migration Caused by Climate Change: How Vulnerable are People in Dryland Areas? (Mitigation and Adaptation Strategies for Global Change 5/4, 379–406, 2000).

151   W.M. Adger and J. Barnett, Compensation for Climate Change Must Meet Needs (Nature 436/7049, 328, 2005); C. Mortreux and J. Barnett, Climate Change, Migration and Adaptation in Funafuti, Tuvalu. Global Environmental Change, in press).

152   N. Myers, Environmental Refugees: A Growing Phenomenon of the 21st century (Philosophical Transactions of the Royal Society 357/1420, 609–613, 2002); Christian Aid, Human Tide: The Real Migration Crisis (Christian Aid: London, 2007); Brown (2008); S. Perch-Nielsen et al., Exploring the Link Between Climate Change and Migration (Climatic Change, 2008); R. McLeman and B. Smit, Migration as a Human Adaptation to Climate Change (Climatic Change 76/1–2, 31–53, 2006); Meze-Hausken (2000).

153   G. Hugo, Environmental Concerns and International Migration (International Migration Review 30/1, 105–131, 1996); S. Castles, Environmental change and forced migration: Making sense of the debate, UNHCR Working Papers 70, 1–14, 2002); S. Lonergan, The Role of Environmental Degradation in Population Displacement (Environmental Change and Security Project Report No. 4, 5–15, 1998).

154   Grothmann and Patt (2005); J. Connell and R. King, Island Migration in a Changing World (In: King, R. and Connell, J. (eds.), Small Worlds, Global Lives: Islands and Migration, Pinter Publications, London, 1–26, 1990.

155   R. Bedford et al., International Migration in New Zealand: Context, Components and Policy Issues (Population Studies Centre Discussion Papers (37), University of Waikato, 2000); C. Macpherson and L. Macpherson, The Changing Contours of Migrant Samoan Kinship (In: King, R. and Connell, J. (eds.), Small Worlds, Global Lives: Islands and Migration. Pinter Publications, London, 277–

296, 1990); A. Ravuvu, Security and Confidence as Basis Factors in Pacific Islanders' Migration (In: Spickard, P., Rondilla, J., and Wright, D. (eds.), Pacific Diaspora: Island Peoples in the United States and Across the Pacific. University of Hawaii Press, Honolulu, 87–98, 2002); R. Oderth, An Introduction to the Study of Human Migration: an Interdisciplinary Perspective (Writers Club Press, Lincoln, 2002).

156   Brown (2008).
157   R. Franke and B. Chasin, Seeds of Famine (New Jersey: Rowman and Allanheld, 1980).
158   K. Smith, Environmental Hazards: Assessing Risk and Reducing Disaster (Routledge, London, 1996).
159   S. Rahmstorf et al., Recent climate observations compared to projections (Science 316/5825, 709, 2007); S.D. Donner et al. (2005).
160   J. Barnett and W.N. Adger, Climate Dangers and Atoll Countries (Climatic Change 61, 321–337, 2003).
161   ACIA, Impacts of a warming Arctic (Arctic Climate Impact Assessment, Cambridge University Press, 2004); E.J. Keskitalo, Climate change and globalization in the Arctic: An integrated approach to vulnerability assessment (London, Earthscan, 2008).
162   W.N. Adger et al. (eds.), Adapting to Climate Change: Thresholds, Values, Governance (Cambridge: Cambridge University Press (in press)).
163   CSIS, The Age of Consequences: The Foreign Policy and National Security Implications of Global Climate Change (Washington D.C., CSIS, 2007); A. Dupont and G. Pearman, Heating Up the Planet: Climate Change and Security (Lowy Institute Paper 12. Double Bay, The Lowy Institute, 2006); C. Jasparro and J. Taylor, Climate Change and Regional Vulnerability to Transnational Security Threats in Southeast Asia (Geopolitics 13/2, 232–256, 2008); N. Myers, Environmental Refugees (Population and Environment 19/2, 167–182, 1997); P. Schwartz and D. Randall, An Abrupt Climate Change Scenario and its Implications for United States National Security (San Franciscio, Global Business Network, 2003). http://www.gbn.com/ArticleDisplayServlet.srv?aid=26231, cited April 8 2007; German Advisory Council on Global Change (WGBU), World in Transition: Climate Change as a Security Risk (Summary for Policy Makers. Berlin, WBGU, 2007). http://www.wbgu.de/wbgu_jg2007_engl.html, cited November 29 2007.
164   J. Barnett, Security and Climate Change (Global Environmental Change Volume, Issue 1, 7–17, 2003); J. Barnett and N. Adger, Climate Change, Human Security and Violent Conflict (Political Geography 26/6, 639–655, 2007); H. Buhaug et al., Implications of climate change for armed conflict (Paper presented at the World Bank Workshop on the Social Dimensions of Climate Change, The World Bank, Washington, 5–6 March 2008); R. Nordås and N. Gleditsch, Climate conflict: common sense or nonsense? (Political Geography 26/6, 627–638, 2007); J. Scheffran, Climate Change and Security (Bulletin of the Atomic Scientists 64/2, 19–25, 59–60, 2008); I. Salehyan, From Climate Change to Conflict? No Consensus Yet (Journal of Peace Research 45/3, 315–326, 2008).
165   E. Miguel et al., Economic Shocks and Civil Conflict: An Instrumental Variables Approach (Journal of Political Economy 112–4, 725–753, 2004).
166   C. Hendrix and S. Glaser, Trends and Triggers: Climate, Climate Change and Civil Conflict in Sub-Saharan Africa (Political Geography 26/6, 695–715, 2007); P. Meier et al., Environmental Influence on Pastoral Conflict in the Horn of Africa (Political Geography 26/6, 716–735, 2007). GECHS: Disaster Risk Reduction, Climate Change Adaption and Human Security, Report 2008.
167   E. Nel and M. Ringharts, Natural Disasters and the Risk of Violent Civil Conflict (International Studies Quarterly 52/1, 159–185, 2008).
168   H. Buhaug et al., Implications of climate change for armed conflict (Paper presented at the World Bank Workshop on the Social Dimensions of Climate Change, The World Bank, Washington, 5–6 March 2008).
169   Buhaug et al., (2008); J. Barnett and W.N Adger, Climate change, human security and violent conflict (Political Geography 26/6, 639–655, 2007); C. Kahl, States, Scarcity, and Civil Strife in the Developing World (Princeton University Press: New Jersey, 2006).
170   G. Baechler, Violence Through Environmental Discrimination: Causes, Rwanda Arena and Conflict Model (Kluwer: Dordrecht, 1999). N. Gleditsch et al., Climate Change and Conflict: The Migration

Link. International Peace Academy Coping With Crisis (Working Paper: International Peace Academy: New York, 2007). R, Reuveny, Climate Change Induced Migration and Violent Conflict (Political Geography 26/6, 656–673, 2007).

171 A.Carius et al., Climate Change and Security: Challenges for German Development Cooperation (Deutsche Gesellschaft für Technische Zusammenarbeit (GTZ) GmbH: Eschborn, 2008).

172 A.T. Wolf, A. Kramer, A. Carius, and G.D. Dabelko, Navigating peace: Water can be a pathway to peace, not war (Woodrow Wilson International Center for Scholars, July 2006).

173 Buhaug et al. (2008).

174 Sen (1999).

175 UNDP (2007/2008, 39).

176 IPCC (2007a).

177 IPCC, IPCC Workshop on Changes in Extreme Weather and Climate Events Beijing (Workshop Report China 11–13 June, 2002; R.E. Benestad and J.E. Haugen, On complex extremes: flood hazards and combined high spring-time precipitation and temperature in Norway (Climate Change 85/3–4, 381–406, 2007).

178 UN General Assembly, International cooperation on humanitarian assistance in the field of natural disasters, from relief to development (Report of the Secretary-General. No. A/62/323. New York, 2007); United Nations Economic and Social Council, 2008 (ECOSOC/6345, Background Release, 27 June 2008).

179 Schipper (2008).

180 O'Brien and Leichenko (2007).

181 O'Brien and Leichenko (2007).

Chapter 3

Vikram Kolmannskog

CLIMATE CHANGED: PEOPLE DISPLACED

[...]

Answering the Basics: The How, Who, Where, and How Many

As early as in the First Assessment Report in 1990, the Intergovernmental Panel on Climate Change (IPCC) stated that the gravest effects of climate change may be those on human mobility. Yet, there is still a lack of research relating to this issue. Over the last year, some progress has been made within the humanitarian community in seeking answers to basic questions, like how and where people are displaced, who the displaced people are, and how many.

More Frequent and Severe Disasters

We know that climate change has effects here and now. In 2007, the Fourth Assessment Report of the IPCC authoritatively established that human-induced climate change is accelerating and already has severe impacts on the environment and human lives.[1] A significant impact of climate change is the increase in the frequency and severity of certain hazards. Hazards combined with vulnerability can result in disasters. The UN International Strategy for Disaster Reduction has the following definition of a disaster:

> A serious disruption of the functioning of a community or a society causing widespread human, material, economic or environmental losses which exceed the ability of the affected community or society to cope using its own resources. A disaster is a function of the risk process. It results from the combination of hazards, conditions of vulnerability and insufficient capacity or measures to reduce the potential negative consequences of risk.[2]

The overall trend shows that the number of recorded natural disasters has doubled from approximately 200 to over 400 per year over the past two decades.[3] The majority are climate-related disasters – that is, disasters which climate change can influence both in terms of frequency and severity. These include the meteorological (for example storm), the hydrological (for example flood), and the climatological (for example drought). According to the UN Emergency Relief Coordinator, this situation of more frequent and severe disasters may be "the new normal".[4]

Linking Climate Change, Disasters and Displacement

While there is not a mono-causal relation between climate change, disasters and displacement, the existence of a clear link between the phenomena is increasingly recognised.[5] Voluntary migration can be a form of coping or adaptation, but climate change and disasters also contribute to forced displacement as a survival strategy.

The current projections for the number of people who will be displaced in the context of climate change vary greatly. For example, the last IPCC report quotes estimates that 150 million people may be displaced by 2050, and the Stern Review of the Economics of Climate Change cites estimates of 200 million displaced by 2050.[6] These estimates are generally accepted to be subject to high degrees of uncertainty, primarily because there is no baseline information on current levels of disaster-related displacement.

To address the need for basic answers to inform policy, advocacy and operations, an expert group was established in 2008 under the humanitarian forum Inter-Agency Standing Committee (IASC). A typology based on the work of the Representative to the UN Secretary General on the Human Rights of Internally Displaced Persons, was further developed to clarify how people can be displaced in the context of climate change and their protection status.[7] Based on this typology, The UN Office for the Coordination of Humanitarian Affairs (OCHA) and the Internal Displacement Monitoring Centre of NRC (IDMC) carried out a study to start addressing the question of how many people are displaced.[8] The results indicate that millions are already displaced due to climate-related disasters each year.

How and How Many?

The first category of the IASC typology concerns displacement linked to sudden-onset disasters, such as floods and storms. According to the OCHA-IDMC study, approximately 36 million people were displaced as a result of sudden-onset natural disasters in 2008. The Sichuan earthquake in China alone displaced 15 million people.

Had it not been for this earthquake, climate-related disasters would have been responsible for over 90 percent of disaster-related displacement in 2008. More than 20 million people were displaced due to climate-related sudden-onset disasters.

The second category concerns displacement linked to slow-onset disasters, such as drought, which can seriously impact on people's livelihoods. According to the OCHA-IDMC study, more than 26.5 million people were reported affected by 12 droughts in 2008, but estimates for displacement are not readily available. Determining the element of force and ascribing causation is much more complex than in sudden-onset disasters. A particular slow-onset disaster case, which is separated out as a third category in the IASC typology, is that linked to sea-level rise and resulting in loss of state territory, as in the case of small island states. As of 2008, the only found permanent relocation plans identified in the OCHA-IDMC study concerned the forced displacement of the 2000 inhabitants of the Tulun (Carteret) and 400 of the Takuu (Mortlock) Islands in Papua New Guinea. However, according to current IPCC findings, this trend is likely to substantially accelerate in the future.

The final IASC category concerns displacement linked to conflict. According to the OCHA-IDMC study, 4.6 million were newly internally displaced and 42 million people were living in forced displacement due to conflict and persecution in 2008. According to some researchers, climate change impacts such as drought may have consequences for conflict, for by example by making resources scarcer and increasing competition.[9]

There are many interlinkages that also need consideration: Disasters and degradation can trigger displacement and conflicts; and conflicts and displacement, in turn, often cause further environmental degradation.

The IASC typology can be considered a work in progress. There are other effects of climate change not explicitly dealt with, such as increases in certain diseases and epidemics. Some of these effects are related to the "natural" disasters while others can perhaps be considered either sudden-onset or slow-onset disasters in themselves.[10] We could also add another category, namely displacement linked to measures to mitigate or adapt to climate change. For example, biofuel projects and forest conservation could lead to displacement if not carried out with full respect for the rights of indigenous and local people.[11]

Finally, it is important not to overlook those who are not displaced. While some remain because of resilient capacity, others may in fact be forced to stay. They do not have the resources to move.[12] In general, the most vulnerable people are often the most exposed and affected by disasters. Displacement will result in particular needs, but it is important to stress that many of those left behind may also have very serious protection concerns and there is a need for an inclusive approach to all affected.

Hotspots of Displacement

In one way or another, all countries will eventually be affected by climate change, but some are more immediately and particularly exposed. The IPCC report highlights the following groups of countries: Small Island Developing States (SIDS), Africa, Mega-deltas (particularly in Asia), and the polar regions. As already mentioned, the impacts of climate change such as disasters depend on exposure, hazard, vulnerability and coping capacity. The locations of the disasters are not predicted to change much in the near future. Of the 20 disasters that caused most displacement in 2008, 17 were in Asia.

In all the categories of displacement discussed above, the displacement may be internal or cross-border, temporary or permanent, but it is likely that the majority of the displaced remain within their country of origin. Much of the sudden-onset disaster displacement is temporary and short-distance. The effectiveness and success of response, recovery and rehabilitation efforts largely determine how long people are displaced. Countries may be more or less willing and able to address these challenges, and the international community has a responsibility to support them.

The Need for Research and Action

Climate change is likely to lead to increasing rates of displacement, and it is vital that evolving frameworks for climate change adaptation address displacement issues. Consistent application of a natural disaster displacement monitoring methodology would provide a baseline for informed estimates as to how current trends may be affected by climate change in the future, and would be a necessary element for any improvement in the response for the displaced. In addition, data should be collected on related factors, including the duration of displacement and the needs of displaced populations. There is also a need for further research on displacement related to slow-onset disasters and sea level rise, the links between climate change, conflict and displacement, and climate change impacts on those who already are displaced. But, as should be clear by now, a lack of information can no longer be used as an excuse to delay action.

[...]

Internal Displacement: International Law and Protection[14]

As the IASC expert group as well as others emphasize, it is likely that the majority of people displaced in the context of climate change in the near future remain within the borders of their country of origin. Today, about two thirds of all people displaced by persecution, war and conflict remain within their own countries.[15] People's mobility largely depends on resources and networks, and climate change is likely to negatively affect people's resources, increase their vulnerability and thereby reduce their mobility.[16] Many of those affected by climate change are likely to be internally displaced persons.

The Guiding Principles on Internal Displacement

Several rights and systems of law have eventually been identified as relevant for internally displaced persons (IDPs). Rather than creating a separate convention, the 1998 UN Guiding Principles on Internal Displacement (UN Doc. E/CN.4/1998/53/Add.2) is a synthesis of such law, in particular human rights law. The Guiding Principles were recognized by states as "an important international framework for the protection of internally displaced persons" at the World Summit in 2005 as well as in several UN General Assembly Resolutions (see for example UN Doc A/60/L.1). There are several examples of national and regional IDP legislation and policy inspired by the Guiding Principles. The latest major development is the adoption of an African Union convention on internal displacement in Kampala, October 2009.[17]

The Guiding Principles provide the normative framework for addressing all displacement occurring within a country. According to the broad and descriptive definition in the Introduction of the Guiding Principles, "internally displaced persons" are:

> persons or groups of persons who have been forced or obliged to flee or to leave their homes or places of habitual residence, in particular as a result of, or in order to avoid, the effects of armed conflict, situations of generalized violence, violations of human rights or natural or human-made disasters, and who have not crossed an internationally recognized State border.

The UN and the Representative of the UN Secretary General on the Human Rights of Internally Displaced Persons in particular, have also clarified that this definition covers all climate change-related displacement, including slow-onset disaster displacement.[18] The explicit inclusion of disasters in the definition is a recognition that persons displaced by disasters also have human rights and protection needs requiring international attention. In the aftermath of the 2004 Asian Tsunami, there was increased recognition of the importance of a human rights approach in dealing with those affected by disasters. Operational Guidelines on Human Rights and Natural Disasters have been developed, which address the needs of all affected, including the displaced.[19]

Operational Challenges

Although there is a normative framework for climate change-related internal displacement, there are still serious challenges on the ground. The states, which have the primary responsibility for people on their territory, are sometimes unwilling or unable to protect displaced people, and in some cases even deny the entry of international protection and assistance agencies referring to the principle of national sovereignty and non-interference.

There are also challenges relating to the roles and responsibilities of the humanitarian agencies. As part of UN humanitarian reform, a division of labour has been established known as the Cluster Approach. In conflict-related internal displacement, UNHCR has assumed global leadership of the Protection Cluster, and co-leads the global Camp Coordination and Camp Management Cluster with the International Organization for Migration (IOM), and the Emergency Shelter Cluster with the International Federation of Red Cross and Red Crescent Societies (IFRC). The Cluster Approach is also relevant in natural disaster scenarios. At country level, the leadership role for protection in natural disaster situations is decided upon by UNHCR, the Office of the UN High Commissioner for Human Rights and UNICEF, on a case-by-case basis. This arrangement has come under criticism, as it does not bring about the necessary predictability or rapidity of response, and the protection agencies are currently engaged in a thorough review. In addition, what is known as the One UN initiative aims at enhancing system-wide UN coherence at the country level and encompasses a number of dimensions: One Leader, One Programme, One Budget and, where appropriate, One Office. It is likely that climate change and the more frequent and severe natural disasters will test the capacities of humanitarian actors, and may call for a new distribution of roles and/or new models of cooperation.[20]

Protection and Natural Disaster

The protection dimension in natural disaster response is still poorly understood at the national level, particularly by state agencies, which after all have the primary responsibility to protect.[21] The Human Rights Committee has stressed the need to review relevant "practices and policies to ensure the full implementation of [the] obligation to protect life and of the prohibition of discrimination, whether direct or indirect, as well as of the United Nations Guiding Principles on Internal Displacement, in matters related to disaster prevention and preparedness, emergency assistance and relief measures" (CCPR/C/USA/CO/3, para. 26).

Particular challenges for the displaced, as well as for the authorities concerned, arise in the context of evacuations, relocations, resettlements and, more generally, the need to find durable solutions for those among the displaced who cannot return. Of the approximately 20 million people displaced by climate-related sudden-onset disasters in 2008, approximately 7.4 million were evacuated either as a preventive measure before a disaster or as palliative measure after a disaster, according to the OCHA-IDMC study.

Evacuation and Relocation

The state's duty to protect people entails an obligation to help people move from zones where they face a danger. A failure to assist people who cannot leave such zones on their own may amount to a human rights violation if competent authorities knew or should have known about the danger and had the capacity to act. For example, the European Court of Human Rights found a breach of the right to life because authorities had not acted adequately in preventing a mudslide (Budayeva and others v. Russia, app. nos. 15339/02, 21166/02, 20058/02, 11673/02 and 15343/02, 30 March 2008). In many areas it is foreseeable that climate change and disasters may result in the need for relocation. For example, accelerating coastal erosion in Alaska is likely to force several communities to relocate; yet no process has been established to address this.[22]

Participation and Non-Discrimination

Participatory, rights-based planning is required. According to the Representative of the UN Secretary General on the Human Rights of Internally Displaced Persons, an approach that treats people as objects of state care rather than with dignity is still prevalent in disasters.[23] Human dignity involves, as a minimum, a right to be heard. Thus, the person participates, and is not merely a means, in decisions concerning herself. According to the Guiding Principles, internally displaced people have a right to be informed, consulted and to participate in decisions affecting them (see in particular Guiding Principles 7.3.c and d, 18.3 and 28.2). The affected and the population at risk should be consulted and invited to participate in the process from the start, including in exploring intervention measures to help them remain or move, and in evacuation, relocation, resettlement and return decisions and design. According to the Operational Guidelines on Human Rights and Natural Disasters, rights are often violated, not because of conscious intention, but because of the lack of rights-based planning.

Non-discrimination is another paramount element in human dignity and human rights; all humans have dignity and certain rights. Evacuation and relocation measures must not discriminate against certain ethnic, religious or other groups. In some of the countries affected by the Asian Tsunami there were reports that buffer zones were being established in a discriminatory manner allowing construction of tourism facilities while local residents were not allowed to return and reconstruct their homes.[24]

Non-discrimination may require that distinctions are made to take into account special protection needs. After Hurricane Katrina hit New Orleans, the Human Rights Committee received reports that the poor, and in particular African-Americans, were disadvantaged because the rescue and evacuation plans were based on the assumption that people would use their private vehicles, thus disadvantaging those not owning a car (CCPR/C/USA/CO/3). During displacement, pre-existing patterns of discrimination are often exacerbated, putting vulnerable groups at further risk of human rights abuses. In the context of Hurricane Katrina, the Human

Rights Committee highlighted the importance of ensuring that the rights of the poor, and in particular African-Americans, are fully taken into consideration in the reconstruction plans with regard to access to housing, education and healthcare (CCPR/C/USA/CO/3).

Use of Force

Sometimes, populations at risk refuse to evacuate, either out of fear of losing livestock or other property or because they do not take warnings seriously. There is also a real risk that states use climate change and disasters as a pretext, while in reality having other reasons for relocating certain groups from certain areas, as we saw examples of after the Asian Tsunami. While evacuation or relocation to safer areas in consultation with the affected can and should be encouraged and facilitated, forced evacuation or relocation is only permissible on certain conditions.

Freedom of movement is prominently enshrined in the 1966 International Covenant on Civil and Political Rights article 12, and includes the right to remain at the place of habitual residence. In the context of climate change and disasters, the prohibition on arbitrary displacement (Guiding Principle 6.1) means that the safety and health of those affected must require the forced evacuation or relocation (Guiding Principle 6.2.d). The evacuation or relocation must be a measure of last resort; all feasible alternatives must have been explored in order to avoid displacement altogether (Guiding Principle 7.1). In the context of climate change and disasters, this means that the authorities must make real efforts to implement other adaptation and disaster risk reduction measures to make the areas safe enough so people can choose to stay. Involuntary relocation and resettlement rarely leads to improvements in the quality of life of those who are moved, so moving communities in anticipation of climate change may precipitate vulnerability more than it avoids it.[25]

Return and Resettlement

While consultation and participation of affected people may be difficult or even impossible during the emergency phase of a disaster, it is particularly important with regard to return, resettlement and recovery (Guiding Principle 28.2). On the basis of freedom of movement, Guiding Principle 28 spells out three solutions which all IDPs have the right to choose between: return to the place of origin, local integration in the place of displacement, and resettlement in another part of the country. As in the case of initial relocation, it can only be for the safety and health of those affected (Guiding Principle 6.2.d) that a state prohibits return to an area.

As a corollary to Guiding Principle 28, Guiding Principle 15 d stipulates the right of internally displaced persons to be protected against forcible return or resettlement to places where their life, safety, liberty or health would be at risk. Returning a person to an area that was recently struck by disaster, is generally disaster-prone or severely environmentally degraded, may be in breach of the prohibition on forced return reflected in Guiding Principle 15 d. In the annotations to the definition of IDPs in the Guiding Principles, people who have moved voluntarily to another part of their country but cannot return to their homes because of events occurring during their absence that make return impossible or unreasonable, are also considered displaced.[26]

Even when it is permissible to enforce return, it may be unreasonable to expect the person to return, and some places of origin may become uninhabitable in the near future and return may be inadvisable. If there is an increase in the population living in the area, the process of environmental degradation and disaster could be further accelerated. What has been said about the possibility, permissibility and reasonableness of forced return, applies similarly to forced resettlement to a third area.

Housing, Property and Livelihoods

The affected should receive compensation for property and land lost, as well as assistance in resettling and re-establishing their livelihoods and residence elsewhere (Guiding Principles 28 and 29). If it is a permanent relocation, some development principles could also apply by analogy. The World Bank's Operational Policy 4.12 on Involuntary Resettlement of January 2002 requires that all the affected people should have incomes and standards of living that are at least equivalent to their pre-project condition.

Experience indicates that while there is investment in building houses at the relocation and resettlement sites, there is often not sufficient focus on the creation of livelihood opportunities and the provision of basic services such as water, sanitation, education and health. In some tsunami-affected countries, fishing families relocated or resettled in-land sometimes ended up destitute, because adequate livelihood alternatives were not available or made available. Where people are unable to return to previous sources of livelihood, appropriate measures, including provision of re-training opportunities, should be taken (see also the Operational Guidelines C 4.2). According to the Representative of the UN Secretary General on the Human Rights of Internally Displaced Persons, problems relating to housing, property and livelihoods result in most cases from an inadequate legal and budgetary framework, and the fact that the affected often come from marginalized sectors of society that continue to be marginalized.[27]

States and humanitarian agencies should review policies, laws and institutional arrangements and take a rights approach when addressing climate change, disasters and displacement. Considering that climate change is a global process, the international community also has a responsibility to support and strengthen different states' ability to provide protection from displacement, during displacement and to end displacement.

Cross-Border Displacement: International Law and Protection[28]

While it is likely that the majority of the displaced remain within their country of origin, some may cross internationally recognised borders and have an uncertain legal status. While new legal and governance solutions may be needed, developing these will take time, and it is also important to look at how existing international law already applies. Some of the displaced should indeed be considered refugees, and considerations relating to the possibility, permissibility and reasonableness of return may provide a starting point to strengthen or even expand existing instruments and mechanisms to address the cross-border protection gap.

Relocation, Resettlement and Statelessness

In some extreme cases, such as in the case of the "sinking" island states and potential statelessness, there may be a need for a cross-border relocation. The President of the Maldives announced late last year that they want to buy land in another country.[29] The government of Kiribati is trying to secure enhanced labour migration options to Australia and New Zealand, but they also recognise that migration schemes will eventually need to be accompanied by humanitarian options and are keen to secure international agreements in which other governments recognise that climate change has contributed to their predicament and acknowledge relocation as part of their obligation to assist. The government of Tuvalu, on the other hand, does not want relocation to feature in international agreements, because of its fear that if it does, industrialized countries may simply think that they can solve problems like rising sea levels by relocating affected populations rather than reducing greenhouse gas emissions.[30]

It is still unclear whether people who lose their state due to climate change impacts, such as the "sinking" island state citizens, would be considered stateless. According to the 1954 Convention Relating to the Status of Stateless Persons article 1, a stateless person is "a person who is not considered as a national by any state under the operation of its law." According to McAdam, the "sinking" island citizens would not be protected because the definition of statelessness is premised on the denial of nationality through the operation of the law of a particular state, rather than through the disappearance of a state altogether.[31] Furthermore, current legal regimes are hardly sufficient to address their very specific needs, including relocation.

UNHCR has been mandated to engage in preventing and reducing statelessness as well as to protect stateless persons (see GA/RES/50/152, 9 February 1996, paras.14–15). In a recent submission to the climate change negotiations, UNHCR, supported by IOM and NRC, recommends multilateral comprehensive agreements that could provide where and on what legal basis populations affected by climate change would be permitted to move, and their status.[32] Stateless refugees are protected in the 1951 Convention, article 33(1).

The Refugee Regime

According to article 1A of the 1951 Convention relating to the Status of Refugees, as modified by the 1967 Protocol, a refugee is a person who owing to a well-founded fear of being persecuted for reasons of race, religion, nationality, membership of a particular social group, or political opinion, is outside the country of his nationality, and is unable, or owing to such fear, is unwilling to avail himself of the protection of that country or who, not having a nationality and being outside the country of his former habitual residence as a result of such events, is unable or, owing to such fear, is unwilling to return to it.

Displacement in the context of climate change and disasters was not part of the drafters' considerations when this definition was formulated. Nonetheless, some people displaced across borders in the context of climate change could qualify for refugee status and protection. Serious or systematic human rights violations are normally considered to amount to persecution.[33] Experience shows that situations of both natural disasters and conflict are prone to human rights violations. The 1951 Convention, as well as UNHCR's mandate, will as a minimum be applicable in situations where the victims of natural disasters flee because their government has consciously withheld or obstructed assistance in order to punish or marginalize them on one of the five grounds in the definition.[34] In addition, there are often several reasons why a person moves, and convention refugees may flee in the context of disasters while the well-founded fear of persecution exists independently.[35]

There are regional instruments with broader definitions, but none explicitly mention climate change or disasters as a reason to grant refugee status. The 1969 OAU Convention Governing the Specific Aspects of Refugee Problems in Africa, article 1.2, includes as refugees persons forced to flee due to "events seriously disturbing public order." Although there have been examples of practice to permit people displaced by disasters across borders to remain temporarily, it seems that in most cases, African governments have not characterised this as an obligation arising under the OAU Convention.[36] In Latin America, the 1984 Cartagena Declaration on Refugees, which has inspired the legislation of many states in the region, also includes as refugees in article 3 persons forced to flee due to "other circumstances which have seriously disturbed public order." However, the International Conference on Central American Refugees does not understand the "other circumstances" to include natural disasters.[37] Jurisprudence based on these regional definitions is scarce, however, and there is a need to develop doctrine and guidance to states on the interpretation of these criteria. We may also see a change in practice and interpretation with the increasing frequency and severity of disasters and ensuing displacement.

The Possibility, Permissibility and Reasonableness of Return

Since many of the cross-border displaced persons will not qualify as either stateless persons or refugees, some advocates for their protection have suggested amending the 1951 Convention. But any initiative to modify the refugee definition would risk a renegotiation of the Convention, which, in the current political situation, may undermine the international refugee protection regime altogether.[38] Some solution to the normative protection gap may be found in the broader human rights law and considerations of the possibility, permissibility and reasonableness of return.

We may see cases where return of a person to his or her place of origin at some point becomes impossible due to climate change and disasters. The "sinking" island states may be an extreme example. In other cases disasters are likely to affect infrastructure required to effectuate a return.

Forced return may also be impermissible, either because it is considered a direct breach of a fundamental right or considered to be a more indirect breach of such a right. The principle of non-refoulement in the Convention relating to the Status of Refugees article 33(1) stipulates a prohibition of expelling or returning ("refouler") a refugee "in any manner whatsoever to the frontiers of territories where his life or freedom would be threatened on account of his race, religion, nationality, membership of a social group or political opinion." This fundamental principle is widely regarded as being a part of customary international law and has counterparts in human rights law. In human rights law, non-refoulement is an absolute and general ban on sending a person, independent of conduct or status, to places where he or she risks certain rights violations.

It may also be considered that return in some cases is unreasonable. Not only strict permissibility, but also a more discretionary reasonableness of return, would be relevant for states to consider in the context of climate change.

Protection against Return to Torture Inhuman and Degrading Treatment

Most agree that the prohibition on torture is a peremptory norm, but there is disagreement regarding the extent to which one is protected by customary law against lesser ill-treatment and human rights violations.

No matter how much a disaster has been induced or created by humans, it is doubtful, to say the least, if it can meet the international definition of torture as the infliction of severe pain or suffering by a public official for an enumerated purpose such as punishment or obtaining a confession. It could also seem far-fetched to call a disaster cruel, inhuman or degrading treatment. In some cases, rather than claiming that a person is returned to ill-treatment, the return itself could arguably constitute the ill-treatment and perhaps even torture. Let us illustrate with a rather extreme example. How should we consider a case where a public official leaves a person to cope by himself with hardly any means in the middle of a desert? There is a continuum between direct and indirect human rights violations.

Generally, courts have carefully circumscribed the meaning of "inhuman or degrading treatment", but there are cases where the concept of "inhuman treatment" has been interpreted rather progressively. In the case of D v. the United Kingdom (application number 30240/96, 2 May 1997), the European Court of Human Rights considered that returning an HIV-infected person to St. Kitts would amount to "inhuman treatment," due to inter alia the lack of sufficient medical treatment, social network, a home or any prospect of income. During and after disasters, such as the hurricane Mitch in Central-America in 1998 and the cyclone Nargis in Myanmar in 2008, homes and vital infrastructure is destroyed or damaged, hindering the provision of basic services such as clean water, electricity and food. One could consider that persons with particular vulnerabilities are protected against return to such circumstances. Clearly, law relating to the permissibility of return is relevant in a climate change context.

Other Rights and Principles Relevant To Return

Climate change and disasters have negative effects on the realisation of several human rights.[39] In theory, any human rights violation under systems such as the European Convention on Human Rights could give rise to a non-refoulement obligation (R v. Special Adjudicator ex parte Ullah, 2004 UKHL 26, paras 24–25). Importantly, the right to life is non-derogable and has very limited exceptions (article 2(2) and article 15(2)). Hence, a person should not be sent back where there is a danger to her life. In addition, one could apply the non-refoulement of refugee law (which includes protection of life) by analogy. Climate change and disasters also affect other human rights, such as the right to food, the right to water, the right to health and the right to adequate housing. Except for absolute rights, such as the right to life and the ban on torture and certain ill-treatment, most human rights provisions permit a balancing test between the interests of the individual and the state. The "new normal" of climate change with more frequent and severe disasters must weigh heavily. These rights will often also be linked to the right to life, and could arguably also be linked to the ban on inhuman and degrading treatment.

Related to the question of permissibility and reasonableness of return, is the principle of return in safety and dignity. International treaties, UN resolutions, UNHCR handbooks and the High Commissioners' speeches, indicate that important elements of the norm include participation, voluntariness, restoration of rights and sustainability of returns.[40]

A Protection Status

If return is not possible, permissible or reasonable due to circumstances in the place of origin and personal conditions, a person should receive protection regardless of the initial cause of movement. The Representative of the UN Secretary General on the Human Rights of Internally Displaced Persons has argued that in the context of climate change, such persons could in fact be considered displaced.[41] In cases of slow-onset disasters, it would not be so much a question of why someone left initially, but rather whether the gradual degradation has reached a critical point where they cannot be expected to return now. As already mentioned, the annotations to the Guiding Principles on Internal Displacement include as displaced persons those who have moved voluntarily to another part of their country, but cannot return to their homes because of events occurring during their absence that make return impossible or unreasonable. To a certain degree, this line of thinking is also acknowledged in traditional refugee law with the recognition of surplus refugees, who were not refugees when they left their country, but who became refugees at a later date due to circumstances arising in the country of origin or as a result of their own actions.

Naturally, it is the present and future risk of rights violations, rather than the past, which is crucial in determining protection need. Where this need is acknowledged, a clear protection status should also be granted. Existing human rights law, including the non-refoulement principle, does not provide for a right to stay nor dictate the content of any protection, but it must include non-rejection at the border to be effective, and can provide a basis for some form of complementary protection.

State Practice and Complementary Protection in Natural Disaster Cases

Complementary forms of protection have been granted to persons who do not fit so well in the refugee definition, but nonetheless are considered to be in need of substitute protection.

We should not assume that people displaced by climate change and disasters will automatically and permanently lose the protection of their state of origin. The responsibility of neighbouring and more distant states receiving the displaced should come in support of that of the state of nationality. The American Temporary Protected Status mechanism seems to reflect such thinking. According to the 1965 Immigration and Nationality Act, section 244, the nationals of a foreign state can be designated for Temporary Protected Status (TPS) on condition that:

1.   There has been an environmental disaster in the foreign state resulting in a substantial, but temporary, disruption of living conditions;
2.   The foreign state is unable, temporarily, to handle adequately the return of its own nationals; and
3.   The foreign state officially has requested such designation.

In the aftermath of the hurricane Mitch in 1998, the United States made the unprecedented decision to grant TPS to Hondurans and Nicaraguans and other Central Americans. The repeated US extensions of TPS for Hondurans and Nicaraguans is commendable, but it does not change the fact that the individuals in question are still residing in the country on a temporary basis, more than ten years after the disaster struck. Only a few of the other nationalities that appear to qualify for TPS have been accepted. The wide discretion in designating countries for TPS raises a concern that the failure to designate a country may be due to domestic politics, ideology, geographical proximity to the United States, foreign policy interests, the number of nationals present in the United States who would benefit from a designation, and other factors unrelated to human rights protection.[42] Furthermore, in extreme disaster scenarios, the state of origin may be unable to even advocate with other states on behalf of its citizens in distress. There are also cases in which displacement relates to a certain unwillingness to protect, or even active human rights violations, in the state of origin. While the American model recognises a role for the state of origin, it is not a strong, legal obligation to protect the individual.

In Finland and Sweden, another model has been chosen.[43] While they emphasise that the first alternative in natural disasters is internal flight and international humanitarian help, the countries also recognise that complementary protection may be necessary. There are provisions in both countries' Aliens Acts to extend either temporary or permanent protection to foreign nationals who cannot return safely to their home country because of an "environmental disaster" (see for example the Swedish Aliens Act 2005:716, Chapter 4, Section 2).

While other countries may not have an explicit recognition of such displacement in their legislation, some have an inclusive practice of temporary or discretionary "humanitarian" stay.[44] From 2001 to 2006, there was a presumption in Denmark that families with young children, and eventually also landless people, should not be returned to Afghanistan due to the drought there. In non-EU countries there is also increasing attention being paid to the topic.

State Practice and Complementary Protection in Conflict Cases

Climate change and disasters could also contribute to increasing conflict and related cross-border displacement. Regional instruments like the OAU Convention and the Cartagena Declaration include as refugees persons fleeing from "generalised violence." The EU Temporary Protection Directive provides for temporary protection in mass-influx situations of persons fleeing armed conflict, and the EU Qualification Directive extends subsidiary protection if there is "a serious and individual threat to a civilian's life or person by reason of indiscriminate violence in situations of international or internal armed conflict" (article 2 e, cf. 15 c). But, apart from those that have adopted the OAU Convention or the EU directives, many countries do not yet recognize people fleeing generalised violence as refugees or persons qualifying for complementary protection. This area of law therefore also needs further harmonisation and binding force. As in the disaster displacement cases, one could build further on human rights and non-refoulement, to support existing law with guidelines on when return is permissible and reasonable and when protection status should be granted.

In sum, some displaced persons may qualify as either stateless persons or refugees, and states should recognise them as such, but states should also ensure that migration management systems provide for the entry and protection of others in need. The human rights regime and complementary protection mechanisms can be built on for such solutions. While bilateral deals such as those under the American TPS is one option, the receiving states must also use their sovereign right to grant safe haven in accordance with basic human rights commitments. If return is not possible, permissible or reasonable due to circumstances in the place of origin and personal conditions, a person should receive protection. Temporary or more permanent protection would of course also alleviate pressure on a state struggling with disasters or violent conflicts. As many of the domestic approaches are discretionary and vary greatly, there is a need to address these questions at a regional and international level, but states should also already start adapting their national laws to better respond to climate change and cross-border displacement.

[…]

Climate Change Mitigation and Adaptation

Both climate change mitigation and adaptation are relevant to the obligation to prevent arbitrary displacement from happening in the first place. Even the slightest increase in global mean temperature can lead to unmanageable humanitarian disasters and displacement of people. With a four degrees increase, we risk inter alia one meter sea level rise within 2100, according to the latest scientific conferences. One tenth of the world's population lives in coastal areas less than one meter above sea level. The IPCC recommends that the increase in global mean temperature should not exceed 2 to 2.4 degrees Celsius. Such climate change mitigation would require enormous and fast reductions in greenhouse gas emissions. Who is going to cut, and how much, is a contentious issue.

It is also important that mitigation measures, such as biofuel projects and forest conservation, are carried out with full respect for the rights of indigenous and local people so they do not lead to displacement. While mitigation is a must, there is also a need to adapt to the current and unavoidable impacts of climate change. This includes reducing the risk of climate-related disasters and thereby also the need to move. The measures range from flood defence infrastructure to education and livelihood diversification. How and how much money will be made available for these purposes is still unclear.

Getting Recognition for Migration and Displacement Issues

There are also situations for which mitigation and anticipatory adaptation has not been sufficient to prevent disasters and displacement. Even though the IPCC highlighted migration and displacement already in 1990, this was never addressed in the climate change negotiations and agreements. A sub-group of the IASC task force on climate change has focused on getting recognition for migration and displacement issues in the current negotiations.[47]

Climate change-related migration and displacement was highlighted in some statements during the Conference of the Parties 14 in Poznan, December 2008, most prominently in the opening statement by the Minister of the Environment of Poland and President of COP 14, and in the statement delivered by the Ambassador of Algeria on behalf of the Africa Group. In the risk management workshop, the states expressed further their support and willingness to build on and coordinate with existing institutions and mechanisms such as the Hyogo Framework for Action.[48] The issue of migration and displacement later figured in the assembly document of ideas and proposals where Bangladesh referred to "climate refugees."[49] The particular challenge of relocation has also been mentioned, inter alia by Mexico during the risk management workshop held in Bonn in April 2009.

A Reference in the Draft Agreement

In May 2009 a draft negotiation text had been prepared and was made public.[50] It was a 200-page document based on literally hundreds of submissions. Many text proposals suggested by humanitarian agencies were included and had support from both Annex 1 (mainly industrialized) and Annex 2 (mainly developing) countries, but much of the text was in brackets which meant it required further negotiation.

Paragraph 25 (e) of the first draft negotiation text included as adaptation actions, "activities related to national and international migration/planned relocation." The reference ensured by the Bangladeshis was well received by many states at the first reading in Bonn in June 2009. Humanitarian agencies present, such as UNHCR, IOM, the Representative to the Secretary-General on the Human Rights of IDPs and NRC, welcomed the reference and offered some advice on how the text could be modified.[51] At a second reading of the draft text, suggested modifications had been included, and a revised negotiation text was made available.[52]

Since then, negotiation sessions have been held in Bonn in August, Bangkok in September and Barcelona in October 2009. Different groups were set up to edit and shorten the 200 pages initial negotiation draft. The inclusion of migration and displacement issues received explicit support by the G77 and China during the Bangkok session. Significantly, no states have opposed it. After Barcelona, the reference to migration and displacement featured in the non-paper 41 on adaptation in paragraph 13:[53]

> All Parties [shall] [should] jointly undertake action under the Convention to enhance adaptation at the international level, including through:
>
> (b) Activities related to migration and displacement or planned relocation of persons affected by climate change, while acknowledging the need to identify modalities of interstate cooperation to respond to the needs of affected populations who either cross an international frontier as a result of, or find themselves abroad and are unable to return owing to, the effects of climate change.

It is important that the reference to "the international level" is not interpreted to mean that only cross-border movements are addressed. It should rather be interpreted to encompass international cooperation as well as international standards to address, inter alia, migration and displacement, whether such movements are internal or cross-border. Much of the movement is after all expected to be within countries, and the plight of IDPs is a matter of international concern. For clarity, however, it could be useful to modify litra b slightly to read "Activities related to internal and cross-border migration and displacement."

A Strong, Binding Legal Agreement

In addition to the displacement-specific text, key language on risk management and disaster risk reduction was prominent. While these were significant steps in the right direction, the overall negotiation climate was not very good. Little progress was made on the more contentious issues such as greenhouse gas emission targets and funding, and there was a widening gap between developing countries and developed countries. Hopefully, Copenhagen results in a political agreement with clear commitments on all the key elements, and we will see a binding legal agreement within 2010. It remains to be seen whether and how this agreement incorporates displacement and other humanitarian issues.

[…]

Vikram Kolmannskog, '*Climate Changed: People Displaced*', Norwegian Refugee Council, 2009, pp. 5–23.

Notes

1   The Fourth Assessment Report of the Intergovernmental Panel on Climate Change, available at www.ipcc.ch.
2   United Nations International Strategy for Disaster Reduction. Terminology: Basic terms of disaster risk reduction, available at www.unisdr.org/eng/library/lib-terminology-eng%20home.htm.
3   Emergency Events Database available at www.em-dat.be.
4   Under-Secretary-General for Humanitarian Affairs and Emergency Relief Coordinator, Opening Remarks at the Dubai International Humanitarian Aid and Development Conference and Exhibition "DIHAD 2008 Conference", 8 April 2008, available at: http://www.reliefweb.int/rw/rwb.nsf/db900sid/YSAR-7DHL88?OpenDocument.
5   See for example Kolmannskog, 2008, Future floods of refugees, Oslo: Norwegian Refugee Council, available at http://www.nrc.no/arch/_img/ 9268480.pdf.
6   UK Treasury. 2005. Stern Review on the Economics of Climate Change, available at http://webarchive.nationalarchives.gov.uk/+/http://www.hmtreasury.gov.uk/independent_reviews/stern_review_economics_climate_ change/sternreview_index.cfm.
7   Informal Group on Migration/Displacement and Climate Change of the IASC, 2008, Climate Change, Migration and Displacement: Who Will Be Affected?: Working paper submitted by the informal group on Migration/Displacement and Climate Change of the IASC, 31 October 2008, available at http://unfccc.int/resource/docs/2008/smsn/igo/022.pdf.
8   OCHA and IDMC/NRC, 2009, Monitoring Disaster Displacement in the Context of Climate Change, available at http://www.internal-displacement. org/8025708F004BE3B1/(httpInfoFiles)/12E8C7224C2A6A9EC125763900315 AD4/$file/monitoring-disaster-displacement.pdf.
9   See for example Kolmannskog, 2008, Future floods of refugees, Oslo: Norwegian Refugee Council, available at http://www.nrc.no/arch/_ img/9268480.pdf.
10 See for example the Emergency Events Database categories at www.em-dat.be.
11 See for example Resisting Displacement by Combatants and Developers: Humanitarian Zones in North-West Colombia, Geneva: Internal Displacement Monitoring Centre, 2007.
12 Black, R. et al., 2008, Demographics and Climate Change: Future Trends and their Policy Implications

for Migration, Development Research Centre on Migration, Globalisation and Poverty, University of Sussex, Brighton, available at http://www.unicef.org/socialpolicy/files/Demographics_and_Climate_Change.pdf.

14   This article is based on Kolmannskog, V, 2009, Dignity in Disasters and Displacement, to be published in the book The Changing Environment for Human Security: New Agendas for Research, Policy, and Action, Global Environmental Change and Human Security Project; and the joint submission to the climate change negotiations Forced Displacement in the Context of Climate Change: Challenges for States under International Law, 15 May 2009, available at http:// unfccc.int/resource/docs/2009/smsn/ igo/049.pdf.

15   NRC, Flyktningregnskapet 2009, available in Norwegian at http://www. flyktninghjelpen.no/?did=9408606.

16   Black R et al., 2008, Demographics and Climate Change: Future Trends and their Policy Implications for Migration, Development Research Centre on Migration, Globalisation and Poverty, University of Sussex, Brighton, available at http://www.unicef.org/socialpolicy/files/Demographics_and_Climate_Change.pdf.

17   Available at http://www.reliefweb.int/rw/lib.nsf/db900SID/SNAA- 7X73KL?OpenDocument.

18   See for example Informal Group on Migration/Displacement and Climate Change of the IASC, 2008, Climate Change, Migration and Displacement: Who will be affected?: Working paper submitted by the informal group on Migration/ Displacement and Climate Change of the IASC – 31 October 2008, available at http://unfccc.int/ resource/docs/2008/smsn/igo/022.pdf.

19   Available   at   http://www.brookings.edu/~/media/Files/rc/reports/2008/spring_natural_disasters/spring_natural_disasters.pdf.

20   See also Climate change, natural disasters and human displacement: a UNHCR perspective, 14 August 2009, available at http://www.unhcr.org/cgi- bin/texis/vtx/search?page=search&docid=4901e81a4&query=climate%20 change.

21   Albuja S and Cavelier I, 2009, Work in progress towards protection in natural disasters: challenges from a conflict/ disaster context, Paper prepared for the Refugee Studies Centre/Humanitarian Policy Group's conference 'Protecting People in Conflict and Crisis: Responding to the Challenges of a Changing World,' University of Oxford. September 2009, on file with author.

22   Warner K 2009 Global Environmental Change and Migration: Governance challenges, submitted to Global Environmental Change, on file with author.

23   Protection of Internally Displaced Persons in Situations of Natural Disasters, Report of the Representative of the Secretary-General on the Human Rights of Internally Displaced Persons, Walter Kälin, to the Human Rights Council, tenth session, A/HRC/10/13/ Add.1, 5 March 2009.

24   Protection of Internally Displaced Persons in Situations of Natural Disasters, Report of the Representative of the Secretary-General on the Human Rights of Internally Displaced Persons, Walter Kälin, to the Human Rights Council, tenth session, A/HRC/10/13/ Add.1, 5 March 2009.

25   Barnett J and Webber M, 2009, Accommodating Migration to Promote Adaptation to Climate Change, a policy brief prepared for the Secretariat of the Swedish Commission on Climate Change and Development and the World Bank World Development Report 2010 team, available on www.ccdcommission.org/Filer/documents/Accommodating%20Migration.pdf.

26   Kälin W 2008 Guiding Principles on Internal Displacement: Annotations, Studies in Transnational Legal Policy, No. 38. Washington DC: The American Society of International Law and The Brookings Institution – University of Bern Project on Internal Displacement, available at www.asil.org/pdfs/stlp.pdf.

27   Protection of Internally Displaced Persons in Situations of Natural Disasters, Report of the Representative of the Secretary-General on the Human Rights of Internally Displaced Persons, Walter Kälin, to the Human Rights Council, tenth session, A/HRC/10/13/ Add.1, 5 March 2009.

28   This article is based on Kolmannskog V, The Point of No Return, Refugee Watch, December 2009; Kolmannskog V and Myrstad F, Environmental Displacement in European Asylum Law, European Journal of Migration and Law, Volume 11, Issue 4, 2009, pages 313–326; and the joint submission to the climate change negotiations Forced Displacement in the Context of Climate Change: Challenges

for States under International Law, 15 May 2009, available at http://unfccc.int/resource/docs/2009/smsn/igo/049.pdf.

29   Guardian 2008 Paradise almost lost: Maldives seek to buy a new homeland, 10 November 2008, available at http:// www.guardian.co.uk/environment/2008/ nov/10/maldives-climate-change.

30   Inside Story 2009 We aren't refugees, 29 June 2009, available at http://inside. org.au/we-arent-refugees.

31   McAdam and Sau 2008 An Insecure Climate for Human Security? Climate-Induced Displacement and International Law, Sydney Centre Working Paper No. 4. University of Sydney, available at http://www.law.usyd.edu.au/scil/ pdf/SCIL%20WP%204%20Final.pdf.

32   Climate Change and Statelessness: An Overview, submitted by UNHCR supported by IOM and NRC to the UNFCCC, 15 May 2009, available at http://www.unhcr.org/refworld/ docid/4a2d189d3. html.

33   Handbook on Procedures and Criteria for Determining Refugee Status under the 1951 Convention and the 1967 Protocol relating to the Status of Refugees, para. 53, available at http://www.unhcr. org/publ/PUBL/3d58e13b4.pdf.

34   Climate Change, Natural Disasters and Human Displacement: A UNHCR Perspective, 14 August 2009, available at http://www.unhcr.org/refworld/ docid/4a8e4f8b2.html.

35   See for example Kolmannskog, 2008, Future floods of refugees, Oslo: Norwegian Refugee Council, available at http://www.nrc.no/arch/_ img/9268480.pdf.

36   Edwards A, 2006, Refugee Status Determination in Africa, RADIC, 14.

37   International Conference on Central American Refugees, United Nations High Commissioner for Refugees 1989, Document CIREFCA/89/9.

38   See for example Kolmannskog, 2008, Future floods of refugees, Oslo: Norwegian Refugee Council, available at http://www.nrc.no/arch/_img/9268480.pdf; and Climate Change, Natural Disasters and Human Displacement: A UNHCR Perspective, 14 August 2009, available at http://www. unhcr. org/refworld/docid/4a8e4f8b2. Html.

39   Report of the Office of the United Nations High Commissioner for Human Rights on the relationship between climate change and human rights, UN Doc A/HRC/10/61, New York: United Nations, available at http://ochaonline.un.org/OchaLink- Click.aspx?link=ocha&docId=1116540.

40   Bradley, 2007, Return in Dignity: A Neglected Protection Challenge, RSC Working Paper No. 40, University of Oxford, available at http://www.rsc. ox.ac.uk/PDFs/RSCworkingpaper40.Pdf.

41   Kälin, W, 2008, Displacement Caused by the Effects of Climate Change: Who will be affected and what are the gaps in the normative framework for their protection? available at: http://www. brookings.edu/papers/2008/1016_climate_change_kalin.aspx.

42   Frelick and Kohnen, 1995, Filling the gap: Temporary Protected Status, Journal of Refugee Studies, 8(4), p 339–363.

43   Kolmannskog V and Myrstad F, Environmental Displacement in European Asylum Law, European Journal of Migration and Law, Volume 11, Issue 4, 2009, pages 313–326.

44   Kolmannskog V and Myrstad F, Environmental Displacement in European Asylum Law, European Journal of Migration and Law, Volume 11, Issue 4, 2009, pages 313–326.

[…]

47   The following submissions from humanitarian agencies to the UNFCCC specifically address migration and displacement:
1) Change, Migration and Displacement: Who will be affected? Working paper submitted by the informal group on Migration/Displacement and Climate Change of the IASC – 31 October 2008 to the UNFCCC Secretariat, available at http://unfccc.int/resource/ docs/2008/smsn/igo/022.pdf.
2) Climate change, migration and displacement: impacts, vulnerability and adaptation options, Submission by the IOM, UNHCR and UNU, in cooperation with NRC and the RSG on the Human Rights of IDPs, 6 February 2009, available at http:// unfccc.int/resource/docs/2008/smsn/ igo/031.pdf.
3) Forced displacement in the context of climate change: Challenges for states under international law, Submission by UNHCR in cooperation with NRC, the RSG on the Human Rights of IDPs and

UNU, 15 May 2009, available at http://unfccc.int/resource/docs/2009/ smsn/igo/049.pdf.

4) Climate change and statelessness: An overview, Submission by UNHCR supported by IOM and NRC 15 May 2009, available at http://unfccc.int/ resource/docs/2009/smsn/igo/048.pdf.

48   Available at http://www.unisdr.org/ eng/hfa/hfa.htm.

49   FCCC/AWGLCA/2008/16/Rev.1, available at: http://unfccc.int/resource/ docs/2008/awglca4/ eng/16r01.pdf.

50   The first draft negotiation text is available at: http://unfccc.int/documentation/documents/ advanced_search/items/3594. php?rec=j&priref=600005243#beg

51   "Comments and Proposed Revisions to the Negotiating Text Prepared by the Chair of the UNFCCC Ad Hoc Working Group on Long-Term Cooperative Action", available at http:// www.unhcr. org/4a408cc19.html.

52   The revised draft negotiation text is available at: http://unfccc.int/resource/docs/2009/awglca6/ eng/inf01.pdf.

53   The adaptation non-paper 53 of 6th November 2009 is available at http://unfccc.int/files/meetings/ ad_hoc_working_groups/lca/application/pdf/awglcaadaptnp53061109.pdf.

Chapter 4

Steve Lonergan

THE ROLE OF ENVIRONMENTAL DEGRADATION IN POPULATION DISPLACEMENT

[...]

Introduction

The UNHCR in the 1993 State of the World's Refugees identified four root causes of refugee flows. These were: political instability; economic tensions; ethnic conflict; and environmental degradation. The claim that environmental degradation was a root cause of refugee flows was a direct response to a growing number of articles positing a link between environmental degradation and population movement, and a recognition that the numbers of displaced persons internationally was much larger than indicated by the statistics on refugee flows.

According to many writers, the number of people who have been displaced by environmental degradation is immense. Jacobson (1988) notes that, "environmental refugees have become the single largest class of displaced persons in the world."

Homer-Dixon (1991) further notes that environmental degradation is likely to produce "waves of environmental refugees that spill across borders with destabilizing effects" on domestic order and international relations. Speaking of displaced persons unaccounted for in official refugee figures, the Executive Director of UNEP at the time, Mustafa Tolba (1985), stated that "these people are the millions fleeing the droughts of northern Africa, the victims of Bhopal and the thousands made homeless by the Mexico earthquake. They are environmental refugees." Estimates of the number of environmental refugees start at 10 million (compared to 17 million official refugees); more than half of these are believed to be in Sub-Saharan Africa (Jacobson, 1988; Trolldalen, et. al., 1992; Westing, 1992). Because governments generally take little official account of this unconventional category, Myers (1992) estimates that the numbers may be as high as 25 million. It is also claimed that the numbers are increasing rapidly. The Intergovernmental Panel on Climate Change (IPCC, 1990) noted that the greatest effect of climate change may be on human migration as millions of people will be displaced due to shoreline erosion, coastal flooding and agricultural disruption. Following from this, Myers (1992) projected environmental refugees in a greenhouse-affected world (in yr. 2050) at 150 million persons. Westing (1992) further documented displaced persons throughout the world in 1990 (using UN data), including officially recognized refugees (16.7 million), unrecognized, cross-border "refugees" (3.5 million), and unrecognized, internal "refugees" (21.3 million). He sums these into a category of "total national refugees" with 41.5 million persons. In 1986, the total was only 26.4 million, and he speculates that the growth is due to the addition of "environmental refugees."

The consideration for people who may have been displaced by environmental degradation has reached far beyond a humanitarian concern for a disenfranchised population; in some quarters, it is being considered a "threat to security." Betterton (1992, as cited in Honebrink, 1993) noted that the U.S. military may be needed "to guard the border with Mexico, as it is expected that problems may result from environmental refugees fleeing the Third World."

Indeed, the anti-immigration literature in the United States and Europe often claims that immigration is a cause of environmental degradation, thereby bringing the links full circle (see, for example, Beck, 1996; Williamson, 1996; and the literature distributed by FAIR, the Federation of Americans for Immigration Reform). Quotes like the ones below are becoming increasingly prevalent in the popular literature.

> It is not antihuman or antisocial to say that too many people can be a problem…. People pollute, and too many people living in an area can degrade that area irrevocably. Immigration at high levels exacerbates our resource and environmental problems. It will leave a poorer, more crowded, more divided country for our children (Lamm and Imhoff, 1985).

> …Immigration has been a substantial cause of the negative environmental news that must be mixed among all the good…. Thus, to what extent environmental problems can be blamed on U.S. population growth, the preponderance of that blame rests on U.S. immigration policy. Only a reduction in numbers will deal with the environmental problem. (Beck, 1996).

While some may feel that such claims are little more than disguised racism—a "greening of hate" might be a better term—it is important to accept that the issue of environmental degradation and population displacement has reached a level of "high politics" discourse. This is true whether viewing environmental degradation as a "cause" or an "effect."

The purpose of this paper is to clarify the myriad of issues surrounding the linkage between environmental degradation and population displacement. The presentation on the following pages adopts a problem-based approach, attempting to answer crucial questions regarding, for example, the evidence of a link and the potential policy implications of the existing research. In addition, the concern is only with environment as a possible cause of, or contributor to, population movement, as opposed to the potential environmental repercussions associated with population movement. The latter concern, while very much in the public debate, has been addressed elsewhere (see Li and Lonergan, forthcoming).

The Role of Environment in Migration Movements

Migration has been described as "an extremely varied and complex manifestation and component of equally complex economic, social, cultural, demographic, and political processes operating at the local, regional, national, and international levels" (Castles and Miller, 1993). As complex as migration is, the environment is equally so. And it is similarly problematic to remove environmental processes from the social, economic, political and institutional structures of which they are a part. Therefore, drawing a linear, deterministic relationship between environmental degradation and migration (and security) is not only inappropriate, but impossible, despite the claims of some authors. Nevertheless, we can try to identify certain cases where environment plays an important role as a contributor to population movement, and attempt to design interventions to minimize the negative impacts associated with such cases.

1. How many refugees and migrants are there?

This is an almost impossible question to answer. The International Organization of Migration estimated that there were over 80 million migrants in 1990 (IOM, 1990). Fifteen million of these were refugees and asylum seekers. By 1992, estimates put the total number of migrants at over 100 million, of whom 20 million were refugees and asylum seekers (Castles and Miller, 1993). However, UNHCR (1995) acknowledges that collecting accurate statistical data on refugees and asylum-seekers is "one of the most problematic issues" confronting the agency, and these figures, indeed all figures cited in this article, must be treated with suspicion.

Nevertheless, rough estimates of the total number of displaced persons are often presented with abandon, either for shock value or for political reasons. Myers (1995) states that China has "120 million internal migrants, and at least...six million deserve to be regarded as environmental refugees." He goes on to say that there are now at least 25 million "environmental refugees" (Myers, 1995: 15). The International Organization for Migration (IOM, 1992) goes farther, noting that by the turn of the century there may be one billion persons who have been "environmentally displaced from their original habitat." Such claims lead to much confusion and fear on the part of many, and provide ample "evidence" for those wishing to promote anti-immigration rhetoric in the North.

2. Even if we cannot accurately estimate the number of migrants, what have traditionally been presented as the causes of migration flows?

The literature on migration is voluminous, and there will be no attempt to repeat this information here. Theories on the causes of migration flows can generally be categorized into two broad perspectives. The first is a "neo-classical economics equilibrium approach," which suggests that population movement is a "natural" response to interregional differences in social and economic opportunities, and people generally move from where labour is plentiful and capital is scarce to labour deficit and capital rich areas. Thus, the level of development in various regions of the globe is seen as determining the magnitude and direction of migratory streams. Extensions to the neo-classical approach explain population movements based on a combination of "push" and "pull" factors; existing conditions at the place of origin may motivate an individual to leave, or qualities of the area of destination may attract a potential migrant. Demographic pressures, political instability, lack of economic opportunities and, more recently, environmental degradation have been posed as possible "push" factors.

The second approach criticizes the neo-classical economic perspective for placing too much emphasis on the free choice of individuals, and for neglecting the macro-structural forces which lie at the base of the regional disparities to which people respond. Population movements are not unique or isolated events, but are related to the international power structure and institutional organization. According to this "structuralist" approach, the explanation for population movements lies in the deeper, underlying forces which structure the unequal distribution of opportunities between regions. Population movements, then, are a response to broader structural forces in society, in particular those associated with the uneven penetration of capitalism which has created substantial spatial inequalities.

The difference between neo-classical economic theories of population movements and the structuralist approach influences all aspects of any discussion regarding the issue. Not only do the theories offer opposing views of the causes of refugee movements, but they also imply very different outcomes. The neo-classical approach, arguing that population displacements are natural occurrences, suggests that they are positive events and that policy development should reflect and reinforce the beneficial aspects of these movements. The structuralist approach, however, emphasizes that population movements are a response to unnatural imbalances in power and opportunities. Consequently, the negative aspects of population displacements are a function of inequities in development, and policy should be developed to address these imbalances and attempt to stem what must be viewed as a consequence of the inequitable distribution of resources in society.

3. What role does the environment play as a contributor to population movement?

(a) The Advocates

Although there is growing awareness of, and interest in, the relationship between environmental change and population movement, the traditional literature on migration has largely ignored the connection. In their report to the Trilateral Commission (International Migration: Challenges in a New Era), Meissner et al, (1993) never once mention environment or resources. Rogers (1992) in his discussion on migration presents four key indicators of "migration potential:"

- population growth;
- economic restructuring;
- increasing economic disparities; and
- increased refugee flows.

Again, environment is not mentioned. Other recent reviews on the causes of migration which fail to include environmental degradation or resource depletion as factors include Appleyard, 1991 and Massey et al., 1993). This stands in stark contrast to the statements in The State of the World's Refugees (UNHCR, 1993), which clearly identify environmental degradation as a root cause of population displacement, as mentioned above (it is worth noting, however, that the 1995 volume by UNHCR does not make a similar claim).

Countering the traditional perspective on migration is a growing literature which claims that traditional theories fail to recognize the true extent and complexity of migratory responses to environmental degradation (cf. Hall and Hanson, 1992; Kavanagh and Lonergan, 1992; Fornos,

1993; Stoett, 1993; Lee, 1996; Suhrke, 1992, 1996; Vlachos, 1996). Most attention has focused on the plight of "ecological refugees" or "environmental refugees" (El-Hinnawi, 1985; UNHCR, 1993). While the World Commission on Environment and Development (WCED) identified environmentally-induced population displacement as a "recent phenomenon" (WCED, 1987), there is little doubt that throughout history people have had to move from their land because it has become degraded through natural disasters, warfare or over-exploitation. Intuitively, it makes sense that environmental change may affect socio-economic conditions which, in turn, could lead to out-migration. Indeed, recurrent droughts and extreme flooding have uprooted millions of people, although whether environmental catastrophes were the root cause of such movement is unclear.

The concern that environmental degradation will produce "waves of refugees," however, is more recent, based largely on the writings of El-Hinnawi (1985), Jacobson (1988) and Myers (1993; 1995). Suhrke (1992) labels this group the "maximalists." Supporting their arguments is the fact that environmental disasters such as floods, droughts and earthquakes are displacing ever larger numbers of people, not necessarily because the severity of these events is becoming greater,[2] but because population density, especially in regions which are prone to disaster, is increasing rapidly. Land and resource scarcity elsewhere may also be a strong contributor to these increases in density in vulnerable areas.

Since its first official use in 1985 by El-Hinnawi in his United Nations Environment Programme (UNEP) report, the phrase "environmental refugee" has appeared with increasing frequency in the literature on environment and development. "Environmental refugees" are defined by El-Hinnawi as:

> ...those people who have been forced to leave their traditional habitat, temporarily or permanently, because of a marked environmental disruption (natural and/or triggered by people) that jeopardized their existence and/or seriously affected the quality of their life. (El-Hinnawi, 1985, p.4)

Jacobson (1988) notes that "environmental refugees have become the single largest class of displaced persons in the world," with an estimated 10 million environmentally-displaced persons in the late 1980s, compared with 17 million official political refugees displaced by warfare, strife and persecution. And the conclusion by the UNHCR is unequivocal: "There are, nevertheless, clear links between environmental degradation and refugee flows" (UNHCR, 1993, p. 18). While the UNHCR claim may be true, it does not necessarily follow that environmental degradation has been the cause of a majority of "refugee" flows.

(b) The Contrarians

Despite these claims, it remains that there has been little substantive research directed at the question of the role of environmental change in population movement. Considerable confusion has arisen over definitions, the size of these "refugee" flows and whether one, indeed, can isolate environmental causes from the complex set of variables affecting population movement. While there is a sense that drastic environmental change may affect the structural forces which, in turn, link to population movement, the environment is seen as little more than a "contextual factor" which is taken into consideration in decision-making (Suhrke, 1992, labels this perspective the "minimalist"). The arguments presented by the "maximalists" (it is claimed) are ill-founded, and based on anecdotal information.

[...]

For example, Myers (1993) estimates that for every person who moves across an international boundary to escape environmental pressures there may be two or three similarly displaced people who move within their territory of nationhood—so-called "internally displaced persons." Myers adds these two categories of population movement together and estimates the total number of "environmentally displaced" persons to be as high as 25 million (he further predicts, as a worst case scenario, that this figure may increase to 150 million by the year 2050 as a result of the "greenhouse effect" and rising sea-levels). Westing (1992) speculates that the growth in the world's refugee and internally displaced population from 26.4 million in 1986 to 41.5 million in 1990 may have been attributable to environmental degradation, which has forced people from their land.

The writings noted above which have popularized the phenomenon of "environmental refugees" are problematic for reasons which are both definitional and substantive. First, the words "estimate" and "speculate" above are used advisedly: in most cases these figures are little more than educated guesswork—there is little empirical evidence with which to authenticate these authors' claims (Mougeot, 1992).

Second, there is too often an uncritical acceptance of a direct causal link between environmental degradation and population displacement. Implicit in these writings is the belief that environmental degradation—as a possible cause of population displacement—can be separated from other social, economic or political causes. It must be recognized that the degradation of the environment is socially and spatially constructed; only through a structural understanding of the environment in the broader political and cultural context of a region or country can one begin to understand the "role" it plays as a factor in population movement.

Third, not only are the definitions offered for environmental refugees ambiguous and inconsistent, the projections of future numbers do not take into consideration adaptation, there is no discussion of the role of public policy—or other factors—in the increase in the numbers of displaced people, and the analyses are, in most cases, quite superficial. Why do people continue to move into Mexico City and Chongqing, China, two of the most polluted places on Earth? Why does severe environmental degradation not generate large out-migration in many cases?

Last, some authors are concerned that there is no legal basis for the definition of "environmental refugee." Not only does this conflict with the standard definition of refugees which was codified in the 1951 Convention and 1967 Protocol relating to the Status of Refugees, but it may undermine current work towards using broader human rights criteria to determine refugee status (McGregor, 1993).

Despite these criticisms, it is important not to trivialize the potential role environmental change may play in population movement. It is entirely possible that the impact of environmental degradation and resource depletion on population movement may be even more important than these authors suggest.

4. What does "environment" mean in the context of migration?

Part of the difficulty in determining what role the "environment" plays as a cause of, or contributor to, population movement is that authors interpret "environment" quite broadly, or keep it ill-defined. El-Hinnawi (1985), for example, notes three categories of "environmental refugees:

- Those temporarily displaced because of an environmental stress such as an earthquake, or cyclone, and who will likely return to their original habitat;
- Those permanently displaced because of permanent changes to their habitat, such as dams or lakes; and
- Those who are permanently displaced desiring an improved quality of life because their original habitat can no longer provide for their basic needs.

In these three categories, El-Hinnawi has incorporated three very different groups of migrants. In the first case, there is a temporary movement from physical danger; the second category involves development projects where individuals are forced to resettle within a region (and there is a question how many "internal" refugees are generated by these processes); and the third reflects a voluntary movement based on the "push-pull" model noted above.

It is useful to categorize environmental stress, as follows (Lonergan, 1994):

NATURAL DISASTERS
Natural disasters include floods, volcanoes and earthquakes. They are usually characterized by a rapid onset, and their impact (destructiveness) is a function of the number of vulnerable people in the region rather than the severity of the disaster, per se. Poor people in developing countries are the most affected because they are the most vulnerable. (Droughts, despite a slower onset, are also included in this category.) Recent earthquakes in Pakistan and flooding in many regions of the world indicate not only the destructiveness of disasters, but their ability to displace large numbers of people.

CUMULATIVE CHANGES OR "SLOW-ONSET CHANGES"
Cumulative changes are generally natural processes occurring at a slower rate which interact with—and are advanced by—human activities. The processes include deforestation, land degradation, erosion, salinity, siltation, waterlogging, desertification and climate warming. Human-induced soil degradation is one factor which directly affects economic sufficiency in rural areas. Water availability is another factor which may affect sustainable livelihoods. Do factors such as water scarcity and human-induced soil degradation in and of themselves cause population displacement? The linkage is much more indirect; in most cases, one or more of rapid population growth, economic decline, inequitable distribution of resources, lack of institutional support and political repression are also present.

ACCIDENTAL DISRUPTIONS OR INDUSTRIAL ACCIDENTS
This category includes chemical manufacture and transport and nuclear reactor accidents. The two most obvious examples are the nuclear accident at Chernobyl, in the former USSR in 1986, and the Union Carbide accident in Bhopal, India, in 1987. Between 1986 and 1992, there were

over 75 major chemical accidents which killed almost 4,000 persons worldwide, injured another 62,000, and displaced over 2 million (UNEP, 1993). Most of these displacements, however, were temporary. In the case of the accident at Bhopal, despite the death of 2,800 people and illnesses to 200,000 more, there was virtually no mass movement of population out of the region.

DEVELOPMENT PROJECTS

Development projects which involve forced resettlement include dams and irrigation projects. In India, for example, it has been estimated that over 20 million persons have been uprooted by development projects in the past three decades (Fornos, 1992). The Three Gorges Dam project in China—expected to displace over 1 million persons—and the Sardar Sarovar Dam project in India are the most notable present examples. Rapid urbanization in some regions of the world is also forcing people from their land; conversion of agricultural land to urban uses has long been a phenomenon in the North, and increasingly this is the case in the South as well.

CONFLICT AND WARFARE

Environmental degradation is considered by many to be both a cause and effect of armed conflict. Although the evidence of wars being fought over the environment is weak (except, of course, over land), there is an increasing use of the environment as a "weapon" of war or, as Gleick (1990) notes, as a "strategic tool." One obvious example in this category was the threat by then President Ozal of Turkey to restrict the flow of the Euphrates to Syria and Iraq in order to pressure Syria to discontinue its support of Kurdish separatists in Turkey. Other examples include the purposeful discharge of oil into the Persian Gulf during the Gulf War and the destruction of irrigation systems during conflicts in Somalia. Such activities have similar and, indeed, more immediate consequences as the slow-onset changes noted above. But in these cases, it seems clear that the "environment" is merely a symptom of a larger conflict, and the root cause of any population movement is the conflict itself, and the reasons behind it.

5. How does one reconcile these different aspects of environment?

Collectively, it is claimed that these "environmental" changes have resulted in millions of displaced persons. The global deterioration of the environment, continued population growth, and increasing resource scarcity will likely play an increasing role in population movement in the future. But are these factors all "environmental?" And what are the links to migration?

To understand causal relationships, and to better design policy interventions, it is imperative that these five categories be treated separately, and not considered collectively as "environmental degradation." In some cases, there is minimal impact on population movement, while in others, the role of "environment" is extremely difficult to ascertain. It is clear, for example, that industrial accidents have had relatively little impact on migration, with the exception of Chernobyl. Most accidents have resulted in a short-term relocation, but very few (of the more than 2 million cited above) have been displaced permanently from their homes. In the context of other changes, this is a relatively minor concern.

Development projects, while there is little question that they displace large populations, should also be treated separately from other categories. The magnitude of some of the projects is, indeed, daunting, and it has caused the World Bank to avoid any projects which involve major resettlement programs (such as Sardar Sarovar in India). In theory, these projects include a resettlement component, and are unlikely to produce the "waves of environmental refugees" that Homer-Dixon cautions about.

The links between natural disasters and population displacement are also problematic. Sadako Ogata, the UN High Commissioner on Refugees, stated in 1992 that the "majority of refugees are found in arid and semi arid areas of the poorest countries of the world." Examples of the devastating impact of natural disasters, however, generally come from Bangladesh, Central America, Haiti and South Korea. There is little question that the number of people affected by natural disasters has increased markedly over the past three decades (from 28 million in the 1960s to 64 million in the 1980s). Population growth—particularly in vulnerable areas—and poverty have combined to make larger numbers of people susceptible to environmental disasters. And while the number of homeless is significant, it does not imply that these people migrated to different regions or countries.

Indeed, some authors claim that sudden onset disasters have resulted more in increased death rather than increased flight (Lee, 1996). The category of cumulative, or slow-onset change, may well be the most important in terms of being a force in population migration, but it is also the most difficult to measure. Environmental changes such as increased water scarcity and soil degradation may be one factor among many facing a potential migrant. As was noted before, removing environmental processes from the social, economic and political processes in which they are embedded is virtually impossible.

6. What is the evidence presented for a link between environment and migration?

Numerous examples are presented to substantiate the link between environmental change and population movement, but the most common are the Sahel in Africa, El Salvador, Haiti, and Bangladesh (El Hinnawi, 1985; Hall and Hanson, 1992; Surhke, 1992; Myers, 1995). There is little doubt that each of these regions/countries has experienced significant environmental stress: droughts, deforestation, soil degradation, and flooding are the most notable. But it is also clear that there are a myriad of other social, economic and institutional processes which are present. Rapid population growth, inequitable land distribution, civil war, extreme poverty, and so on. For example, the Kissinger Report of 1984 attributed the conflict in El Salvador to poverty and inequality; the conflict in the country has resulted in over a million people displaced. But what role did the environment play? Deforestation, exploitation of coastal resources, and the civil war have resulted in substantial environmental damage in the country (Hall and Hanson, 1992). In turn, as Leonard (1989) notes,

> If deterioration of these natural resource systems continues, political and social instability will be exacerbated as will economic stagnation and rural poverty. This phenomenon in turn will constrain future economic and social development in all seven countries of greater Central America.

Is environmental degradation a root cause of population movement in El Salvador? It likely played a role, but it was certainly not a root cause.

Another often used example is the Sahel, where droughts and famine have severely impacted people in almost every country in the region. But poverty, marginal agricultural land, institutional constraints, war, inflation and landlessness not only increased the vulnerability of the population to climate variation, but affected the ability of individuals and communities to adapt to a changing environment. The people became more vulnerable, not because of environmental degradation, per se, but because of a host of other social, economic and institutional factors. The same is true in all cases which are used as "evidence" of environmental refugees. The key factor is that certain populations are becoming more vulnerable to environmental change because of other factors; primary among these are poverty and resource inequality, coupled with population growth, institutional constraints, and economic insufficiency.

7. Is there evidence to the contrary? That environmental change is not linked to migration?

This question is equally problematic. Direct evidence refuting the claim that environmental factors influence population migration suffers the same difficulties of isolating one factor as all studies. Mougeot (1992) did review World Bank projects to determine if environment was a proximate cause of population movement and found no evidence of a connection, but the scope of this study was very limited. It is clear that there remains a need to better understand the linkages between environmental change and population displacement, to identify regions and populations most vulnerable to environmental degradation, and to lend support to the populations at risk. And despite the fact that evidence provided to identify the link between environmental degradation and population displacement is highly speculative, it is important not to trivialize the role the above factors increasingly may play in population movements. Individuals, families and communities have a remarkable ability to adapt to changing and distressed conditions, and the initial response is to develop stronger safety and coping mechanisms to deal with adverse ecological and economic circumstances. But continued environmental degradation and resource depletion coupled with increasing impoverishment in certain regions is placing a heavy burden on these adaptation responses, and they are becoming powerful impelling factors in population displacement.

8. What types of environmental problems might there be in the future which could affect migration?

The Intergovernmental Panel on Climate Change (IPCC) noted in 1990 that the greatest effect of climate change may be on human migration as millions of people will be displaced due to shoreline erosion, coastal flooding and agricultural disruption. Based on this, Myers (1992) projects "environmental refugees" in a greenhouse-affected world (by the middle of the next century) at 150 million persons. While this may be an overstatement, it is true that sea-level rise and coastal flooding will require significant adaptation on the part of some countries, particularly those which have large populations living within a meter of sea-level. The IPCC adds that up to 360,000 km of coastline might be affected. None of the estimates of migration associated with global warming gives any consideration to adaptation mechanisms. While there may be significant implications for some regions, these changes will occur slowly, and by all accounts, most communities and regions will be able to adapt without substantial social or economic cost. Again, the most vulnerable will be the poor, with few options in the face of environmental change.

Water scarcity and poor air quality are other problems which come to mind. But Amman, Jordan—with severe water scarcity—and Mexico City—with the world's worst levels of air pollution—are both growing very rapidly from immigration. Indeed, in some instances, it is easier to find cases of people moving to regions which have suffered environmental degradation than moving away from those regions.

Likely the greatest impact on people's decision to move will be degradation of the land, through deforestation and inappropriate agricultural practices. Salinization and waterlogging of irrigated land will reduce output and increase the economic discrepancies between regions. However, even land degradation is a gradual process, which allows for adaptation.

There is a need for further study of the adaptation mechanisms available to individuals and communities. How have regions coped with environmental stress? Why hasn't resource scarcity resulted in major migrations? What types of adaptation mechanisms can do nor nations assist with?

9. What conclusions can be drawn from the above information?

The four general conclusions below (some of which are adapted from Lonergan and Parnwell, forthcoming), reflect the answers to the questions above.

- Generalizations about the relationship between environmental degradation and population movement mask a great deal of the complexity which characterizes migration decision-making. Much of the literature suggests a deterministic cause and effect model where a set of environmental stresses will result in a similar response—migration— from individuals and communities. This may occur with certain forms of environmental catastrophe, where there is no option but to move. But in general such a model is very misleading. Levels of internal differentiation within communities are typically high, and thus people will have different levels of ability to cope with environmental stresses. Furthermore, people's "tolerance thresholds" are highly variable, being surpassed very readily in some (perhaps the more footloose members of a rural community), and being almost insurmountable in others (for instance, older residents who have a strong attachment to the home area and thus a built-in inertia). A proper appreciation and understanding of the complexity and diversity of human responses to environmental degradation is essential if we are to identify the full extent of the phenomenon and plan accordingly.
- It is extremely difficult to isolate the specific contribution of environmental change in many forms of population movement, especially those which are more "voluntary" in nature. It may be relatively easy to identify the parallel occurrence of environmental degradation and population movement, but assuming a causal link may be misleading and dangerous. In reality, movement takes place in response to a combination of environmental, economic, social and

political (including armed conflict) stimuli. Thus separating environmental processes from the structures within which they are embedded is both difficult and a distortion of reality.

- There is also an implicit assumption in the literature that movement is an assured means of obtaining relief from environmental pressures. Despite the ancient Chinese proverb that states "Of thirty ways to escape danger, running away is the best" (from El Hinnawi, 1985), it is not necessarily the case that movement always reduces environmental—or other—stress. In reality, movement may lead to the substitution of one set of stresses (environmental) for another (economic, social, political and/or further environmental stresses). Movers may have to accept whatever opportunities come their way in the new location.

- An important question—often overlooked where the central preoccupation is with identifying the volume of the migratory movement—concerns the future intentions of environmentally-displaced persons, not least with regard to the duration of their sojourn. Do migrants intend to return to their home area, if that option is available, or remain in their new location? The answer to this question will have a significant bearing upon their actions and behaviour in their place of refuge, and is also crucial to the planning pro reaching any one of these stages will be a function of the severity of the environmental crisis and the opportunities which become available to the displaced through movement.

Policy Recommendations

These four general conclusions underscore the difficulty in developing policy prescriptions to deal with the issue of environmental degradation and population movement. Migration is a complex phenomenon, and it is not clear what role environmental degradation plays in influencing a person's decision to migrate. It is also difficult, if not impossible, to isolate environment from other social, economic, and political factors. And there has been a dearth of research that focuses on individual or collective human perceptions and evaluations of actual and expected conditions of the environment as a source of insecurity and migration stress. Developing policy prescriptions in this context, therefore, is a risky enterprise, at best. However, accepting these difficulties, two sets of recommendations are presented below. The first set presents general policy recommendations for assisting communities and regions under environmental stress, particularly where that stress may contribute to population movement. The second set provides specific policy recommendations for agencies involved in setting refugee policy.

What types of policy recommendations can one make globally?

Despite the complex nature of migration flows, and the ongoing debate on the role of environmental degradation as a cause of, or contributor to, migration, there is little doubt that we need to give greater consideration to environmental deterioration and resource scarcity in our development assistance activities. This implies a major emphasis on promoting sustainable development and its ecological, economic and social manifestations, and ensuring human security. More specific recommendations include:

- Develop a system to help anticipate migrations which might be triggered by environmental disruptions;

- Focus efforts on identifying adaptation mechanisms, and how these mechanisms might be reinforced in vulnerable communities and regions;
- Develop case studies of how environmental degradation influences migration, with specific consideration of developing procedures to assist those affected by environmental disruptions;
- Develop better working relationships among human rights, environment, population and migration organizations;
- Involve migrants and refugees directly in the development of programs to assist those affected by environmental deterioration;
- Recognize the cumulative causality of environmental degradation and population movement, and assist receiving regions to ensure minimal environmental impacts of the migration flows;
- Provide assistance to countries most vulnerable to future environmental change; and
- Recognize that human rights and the environment—indeed, human security and all its components—should be the cornerstone of any assistance policies.

Can we make more specific policy recommendations that are relevant to government agencies?

As noted above, environmental degradation and resource depletion are only two of many factors that may contribute to insecurity and, as a response, population movement. Other key factors surely include population growth and an inequitable distribution of income and/or resources (often linked to impoverishment). The following quote from the World Commission on Environment and Development (WCED, 1987) is telling:

> ... Poverty is a major cause and effect of global environmental problems. It is therefore futile to attempt to deal with environmental problems without a broader perspective that encompasses the factors underlying world poverty and international equality.

This implies that policy prescriptions should focus on promoting sustainability in resource use, reducing rates of population growth, and addressing the inequitable distribution of income and access to resources between and within countries. Such policies should also incorporate activities which will assist in reducing both the biophysical and social vulnerability of individuals and communities to environmental change.
Examples include:

- An increase in support for family planning in developing countries. Since population growth is a threat to the environment and to the economic livelihood of many people, it is imperative that birth rates are brought down.
- There must be greater focus on agricultural activities in developing countries. This should focus on reducing erosion and deforestation, and increasing the sustainability of small farms in marginal areas.
- Greater effort should be made to improve education and awareness with respect to the environment. This includes care for the environment and sustainable resource use.
- In this context, an adequate supply of freshwater is crucial. It is also imperative that treated water be recycled to agricultural uses. Inefficient use of water, water loss in urban areas, and the lack of systems to use recycled water greatly affect social welfare.

• There must be greater capacity building in the administration of environmental programs. This ranges from increased support for NGOs in the environmental field to the development of government agencies that can participate in international environmental work.

The complex nature of environment – population linkages makes it difficult to develop policy recommendations that are as concrete as many would like. However, it is apparent that environmental degradation and resource depletion may play a contributing role in affecting population movement, often filtered through contexts of poverty and inequity. In turn, it is clear that some population movements—particularly large scale, mass movements—have a negative impact on the natural environment of receiving regions. In order to develop a more concise policy agenda, it is imperative that further attention be given to the links among environment, population and poverty; to which groups are most vulnerable to environmental change; and to identifying vulnerable regions and future "hot spots" of insecurity and potential migration / refugee pressure.

[…]

From Steve Lonergan, '*The Role of Environmental Degradation in Population Displacement'*, Environmental Change and Security Project Report, Issue 4, Spring 1998, pp. 5–15.

Note

1. Steve Lonergan is Chair of the Global Environmental Change and Human Security Project (GECHS) of the International Human Dimensions Programme on Global Environmental Change, and Professor of Geography, University of Victoria, Canada. Excerpted with permission from The Role of Environmental Degradation in Population Displacement, of the GECHS Project, Research Report 1, 1998.

Chapter 5

Oli Brown

CLIMATE CHANGE AND FORCED MIGRATION
Observations, Projections and Implications

Summary

In 1990, the Intergovernmental Panel on Climate Change (IPCC) noted that the greatest single impact of climate change could be on human migration—with millions of people displaced by shoreline erosion, coastal flooding and agricultural disruption. Since then various analysts have tried to put numbers on future flows of climate migrants (sometimes called 'climate refugees')—the most widely repeated prediction being 200 million forced climate migrants by 2050.

But repetition does not make the figure any more accurate. While the scientific argument for climate change is increasingly confident, the consequences of climate change for human population distribution are unclear and unpredictable. With so many other social, economic and environmental factors at work, establishing a linear, causative relationship between anthropogenic climate change and forced migration has, to date, been difficult.

This may change in future. The available science, summarised in the latest assessment report of the IPCC, translates into a simple fact: on current predictions the 'carrying capacity' of large parts of the world will be compromised by climate change.

The meteorological impact of climate change can be divided into two distinct drivers of migration: *climate processes* such as sea-level rise, salinisation of agricultural land, desertification and growing water scarcity, and *climate events* such as flooding, storms and glacial lake outburst floods. But non-climate drivers, such as government policy, population growth and community-level resilience to natural disaster, are also important. All contribute to more vulnerable people living in more marginal areas.

The problem is one of time (the speed of change) and scale (the number of people it will affect). But the simplistic image of a coastal farmer being forced to pack up and move to a rich country is not typical. On the contrary, as is already the case with political refugees, it is likely that the burden of providing for climate migrants will be born by the poorest countries—those least responsible for emissions of greenhouse gases.

Temporary migration as an adaptive response to climate stress is already apparent in many areas. But the picture is nuanced; the ability to migrate is a function of mobility and resources (both financial and social). In other words, the people most vulnerable to climate change are not necessarily the ones most likely to migrate.

Predicting future flows of forced climate migrants is complex; stymied by a lack of baseline data, distorted by population growth and reliant on the evolution of climate change as well as the quantity of future emissions. Nonetheless this paper sets out three broad scenarios, based on differing emissions forecasts, for what we might expect. These range from the best case scenario where serious emissions reduction takes place and a 'Marshall Plan' for adaptation is put in place, to the 'business as usual' scenario where the large scale migration foreseen by some analysts comes true, or is exceeded.

Forced migration hinders development in at least four ways: by increasing pressure on urban infrastructure and services, by undermining economic growth, by increasing the risk of conflict and by leading to worse health, educational and social indicators among migrants themselves.

There has been a collective, and rather successful, attempt to ignore the scale of the problem. Forced climate migrants fall through the cracks of international refugee and immigration policy—and there is considerable resistance to the idea of expanding the definition of political refugees to incorporate climate 'refugees'. Meanwhile, large scale migration is not taken into account in national adaptation strategies which tend to see migration as a 'failure of adaptation'. So far there is no 'home' for forced climate migrants in the international community, both literally and figuratively.

1. Introduction

A growing crisis

As early as 1990 the Intergovernmental Panel on Climate Change (IPCC) noted that the greatest single impact of climate change might be on human migration—with millions of people displaced by shoreline erosion, coastal flooding and agricultural disruption.[2] Since then, successive reports have argued that environmental degradation, and in particular climate change, is poised to become a major driver of population displacement—a crisis in the making.

In the mid 1990s it was widely reported that up to 25 million people had been forced from their homes and off their land by a range of serious environmental pressures including pollution, land degradation, droughts and natural disasters. At the time it was declared that these 'environmental refugees', as they were called (see Box 1), exceeded all documented refugees from war and political persecution put together.[3]

The 2001 World Disasters Report of the Red Cross and Red Crescent Societies repeated the estimate of 25 million current 'environmental refugees'. And in October 2005 the UN University's Institute for Environment and Human Security warned that the international community should prepare for 50 million environmental refugees by 2010.[4]

A few analysts, of whom Norman Myers of Oxford University is perhaps the best known, have tried to estimate the numbers of people who will be forced to move over the long term as a direct result of climate change. "When global warming takes holds" Professor Myers argues, "there could be as many as 200 million people overtaken by disruptions of monsoon systems and other rainfall regimes, by droughts of unprecedented severity and duration, and by sea-level rise and coastal flooding".[5]

Box 1. Refugee or migrant?

Labels are important. One immediately contentious issue is whether people displaced by climate change should be defined as 'climate refugees' or as 'climate migrants'. This is not just semantics—which definition becomes generally accepted will have very real implications for the obligations of the international community under international law. Campaigners have long used the phrase 'environmental refugee' or 'climate refugee' to convey added urgency to the issue. They argue that, in the most literal sense of the words, such people need to 'seek refuge' from the impacts of climate change. Any other terminology, they maintain, would downplay the seriousness of these people's situation. The word 'refugee' resonates with the general public who can sympathise with the implied sense of duress. It also carries fewer negative connotations than 'migrant' which tends to imply a voluntary move towards a more attractive lifestyle. However, the use of the word 'refugee' to describe those fleeing from environmental pressures is not strictly accurate under current international law. The United Nations' 1951 Convention and 1967 Protocol relating to the status of refugees are clear that the term should be restricted to those fleeing persecution: "a refugee is a person who owing to a well-founded fear of being persecuted for reasons of race, religion, nationality, membership of a particular social group, or political opinion, is outside the country of his nationality, and is unable to or, owing to such fear, is unwilling to avail himself of the protection of that country".[12]

There are other problems with using the term 'refugee'. Strictly speaking, categorization as a refugee is reliant on crossing an internationally recognised border: someone displaced within their own country is an 'internally displaced person' (IDP). Given that on current predictions the majority of people displaced by climate change will stay within their own borders, restricting the definition to those who cross international borders may seriously understate the extent of the problem. Second, the concept of a 'refugee' tends to imply a right of return once the persecution that triggered the original flight has ceased. This is, of course, impossible in the case of sea level rise and so again the term distorts the nature of the problem. Third, and perhaps most importantly, there is concern that expanding the definition of a refugee from political persecution to encompass environmental stressors would dilute the available international mechanisms and goodwill to cater for existing refugees.

The question of definition makes for a hotly contested debate amongst international human rights lawyers.[13] However, in practice there is considerable resistance among the international community to any expansion of the definition of a 'refugee'. Developed countries fear that accepting the term refugee would compel them to offer the same protections as political refugees; a precedent that no country has yet been willing to set.[14] Meanwhile, the international institutions currently charged with providing for refugees, principally the office of the United Nations High Commissioner for Refugees (UNHCR), are already overstretched and are unable to cope with their current 'stock' of refugees.[15]

The UNHCR itself is taking on an expanded role in the provision of care to IDPs and so is highly resistant to any further expansion of its mandate.[16,17]

If the term 'climate refugee' is problematic it is still used, in part, for lack of a good alternative. 'Climate evacuee' implies temporary movement within national borders (as was the case with Hurricane Katrina). 'Climate migrant' implies the 'pull' of the destination more than the 'push' of the source country and carries negative connotations which reduce the implied responsibility of the international community for their welfare. But for lack of an adequate definition under international law, environmental migrants are almost invisible in the international system: no institution is responsible for collecting data on their numbers, let alone providing them with basic services. Unable to prove political persecution in their country of origin they fall through the cracks in asylum law.

How then should we categorize these people? One proposed definition from Jeff Crisp of the UNHCR is, "People who are displaced from or who feel obliged to leave their usual place of residence, because their lives, livelihoods and welfare have been placed at serious risk as a result of adverse environmental, ecological or climatic processes and events".[18] This definition makes no reference to cross-border movement or whether the movement is temporary or permanent but does describe an element of compulsion with varying degrees of threat (to people's welfare, livelihoods or lives).

This study uses the term 'forced migrant' in the knowledge that it is not a universally accepted term but in the hope that it conveys a reasonably accurate impression of the increasing phenomenon of non-voluntary population displacement likely as the impacts of climate change grow and accumulate.

200 million climate migrants by 2050?

Professor Myers' estimate of 200 million climate migrants by 2050 has become the accepted figure cited in respected publications from the IPCC to the Stern Review on the Economics of Climate Change.[6]

This is a daunting figure; representing a ten-fold increase over today's entire documented refugee and internally displaced populations.[7] To put the number in perspective it would mean that by 2050 one in every forty-five people in the world will have been displaced by climate change. It would also exceed the current global migrant population: according to the International Organization for Migration about 192 million people, or 3 percent of the world's population, now live outside their place of birth.[8]

But this prediction is still very tentative. Professor Myers himself admits that his estimate, although calculated from the best available data, required some 'heroic extrapolations'.[9] Not that any criticism is implied; the simple fact is that nobody really knows with any certainty what climate change will mean for human population distribution.

A complex, unpredictable relationship

The scientific basis for climate change is increasingly well established. An enormous amount of time and energy have gone into determining the meteorological impacts of climate change in terms of raised sea levels, altered precipitation patterns and more frequent and fierce storms. Much less time, energy and resources, however, have been spent on empirical analysis of the impacts of climate change on human populations.

Partly, this is because the relationship is so unpredictable: the science of climate change is complex enough – let alone its impact on societies of differing resources and varied capacity to adapt to external shocks. Partly, it is because individual migrants' decisions to leave their homes vary so widely: deciding causality between economic 'pull' and environmental 'push' is often highly subjective. And finally, disaggregating the role of climate change from other environmental, economic and social factors requires an ambitious analytical step into the dark. In short, drawing a causative, linear line between climate change and forced migration is very difficult.

For example, Hurricane Katrina, which lashed the Gulf Coast of the United States in August 2005 and temporarily displaced over a million people,[10] is often presented (quite rightly) as a preview of the kind of more intense and frequent extreme weather events we can expect from climate change. But the hurricane was more than just a meteorological event: the damage it caused was a product of poor disaster planning, consistent underinvestment in the city's protective levees as well the systematic destruction of the wetlands in the Mississippi delta that might have lessened the force of the storm. Labelling it a 'climate change event' over-simplifies both its causes and its effects.

Nevertheless, estimates of future numbers of climate change migrants are repeated almost glibly, either for shock value or for want of a better figure.[11] This paper sets out to challenge the predictions by trying to pick apart the terminology, the time frame and the degree of uncertainty implicit in them.

Section 2 looks at the ways that climate change might lead to increased forced migration. Section 3 then analyses some predictions for numbers of future climate migrants, examines some of the uncertainties with these predictions and lays out three different tentative scenarios on future numbers of forced migrants. Which (if any) of these comes to pass depends on future population growth, distribution and resilience to environmental pressures as well as the ability of the international community to curb greenhouse gas emissions and help the poorest countries adapt to the impacts of climate change. Section 4 assesses the development implications of forced migration within countries and across borders. Finally, section 5 investigates a variety of international and domestic policy responses to the prospect of large scale population movements caused by climate change.

2. Climate change and forced migration

Not such a wonderful world

Put simply: climate change will cause population movement by making certain parts of the world much less viable places to live; by causing food and water supplies to become more unreliable and increasing the frequency and severity of floods and storms. Recent reports from the IPCC and elsewhere set out the parameters for what we can expect:

By 2099 the world is expected to be on average between 1.8 and 4°C hotter than it is now.[19] Large areas are expected to become drier—the proportion of land in constant drought expected to increase from 2 percent to 10 percent by 2050.[20] Meanwhile, the proportion of land suffering extreme drought is predicted to increase from 1 percent at present to 30 percent by the end of the 21st century.[21] Rainfall patterns will change as the hydrological cycle becomes more intense. In some places this means that rain will be more likely to fall in deluges (washing away top-soil and causing flooding).

Changed rainfall patterns and a more intense hydrological cycle mean that extreme weather events such as droughts, storms and floods are expected to become increasingly frequent and severe.[22] For example, it is estimated that the South Asian monsoon will become stronger with up to 20 percent more rain falling on eastern India and Bangladesh by 2050.[23] Conversely, less rain is expected at low to mid-latitudes; by 2050 sub-Saharan Africa is predicted to have up to 10 percent less annual rainfall in its interior.[24]

Less rain would have particularly serious impacts for sub-Saharan African agriculture which is largely rain-fed: the 2007 IPCC report of the Second Working Group estimates that yields from rain fed agriculture could fall by up to 50 percent by 2020.[25] "Agricultural production, including access to food, in many African countries and regions is projected to be severely compromised by climate variability and change" the report notes.[26]

According to the same report, crop yields in central and south Asia could fall by 30 percent by the middle of the twenty-first century.[27] Some fish stocks will migrate towards the poles and colder waters and may deplete as surface water run-off and higher sea temperatures lead to more frequent hazardous algal blooms and coral bleaching.[28] Compounding this, climate change is predicted to worsen a variety of health problems leading to more widespread malnutrition and diarrhoeal diseases, and altered distribution of some vectors of disease transmission such as the malarial mosquito.[29]

Meanwhile, melting glaciers will increase the risk of flooding during the wet season and reduce dry-season water supplies to one-sixth of the world's population, predominantly in the Indian sub-continent, parts of China and the Andes.[30] Melting glaciers will increase the risk of glacial lake outburst floods particularly in mountainous countries like Nepal, Peru and Bhutan.

Global average sea level, after accounting for coastal land uplift and subsidence, is projected to rise between 8cms and 13cms by 2030, between 17cms and 29cms by 2050, and between 35cms and 82cm by 2100 (depending on the model and scenario used).[31] Thermal expansion of sea water accounts for nearly two-thirds of this rise with glacial melt providing the rest.[32] Large delta systems are at particular risk of flooding.[33]

According to Nicholls and Lowe, using a mid-range climate sensitivity projection, the number of people flooded per year is expected to increase by between 10 and 25 million per year by the 2050s and between 40 and 140 million per year by 2100s, depending on the future emissions scenario.[34]

The area of coastal wetlands is projected to decrease as a result of sea level rise. For a high emissions scenario and high climate sensitivity, wetland loss could be as high as 25% and 42% of the world's existing coastal wetlands by the 2050s and 2100s respectively.[35]

The avalanche of statistics above translates into a simple fact—that on current trends the 'carrying capacity' of large parts of the world, i.e. the ability of different ecosystems to provide food, water and shelter for human populations, will be compromised by climate change.

Climate processes and climate events

Robert McLeman of the University of Ottawa, unpacks the drivers of forced migration into two distinct groups.[36] First, there are the climate drivers. These themselves are of two types—climate processes and climate events. *Climate processes* are slow-onset changes such as sea-level rise, salinisation of agricultural land, desertification, growing water scarcity and food insecurity. Sea-level rise patently makes certain coastal areas and small island states uninhabitable. Cumulatively they erode livelihoods and change the incentives to 'stick it out' in a particular location. Some women in the Sahel, for example, already have to walk up to 25 kilometres a day to fetch water. If their journey gets longer they will simply have to move permanently.[37]

On a national level, sea level rise could have serious implications for food security and economic growth. This is a particular concern in countries that have a large part of their industrial capacity under the 'one metre' zone. Bangladesh's Gangetic plain and the Nile Delta in Egypt, which are breadbaskets for both countries, are two such examples. Egypt's Nile Delta is one of the most densely populated areas of the world and is extremely vulnerable to sea level rise. A rise of just 1 metre would displace at least 6 million people and flood 4,500 km^2 of farmland.[38]

Climate events, on the other hand, are sudden and dramatic hazards such monsoon floods, glacial lake outburst floods, storms, hurricanes and typhoons. These force people off their land much more quickly and dramatically. Hurricanes Katrina and Rita, for example, which lashed the Gulf Coast of the United States in August and September 2005 left an estimated 2 million people homeless.[39] The 2000 World Disasters Report estimated that 256 million people were affected by disasters (both weather-related and geo-physical) in the year 2000, up from an average of 211 million per year during the 1990s—an increase the Red Cross attributes to increased 'hydro-meteorological' events.[40]

Non-climate drivers

Equally important though are the non-climate drivers. It is clear that many natural disasters are, at least in part, 'man-made'. A natural hazard (such as an approaching storm) only becomes a 'natural disaster' if a community is particularly *vulnerable* to its impacts. A tropical typhoon, for example, becomes a disaster if there is no early-warning system, the houses are poorly built and people are unaware of what to do in the event of a storm. A community's vulnerability, then, is a function of its *exposure* to climatic conditions (such as a coastal location) and the community's *adaptive capacity* (the capacity of a particular community to weather the worst of the storm and recover after it).

Different regions, countries and communities have very different adaptive capacities: pastoralist groups in the Sahel, for example, are socially, culturally and technically equipped to deal with a different range of natural hazards than, say, mountain dwellers in the Himalayas.[41] National and individual wealth is one clear determinant of vulnerability—enabling better disaster risk reduction, disaster education and speedier responses. In the decade from 1994 to 2003 natural disasters in countries of high human development killed an average of 44 people per event, while disasters in countries of low human development killed an average of 300 people each.[42]

On a national scale, Bangladesh has very different adaptive capacities and disaster resilience to the United States. In April 1991 Tropical Cyclone Gorky hit the Chittagong district of south-eastern Bangladesh. Winds of up to 260 kilometres per hour and a six metre high storm surge battered much of the country killing at least 138,000 people and leaving as many as 10 million people homeless.[43] But the following year in August 1992, a stronger storm, the category five Hurricane Andrew, hit Florida and Louisiana with winds of 280 kilometres per hour and a 5.2 metre storm surge but, while it left $43 billion in damages in its wake, it caused only 65 deaths.[44]

Climate change will challenge the adaptive capacities of many different communities, and overwhelm some, by interacting with and exacerbating existing problems of food security, water scarcity and the scant protection afforded by marginal lands. At some point that land becomes no longer capable of sustaining livelihoods and people will be forced to migrate to areas that present better opportunities. The 'tipping points' will vary from place to place and from individual to individual. Natural disasters might displace large numbers of people for relatively short periods of time, but the slow-onset drivers are likely to displace permanently many more people in a less headline grabbing way.

Population, poverty and governance are key variables

Migration, even forced migration, is not usually just a product of an environmental 'push' from a climate process like sea level rise. Except in cases of climate events, where people flee for their lives, it does require some kind of 'pull': be it environmental, social or economic. There has to be the hope of a better life elsewhere, however much of a gamble it might be. Past environmental migratory movements, such as in the US Dust Bowl years in the 1930s, suggest that being able to migrate away from severe climatic conditions, in this case prolonged drought, requires would-be migrants to have some 'social and financial capital' such as existing support networks in the destination area and the funds to be able to move.[45]

It also should be mentioned, and this is absent from much of the campaigning literature, that climate change will make some places better able to sustain larger populations. This is particularly reflected in predictions for less severe total temperature rises, i.e. 2–3 degrees celsius over the 21st century rather than rise of 4–5 degrees or more. This is for three main reasons. First, higher temperatures will likely extend growing seasons and reduce frost risk in mid- to high-latitude areas such as Europe, Australia and New Zealand and make new crops viable (already vineyards are spreading north in Britain).[46] Second, the 'fertilization effect' of more CO_2 in the atmosphere is predicted to increase crop yields and the density of vegetation in some areas.[47] And third, altered rainfall patterns mean that rain might increase in areas previously suffering water stress. A 2005 study, for example, predicts that a warmer north Atlantic and hotter Sahara will trigger more rain for the Sahel.[48]

In other words, climate change might provide both 'push' and 'pull' for some population displacement. This is not to downplay the seriousness of climate change: above 4 or 5 degrees celsius, the predicted impacts of climate change become almost universally negative.[49] But it is to make that point that the role of climate change in population displacement is not a linear relationship of cause and effect, of environmental 'push' and economic 'pull'.

Non-climatic drivers remain a key variable. It is, after all, population growth, income distribution and government policy that push people to live on marginal lands in the first place. In other words a community's vulnerability to climate change is not a constant—it can be increased or decreased for reasons that have nothing to do with greenhouse gas emissions.[50] In this sense the non-climatic drivers (that put vulnerable people in marginal situations) can be as important a determinant of the problem as the strength of the 'climate signal' itself.

As Steve Lonergan of the University of Victoria, Canada, noted in 1998, "there is too often an uncritical acceptance of a direct causal link between environmental degradation and population displacement. Implicit in these writings is the belief that environmental degradation—as a possible cause of population displacement—can be separated from other social, economic or political causes. It must be recognized that the degradation of the environment is socially and spatially constructed; only through a structural understanding of the environment in the broader political and cultural context of a region or country can one begin to understand the "role" it plays as a factor in population movement".[51]

Intuitively we can see that climate change will play a role in future movements of people. But putting empirically sound figures on the extent of the problem is complex. And it is hard to persuade decision makers to take the issue seriously without being able to wave concrete figures in front of them. This is the subject of the next section.

3. Predictions

Prediction is very difficult, especially about the future.

(Niels Bohr, Danish physicist (1885–1962))

Climate migration is not new

Archaeological evidence suggests that human settlement patterns have responded repeatedly to changes in the climate.[52,53] There is evidence that the emergence of the first large, urban societies was driven by a combination of climatic and environmental desiccation. The complex societies of Egypt and Mesopotomia, for example, emerged as people migrated away from desiccating rangelands and into riverine areas. The resulting need to organise densely packed populations in order to manage scarce resources in restricted areas has been identified as one of the main driving forces behind the development of the first civilisations.[54]

Much later, during the fourth century CE, growing aridity and frigid temperatures from a prolonged cold snap caused the Hun and German hordes to surge across the Volga and Rhine into milder Gaul and eventually led to the sack of Rome by the Visigoths. Likewise, the eight-century Muslim expansion into the Mediterranean and southern Europe was, to some extent, driven by drought in the Middle East.[55]

Existing patterns of climate migration

Migration is (and always has been) an important mechanism to deal with climate stress. Pastoralist societies have of course habitually migrated, with their animals, from water source to grazing lands in response to drought as well as part of their normal mode of life. But it is becoming apparent that migration as a response to environmental change is not limited to nomadic societies.

In western Sudan, for example, studies have shown that one adaptive response to drought is to send an older male family member to Khartoum to try and find paid labour to tide the family over until after the drought.[56] Temporary migration in times of climate stress can help top-up a family's income (through remittances from paid work elsewhere) and reduce the draw on local resources (fewer mouths to feed).

When climate stresses coincide with economic or social stresses, the potential for forced migration from rural areas increases significantly. But the picture is nuanced. In West Africa, the distance that people migrate is a function of their family's resources; in really bad drought years they can not afford to travel far and instead try to find paid work in local cities (see Box 2). Known locally as "eating the dry season" it occurs today in many parts of drought-stricken West Africa.

The ability to migrate is, almost by definition, a function of mobility. In the 1930s Dustbowl Years in the US migrants from the Great Plains tended to be tenant farmers without strong ancestral or financial ties to the land.[57] The decision to migrate is normally taken at a household level (unless the state is clearing an area) — and relies on individual calculations of social and financial capital. Migration is typically not the first adaptive response households take when confronted by climate stress; rather it resorted to when other means of adaptation (such as selling livestock) are insufficient to meet their immediate needs and often when their communities or governments have proven incapable of giving assistance.

Migration, especially when it is a response to slower acting climate processes (rather than a sudden climatic event like a hurricane), typically requires access to money, family networks and contacts in the destination country. Even in the most extreme, unanticipated natural

Box 2. "Eating the dry season" – temporary labour migration in West Africa

In the West African Sahel recent studies have cast light on the use of temporary migration as an adaptive mechanism to climate change. The region has suffered a prolonged drought for much of the past three decades and one way that households have adapted is by sending their young men and women in search of wage labour after each harvest.[58] But how far they travel depends, in part, on the success of the harvest. A good harvest might give the family sufficient resources to send a member to Europe in search of work. While the potential rewards in terms of remittances are high, it is a highly speculative gamble—in addition to dangerous journey, the rewards are uncertain. In addition the chances are the migrant will not be back in time for the next year's planting. But in a drought year, when harvests are poor, the young men and women tend to stay much closer to home, instead travelling to nearby cities for paid work so as to reduce the drain on the household's food reserves and top-up household income. In such years the risk of losing the 'migration gamble' is simply too great.[59]

disasters, migrants, if they have any choice, tend to travel along pre-existing paths—to places where they have family, support networks, historical ties and so on. Most people displaced by environmental causes will find new homes within the boundaries of their own countries. Evacuees from Hurricanes Rita and Katrina did not stream across the border to Mexico but typically found temporary refuge with family members elsewhere in the country.[60]

[...]

The image of a coastal farmer getting inundated by rising sea levels and being forced to pack up and move to a rich country simply is not borne out by experience. The 2004 Asian Tsunami, for example, killed more than 200,000 people and displaced twice as many. But those people were largely not displaced to OECD countries. Instead the burden of displacement (and of providing for evacuees) is overwhelmingly born by the local region.

Those who cannot, or choose not to, find new homes within their own country tend to seek refuge in places where they have existing cultural or ethnic ties. So Bangladeshis would seek refuge in India or Pakistan, Indonesians from Sumatra would look to Malaysia and so on.[63] Likewise, intercontinental migration is most likely to follow pre-existing paths and old colonial relationships. So the United Kingdom might be an obvious destination for Pakistanis and Bangladeshis, France for would-be migrants from Francophone West Africa, and Australia and New Zealand for some groups in the South Pacific.

In short, people have had to move for environmental reasons for thousands of years. But recent examples provide useful, albeit sobering, analogues for the likely impact of future climate change. The 1998 monsoon floods in Bangladesh brought some of the worst flooding in living memory, inundating two-thirds of the country for two months, devastating its infrastructure and agricultural base and leading to fears about the country's long-term future in a world of higher ocean levels and more intense cyclones.[64] The floods left an estimated 21 million people homeless.[65] Meanwhile the Yangtze floods of the same year temporarily displaced an estimated 14 million people and triggered the largest ever peace-time deployment of the People's Liberation Army to provide humanitarian aid and rebuild critical infrastructure.[66] However, it is one thing to reflect on past and present climate-triggered population movements and quite another to predict accurate figures for future population displacement.

The problem of prediction

Although meteorological science and climate modelling techniques have progressed dramatically over the past decade, we still cannot accurately predict the impact of climate change on our weather systems. Amongst much else there is uncertainty about the way rainfall patterns will change and continuing debate on whether global warming will lead to more frequent and fierce hurricanes.[67]

So far, and quite understandably, the focus of the scientific community has been on establishing the extent and nature of anthropogenic climate change. Less time and energy have gone into predicting the impact of future climate change on human societies in any more than the most general terms. The complex interactions between different meteorological and social factors make cause and effect models tricky and often inappropriate. Consequently, the figures that analysts have produced to date are little more than well-educated guesswork. Developing more solid predictions will require a lot of hard number-crunching that is only really starting now.[68]

These predictions are complicated by three factors:

- First, forced climate migration will take place against a background of unprecedented changes in the number and distribution of the world's population. The global population is currently growing at a rate of 1.1 percent and is predicted to reach 9.075 billion by 2050 (from its 2005 level of 6.54 billion). Meanwhile, there is an accelerating move to urban areas. Already 49 percent of the world's population lives in cities, and the growth rate of the urban population is nearly double (2 percent) that of total population growth.[69]

  These trends are even more pronounced in low and middle-income countries. Between 2005 and 2010 Burundi, for example, is expected to have a population growth rate of 3.7 percent and an urban growth rate of 6.8 percent.[70] Meanwhile, the Sahelian region of Northern Nigeria, perhaps the area of the country most susceptible to climate change, is already characterised by high population growth (about 3.1 percent) and rapid urbanisation (about 7 percent).[71] Clearly it would be absurd to attribute the entire urban drift to climate change, but disaggregating what role climate change might play in added rural-urban migration is very hard.

- Second, we have no real base-line figure for current migratory movements. Nor is there much capacity in developing countries or the international community to gather this sort of data, particularly for internal migration. What limited capacity exists is focused on tracking cross-border migration. Given that a majority of forced climate migrants will stay within their own borders the machinery to collect data on these movements simply does not yet exist.

- Third, what happens in the second half of the twenty-first century depends to a great extent on what we do today. Until 2050 the degree of inertia in the climate system that means that climate change over the next fifty years is largely predetermined.[72] However, the extent and nature of climate change after then is reliant on current emissions. Consequently, many analysts think that it is highly speculative to try to push predictions past 2050.[73]

The Climate Canaries

Nonetheless, there has been a somewhat breathless competition in the world's media to find the first conclusive 'victims' of climate change—who, like a miner's canary, will mark the beginning of a period of irreversible climate impacts. Four cases have been quite extensively highlighted in the past few years: the Cartaret islands in Papua New Guinea, the residents of Lateu village in Vanuatu, the relocation of Shishmaref village on Sarichef island in Alaska, and the submergence of Lohachara island in India's Hooghly river.

In 2005 it was officially decided to evacuate the 1,000 residents of the Carteret Islands, a group of small and low-lying coral atolls administered by Papua New Guinea. Storm-related erosion and salt water intrusion had rendered the population almost entirely dependent on outside aid. Ten families at a time will now be moved to the larger island of Bougainville, 100 kilometres away.[74]

A second group of about a hundred residents of Lateu, on the island of Tegua on Vanuatu, were relocated farther inland, again following storm-damage, erosion and salt damage to their original village. In both cases, the declaration of their status as 'the first climate change refugees' was timed to coincide with the United Nations Climate Convention meeting in November 2005.[75]

Shishmaref village lies on Sarichef island just north of the Bering strait. A combination of melting permafrost and sea-shore erosion, at a rate of up to 3.3 metres a year, have forced the inhabitants to relocate their village several kilometres to the south.[76] It is thought that climate change has directly exacerbated the sea-erosion by thinning the sea ice which used to reduce the force of local tides and currents.

In December 2006 there were widespread reports of the first submergence of an inhabited island due to climate change. Researchers reported that Lohachara island in the Hooghly river delta, once home to 10,000 people, and which had first started flooding 20 years ago, had finally been entirely submerged. One of a number of vanishing islands in the delta, the loss of the islands and other coastal land in the delta has left thousands of people homeless.[77]

However, in the interests of balance, it's worth noting that there is little scientific consensus that these four cases are definitively the result of anthropogenic climate change. Fred Terry, director of the UNDP's programme in Bougainville argues that in the case of the Carteret Islands dynamite fishing has destroyed the natural protection offered by the reef, whilst natural subsidence and tectonic movement might also explain the islands' inundation. In fact plans to evacuate the residents have been discussed since the early 1980s, but were interrupted by the war on the neighbouring Papua New Guinean island of Bougainville.[78] Likewise, Lohachara island, a sandbar in the Hooghly delta (and so inherently unstable), was eroded by river currents, weakened by mangrove destruction, and submerged by tectonic titling and local subsidence.[79]

So far the publicised examples of forced migration caused by anthropogenic climate change are more anecdotal than empirical, affecting a few hundred or thousand people at a time. The urge to grab the headlines has tended to obscure the fact that we know that climate variation has influenced human population distribution for thousands of years. But while the evidence for a distinctively anthropogenic 'climate change signal' in forced migration so far is circumstantial, it is mounting. And with all available scenarios predicting accelerating climate change impacting growing populations and more people living on marginal land, forced climate migration is certain to increase. The important questions are: By how much? And with what implications for development?

The Good, the Bad and the (very) Ugly: Climate migrant scenarios

The impact of climate change as a driver of future forced migration depends on several factors:

1. The quantity of future greenhouse gas emissions;
2. The rate of future population growth and distribution;
3. The meteorological evolution of climate change;
4. The effectiveness of local and national adaptation strategies.

The IPCC has devised a series of scenarios, called the Emissions Scenarios of the IPCC Special Report on Emission Scenarios (or SRES for short), which set out a range of different future emissions scenarios varied according to demographic, technological and economic developments. There are six basic 'storylines', each of which aggregates different rates of population and economic growth as well as the future 'energy mix'. They range from the most greenhouse gas intensive (A1F1—where energy is mostly derived from fossil fuels and economic growth is rapid) to the less intensive B1 storyline (where the world economy moves towards less resource intensity and cleaner technology). All the scenarios assume no additional climate change initiatives such as the emissions targets under the Kyoto protocol. Three of the SRES scenarios are used here as starting points to imagine three highly speculative scenarios for future climate-induced migration:[80]

The Good

The first (B1) is the best case scenario. Its impact is relatively low but so also is its likelihood. The B1 storyline describes a world whose population peaks mid-century around 9 billion and declines thereafter towards 7 billion. There is a rapid change in economic structures towards a service and information economy with a reduction in material intensity and the introduction of clean and resource efficient technologies. "The emphasis is on global solutions to economic, social and environmental sustainability, including improved equity, but without additional climate initiatives".[81]

In addition (and this is where this scenario diverges from the B1 storyline) we can imagine that a serious post 2012 regime is put in place by the international community to reduce carbon emissions. The BRIC countries (Brazil, Russia, India and China) join as full members and work to cut their own emissions. Atmospheric concentrations of CO_2 stabilise around 600ppm by end of century leading to temperature rise over the century of around 1.8 degrees and sea level rise of from 18 to 38 cms.[82] In addition a 'Marshall plan' for adaptation helps countries deal with the worst impacts of climate change.

Nonetheless, according to the Stern report, such a temperature rise would still lead to a 20–30 percent decrease in water availability in some vulnerable regions such as Southern Africa and the Mediterranean countries. It would also result in declining crop yields in tropical regions. In Africa crop yields could be cut by between 5 to 10 percent.[83] Meanwhile up to 10 million more people would be affected by coastal flooding each year.[84]

In this case the headline figure for climate migration (the 200 million 'climate refugees' by 2050) might, in hindsight, seem like an exaggeration. Instead we could expect increased migration of between five and ten percent along existing routes. There would be increased rural to urban migration but it would prove largely manageable, if not indistinguishable, within existing patterns of migration.

The Bad

Our second scenario uses the 'A1B' storyline as its starting point. A1B envisages a world of very rapid economic growth, with a global population that peaks mid century and declines thereafter, as well as the swift up-take of new and more efficient technologies. The scenario predicts economic convergence among regions, increased social and cultural interactions and a substantial reduction in regional differences in per capita income. In this scenario the world's energy is sourced from a balance between fossil intensive and non-fossil energy sources.[85] We can imagine that international efforts to reduce greenhouse gas emissions are delayed, patchy and not particularly effective. Some effort and funds are invested into adaptation, but not enough.

The estimate for temperature rise over the 21st century for the A1B storyline is 2.4°C (with a likely range from 1.7°C to 4.4°C). Atmospheric concentrations of CO_2 by the end of the century are 850 ppm (three times pre-industrial levels).[86] With higher temperatures the practical implications of climate change are much greater. Under this scenario sea level rise would be between 21cm and 48cm and precipitation in sub-tropical areas would fall by up to 20 percent.[87] According to the Stern report, a 3°C temperature rise would mean one to four billion people would suffer water shortages and between 150 to 550 additional million people would be at risk of hunger. Conversely other areas would gain unwelcome water with coastal flooding affecting between 11 and 170 million additional people each year.[88] Marginal lands would become increasingly uninhabitable, with dramatic increases in internal rural to urban migration and also emigration to richer countries, particularly of young, skilled people. Meanwhile, millions of people would be temporarily displaced by individual extreme weather events.

The Ugly

The third scenario uses the A1F1 storyline as its starting point. A1F1 is similar to A1B in that it forecasts rapid economic growth and a global population that peaks mid-century and falls thereafter. However, unlike A1B, energy in the A1F1 world continues to be overwhelmingly sourced from fossil-fuel supplies—and is a 'business as usual scenario' without any Kyoto emission reductions or serious attempts at adaptation.[89] On this trend, atmospheric concentration of CO_2 by 2099 will be 1,550ppm: five times pre-industrial levels and four times current levels.

Such CO_2 levels would result in a temperature rise over the century of 4.0°C (with a likely range from 2.4°C to 6.4°C) and sea level rise from 29cm to 59cm.[90] According to the Stern report a temperature rise of 4.0°C would result in a 30 to 50 percent decrease in water availability in Southern Africa and Mediterranean. Agricultural yields would decline by 15 to 35 percent in Africa and entire regions, such as parts of Australia, would fall out of production.[91] With high climate sensitivity, the number of people flooded per year could be as many as 160 million by the 2050s and 420 million by the 2100s.[92]

Under this scenario, predictions of 200 million people displaced by climate change might easily be exceeded. Large areas of Southern China, South Asia and the Sahelian region of sub-Saharan Africa could become uninhabitable on a permanent basis. Climate forced migration would be unmistakeable with tens of millions of people at a time displaced by extreme weather events, such as floods, storms and glacial lake outburst floods, and many millions more displaced by climate processes like desertification, salinisation of agricultural land and sea-level rise.

The above scenarios all assume a roughly linear evolution of climate change. But the picture would change again in the case of abrupt climate change such as the collapse of the Gulf Stream or melting of the Greenland or Antarctic ice sheets. The IPCC estimates that the elimination of the Greenland ice sheet would lead to a contribution to a sea level rise of about 7m.[93] The Stern report estimated that the melting or collapse of the ice sheets would raise sea levels and eventually threaten 4 million km² of land which is currently home to 5 percent (around 310 million people) of the world's population.[94]

4. Development implications

There is irony in the fact that it is the developing countries—the least responsible for emissions of greenhouse gases—that will be the most affected by climate change. If the situation with refugees from war and political persecution is any indication, they will also bear the greatest burden of providing for forced climate migrants. For example, in 2000, the twenty countries with the highest ratios of official refugees had an annual per-capita income of just US$850.[95]

Assessing regional vulnerabilities

Numerically and geographically, South and East Asia are particularly vulnerable to large scale forced migration. This is because sea-level rise will have a disproportionate effect on their large populations living in low-lying areas. Six of Asia's ten mega-cities are located on the coast (Jakarta, Shanghai, Tokyo, Manila, Bangkok and Mumbai).[96] China, meanwhile, has 41 percent of its population, 60 percent of its wealth and seventy percent of its megacities in coastal areas.[97]

Millions more are vulnerable in Africa, particularly around the Nile Delta and along the west coast of Africa. Changed patterns of rainfall would have particularly serious impacts for food security in sub-Saharan Africa. According to the latest IPCC report reduced rainfall could lower crop yields by as much as 20 percent by 2020, leading to increased malnutrition.[98]

Small island states around the world are particularly vulnerable to sea level rise because in many cases (the Bahamas, Kiribati, the Maldives and the Marshall Islands) much of their land is less than three or four metres above present sea level.[99] One 1999 analysis estimated that, by 2080, flood risk for people living in small island states will be 200 times greater than if there had been no global warming.[100] Other island states tend to have high levels of development and high density population around their coasts. Half the population of the Caribbean, for example, lives within 1.5km of the shoreline.[101]

Forced migration and development

Over the short term, climate-change forced migration will make the Millennium Development Goals (MDGs) harder to achieve.[102] Over the long term, large scale climate change migration could roll back much of the progress that has been made so far. Particularly threatened is the uninterrupted provision of the education and health services that underlie goals 2 (universal primary education) and goals 4 and 5 (reducing child and maternal mortality and combating HIV/Aids, malaria and other diseases).

Forced migration hinders development in at least four ways; by increasing pressure on urban infrastructure and services, undermining economic growth, increasing the risk of conflict and leading to worse health, educational and social indicators among migrants themselves.

What impact climate change migration ultimately has on development depends, of course, on which of the above storylines plays out: it is clear that 200 million people displaced by climate change would be much more detrimental to development than 10 million. There is also a large difference in development outcomes between those displaced by long-term climate processes (sea level rise) and short-term climate events (storms). Aggregated figures for forced climate migration mask this distinction.

4.1 The urban flood

Increasing food and water scarcity due to climate change in rural areas will accelerate the dramatic rural-urban drift in the developing world. Urban areas offer access to the cash economy (rather than subsistence farming) and can make it easier to provide services. However, rapid and unplanned urbanisation has serious implications for urban welfare and urban service provision.

Already, one third of the world's urban population, about one billion people, lives in slums: in poor quality housing with limited clean water, sanitation and education services.[103] By 2030 it is estimated that this number will rise to 1.7 billion people.[104] High population densities and high contact rates help to spread disease, while health and education services are often inadequate. In India, for example, unplanned urbanisation has been associated with the spread of dengue fever.[105]

4.2 Hollowed economies

Mass migration disrupts production systems and undermines domestic markets. In addition, the loss of 'human capital' in the form of the labour force and investment in education undermines economic growth. This can establish a self-reinforcing of limited economic opportunity that contributes to future migration.

The 'brain drain' effect from developing countries is already a serious problem. For example, in 2006, 926 Ghanaian doctors were practising in the OECD alone, representing a much-needed 29 percent of those still practising in Ghana.[106] Climate change could accelerate the brain drain as it is typically those with larger reserves of financial and social capital who are able to move away.

One of the legacies of the 1930s Dust Bowl case was that those who fled the drought were young, skilled families with some money and strong social networks—the very kind of people that are essential components of successful communities. "The places they left behind", says Ottawa University's Robert McLeman, "became increasingly polarized between affluent property owners and an impoverished underclass, a downward spiral from which some communities never recovered.

Future climate-migration holds a similar potential to have negative long-term consequences for socio-economic stability in affected areas".[107]

4.3 Political instability and ethnic conflict

Large-scale population displacement will redraw the ethnic map of many countries, bringing previously separate groups into close proximity with each other and in competition for the same resources. In the context of poor governance, poverty and easy access to small arms, these situations can easily turn violent. In Nigeria, 3,500 square kilometres (1,350 square miles) of land are turning into desert every year, making desertification the country's leading problem. As the desert advances, farmers and herdsmen are forced to move, either squeezing into the shrinking area of habitable land or forced into the already overcrowded cities.[108] There is also a fairly widely-held belief that the current crisis in Darfur has its origins in the extended drought that brought pastoralists into competition with farmers.[109]

Large population movements are already recognised by the UN Security Council as constituting a potential threat to international peace and security, particularly if there are existing ethnic and social tensions.[110] According to John Ashton, the UK's climate change envoy, "Massive migrations, particularly in the arid or semi-arid areas in which more than a third of the world's people live, will turn fragile states into failed states and increase the pressure on regional neighbours—a dynamic that is already apparent in Africa".[111]

4.4 Health impacts and welfare of forced migrants

Population displacement undermines the provision of medical care and vaccination programmes; making infectious diseases harder to deal with and more deadly. It is well documented that refugee populations suffer worse health outcomes than settled populations. Forced migrants, especially forced to flee quickly from climate events, are also at greater risk of sexual exploitation, human trafficking and sexual and gender-based violence.[112]

Forced migration in response to climate stresses can also spread epidemic disease. Visceral leishmaniasis (VL) is one example. VL is a widespread parasitic disease with a global incidence of 500,000 new human cases each year. In northeastern Brazil, periodic epidemic waves of VL have been associated with migrations to urban areas after long periods of drought.[113]

5. Policy Responses: Heads in the sand

Despite the serious development implications of large-scale forced climate migration, international capacity and interest in dealing with it is limited.[114] Bold speeches and elaborate commitments to the pursuit of noble goals like refugee rights, environmental protection and sustainable development typically fall prey to narrow geo-political interests when the time for action comes. The result is that forced climate migrants fall through the cracks of international refugee and immigration policy. There is no 'home' for forced climate migrants, either literally or figuratively.

Instead, there is a collective, and rather successful, attempt to ignore the scale of the problem. Until now the international community has largely focused on mitigating climate change by setting emissions targets for OECD countries and agonising about how to bring it new members to a post-Kyoto 2012 framework. More recently, greater attention has been paid to helping countries adapt to the impacts of climate change. But this approach to adaptation is fundamentally based on the idea of adapting 'in situ'. Migration is seen as a failure of adaptation.

Potential progress can be divided into three, quite distinct, areas. I say 'potential' here as there has not been real progress on any front—yet. First is the legal-political approach to expand the definition of a refugee under current international law. Second is the extent to which forced migration is being incorporated into current domestic plans for climate change adaptation. Third is whether the OECD countries are willing to open their 'immigration gates' to climate migrants.

5.1. Expanding the definition of a 'refugee'

There have been some attempts to broaden the existing definition of a political refugee to include those displaced for environmental reasons or to write a new convention that specifically protects such people.[115] The lack of an accepted definition of an environmental refugee means that, unless they're relocated by extreme weather events, their displacement does not trigger any access to financial grants, food aid, tools, shelter, schools or clinics.

As a result there is no structural capacity in the international system to provide for environmental migrants. Climate migrants are not recognised as a problem in any binding international treaty, nor is there an international body charged with providing for climate migrants, or even counting them. Instead the default response of OECD donor countries to extreme weather events is to give humanitarian aid and invest in early warning systems.

In 2005 the Director of the UN University Institute for Environment and Human Security, Janos Bogardi, argued, "there are well-founded fears that the number of people fleeing untenable environmental conditions may grow exponentially as the world experiences the effects of climate change and other phenomena. This new category of 'refugee' needs to find a place in international agreements. We need to better anticipate support requirements, similar to those of people fleeing other unviable situations".[116]

In August 2006 a meeting of NGOs and some affected countries was held in the Maldives to discuss how an expanded definition might be worked into international law. Inclusion within current refugee law would bring the existing weight of international law and precedent to act on the issue—and would trigger certain obligations on the part of other countries being forced to act on refugees. However, since then the process has faltered and it is hard to foresee any realistic consensus on an expanded definition.[117]

5.2 Adaptation in affected countries

As climate change advances, individual countries will have to make a series of cost-benefit decisions on what they want to protect; building sea walls here, staging managed retreats from eroding shorelines there. The resources and foresight at the disposal of national politicians will define how much each country is affected by climate change, including how many of its population are forced to move.

Domestic policy remains a key variable in disaster risk reduction and population distribution. With the right kind of adaptation, countries can reduce their vulnerability to the impacts of climate events and manage the evolution of climate processes. Cuba, for example, lies directly in a hurricane path but suffers less from hurricanes than its neighbours because of careful preparation, effective early warning systems and widespread storm education.

But few countries are putting any plans in place for the prospect of large-scale forced climate migration. The UNFCCC has supported the development of National Adaptation Programmes of Action (NAPA) which are supposed to help the LDCs identify and rank their priorities for adaptation to climate change.[118] However, none of the fourteen submitted so far (Bangladesh, Bhutan, Burundi, Cambodia, Comoros, Djibouti, Haiti, Kiribati, Madagascar, Malawi, Mauritania, Niger, Samoa, Senegal) mentions migration or population relocation as a possible policy response.[119]

Of course migration may be the only possible adaptive response in the case of some of the Small Island and low lying states where rising seas will eventually flood large parts of the country. Andrew Simms of the New Economics Foundation points out that domestic level responses are, in some cases, an absurd proposition given that the national level might be under water.[120]

Migration is typically seen as a failure of adaptation, not a form of it. There are precedents though. Between 1984–5 the Ethiopian Government resettled tens of thousands of people from drought-stricken areas.[121] Two decades later the Asian Tsunami gave new impetus to plans in the Maldives to organise a 'staged retreat' from their outlying islands. The plan is to concentrate the islands' 290,000 residents on several dozen, slightly higher islands than the 200 islands that the population is currently spread across.[122]

5.3 Immigration policy in less affected countries

Another determinant of forced migration will be immigration policies in countries less affected by climate change, in particular the OECD countries. Some analysts are beginning to argue that immigration is both a necessary element of global redistributive justice and an important response to climate change; that greenhouse gas emitters should take an allocation of climate migrants in proportion to their historical emissions. Andrew Simms argues, "Is it right that while some states are more responsible for creating problems like global climate change, all states should bear equal responsibility to deal with its displaced people?".[123]

It has been widely reported that New Zealand has agreed to accept the inhabitants of the South Pacific Island state of Tuvalu if and when climate change leaves their country uninhabitable.[124] However, this is an urban myth: New Zealand only accepts 75 Tuvaluans each year through the immigration service's Pacific Access Category which makes no reference to environmental degradation. No other country has yet been willing to set a precedent by explicitly accepting climate migrants under a refugee category.

Sweden is the only country even to get close. Swedish immigration policy mentions environmental migrants as a special category as a 'person in need of protection' who is unable to return to his native country because of an environmental disaster. However, the extent to which this includes climate change impacts has not yet been clarified. In the parliamentary text explaining the category, a nuclear disaster is given as an example of an 'environmental disaster' whereas natural disasters are not specifically mentioned.[125]

However, there are increasing examples of immigration concessions for victims of natural disasters – albeit on an ad hoc basis. For example, in 2003 the US immigration service extended for two more years the Temporary Protection Status it granted to 80,000 Hondurans who had fled to the United States after the 1998 Hurricane Mitch which devastated large parts of Central America.[126] After the 2004 Tsunami, Switzerland, Canada and Malaysia temporarily suspended involuntary returns of failed asylum seekers to affected areas of India, Sri Lanka, Thailand and Indonesia. Likewise Australia put a high priority on processing temporary visas for victims and fast-tracking existing applications. The European Union, for its part, proposed offering temporary asylum to child victims of the disaster so as to allow them several months in Europe to recover from the trauma.[127] Whether or not this adds up to an evolving norm of soft law is highly debateable, but it does show some 'greyness' at the edges of immigration policy.

There is a dilemma here. Relaxing immigration rules as part of a concerted policy to 'release the population pressure' in areas affected by climate change could accelerate the brain drain of talented individuals from the developing world to the developed—and worsen the 'hollowing out' of affected economies, which is itself a driver of migration. On the other hand, shutting borders in both source and destination countries undermines remittance economies and denies developing countries the benefits of access to the international labour market.

[...]

6. Conclusions

Environmental, economic and political degradation are connected—though the categories are permeable. One analyst argues, 'One classification may cause the other or, more likely, each drives the other in a vicious cycle of reinforcing degradations".[130] Migration to the United States is an example, "though nominally economic migrants, many of the estimated 1 million people who flood illegally into the United States annually from Mexico are in part driven by declining ecological conditions in a country where 60 percent of the land is classified as severely degraded".[131]

Anthropogenic climate change exacerbates existing environmental, economic and social vulnerabilities. It follows that adaptation to climate change has to be broader than tackling the marginal increased impact of anthropogenic climate change. Focusing on the impacts of climate change without factoring in the local context is leading to some bizarre policy distortions. For example, in the Philippines, policy makers have begun to acknowledge the flood threats posed by a projected annual sea level rise from climate change of 1 to 3 millimetres per year. But at the same time they are oblivious to, or ignore, the main reason for increasing flood risk: excessive ground water extraction which is lowering land surface by several centimetres to more than a decimetre per year.[132]

On current climate change scenarios a certain amount of forced climate migration is 'locked in'. But how much depends on the international community's mitigation and adaptation plans now. It is clear that the international community has to face up to the prospect of large scale displacement caused by climate change.

There is a need for international recognition of the problem, a better understanding of its dimensions and a willingness to tackle it. This should take several forms:

1. The international community needs to acknowledge formally the plight of forced climate migrants. While it is not clear that an expanded definition of a refugee under international law that included environmental degradation as a 'valid' driver of displacement would lead to net benefits for all (traditional and environmental) refugees, some kind of international recognition is required to cement the issue on the international agenda.

2. Development and adaptation policies in potential source countries of forced climate migrants need to focus on reducing people's vulnerability to climate change, moving people away from marginal areas and supporting livelihoods that are more resilient. In particular

more efficient use of existing resources would offset some of the predicted impacts of climate change. In Pakistan, for example, irrigated agriculture uses 85 percent of the country's fresh water supply but leakage and evaporation means that it is only 50 to 65 percent efficient.[133]

3.   A great deal more research is needed to understand the causes and consequences of climate migration and to monitor numbers. Practitioners, meanwhile, should develop better communication and working relationships between the different human rights, population, environmental and migration organizations that share a mandate to respond to population displacement.[134]

Finally, rather than erecting immigration barriers the international community needs to help generate incentives to keep skilled labour in developing countries and capitalise on the benefits that fluid labour markets can bring. The international regulation of labour migration, adaptation to climate change and capacity building in vulnerable countries are inherently intertwined. Migration will be used by some households in vulnerable countries as a means of adapting to climate change and sea level rise. Clearly there has to be a balance of policies that promotes the incentives for workers to stay in their home countries whilst not closing the door of international labour mobility.

From Oli Brown, 'Climate Change and Forced Migration: Observations, Projections and Implications', A Background Paper for the Human Development Report 2007/2008, Fighting Climate Change: Human Solidarity in a Divided World, Human Development Report Office, Occasional Paper, 2007, p. 17

Notes

[…]

2   Steve Lonergan (1998) "The role of environmental degradation in population displacement", Environmental Change and Security Project Report, Issue 4 (Spring 1998): p. 5.

3   Norman Myers (2005) "Environmental refugees: an emergent security issue", 13th Economic Forum, May 2005, Prague.

4   United Nations University (2005) "As ranks of 'Environmental Refugees' swell worldwide, calls grow for better definition, recognition, support", UN Day for Disaster Reduction, 12th October 2005.

5   Norman Myers (2005) "Environmental refugees: An emergent security issue", 13th Economic Forum, May 2005, Prague.

6   Nicholas Stern (ed.) (2006) "The economics of climate change: the Stern review", Cambridge University Press, Cambridge, p. 3.

7   In 1975, there were 2.4 million refugees globally but the number of refugees and people of concern to the UNHCR grew ten-fold in the following two decades, peaking at 27.4 million in 1995. Since 1995, the number of political refugees has declined significantly, mainly due to several ambitious repatriation programmes and an overall decline in new conflicts. Nevertheless, in early 2005, 19.2 million people were still listed as refugees and people of concern to the UNHCR. In Dupont, Alan & Pearman, Graeme (2006) "Heating up the Planet: Climate Change and Security", Lowry Institute for International Policy, Paper 12, Sydney, p. 55.

8   International Organization for Migration, http://www.iom.int/jahia/page3.html, accessed 10th March 2007.

9   Personal communication.

10 Spencer S. Hsu (2006) "2 Million Displaced By Storms", Washington Post, 16th January 2006 http://www.washingtonpost.com/wp-dyn/content/article/2006/01/12/AR2006011201912.html, accessed 3 April 2007.

11 Steve Lonergan (1998) "The role of environmental degradation in population displacement", Environmental Change and Security Project Report, Issue 4 (Spring 1998): p. 6.

12 Resolution 429 of the United Nations General Assembly, 1951, http://www.cas/com/discoveryguides/refugee/review2.php, accessed 14 March 2007.

13 Subsequent actions, conventions and declarations may have nuanced the 1951 Convention and 1967 Protocol – it is the extent of this nuance and the weight of 'soft law' precedence that is the focus of this debate. For example, in 1969, the Organisation of African Unity (now the African Union) released the 'Convention governing the specific aspects of refugee problems in Africa' which cracked open the definition to include "events seriously disrupting public order" (http://www.africa-union.org/Official_documents/Treaties_%20Conventions_%20Protocols/Refugee_Convention.pdf, accessed 4th April 2007). In 1984 the Cartagena Declaration on Refugees expanded it further to encompass "massive violations to human rights and other circumstances which have seriously disturbed public order" and while the Cartagena declaration is not a legally binding document it has heavily influenced domestic law. http://www.asylumlaw.org/docs/international/CentralAmerica.PDF, accessed 4th April 2007.

14 Reports that New Zealand has agreed to accept the entire Tuvaluan population once climate change makes their islands uninhabitable (thereby setting such a precedent) are false.

15 Personal communication with the author.

16 Personal communications with the author. Also in Lonergan S. (1998) "The role of environmental degradation in population displacement", Environmental Change and Security Project Report, Issue 4 (Spring 1998): p. 7.

17 UNHCR (2006) "UNHCR's contribution to the inter-agency response to IDP needs – supplementary appeal", UNHCR, May 2006, p. 3.

18 Cited by Jeff Crisp, "Environmental refugees: a UNHCR perspective", 12th June 2006, Lausanne.

19 The temperature spread refers to the current best estimates for 21st average temperature rises under low emission (B1) and high emission (A1F1) IPCC Special Report on Emission Scenarios (SRES). The range across both SRESs is from 1.1°C to 6.4°C – in IPCC (2007) "Climate change 2007: the physical science basis – summary for policy makers", Contribution of Working Group I to the Fourth Assessment Report of the Intergovernmental Panel on Climate Change, Paris, February 2007, p. 10.

20 Tearfund, (2006) "Feeling the heat: why governments must act to tackle the impact of climate change on global water supplies and avert mass movement of climate change refugees", London, p. 5.

21 Eleanor Burke et al. (2006) "Modelling the recent evolution of global drought and projections for the twenty-first century with the Hadley Centre climate model", Journal of Hydrometeorology, Vol. 7, October 2006.

22 John Houghton (2005) "Global warming: the complete briefing", Cambridge University Press, 2005.

23 John Houghton (2005) "Global warming: the complete briefing", Cambridge University Press, 2005.

24 Anthony Nyong (2005) "Impacts of climate change in the tropics – the African experience". Avoiding Dangerous Climate Change Symposium (Met Office, UK, February 2005) keynote presentation.

25 IPCC (2007) "Working Group II Contribution to the Intergovernmental Panel on Climate Change Fourth Assessment Report Climate Change 2007: Climate Change Impacts, Adaptation and Vulnerability", April 2007, p. 10.

26 IPCC (2007) "Working Group II Contribution to the Intergovernmental Panel on Climate Change Fourth Assessment Report Climate Change 2007: Climate Change Impacts, Adaptation and Vulnerability", April 2007, p. 10.

27 IPCC (2007) "Working Group II Contribution to the Intergovernmental Panel on Climate Change Fourth Assessment Report Climate Change 2007: Climate Change Impacts, Adaptation and Vulnerability", April 2007, p. 11.

28 IPCC (2001) "Climate change: Working Group II: Impacts, adaptation and vulnerability" http://

www.grida.no/climate/ipcc_tar/wg2/561.htm, accessed 15 April 2007.

29   IPCC (2007) "Working Group II Contribution to the Intergovernmental Panel on Climate Change Fourth Assessment Report Climate Change 2007: Climate Change Impacts, Adaptation and Vulnerability", April 2007, pp. 9–10.

30   Nicholas Stern (ed.) (2006) "The economics of climate change: the Stern review", Cambridge University Press, Cambridge, p. 56.

31   Debbie Hemming et al. (2007) "Impacts of mean sea level rise based on current state-of-the-art modelling", Hadley Centre for Climate Prediction and Research, Exeter.

32   Debbie Hemming et al. (2007) "Impacts of mean sea level rise based on current state-of-the-art modelling", Hadley Centre for Climate Prediction and Research, Exeter.

33   Debbie Hemming et al. (2007) "Impacts of mean sea level rise based on current state-of-the-art modelling", Hadley Centre for Climate Prediction and Research, Exeter.

34   Robert J. Nicholls & Jason Lowe (2004) "Benefits of mitigation of climate change for coastal areas", Global Environmental Change, 14.

35   Ibid.

36   Personal communication.

37   De Wit, M. & Stanjiewicz, J. (2006) "Changes in surface water supply across Africa with predicted climate change", Science, Vol. 311, 31 March 2006, cited in Tearfund (2006), p. 15.

38   Nicholas Stern (ed.) (2006) "The economics of climate change: the Stern review", Cambridge University Press, Cambridge.

39   Spencer Hsu (2006) "2 Million Displaced By Storms", Washington Post, 16 January 2006, http://www.washingtonpost.com/wp-dyn/content/article/2006/01/12/AR2006011201912.html, accessed 3 April 2007.

40   Christina Ward (2001) "World Disasters Report Calls for Improved Aid Programs", http://www.redcross.org/news/in/ifrc/010702disreport.html, accessed 2007.

41   Ced Hesse & Lorenzo Cotula (2006) "Climate change and pastoralists: investing in people to respond to adversity", Sustainable Development Opinion, IIED, London.

42   Natural disasters here include both hydro-meteorological disasters and geo-physical ones. However the former outnumber the latter nine to one in frequency. IFRC (2004) "World Disasters Report 2004: focus on community resilience", chapter 8, http://www.ifrc.org/publicat/wdr2004/chapter8.asp, accessed 20 April 2007.

43   National Oceanic and Atmospheric Administration, "NOAA's Top Global Weather, Water and Climate Events of the 20th Century", http://www.noaanews.noaa.gov/stories/images/global.pdf, accessed 20 April 2007.

44   In inflation adjusted and wealth normalised 2004 USD. National Oceanic and Atmospheric Administration, http://www.aoml.noaa.gov/hrd/tcfaq/costliesttable3.html, accessed 21 April 2007.

45   Robert McLeman & Barry Smit (2005) "Assessing the security implications of climate change-related migration", presentation to workshop on Human Security and Climate Change, 21–23 June 2005, Oslo, pp. 8–9.

46   IPCC (2007) "Working Group II Contribution to the Intergovernmental Panel on Climate Change Fourth Assessment Report Climate Change 2007: Climate Change Impacts, Adaptation and Vulnerability", April 2007, p. 8.

47   USGCRP (2000) "Climate Change Impacts on the United States: The Potential Consequences of Climate Variability and Change. Overview: Agriculture", US Global Change Research Program, http://www.usgcrp.gov/usgcrp/Library/nationalassessment/overviewagriculture.htm, accessed 2007.

48   Martin Hoerling et al. (2006) "Detection and Attribution of Twentieth-century Northern and Southern African rainfall change", Journal of Climate, Volume 19, Issue 16, August 2006, pp. 3989–4008.

49   IPCC (2007) "Working Group II Contribution to the Intergovernmental Panel on Climate Change Fourth Assessment Report Climate Change 2007: Climate Change Impacts, Adaptation and

Vulnerability", April 2007, p. 8.

50   Roger Pielke, Gwyn Prins, Steve Rayner & Daniel Sarewitz (2007) "Lifting the taboo on adaptation: renewed attention to policies for adapting to climate change cannot come too soon", in Nature, Vol. 445, 8 February 2007, p. 597.

51   Steve Lonergan (1998) "The role of environmental degradation in population displacement", Environmental Change and Security Project Report, Issue 4 (Spring 1998): p. 8.

52   Robert McLeman & Barry Smit (2006) "Migration as an adaptation to Climate Change", Climate Change, 2006.

53   Nick Brooks (2006) "Climate Change, drought and pastoralism in the Sahel", Discussion note for the World Initiative on Sustainable Pastoralism, November 2006.

54   Nick Brooks (2006) "Climate Change, drought and pastoralism in the Sahel", Discussion note for the World Initiative on Sustainable Pastoralism, November 2006.

55   Alan Dupont & Graeme Pearman (2006) "Heating up the Planet: Climate Change and Security", Lowry Institute for International Policy, Paper 12, Sydney, p. 1.

56   Robert McLeman & Barry Smit (2004) "Climate change, migration and security", Canadian Security Intelligence Service, Commentary No. 86, Ottawa, p. 8.

57   Robert McLeman & Barry Smit (2005) "Assessing the security implications of climate change-related migration", presentation to workshop on Human Security and Climate Change, 21–23 June 2005, Oslo, pp. 8–9.

58   Science Daily (1999) "Jet Stream Studied in West African Drought", Penn State, 9 June 1999, http://www.sciencedaily.com/releases/1999/06/990607072120.htm accessed, 21 April 2007.

59   Robert McLeman (2006) "Global Warming's huddled masses", The Ottawa Citizen, 23 November 2006.

60   Peter Grier (2005) "The Great Katrina Migration", The Christian Science Monitor, 12 September 2005, http://www.csmonitor.com/2005/0912/p01s01-ussc.html, accessed 3 April 2007.
       [...]

63   Alan Dupont & Graeme Pearman (2006) "Heating up the Planet: Climate Change and Security", Lowry Institute for International Policy, Paper 12, Sydney, p. 59.

64   Alan Dupont & Graeme Pearman (2006) "Heating up the Planet: Climate Change and Security", Lowry Institute for International Policy, Paper 12, Australia, pp. 45–46.

65   UNICEF (2004) http://www.unicef.org/infobycountry/bangladesh_22473.html, accessed 10 April 2007.

66   Alan Dupont & Graeme Pearman (2006) "Heating up the Planet: Climate Change and Security", Lowry Institute for International Policy, Paper 12, Sydney, p. 45.

67   Pew Centre on Global Climate Change, "Hurricanes and global warming", http://www.pewclimate.org/hurricanes.cfm, accessed 10 April 2007.

68   Debbie Hemming, Hadley Centre for Climate Change, personal communication.

69   UNFPA (2006) "State of the World Population, 2006", United Nations Population Fund, New York, p. 98.

70   UNFPA (2006) "State of the World Population, 2006", United Nations Population Fund, New York, p. 98.

71   Anthony Nyong, Charles Fiki & Robert McLeman (2006) "Drought-related conflicts, management and resolution in the West African Sahel: considerations for climate change: considerations for climate change research", in Die Erde, vol. 137, issue 3, p. 229.

72   IISD (2001) "Summary of the eighteenth session of the intergovernmental panel on climate change: 24–29 September 2001", Vol. 12 No. 177, 2 October 2001, http://www.iisd.ca/vol12/enb12177e.html, accessed 15 April 2007.

73   Norman Myers, personal communication.

74   John Vidal "Pacific Atlantis: first climate change refugees", The Guardian, London, 25 November 2005

75   Robert McLeman & Barry Smit (2006) "Changement climatique, migrations et sécurité", Les Cahiers de la sécurité 63(4), 95–120.

76   Arctic change "Human and economic indicators – Shishmaref island", http://www.arctic.noaa.gov/

detect/human-shishmaref.shtml, accessed 10 April 2007.

77   http://news.independent.co.uk/environment/article2099971.ece, accessed 15 April 2007.

78   John Vidal (2005) "Pacific Atlantis: first climate change refugees", The Guardian, 25 November 2005, http://www.countercurrents.org/cc-vidal251105.htm, accessed 18 April 2007.

79   http://timblair.net/ee/index.php/weblog/island_erased, accessed 18 April 2007.

80   Warning: the following scenarios aggregate the scientific models for future climate change with international action on emissions reduction and some of the emerging models on specific impacts on coastal flooding and crop yields. They make no pretence at being scientific and are meant only to be indicative. They are compiled from existing models and interviews with analysts and experts. An important caveat that should be borne in mind with the SRES storylines and many of the numerical predictions of future climate change migrants is that, for clarity, they generally assume that people do nothing to manage long-term climate processes or build resilience to short-term climate events (see Annex 2, p. 6).

81   IPCC (2007) "Climate Change 2007: The physical science basis – Summary for Policy Makers", Contribution of Working Group I to the Fourth Assessment Report of the Intergovernmental Panel on Climate Change, Paris, February 2007, p. 14.

82   IPCC (2007) "Climate Change 2007: The physical science basis – Summary for Policy Makers", Contribution of Working Group I to the Fourth Assessment Report of the Intergovernmental Panel on Climate Change, Paris, February 2007, p. 10.

83   Nicholas Stern (ed.) (2006) "The economics of climate change: the Stern review", Cambridge University Press, Cambridge, p. 57.

84   Nicholas Stern (ed.) (2006) "The economics of climate change: the Stern review", Cambridge University Press, Cambridge, p. 57.

85   IPCC (2007) "Climate Change 2007: The physical science basis – Summary for Policy Makers", Contribution of Working Group I to the Fourth Assessment Report of the Intergovernmental Panel on Climate Change, Paris, February 2007, p. 14.

86   IPCC (2007) "Climate Change 2007: The physical science basis – Summary for Policy Makers", Contribution of Working Group I to the Fourth Assessment Report of the Intergovernmental Panel on Climate Change, Paris, February 2007, p. 14.

87   IPCC (2007) "Climate Change 2007: The physical science basis – Summary for Policy Makers", Contribution of Working Group I to the Fourth Assessment Report of the Intergovernmental Panel on Climate Change, Paris, February 2007, pp. 11–12.

88   Nicholas Stern (ed.) (2006) "The economics of climate change: the Stern review", Cambridge University Press, Cambridge, p. 57.

89   IPCC (2007) "Climate Change 2007: The physical science basis – Summary for Policy Makers", Contribution of Working Group I to the Fourth Assessment Report of the Intergovernmental Panel on Climate Change, Paris, February 2007, p. 14.

90   All projections do not include uncertainties in carbon-cycle feedbacks which could increase or decrease the upper bounds of sea level rise: IPCC (2007) "Climate Change 2007: The physical science basis – Summary for Policy Makers", Contribution of Working Group I to the Fourth Assessment Report of the Intergovernmental Panel on Climate Change, Paris, February 2007, p. 11.

91   Nicholas Stern (ed.) (2006) "The Economics of Climate change: the Stern review", Cambridge University Press, Cambridge, p. 57.

92   Robert J. Nicholls & Jason Lowe (2004) "Benefits of mitigation of climate change for coastal areas", Global Environmental Change, 14.

93   IPCC (2007) "Climate Change 2007: The physical science basis – Summary for Policy Makers", Contribution of Working Group I to the Fourth Assessment Report of the Intergovernmental Panel on Climate Change, Paris, February 2007, p. 13.

94   Nicholas Stern (ed.) (2006) "The economics of climate change: the Stern review", Cambridge University Press, Cambridge, p. 56.

95   Norman Myers (2005) "Environmental refugees: an emergent security issue", 13th Economic Forum, May 2005, Prague 23–27 May 2005.

96   Dupont, Alan & Pearman, Graeme (2006) "Heating up the Planet: Climate Change and Security", Lowry Institute for International Policy, Paper 12, Sydney, p. 58.

97   Zhao Xinhgshu (2007) "Climate security in Asia: perception and reality", presentation at the RUSI Conference on Climate Security in Asia, 24 April 2007.

98   IPCC (2007) "Working Group II Contribution to the Intergovernmental Panel on Climate Change Fourth Assessment Report Climate Change 2007: Climate Change Impacts, Adaptation and Vulnerability", April 2007, p. 10.

99   IPCC (1997) "Regional impacts of climate change: summary for policy makers", http://www.grida.no/climate/ipcc/regional/513.htm accessed 20 March 2007.

100   R. Nicholls, F. Hoozemans & M. Marchand, (1999) "Increasing flood risk and wetland looses due to global sea-level rise: regional and global analyses", Global Environmental Change 9 (suppl) cited in A. Dupont & G. Pearman (2006) "Heating up the Planet: Climate Change and Security", Lowry Institute for International Policy, Paper 12, Sydney, p. 47.

101   Nicholas Stern (ed.) (2006) "The economics of climate change: the Stern review", Cambridge University Press, Cambridge, p. 77.

102   UNDP "Climate change and the MDGs", http://www.undp.org/gef/adaptation/dev/02a.htm accessed 3 April 2007.

103   UN-HABITAT. "Urbanization: Facts and Figures", http://www.unhabitat.org/mediacentre/documents/backgrounder5.doc, accessed 22 March 2007.

104   Elliott D. Sclar, Pietro Garau & Gabriella Carolini "The 21st Century Health Challenge of Slums and Cities", The Lancet, Vol. 365, 5 March 2005, http://www.unmillenniumproject.org/documents/TheLancetSlums.pdf accessed 16 April 2007.

105   I. Shah, G.C. Deshpande & P.N. Tardeja, 2004, "Outbreak of dengue in Mumbai and predictive markers for dengue shock syndrome", J. Trop. Pediatrics, 50, 301–305.

106   UNECA (2006) "International Migration and Development: implications for Africa", A background document for the High Level Dialogue on Migration and Development, UN General Assembly, 14–15 September 2006, p. 2.

107   Robert McLeman & Barry Smit (2006) "Changement climatique, migrations et sécurité", Les Cahiers de la sécurité 63(4), pp. 95–120.

108   Lester B. Brown (2004) "Troubling New Flows of Environmental Refugees", Earth Policy Institute, January 28 2004.

109   University for Peace, Africa Programme (2004) http://www.africa.upeace.org/news.cfm?id_activity=301&actual=2004.

110   Francesco Sindico (2005) "Ex-post and ex-ante [Legal] Approaches to Climate Change – threats to the international community", New Zealand Journal of Environmental Law, Vol. 9: pp. 209–238.

111   John Ashton & Tom Burke (2005) "Climate change and global security", www.opendemocracy.net, 21 April 2005, pp. 1–2.

112   UNHCR (2003) "Sexual and Gender-Based Violence against Refugees, Returnees and Internally Displaced Persons: Guidelines for Prevention and Response", Geneva, 2003, p. 1.

113   Carlos Roberto Franke, Mario Ziller, Christoph Staubach & Mojib Latif (2002) "Impact of the El Niño Oscillation on Visceral Leishmaniasis, Brazil", Emerging Infectious Diseases, Vol. 8, No. 9, September 2002: pp. 914–7.

114   Robert McLeman & Barry Smit (2006) "Changement climatique, migrations et sécurité", Les Cahiers de la sécurité 63(4), 95–120.

115   See, for example, Friends of the Earth, Australia, (2005) "A citizen's guide to climate refugees", FOE, Melbourne.

116   United Nations University (2005) "As ranks of 'Environmental Refugees' swell worldwide, calls grow for better definition, recognition, support", UN Day for Disaster Reduction, 12 October 2005.

117   Personal communication.

118   UNFCC "National Adaptation Programmes of Action", http://unfccc.int/adaptation/napas/items/2679.php, accessed 10 March 2007.

119   As of April 2007. Links to all the submitted NAPAs can be found here: http://unfccc.int/national_

reports/napa/items/2719.php, accessed 10 March 2007.

120   Cited in Saleemul Huq & Hannah Reid (2004) "Climate Change and Development – consultation on key researchable issues", IIED, London, p. 6.

121   Helmut Kloos & Adugna Aynalem (1989) "Settler migration during the 1984/85 resettlement programme in Ethiopia", GeoJournal, Vol. 19, No. 2, September 1989, pp. 113–127.

122   Christopher Torchia (2005) "Maldives pushes ahead with relocation plan", Associated Press, 28 February 2005.

123   http://eagle1.american.edu/~sj1580a/haiti-hurricane.htm.

124   Tearfund (2006) "Feeling the heat: why governments must act to tackle the impact of climate change on global water supplies and avert mass movement of climate change refugees", London, p.10.

125   Helené Lackenbauer, IFRC, personal communication.

126   Robert McLeman & Barry Smit (2004) "Climate change, migration and security", Canadian Security Intelligence Service, Commentary No. 86, Ottawa, p. 8.

127   Frank Laczko & Elizabeth Collett (2005) "Assessing the Tsunami's Effects on Migration", International Organization for Migration.

[…]

130   Ethan Goffman (2006) "Environmental refugees: how many, how bad?", CSA Discovery Guides http://www.case.com/discoveryguides/discoveryguides-main.php accessed 28 April 2007.

131   Migration and Tourism (2000) Our Planet Magazine, United Nations Environment Program, http://www.ourplanet.com/aaas/pages/population05.html, accessed 28 April 2007.

132   Roger Pielke, Gwyn Prins, Steve Rayner & Daniel Sarewitz (2007) "Lifting the taboo on adaptation: renewed attention to policies for adapting to climate change cannot come too soon", in Nature, Vol. 445, 8 February 2007, p. 598.

133   Aamir Kabir (2002) "Managing the Water Shortages", IUCN, http://www.waterinfo.net.pk/artmw.htm, accessed 30 April 2007.

134   Steve Lonergan (2000) "Environmental Degradation and Migration" and "Sustainable Development: A Southern Perspective", a two-part meeting in the AVISO Policy Briefing Series, 13 April 2000, Senegal.

Chapter 6

Dan Smith and Janani Vivekananda

A CLIMATE OF CONFLICT
The Links between Climate Change, Peace and War

[…]

2. The Double-Headed Problem of Climate Change and Violent Conflict

Climate change and violent conflict present countries and communities with a double-headed problem. The two parts are mutually reinforcing; many of the countries predicted to be worst affected by climate change are also affected or threatened by violence and instability.

The increase in global average temperatures that is already unfolding and is projected to continue will change the climate in many parts of the world. The effects will vary – sea-level rise threatening low-lying small islands and coastal areas, more severe droughts and shorter growing seasons in some places, more storms and floods in others, glaciers melting, deserts forming. These will combine with existing pressures on natural resources and lead in many areas to failing crops, inadequate food supplies and increasingly insecure livelihoods. These further consequences will be especially sharp in countries where poverty, exclusion, inequality and injustice are already entrenched.

From everything we know about how mutually interlocking factors such as poverty, bad governance and the legacy of past conflicts generate risks of new violence, it is safe to predict that the consequences of climate change will combine with other factors to put additional strain on already fragile social and political systems. These are the conditions in which conflicts flourish and cannot be resolved without violence because governments are arbitrary, inept and corrupt. If the relationship between climate change and violent conflict is not addressed, there will be a vicious circle of failure to adapt to climate change, worsening the risk of violent conflict and, in turn, reducing further the ability to adapt.

Risk and risk management

The effects of changing weather patterns will render previous lifestyles and habitats unviable in many places. Some of these changes will be sudden, such as tropical storms and flash floods. Others will be much slower in their onset, such as the steadily falling water levels in the Ganges basin, lengthening droughts on the margins of the Sahel, glacial melting in Peru and Nepal, and rising sea levels. This will lead to increased food insecurity – not just food shortages but uncertainty of supply.

Both sudden shocks and slow onset changes can increase the risk of violent conflict in unstable states because they lack the capacity to respond, adapt and recover. It is likely that the most common way of thinking about how to respond to these problems is through huge humanitarian relief efforts, since such events and the response to them get a great deal of news coverage. But there is a growing awareness that what is really needed is for communities and countries to prepare against sudden shocks, to build their resilience and their adaptive capacity. Where that is possible, communities will not only be better prepared against potential disasters such as floods, but they will, in consequence, also be reducing the risk of conflicts erupting, getting out of control and escalating to violence. Seen in this light, adaptation to the effects of climate change can be a part of peacebuilding, and peacebuilding is a way of increasing adaptive capacity. In the medium to long term, peacebuilding will also increase unstable states' capacities for mitigation.

Vulnerability to climate change is the product of three factors – exposure, sensitivity and adaptive capacity.[3] The first issue is whether a country – or a city, or community, or region – is going to be exposed to physical effects of climate change such as increased frequency of extreme weather. The second issue is how sensitive it is to that exposure – a storm may hit two cities but only cause floods in one of them because it is low lying. And the third issue is whether there is adaptive capacity which, for example, enables city authorities to build flood defences and be ready with quick and safe evacuation plans, while the national government has prepared to care for those who are displaced and can swiftly allocate resources for repair and rebuilding when the floods recede.

This can all be best understood as a matter of identifying and managing risk. Strengthening the capacity to adapt to climate change will not eliminate risk, but it will reduce it. Where there is a risk of violent conflict because of a combination of factors such as poverty, bad governance and a recent history of war, the capacity to manage the risks associated with climate change is also much reduced.

The consequences of consequences

In many countries, one cumulative impact of climate change will be to increase the potential for violent conflict. As we trace this process, we are looking at the consequences of consequences and attempting to track their interactions with other social processes with roots in different aspects of the human condition. Whether countries and communities can adapt so as to cope with the adverse knock-on effects of climate change depends on how a number of variables play out.

It is worth prefacing a brief look at these key variables with two general comments about the causes of violent conflict. It is axiomatic that conflict, as such, is not the central problem – rather, violent conflict is. In other words, conflicts are inevitable, necessary and often productive and key to social progress. What matters is how the conflicts are handled; in particular, whether the participants can reach an acceptable outcome without violence. In this perspective, the consequences of consequences of climate change are bound to include conflict, but need not include violent conflict up to and including the level of war.

Box 2 Darfur: Understanding the causes

The conflict and resulting human tragedy that have unfolded in Darfur since 2003 have grabbed international headlines. As the UN Security Council hammered out a deal to get an international peacekeeping force deployed there, discussions about how to understand the causes of the conflict intensified.

When Darfur first made headlines, the most common way of explaining the context was in terms of ethnic differences between Arabs and Africans. More recently, some have argued – UN Secretary General Ban Ki-Moon among them – that 'the Darfur conflict began as an ecological crisis, arising at least in part from climate change'.[9]

No conflict ever has a single cause. In the case of Sudan, the escalation of violence has been attributed to such factors as: historical grievances; local perceptions of race; demands for a fair distribution of power between different groups; the unfair distribution of economic resources and benefits; disputes over access to and control of increasingly scarce land, livestock and water between pastoralists and agriculturalists; small arms proliferation and the militarisation of youth; and weak state institutions.[10]

Arab–African differences are not as clear cut as some commentators first thought. Political and military alliances frequently shift between ethnic groups, depending on pragmatic considerations. The difference between herders and farmers is also variable. According to the UN Environment Programme,[11] the rural livelihood structures in Sudan are complex and vary from area to area. In many cases, farmers and herders are not separable as some tribes practise both herding and crop cultivation.

The impact of climate change, in particular the 20-year Sahelian drought, played a major role in intensifying grievances in Sudan because it meant there was less land for both farming and herding. These issues played out against a background of economic and political marginalisation, as well as violence. The number of violent conflicts attributable to traditional disputes over the use of land escalated dramatically from the 1970s on.[12] In the mid-1980s, when the north–south Sudanese civil war broke out again after a 10-year hiatus, the government used Arab tribal militias as a means of keeping the southern rebels at bay in Darfur. As a result, ethnic identity started to become more politicised, feeding the escalation of conflicts over land issues with much more destructive fighting than in former times. In 2003 two Darfurian armed groups attacked military installations; the response of local government-backed militias was a further escalation with a campaign of ethnic cleansing, causing over 200,000 deaths and the displacement of over two million.

Thus climate change alone does not explain either the outbreak or the extent of the violence in Darfur. The other 16 countries in the Sahelian belt have felt the impact of global warming, including Mali and Chad, but only Sudan has experienced such devastating conflict. Darfur is, in fact, an exemplary case showing how the physical consequences of climate change interact with other factors to trigger violent conflict.

The conflict itself is taking a further toll of already scarce resources. Militias in Darfur are known for the intentional destruction of villages and forests. The loss of trees in these campaigns reduces the amount of shelter available for livestock and the amount of fuel wood for local communities. This threatens their livelihoods and results in their displacement, while simultaneously worsening the impact of desertification, which makes further conflict over land access more likely.

The massive scale of displacement in Darfur also has a serious impact on the environment. Camps for displaced people mean trees being felled for firewood. The consumption is greater in the many camps where manufacturing bricks is being taken up as a means for people to earn a living, encouraged by development organisations such as DFID. These camps can require up to 200 trees per day for brick-making.[13] Over the weeks and months, combined with the wood needed for domestic use, this adds up to a rate of deforestation that renders the camps unsustainable.

Deforestation already extends as far as 18 kilometres from some camps, as people go further and further afield to find wood. Most of those who go to gather wood in this way are women and children, and this task makes them extremely vulnerable to continuing violence from the militia groups. The incidence of rape has risen as an inevitable result. As the wood runs out, the camps eventually have to move. This is not only hugely disruptive to the hundreds of thousands of camp inhabitants, but it is also detrimental to Darfur's existing problems of drought, desertification and disputes over land-use, which were contributory factors to the conflict from the outset.

[...]

Secondly, when violent conflicts do break out, it is always against the background of a number of different factors interacting with one another. Poverty and poor governance are factors that frequently have a significant role as the background causes of violent conflict; a history of ethno-nationalist politics, environmental degradation and the legacy of previous armed conflicts are further such factors. If these are background causes, in the foreground lie the demands, grievances and positions of the contending parties and the behaviour and credibility of political leaders. It would be misleading to think that climate change alone will cause violent conflict; the problem, rather, lies in the interaction between the effects of climate change and these other factors.

Water

Climate change will significantly affect fresh water supply. Worldwide, over 430 million people currently face water scarcity, and the IPCC predicts that these numbers will rise sharply because climate change will affect surface water levels that are established by rainfall and glacial melting. In some situations, increased glacial melting will cause inland water levels to rise in the short term, followed by a downturn later, but the overall projected impact of climate change is that water scarcity will increase with time.

This will be especially problematic in middle-income countries making the transition from agricultural production to industry. Such states, of which the largest and most advanced in the process are India and China, face an urgent situation as their water resources are already stressed and depleting while demand is growing rapidly.

The conflict risk if water resources are inadequate lies in poor management that either wastes water by inappropriate use of it and inadequate conservation measures, or politicises the issue and seeks a scapegoat on which to blame shortages. Conflicting claims to water resources have been a feature of numerous conflicts as major rivers are very often shared between countries. The situation is particularly problematic when a militarily strong state or region is downstream to a militarily weaker state or region. China, India, Mexico, the Middle East, Southern Africa and Central Asia are among the countries and regions of the world that have been and are likely to be affected by violent conflict over water rights. Tensions over water rights and supply also can be worsened by development programmes that privatise control of the resource without looking after the rights of the poor.[14]

Agriculture

Temperature change and rainfall are decisive for crop and livestock production in the developing world. The IPCC prediction of a temperature rise of 1–3°C in the next 50 years in the global 'business as usual' scenario would mean crop yields falling in mid- to high-altitude regions. If this is borne out by events, regions most likely to be affected by decreasing crop yields include ones that are already prone to food insecurity, such as Southern Africa, Central Asia and South Asia.[15] Studies in India have already seen rice and wheat production decrease as temperature increases, affecting the food security of agriculture-dependent communities.[16] Projected sea-level rise from glacial melting will affect low-lying coastal areas with large populations, reducing the amount of cultivatable land across South Asia and in other areas around the world.

Any disruption in the agricultural sector can massively affect food security, especially for the poorer sections of society. Increased uncertainty about food supply will force communities to find alternative strategies, which often clash with the needs of other communities also facing increased livelihood pressure. In Africa's Sahel region, desertification is reducing the availability of cultivatable land, leading to clashes between herders and farmers. In Northern Nigeria, Sudan and Kenya, these clashes have become violent.[17] The situation in Darfur is most notable (see Box 2).

Energy

Increasing energy consumption is a key reason for global warming but climate change will increase energy requirements in developing countries. Access to reliable, sustainable and affordable energy supplies is vital for development. For example, refrigeration allows local hospitals and clinics to store vital medicines safely; electricity is the basis of modern communications; power is needed to pump water for irrigation and to bring water up from deep wells; and neither industrialisation nor urban development has so far been possible without large-scale energy consumption.

Because energy is such a key development resource, care has to be taken in shaping climate policy. Attempting to develop a strategy to mitigate climate change that includes reduced energy consumption for poor countries would reduce human security, increase poverty and threaten food security. Similarly, reducing energy consumption in middle-income countries would slow economic growth, make poverty reduction much harder to achieve, and generate very high risks of political instability and conflict.

At the same time, of course, meeting increased energy requirements on the basis of business as usual will simply make global warming worse as carbon emissions continue to rise. However, making the transition from fossil fuels to alternative energy sources is proving to be complicated even in rich states with stable, capable governments. It is even more difficult in poor states because the costs of making the transition are relatively higher (i.e., the transition will consume a larger share of scarce economic resources).

Adapting to the energy pressures created by climate change without negative consequences and at affordable costs is a major challenge. Failing to meet it will exacerbate the conflict potential in numerous countries.

[…]

Health

Climate change will pose significant risks to human health. Predicted increases in temperature and rainfall in certain regions are likely to increase the incidence of water-borne diseases such as cholera and malaria which, if unaddressed, could lead to epidemics. Large epidemics could impact the socioeconomic power balance and alter the relations between communities and countries based on availability of material resources to adapt. This could potentially lead to some level of instability or conflict.

Increased natural disasters such as storms and cyclones will lead to increased casualties, putting pressure on already stretched medical resources. Heat waves and water shortages will have an adverse impact on safe drinking water and sanitation that will disproportionately affect the poorest and most marginalised communities, including refugees and internally displaced people.

Failure by the state to provide for basic public health in fragile states is a fundamental factor that erodes the social contract between state and citizens which, in most cases, leads to increased political instability and, often times, violent conflict.

Migration and urbanization

Faced with sudden shocks and with long-term challenges brought about or compounded by climate change, people will move. Taken world-wide, this migration is likely to be on a very large scale, for the basic living conditions of hundreds of millions of people will be influenced by climate change. Stern estimates the scale of migration to reach 200 million by 2050.[22] Some movement will be from one rural community to another, by those hoping to maintain their old lifestyles in a new place. Some movement will be from rural areas where agrarian lifestyles have been overwhelmed by climate change, into urban centres to search for better livelihood options. Others still will cross borders in the hope that a new land will offer better prospects. In each case, those leaving non-viable areas will often migrate to areas that are already only barely viable. A significant part of this new trend of global migration will accelerate urbanisation, adding to urban poverty, conflict and, probably, criminality.

The indirect implications of climate change such as migration and urbanisation present a particular challenge, both to conventional approaches to conflict prevention and to adaptation strategies for climate change. Migration in itself need not be a destabilising factor; it often benefits both those who move and the communities and countries into which they move. But the experience of many countries also shows that there is often great difficulty in accepting immigration. Problems arise particularly when those who already live in an area feel that newcomers are an unwanted burden. This is especially so when communities in search of new livelihood options move to areas that are only just viable. Their presence there can compound social pressures, as it has done in Assam and Bangladesh, for example.[23] In the case of urbanisation, it is noteworthy that even very rapid urbanisation has been managed without violent conflict in prosperous and politically stable nations such as Japan; it is not the process, but the context and the political response to immigration that shape the risks of violent conflict.[24] Nonetheless, that context has so often been conducive to violence and the political response has so often been inflammatory that migration has to be recognised as not only a likely consequence of climate change, but also as a major risk factor in the chain of effects that link climate change and violent conflict.

[...]

Climate change and global insecurity

Failure to help already stressed communities cope with the additional pressure to their livelihoods caused by climate change means that existing grievances will intensify and the risk of violent conflict will increase. Predictions are always uncertain but it is important to identify risks. Our research for the map in this chapter indicates that problems that will be induced or exacerbated by climate change will combine with other factors to create a high risk of armed conflict in 46 conflict-affected states. We identify a further 56 in which the burden of climate change consequences could induce serious political instability, putting them at risk of violent conflict in the long term.

The 46 countries facing a high risk of armed conflict are characterised by some combination of current or recent wars, poverty and inequality, and bad governance. The latter often involves corruption, arbitrary authority, poor systems of justice and weak institutions of government, causing deficiencies in economic regulation and basic services. The combination varies from place to place but all of them suffer from a lethal mix of different types of vulnerability and, consequently, have a high propensity to violent conflict. The armed conflicts that could ensue will probably be fought out with varying degrees of intensity and violence. Some wars kill hundreds of people, others kill hundreds of thousands.

The second group of 56 countries is not so immediately unstable but their government institutions may not be able to take the strain of climate change for a variety of reasons, including a record of arbitrary rule, recent transitions out of dictatorship and war, economic underdevelopment or instability, and lack of technical capacity to handle the issues.

Box 4 Bangladesh: Climate change, migration and conflict risks

The experience of Bangladesh illustrates some of the possible tensions that link climate-related migration to violent conflict. In the recent past, migration has led to violent conflict both within Bangladesh and in neighbouring regions of India.

Bangladesh has a growing population for whom there is not enough land available, and is vulnerable to severe effects from climate change. Part of the country's vulnerability lies in its topography: about half of Bangladesh is located only a few metres above sea level, and about a third is flooded in the rainy season. The Indian Farakka Barrage has made the problem worse over the past 30 years. Completed in 1975, close to the border with Bangladesh, the barrage diverts water from the Ganges River to its Indian tributary, reducing the flow of water in the Bangladeshi tributary. This disturbance to the natural balance of the large Ganges-Brahmaputra delta has caused several severe problems:

- salt water intrusion into Bangladeshi coastal rivers, reaching as far as 100 miles inland on occasion;
- consequent decline in river fishing;
- summer droughts, making the land less productive;
- loss of land to the sea because the reduced river flow meant less sediment was carried into the delta area to give it natural protection against the sea;
- worsened flooding when cyclones hit.

These problems directly affect about 35 million people,[25] exacerbating the effects of other features of rural life – including, not least, poverty, unequal land distribution and, among small farmers, economically inefficient systems of inheritance that divided land among family members into ever smaller plots.

Unable to make a living, many people have migrated. There have been two nearby destinations, as well as others much further afield. Since the 1950s, 12–17 million Bangladeshis have migrated to India (often illegally), attracted by the higher standard of living and lower population density, moving mostly to the adjacent states of Assam and Tripura.[26] And about 400,000–600,000 people have moved within Bangladesh to the Chittagong Hill Tracts (CHT), where they have cleared trees on the steep hillsides and begun farming, resulting in soil erosion and unsustainable livelihoods. In both the neighbouring Indian states and the CHT, there have been conflicts.

Chittagong Hill tribes in Bangladesh were involved in violent conflict with the state for two decades from 1973 until an agreement was reached in 1997. Among the grievances was the influx of people from the plains, whom the Chittagong tribes viewed as a threat. Bangladeshi migration to the north-east Indian region of Assam also contributed to social frictions. The natives resented the newcomers and accused them of stealing land. The immigrants' arrival affected the economy, land distribution and the balance of political power.[27] Violence first erupted in the early 1980s.

These problems continue and further migration as a result of climate change will make them worse. In Bangladesh these pressures combine with persistent political problems that have produced bomb attacks on civilian targets and pressure in some parts of the state for a State of Emergency to be declared. If local and national governments cannot develop measures to cope with the pressures on resources from migration and climate change, the risk of further and more intense violence is very high.

The Double Headed Risk

The consequences of consequences of climate change include a high risk of armed conflict in 46 countries with a total population of 2.7 billion people, and a high risk of political instability in a further 56 countries with a total population of 1.2 billion.

Governance matters

Political stability rests on the strength of the social contract between the government and its citizens. Citizens adhere to the law and pay taxes in return for the state providing for their basic needs, such as security and infrastructure.[30] When the state is perceived to be failing in its basic functions, this contract is eroded.[31] And as the basic problems that government has to solve get deeper, because the demands for resources are becoming more desperate, so the task for government gets more difficult, and the likelihood that it will fail in its basic functions accordingly increases.

[...]

This issue is crucial for two reasons: first, because the decisions that governments take can be extremely important in either moderating or accelerating the social impact of climate change; second, because some state functions are particularly important in relation to the risk of violent conflict. These functions include the provision of primary health care and education, the safeguarding of human rights and democratic systems, and the maintenance of an accountable and effective security sector, including police, army and judiciary.[33] In the event of climate change, an already weak government may find itself unable to meet these basic needs, and one of the consequences of that is an increased risk of violent conflict.

In addition, violent conflict can severely limit the ability of governments to assist in adaptation. Poor governance, combined with other factors, can explain why similarly bad droughts in both Ethiopia and Hungary led to violence only in Ethiopia, and why tropical storms in Haiti and the Dominican Republic led to violence only in Haiti.[34]

It is not poverty alone but uncertainty and the perceived threat of future insecurity that increase the risk of violent conflict.[35] Further, some research indicates that the risk of poverty or its sudden onset also increases the likelihood of individuals joining an armed group.[36] The influence of climate change will be felt as more frequent storms and natural disasters not only cause loss of life and homes, but more generally cause uncertainty and long-term decline in the possibility of maintaining secure livelihoods. In the developed world, these uncertainties and risks can be absorbed by the state's welfare mechanisms and insurance systems. However, in states where such safety nets are already under immense pressure, or do not exist at all due to underdevelopment, weak governance and/or conflict (most notably in countries affected by conflict), the risk of instability cannot be dealt with in this way.

Key risks

This overview of the double-headed problem of climate change and violent conflict reveals a number of key risks that have to be addressed through new policies:

Political instability: Weak governance structures underlie the problem of vulnerability to the impact of climate change. Weak governance is one of the key links in the chain of consequences of consequences. Climate change will put increased pressure on basic state functions such as the provision of basic health care and the guarantee of basic food security. Failed states, fragile states and states in transition, where such institutions either do not exist or are already unable to provide for the basic needs of their citizens, are particularly at risk.

Economic weakness: Economic instability will leave communities highly vulnerable, both to sudden environmental shocks and slow erosion of their livelihood security. The socio-political impacts of climate change will affect poor countries more than further developed states. Poorer countries, which tend to be agrarian states, will be far more susceptible to falling crop yields, extreme weather events and migratory movements. In poorer countries, there is no insurance, either private or state-based, against the effects of crop failure. These impacts of climate change will hinder economic development, and the lack of economic development hinders the ability to adapt to climate change. Empirical studies show that poor countries facing additional pressures are more prone to conflict. Climate change can thus increase obstacles to economic development, worsening poverty and thereby increasing the risk of violent conflict is these states.

Food insecurity: In many areas, the physical effects and the socio-political consequences of climate change will combine to have a profound and destabilising effect on ordinary people's daily lives by reducing food security. The problem here is not simply food shortages but uncertainty of food supply. This may be the result of losing arable land to desert and of shorter growing seasons, but can equally be caused by changes in the food supply chain, such as the loss of roads through flooding (and in other places, the loss of rivers through persistent drought). Political instability and violent conflict also have an effect on food security. Humanitarian assistance can temporarily fill in when there are food shortages but cannot address the underlying problem of lack of food security – and it is only when food security is restored that people can feel safe. In the absence of food security, conflict and migration are almost inevitable consequences.

Demographic changes – migration and urbanisation: Demographic changes always entail a change in power systems and resource allocation. Climate-change-related movements of people will place strain on host communities that already have scarce resources, whether because of population growth, government policy or as an effect of climate change itself. In such circumstances, there is a higher risk of violent conflict. Some of the world's mega-cities are on the coast and are themselves at risk over time from rising sea levels. The combination of population growth, inward migration, declining water supply, other basic shortages and rising sea levels in a city of 15–20 million or more inhabitants adds up to a challenge with which even the most effective city and national government would find hard to cope. Where governance is poor, a social disaster seems close to inevitable.

3. The Unified Solution

The double-headed problem of climate change and violent conflict has a unified solution. The capacities that communities need in order to adapt to the consequences of climate change are very similar to those they need in order to reduce the risk of violent conflict. Addressing one part of the problem in the right way is itself a means of addressing the other part. Indeed, climate change offers an opportunity for peacebuilding: in divided communities, climate change offers a threat to unite against; the need for adaptation offers a task on which to cooperate.

The community is the vital level for action to adapt to and meet climate change but international cooperation is also essential. Climate change and its physical consequences do not respect national borders, so policy and action to address the problem must be developed internationally. This truth has formed the cornerstone of efforts to mitigate climate change for two decades already.

The knock-on socio-economic consequences do not respect national borders either. Large-scale migration, loss of economic output, loss of livelihood security, increased political instability and greater risk of violent conflict will all have consequences that cross national borders. The logic that promotes international cooperation for mitigation works in the same direction when it comes to adaptation.

Furthermore, in many countries that face the double-headed problem, the government is going to be either unwilling or unable – or both – to take on the task of adaptation and peacebuilding. In many of the countries most at risk, the government – and more than that, the system of governance – is part of the problem. The task of helping communities adapt to climate change cannot be left to such governments. There is no alternative except international cooperation to support local action.

Why the international community should act

There are two central motives that should drive international efforts to address the double-headed problem we have identified: the first is to maintain international peace and security; the second, linked to the first, is to support sustainable development.

To maintain international peace and security: The UK government initiated a debate on security and climate change at the UN Security Council in April 2007. There was considerable resistance to this from other governments and it could not be said afterwards that many other governments had been convinced by the UK's arguments. But the very fact that the

UN Security Council was used in this way signalled that climate change is beginning to be considered an issue of international security. The London-based International Institute of Strategic Studies, in its annual Strategic Survey in 2007, similarly identified climate change as a major issue of international security and argued that this would become more widely understood as the effects of climate change begin to bite.[37]

Where the inability to adapt to climate change combines with other stresses to produce violent conflict, neighbouring states and the international community will be affected, not least through the flight of refugees. Even viewed through a narrow economic prism, the cost of a civil war is far higher than the cost of adaptation, so any reluctance in the international community to invest in the adaptation needs of poor communities would be a false economy.

More broadly viewed, a world that is forced into belated efforts to adapt to climate change is almost certainly one in which rivalries between states escalate. Without going into speculative scenarios, the risks that the world faces in relation to climate change will include increased insecurity – unless climate change is treated as an opportunity and becomes the occasion for enhanced cooperation. That is a strong motive for timely international cooperation.

To support sustainable development: The international community has already acknowledged that failure to take climate change into account in development policies and strategies will threaten the achievement of international development goals to reduce poverty and increase literacy and health.[38] Similarly, not paying attention to climate issues in development and peacebuilding can worsen tensions over resources and increase the risk of violent conflict. For example, in Liberia, UN-led programmes are retraining ex-combatants in agricultural skills and reintegrating them into farming communities. According to IPCC projections, however, the region will face a 50 percent cut in crop yields by 2020.[39] Unless the techniques taught are appropriate for the changed environment of the near future – techniques such as half moon planting and water harvesting, for example – the new livelihood opportunities for ex-combatants will be wiped out well within their working lifetime. The existence of unemployed and frustrated ex-combatants is widely regarded as a contributory factor to violent conflict,[40] and violent conflict holds back economic development. But ensuring that development and peacebuilding programmes are sensitive to climate change will bolster or even foster local adaptation and reduce the risk of climate change contributing to violent conflict.

Regional cooperation

It is not only at the level of the UN that international cooperation is relevant. While the world body's role is crucial, it needs to be supplemented by regional and sub-regional bodies such as the African Union and the Organisation of American States, and sub-regional organisations such as the Economic Community of West African States, the Association of South East Asian Nations, and the South Asian Association for Regional Cooperation. Like the EU, but with much less wealth and economic power at their disposal, these bodies represent the common interests of their member states in stability, security and growing trade and prosperity. They can often provide a forum for concerns and a mechanism of support for their members that are closer to the actual concerns of the states involved and less likely to be experienced as an outside threat than, for example, action initiated at the UN level or undertaken by rich northern governments. They could therefore have greater effectiveness and legitimacy in helping develop responses to some of the key risks in the knock-on consequences of climate change.

Some of the measures of adaptation mentioned later in this chapter, such as building stocks of agricultural products as an economic reserve, developing new crop techniques and systems, or identifying post-disaster re-employment opportunities, might be best developed on a regional or subregional basis. Significant numbers of states lack the capacity or the economic resources to make these preparations alone but could play a part in a cooperative system.

Some of the difficult issues of migration could perhaps also be best handled through cooperation at the regional level, developing a framework not only of law, but of interlocking claims and duties on and for one another.

A role for the private sector

The responsibility is not just with governments and inter-governmental organisations. The private sector also has a role to play. International companies operating in at-risk countries have both an interest and a responsibility in safeguarding their investments by working together with governments and communities on adaptation. At a national and local level, again, there is both a company interest and a responsibility to be part of adaptation. Local communities, after all, include small and medium-sized companies, local producers and traders.

Many corporations are already making steps towards sustainable and environmentally friendly business practices. Many companies have developed corporate social responsibility policies that aim to minimise the adverse impacts of the companies on the social environments around them. However, without adequate information on the socio-economic consequences of climate change, some of these well-intentioned policies could actually restrict the adaptation options of some communities in the near future. For example, promoting fair trade coffee is an important step towards generating better conditions for coffee farmers. Yet the predicted increase in temperature of 2°C will dramatically decrease the amount of land suitable for growing coffee.[41] If more farmers were to go into coffee production because they were guaranteed a fair price, and if there were to be no planning for alternative livelihood strategies when climate change strikes, the long-term effect could be harmful.

Well-informed, climate-aware and context-specific business practices, on the other hand, have the scope to provide new adaptation options such as new livelihood opportunities or strengthened infrastructure. For example, if different crops are to be farmed, it is essential that there is efficient distribution of the seeds and of the products – a role for the private sector. Establishing quick re-employment options after drought or extreme weather also offers a role for private companies.

At a different level, business practice should be climate-sensitive, not only in terms of reducing carbon emissions and thus attempting to address the long-term roots of the problem, but also in terms of supporting adaptation to address the short and medium-term consequences. This can involve not only the obvious companies, such as those in energy and transport, but others, such as the finance sector, which is capable of transforming into practical commercial considerations the argument in the 2006 Stern Review that responding constructively to climate change is economically beneficial. Adaptation will require investment, and in some cases will be suitable for private sector investment.

Complexities of cooperation

There is already a considerable international agenda for cooperation on the issue of climate change. For many observers and especially for environmental activists, this agenda does not go nearly far enough on mitigation. But the perspective advanced here is different: important though it is to mitigate global warming, examining the interrelationship between climate change and the risk of armed conflict leads to the conclusion that adaptation needs more attention and more action. Some academic commentators have pointed to 'the long-standing unease in the policy community with regard to adaptation'.[42] Though adaptation does feature on the international agenda, it is mitigation that takes the lion's share of the headlines and the policy initiatives. It is time to recognise that while mitigation is essential, its benefits will come slowly and, in the meantime, adaptation is urgent.

Box 7 Current frameworks and action on climate change adaptation

International Frameworks

UN Framework Convention on Climate Change (UNFCCC): International efforts to tackle climate change are primarily pursued through the UNFCCC. The UNFCCC is an international environmental treaty produced at the United Nations Conference on Environment and Development, known as the Earth Summit, held in Rio de Janeiro in 1992. The parties to the UNFCCC meet annually; the December 2007 meeting in Bali is the 13th Conference of Parties. The UNFCCC acts as an umbrella for international dialogue, policy and funding on climate change. Its overarching mandate, stated in article 2 of the Convention, is to limit greenhouse gas levels to a 'level that would prevent dangerous anthropogenic interference with the climate system'. Under this framework, mitigation of climate change dominates the agenda, with most funding and policy attention geared towards the future of the Kyoto Protocol and a number of separate initiatives.

Intergovernmental Panel on Climate Change (IPCC): The IPCC is the most authoritative source of internationally accepted scientific assessments. These assessments feed into the UNFCCC process and constitute its scientific basis. However, though based on pure science, the reports of the IPCC are produced through intense political negotiation, especially over the confidence with which future effects are predicted, and concerning the analysis of how observed features of climate change are caused. The Fourth Assessment Report (AR4) from the IPCC has come out during 2007 and is more far-reaching in its socio-political analysis of the impacts of climate change than its predecessors.

Working Group II of the IPCC, in particular, has looked more closely at the climate impacts and vulnerabilities of fragile communities than in previous reports. However, it is not the role of the IPCC to provide an assessment of the likely impacts of climate change on violent conflict, so the issue of conflict and peacebuilding potential is not explored in the AR4.

UN International Strategy for Disaster Reduction (ISDR): The ISDR was set up to coordinate approaches at a local, national and international level with the aim of building disaster-resilient communities by promoting increased awareness of the importance of disaster reduction as an integral component of sustainable development.

The Hyogo Framework for Action (HFA): This is a 10-year action framework (2005–2015) for disaster risk reduction. Its three aims are to: integrate disaster risk reduction into sustainable development policies and planning at all levels, with emphasis on disaster planning, mitigation, preparedness and vulnerability reduction; develop and strengthen institutions, mechanisms and capacities at all levels; and to systematically incorporate risk reduction approaches into the implementation of emergency preparedness, response and recovery programmes.

Neither the ISDR nor the HFA was designed to address directly the issues posed by climate change and conflict but they provide useful frameworks to guide and monitor action. However, these frameworks are only as effective as their implementation. NGOs are already finding that action around the HFA is highly top-down and does not sufficiently include local actors.

Global Environment Facility (GEF): Multilateral funding for climate change is mainly channelled through the GEF, a funding agency established in 1991. While most funding for climate change over the last decade has been for mitigation, the GEF has recently set up four new funds for adaptation in developing countries. However, one barrier to using these funds is the GEF rules, which state that they can only be used for the 'incremental costs of global benefits'. While it is relatively easy to calculate the costs of global benefits arising from mitigation projects, it is more difficult to do so for adaptation projects as benefits are usually local rather than global. The four funds are:

- *The Least Developed Countries Fund*: This fund is only for those countries classified as LDCs. It therefore excludes many middle-income countries that also face the risk of instability or violent conflict in the face of climate change. It is reliant on voluntary contributions for funding. Since its launch in 2001, the LDC fund has attracted $120 million in pledges, but only $48 million has been received as of April 2007.
- *The Special Climate Change Fund*: This is for adaptation planning and technology transfer in all developing countries and is reliant on voluntary contributions for funding. As of April 2007, $62 million has been pledged, and $41 million has been received.

- *The Strategic Priority on Adaptation*: A three-year initiative to pilot adaptation capacity-building measures, funded by $50 million from GEF Trust Funds.[43]
- *The Adaptation Fund*: This is intended to fund actual adaptation measures in developing countries. It is not yet operational; the plan is to fund it from CDM credits, amounting to an estimated $1 billion over the next five years. Of the 13 countries that have submitted their NAPAs to the UNFCCC, the total cost of projects proposed to meet only the immediate adaptation needs is $330 million. Factor in the long-term costs, and the 89 other countries in need of assistance, and it is evident that this fund is just a drop in the ocean of what is required.

The cost of adaptation is still hugely under-researched and so an estimate of how much is needed is difficult to discern. However, the World Bank has produced a preliminary estimate that it will cost approximately $10–40 billion to climate-proof investments in the developing world.[44] Even judged against the lower estimates, the pledges received to date are massively inadequate.

At the same time as noting the relative paucity of funds available for adaptation, it is important to add that the international donor community does not only need to spend additional money, it also needs to change the way it meets its current commitments for expenditure on development and peacebuilding. These activities need to be climate-proofed – i.e., the way that development and peacebuilding money is spent has to alter if the challenge of climate change is to be met. This should take an important place on the international agenda, starting with the December 2007 Conference of Parties of the UNFCCC (CoP 13) in Bali.

Organisation for Economic Cooperation & Development (OECD): One forum that has begun looking at integrating the development, peacebuilding and climate adaptation strands is the OECD. The Development Assistance Committee (DAC) is extending the chapter on Environment and Resources in the OECD's Guidelines for Conflict Prevention to take account of climate change. And the DAC Network on Conflict, Peace and Development Cooperation is researching the links between the environment, conflict and peace, issuing briefs and specific assessments on land, water, valuable minerals and forests.[45] The OECD's Working Party on Global and Structural Policy has also recently set up a Climate and Development Project where climate change and conflict are intended to be addressed with strong participation from developing countries.

At The Regional Level

The European Commission (EC): The EC is developing a global monitoring system for environment and security in 2008 as part of the European Strategy for Space.[46] This monitoring measure is intended to oversee implementation of the Kyoto Protocol; it will largely benefit mitigation, rather than adaptation. There are also discussions about the need to link climate change to broader security and development policy strategies and the EC is establishing a new Global Climate Change Alliance between the EU and other vulnerable developing countries.

Apart from the EC, there do not appear to be major regional initiatives addressing adaptation and even the EC is only now coming to this issue.

At The National Level

National Adaptation Programmes of Action (NAPAs): Under the framework of the UNFCCC, the core instrument for addressing adaptation by countries at the national level is through NAPAs. The idea of a NAPA is to provide a process For Least Developed Countries to identify priority activities that respond to their urgent needs for adapting to climate change. To date, 22 states have drawn up a NAPA,[47] and 13 have submitted them to the UNFCCC. In theory, NAPAs take into account existing coping strategies at the grass-roots level, and build upon them to identify priority activities, rather than focusing on scenario-based modelling to assess future vulnerability and long-term policy at state level. However, the process of drafting the NAPAs so far seems to rest more on assistance from donors such as the World Bank and the UN Environmental Programme rather than on participation from community groups and civil society. The NAPAs have an evident potential for integrating peacebuilding and development concerns with adaptation to climate change, but it is too soon to tell whether actual steps are being taken in this direction. In the absence of an effort to integrate the plans and action, the risk is that NAPAs will become just another box for poor governments to tick on the way to getting some funding.

Trade-offs and synergies between adaptation and mitigation

The IPCC's AR4 notes the risk of an unwelcome trade-off between adaptation and mitigation because resources committed to one are not available for the other. As far as the poorest countries are concerned, the fact is that their carbon emissions have been marginal compared to industrial countries and, more recently, the fast developing middle-income countries. Africa as a whole, home to 14 percent of the world's population, is responsible for 3.6 percent of global carbon dioxide emissions while, to take a random example, Australia has 0.32 percent of the world's population, yet produces 1.43 percent of carbon dioxide emissions.[48] With the exceptions of Libya, the Seychelles, Nigeria and South Africa, African countries emit only 0.5 tonnes of carbon dioxide per capita each year. By comparison, as the world's largest emitter, the USA emits over 20 tonnes per capita.[49] For poor and politically unstable countries facing the combined risk of climate change and conflict, therefore, there is not much to gain by concentrating scarce resources onto mitigation, and however heroic their efforts, they will not make much of a dent in global emission levels. From the perspective both of the individual countries and of the international community as a whole, the priority need in the poorest countries is for adaptation.

An additional trade-off between mitigation and adaptation policies has been less discussed. In some circumstances, measures to reduce GHG emissions risk actually hindering adaptation.

For example, among the World Bank's activities in Sri Lanka is the new Renewable Energy for Rural Development Project[50] which aims – among other strategies – to strengthen the national grid through support for privately owned mini-hydroelectricity plants and other renewable energy projects. It is likely to divert scarce water supplies from communities' consumption and agricultural needs. That risks weakening food security at a time when, as part of adaptation to climate change, it should be strengthened. It also risks fostering social tensions because of local resentment towards development initiatives that misfire. Similarly, in Cochabamba, Bolivia, making water into a marketable commodity by contracting water provision out to the private sector pushed up prices and led to violent protests in January to April 2000, with over 100 people injured.[51]

The limits of carbon trading

Carbon trading is one of the key ways in which states are attempting to address the problem of climate change at the international level. Most carbon trading schemes are registered with the Clean Development Mechanism (CDM), an arrangement under the Kyoto Protocol allowing industrialised countries that have committed to reducing greenhouse gas emissions to invest in projects that cut emissions in developing countries as an alternative to more costly emissions reductions in their own countries.

Mitigation is self-evidently important and carbon trading has long been seen as a productive way of doing it but, recently, a number of concerns have arisen around the CDM. A study by Nature in 2007[52] revealed that the CDM was becoming a lucrative industry where companies were paid as much as 50 times more than it cost to reduce emissions. Further investigations have found that there are loopholes allowing for spurious credits to be awarded. There is evidence that the majority of CDM projects would have happened anyway – in other words, companies were simply using the CDM to generate extra income. There were even cases of projects being retrospectively given the CDM tag. Thus, the CDM was not acting as an incentive for new environmentally responsible activities.[53]

Among other problems, the CDM's failure to take account of poverty is concerning. Most CDM projects are in countries undergoing rapid industrialisation and very few are in Africa; in 2005, these accounted for only seven projects in all, 2 percent of the total (and, of these, five were in South Africa).[54] The real problem, however, is the risks entailed in some of the projects. For example, a World Bank landfill gas project in Durban, South Africa, is actively opposed by most local communities because of its adverse health effects.[55] If the effort is made to mitigate climate change in this way, pursued at the expense of the needs and well-being of local communities, there is a risk of social instability. In regions that are already unstable and face a myriad of other pressures, failure to take account of conflict dynamics can contribute to an escalation of such instability into violence.

The problem of maladaptations

The IPCC's Working Group on Impacts, Adaptation and Vulnerability has rightly noted the importance of addressing climate change adaptation in fragile states, especially where these responses are so-called 'no regrets' policies – that is, policies that turn out to be of benefit to a community whether or not the predicted climate change impacts occur.

The IPCC warns against the risks of what it calls maladaptations, which are the result of responses to climate change that lack foresight about climatic or relevant social trends.[56] However, the IPCC draws a problematic conclusion when it argues that this means there should be more emphasis on mitigation to prevent future maladaptations that would increase the costs of climate impacts. It would be equally possible to turn this the other way around and say that the problem of what might be called mismitigation – as outlined above – means there should be less emphasis on mitigation for fear it will go wrong.

The solution is to ensure that maladaptation does not occur at all. In fragile states, this would mean ensuring that policies on climate change are sensitive to conflict risks and, at the same time, ensuring that peacebuilding and development take account of the consequences of the consequences of climate change. In essence, the process entails incorporating adaptation into peacebuilding in a manner that takes account of future vulnerabilities to climate change.

Ensuring the approach is evidence-based

The issue of maladaptation shows that the need is not only to give more attention to adaptation but to make sure it is effective – more of the right kind of adaptation. To this end, examples such as those from Bolivia, Sri Lanka and South Africa show the importance of basing adaptation policies on solid knowledge about local circumstances, including anticipated climate change impact and a thorough contextual analysis.

There are two problems with this – one is time. Such work will take two to five years to complete. Meanwhile, the effects of climate change are already unfolding. The response, therefore, must be incremental. Peacebuilding will help develop the adaptive capacities of communities so they can use the research findings as they come through. In the meantime, peacebuilding and development must be as climate-sensitive as existing knowledge allows, recognising that this knowledge will deepen as time goes by.

The second problem is the risk that the approach to conducting, reporting and using research will be technocratic, top-down and alienating. To ordinary people it will feel like outside experts coming and telling them how things are, how they should live and what they should do. The likelihood is they will ignore this advice or, if necessary, fight it.

A different way of working is possible, grounded in a peacebuilding approach. This emphasises the importance of local knowledge and seeks the active participation of local communities in working out how best to adapt to climate change. While much of the technical knowledge, such as complex climate modelling, would of necessity need to be transferred from states with more advanced research and development capacity, figuring out what to do could and should be an inclusive and participatory process. Where communities are divided because of the experience or risk of violent conflict, addressing these problems could, in fact, provide the occasion for developing a practical, problem-solving dialogue through which cooperative relationships could be established and steadily built up. The aim, in short, is to bring hard science and local knowledge together.

Peacebuilding

'Peacebuilding' means societies equipping themselves to manage conflicts without resorting to violence. It looks different in different contexts – the detailed activities range from local dialogues promoting reconciliation to advocacy that shapes economic policy and business practices. The key is to understand that it is not possible to build peace for people and communities that have been involved in violent conflict; rather, those people and communities must build peace for themselves. It is, however, possible for outsiders to help and participate in that process.

Peace is sustainable only if it is based on a social process in which citizens participate as equals. In general, they will do this only when they see that the peace process offers justice, economic equity and progress, security and good governance. These are the foundations of peacebuilding which, in the long term, are the basis for strong and responsive institutions of government. Peacebuilding is thus holistic, acting on all aspects of a society's security, socio-economic foundations, political frameworks, justice systems, and traditions of reconciliation to strengthen the factors that can contribute to peace. And peacebuilding is also inclusive of all actors and perspectives, including those who are frequently marginalised or excluded.

Peacebuilding works – but it works slowly. It was well over a decade before the peace process in Northern Ireland was regarded by most observers as relatively stable – from the IRA's ceasefire declaration in 1994 to the return to a power-sharing government in 2007 – and the violence in Northern Ireland, though painful and protracted, was relatively low-level by international standards, while the peace process was lavishly funded by comparison with other cases. Peace in Bosnia-Herzegovina has taken a similarly long period to secure, and the process is by no means completed. In Burundi in 2007, there remain elements of risk in the peace process that was initiated in 2000. Peacebuilding takes patience and care and, in its early years, is extremely precarious because it takes far fewer people acting irresponsibly to return a country to violence than are needed to work together to sustain the peace. Yet peacebuilding can transform societies into functional communities that can exist without the threat of violent conflict – a process that we see unfolding in Liberia today.

In Liberia, the key need to which International Alert has been able to contribute is communication as the basis for social participation in the peace process. In Burundi, where Alert has been working since 1995 when civil war was intense, the organisation was able to work at several levels. Alert provided space for political and community leaders to meet, develop mutual confidence and jointly develop ideas for moving the country onto a peaceful path. Alert also worked with civil society activists to help found a national women's peace organisation that, acting as an umbrella for local women's groups, has trained over 10,000 people in conflict resolution, mediation and facilitation. In the Democratic Republic of Congo, Alert has recently developed a programme bringing together people from different regions into a national dialogue on how to sustain peace and human rights in a country that, from the colonial period until recently, knew only dictatorship and war. In the South Caucasus, several Alert projects help people come together and exchange ideas across entrenched lines of conflict, helping to develop social foundations for possible future peace deals.

These activities cannot make peace by themselves, but nor can peace be made without them. A sustainable peace requires a peace agreement between the leaders of the contending parties, their continuing commitment to it after signature, and a social setting to support it and encourage political leaders' continued commitment. The problem that peacebuilding addresses is that, through the experience of violent conflict, societies lose the capacity to resolve difficult issues peacefully. Variously, they lose the institutions that can mediate and negotiate disputes and differences before they get out of hand, and they lose the cultural habits of compromise and tolerance that are required for serious differences to be settled by agreement. Helping societies regain these attributes is what peacebuilding is about.

The way peacebuilding is implemented has to be tailored to the needs of the specific context. Each society and community has its own modes and values. Because the point of peacebuilding is to help societies renew the attributes of a peaceful society, it cannot work on the basis of a top-down recipe. It has to support and enhance the efforts and energies of ordinary people, to develop a process from the ground up so as to ensure that the opportunity offered by a formal peace agreement is seized and lasting peace is created.

Linking peacebuilding and climate adaptation

[…]

In one sense, adaptation to climate change could take many forms, some of them profoundly destructive. If good land for farming or grazing becomes scarce, it could be said that when one group attacks another to drive it away, that is a form of adaptation. Likewise, if the pressures of climate change lead large numbers of people to leave their homes and migrate to urban slums, that also is a form of adaptation. But what people want and need are forms of adaptation that protect human security.

Successful adaptation to climate change will still involve changes in how people live. The key to linking peacebuilding and adaptation with climate change is to ask how people can best change the way they live. What is the best process of change – that is, the process that offers the greatest opportunity to cope peacefully with the challenge of climate change and adapt to it in a way that protects people's well-being? It is, surely, a process that simultaneously meets two objectives: it needs to be based on a proper appreciation of the challenge – i.e., it needs to be scientifically informed; and it needs to be a process that thoroughly involves the people whose lives will change so they shape it and buy into it. For this to be the case, the people involved in the process must understand the problem (so the science must be communicated clearly), see what the options are, gauge the impact of inaction, and choose to change. This approach acknowledges that local knowledge alone is not enough because climate change throws up unprecedented problems, but nor is the best hard science enough by itself, because adaptation needs to be locally grounded and culturally appropriate.

These considerations are even more important when looking at one of the most difficult problems thrown up by climate change and one where some of the greatest fears are likely to reside – migration. As we have argued earlier, as many examples have shown, migration itself does not generate violent conflict, yet it can be an important part of the chain of effects leading to violent conflict because of the responses it so often gets and because of the context. When people find a large number of newcomers arriving, the key issue is to try to develop a common understanding of what the problem is, why it has come about, and then what can be done about it in a way that most meets everybody's needs. The best time to have this discussion is before the pressures of immigration have become intolerable. Research, good information systems and clear government and inter-governmental policies are all essential. But perhaps more important than anything is a commitment to timely dialogue in, with and between the communities that are affected – both those who are forced to migrate by the physical effects of climate change, and those who become hosts to the new migrants. The political issues wrapped up in this part of the climate change problem are extremely tangled, with competing resentments about who benefits from any resources that can be mobilised, and the risk that the question will be politicised in an inflammatory way. To leave such a potentially explosive set of issues alone, however, is to risk that explosion occurring.

The best process of change for a successful adaptation to climate change, in short, is the same as the processes involved in peacebuilding. In both, energies must be engaged in different parts of society – among communities and their leaders, in the private sector, media, political groups, social activists, students and intellectuals – and at different levels – among the elite and among ordinary people. In both, the process must include women as well as men, youth as well as mature adults, minorities as well as majority communities, and it must cross political divides as well. The techniques that will be used are also the same: encouraging dialogue; building confidence; addressing the issues that divide groups and out of which they perceive conflicts to grow; learning; mutual education; developing and strengthening civil society organisations to carry the work forward; strengthening both the capacity and the accountability of the institutions of government.

The processes of peacebuilding and adaptation are not only similar in these ways, they are also synergistic. A society that can develop adaptive strategies for climate change in this way is well equipped to avoid armed conflict. And a society that can manage conflicts and major disagreements over serious issues without a high risk of violence is well equipped to adapt successfully to the challenge of climate change.

There could be a further linkage, because climate change could become a reason for moving on from some of the attitudes and behaviour that were generated by the experience of armed conflict. International Alert has supported dialogues in conflict countries that bring together people who have very different and incompatible perspectives but who share an understanding of the risks and unbearable costs of continuing with (or returning to) open, armed conflict. In the same way, dialogues could bring together people whose different and incompatible perspectives do not prevent them from understanding the common threat of climate change and the shared need to adapt to meet this challenge. It is, seen in one light, no more complicated than adding another crucial item to the agenda of peaceful dialogue. But because it cannot be blamed on one conflicting party over another, and yet it affects all, climate change may have more power for bringing people together than much of the rest of the agenda. Climate change could generate a pragmatic unity because it offers a threat that can put other problems in perspective. And adaptation to climate change offers tasks that can be the object of cooperation between formerly antagonistic groups.

Developing social resilience

Climate adaptation and peacebuilding need comprehensively to address the key risks faced by fragile states affected by climate change. These risks, as outlined at the end of chapter 2 of this report, are political instability, economic weakness, food insecurity and demographic changes such as migration and urbanisation. The measures that are adopted and the way they are implemented have to target not just these four issues but the linkages between them. Awareness of these risks will help national governments and donor agencies develop programmes for the linked goals of development, peacebuilding and adaptation. In so doing, the result will be societies that are increasingly resilient in the face of both short-term shocks and slow onset changes.

One way to gauge this objective is by drawing from the literature on reducing the risk of disaster and looking at the concept of social resilience. This can be understood as the capacity to absorb stress or destructive forces through resistance or adaptation; the capacity to manage or maintain certain basic functions and structures during disastrous events; and the capacity to recover after the event.[57] In principle, the idea of resilience is relevant for thinking about a society's ability to cope with a wide range of problems, from natural calamities, through economic shocks, to invasion, to slow onset changes in the natural environment.

Key characteristics of a resilient society are that it is well governed, understands the risks it faces, can manage those risks and minimise its vulnerability to them, and that it is prepared to respond to unpreventable disasters. Being well governed, the society has clear policies and a strong framework of law and regulation, implemented by capable institutions. It understands the risks it faces because it has the scientific capacities to do so, and can manage them successfully not only because of good planning, but because of public awareness as a result of good communications and sharing of information. It can minimise its vulnerability because it has made provision for social welfare as well as physical protection, and it is well organised with good early warning systems to be able to respond quickly if a natural disaster should strike. Indeed, such a society may experience extreme events such as hurricanes, storms and earthquakes, but its resilience means those events will not actually be disasters.

This depiction of a resilient society is abstract and idealised. It does not describe an existing society – especially not one to be found among the 102 countries that face the double-headed risk of climate change and violent conflict – but it sets objectives to aim for. What it makes clear is that physical protection and preparation for quick response to extreme events are the results of exploring problems and identifying possible solutions, as well as deploying expert knowledge within an open process of information-sharing and discussion. The closer it is possible to get to an inclusive process with the participation of all affected groups, the greater the degree of resilience that can be developed.

Simultaneously addressing peacebuilding needs and climate change adaptation will involve considering how different sectors and actions are connected. For example, building a new road will not only improve transport infrastructure but may also encourage poor communities to settle along the roadside as a means of enhancing their livelihoods through road-side trading. If the road is cutting across a flood plain, those communities also will be increasing their level of vulnerability to climate change. Making a difference in one sector – such as hydropower – without improvements in the provision of other basic services – such as domestic water supply – can fuel new grievances. Developing the resilience of communities so they can adapt successfully to climate change will include developing the capacity to understand these linkages and to act on them.

[...]

The practicalities of adaptation

Adaptation to climate change is already taking place, but it is rarely done in order to build resilience. To date, most adaptation efforts have been initiated within a narrow frame of reference, looking at cost and benefit in terms of narrow economic interests. This, in itself, would not be a bad thing if it were set within a context of social adaptation and building resilience. When it is not, it risks being dysfunctional.

The concrete measures of a successful process of adaptation will emerge from local initiatives and will take a different shape in different contexts as they address different consequences of climate change. There are some examples that can be cited to indicate the practical import of the argument:[69]

- In Mexico and Argentina, in response to increased flooding risks, a number of adjustments have been made: planting dates have been changed and new varieties of crop have been introduced, including drought-resistant plants such as agave and aloe. There also have been changes to overall management systems: stocks of the product have been built up as an economic reserve; farms have diversified by adding livestock operations and the plots used for crops and for grazing have been separated so as to diversify exposure; crop insurance has been set up and local financial pools established as an alternative to commercial crop insurance.
- In Botswana, national government programmes have been set up to recreate employment options after drought. This has entailed working with local authorities and providing assistance to small subsistence farmers to increase crop production.
- In the Philippines, responses to rising sea levels and storm surges include the introduction of participatory risk assessment; provision of grants to strengthen coastal resilience and rehabilitation of infrastructures; construction of cyclone-resistant housing units; retrofit of buildings to improved hazard standards; review of building codes; and reforestation of mangroves.
- And in Bangladesh, where an already rising sea level means that salt water intrusion is a major issue (see Box 4), steps are being taken at the national level, where climate change concerns have been included in the National Water Management Plan and, at local levels, for example, through the use of alternative crops – such as switching from rice production to farming prawns – and the use of low-technology water filters.

Opportunities for coherent adaptation are greater in some sectors, such as agriculture and forestry, buildings and urban infrastructure, but are currently limited in others, such as energy and health. This is only due to a lack of conceptual and empirical knowledge around these areas. There is an evident need to address these research and knowledge gaps, while taking immediate action on areas with stronger existing levels of knowledge and understanding.

The difference in adaptive capacity within and across societies is also a critical issue to be acknowledged in policy and practice. Climate change may not target the marginalised over the affluent, but the differential in capacity to adapt determines who suffers and the extent of that suffering. Communities already facing multiple pressures, such as poor access to economic and natural resources, will face barriers to adaptation. Addressing these barriers will itself be a means of promoting adaptation through bolstering capacity for the process of adaptation.

4. Conclusions and Recommendations

As the IPCC notes, 'societies have a long record of adapting to the impacts of climate through a range of practices…but climate change poses novel risks often outside the range of experience, such as impacts related to drought, heatwaves, accelerated glacier retreat and hurricane intensity.'[70] In short, the future will not be the same as the past. The severity of impacts, both sudden shock and slow onset, will leave some communities unable to adapt to or cope with the physical effects and knock-on consequences of climate change. The most vulnerable communities with the weakest adaptive capacity are in fragile states.

This report has shown that in fragile states the consequences of climate change can interact with existing socio-political and economic tensions, compounding the causal tensions underlying violent conflict. In 46 states already affected by violent conflict, the dual problem of climate change and violent conflict can lock the state into a downward spiral where violent conflict restricts the adaptive capacity and climate change worsens the conflict. In a further 56 states, the consequences of climate change could move them into political instability, creating a high risk of violent conflict further on.

But the potential downward spiral can be transformed into a virtuous circle. The solution to this double-headed problem is a unified one. Essentially, this involves applying the established principles of conflict-sensitive development practices to climate change policies and practice. At the very least, climate change need not increase the risk of violent conflict and, at best, addressing climate change in fragile states can promote peace. By acting together to prevent violent conflict, governments and institutions will be better placed to address the demands of climate change adaptation. In fragile states, therefore, donor governments and institutions must do their best to ensure that climate change strategies are conflict-sensitive, and that peacebuilding and development activities are climate-sensitive.

[…]

Dan Smith and Janani Vivekananda, 'A Climate of Conflict: The Links between Climate Change, Peace and War', International Alert, November 2007, pp. 9–37.

Notes

3   Paavola, Jouni & W. Neil Adger. 'Fair adaptation to climate change,' Ecological Economics, 56, 2006, pp 594–609, p.604. Available at: www.elsevier.com/locate/ecolecon.
      […]
9   Ban Ki-Moon. 'A Climate Culprit in Darfur,' in The Washington Post, 16 June, 2007.
10   O'Callaghan, Sorcha. Presentation at ODI event: 'Environment, Relief and Conflict in Darfur,' 2 August, 2007.
11   UNEP. Sudan Post-Conflict Environmental Assessment, United Nations Environment Programme: Kenya, 2007.
12   'Darfur Rising: Sudan's New Crisis,' ICG Africa Report No. 76, March 2004, p5.
13   Alam, Syed Ashraful. 'Use of biomass fuels in the brick-making industries of Sudan: Implications for deforestation and greenhouse gas emission,' 2006. Available at: http://ethesis.helsinki.fi/julkaisut/

maa/mekol/pg/alam/useofbi o.pdf.

14 Levina, Ellina. 'Domestic Policy Frameworks for Adaptation to Climate Change in the Water Sector,' OECD, 2006; also, see the experience of privatisation of water in Latin American states such as the 'Cochabamba Water Wars' that took place in Bolivia between January and April 2000, due to privatisation of the municipal water supply. See: http://ipsnews.net/news.asp?idnews=35418.

15 Intergovernmental Panel on Climate Change. Climate Change 2007: Impacts, Adaptation, and Vulnerability. Contribution of Working Group II to the Fourth Assessment Report of the Intergovernmental Panel on Climate Change. Cambridge University Press: Cambridge, 2007.

16 'Climate Change Impacts on Agriculture in India,' UK DEFRA, 2007. Available at: http://www.defra. gov.uk/ENVIRONMENT/climatechange/internat/devcountry/pdf/india-climate-6-agriculture. pdf; and research conducted by Indian Agricultural Research Institute available at: http://www.iari. res.in/

17 Hussein, Karim, James Sumberg & David Seddon. 'Increasing Violent Conflict between Herders and Farmers in Africa: Claims and Evidence,' Development Policy Review, 1999.
    [...]

22 Stern, Nicholas et al. The Economics of Climate Change (The Stern Review). Cambridge University Press: Cambridge, 2007. Available at: http://www.hmtreasury.gov.uk/independent_reviews/ stern_review_economics_climate_change/stern_review_report.cfm.

23 Baechler, G. 'Environmental degradation in the south as a cause of armed conflict,' In: Carius, A. and K. Lietzmann, Environmental change and security: A European perspective, Berlin: Springer-Verlag, 1999; Van Ireland, E., M. Klaassen, T. Nierop, & H. van der Wusten, Climate change: Socio-economic impacts and violent conflict, Wageningen: Dutch National Research Programme on Global Air Pollution and Climate Change, Report No. 410200006, 1996.

24 Barnett, Jon. 'Security and Climate change,' Global Environmental Change, 13, 2003, pp7–17.

25 Reuveny, Rafael. 'Environmental Change, Migration and Conflict: Theoretical Analysis and Empirical Explorations,' GECHS, June 2005.

26 Reuveny, Rafael. 'Environmental Change, Migration and Conflict: Theoretical Analysis and Empirical Explorations,' GECHS, June 2005.

27 Reuveny, Rafael. 'Environmental Change, Migration and Conflict: Theoretical Analysis and Empirical Explorations,' GECHS, June 2005.

30 Dworkin, Ronald. Law's Empire. Oxford: Hart Publishing, 1998; Raz, Joseph. Authority, Law and Morality. Monist, 68, 1985, pp295–314.

31 Raz, Joseph. Authority, Law and Morality. Monist, 68, 1985, pp295–314.

32 'Drought Cuts Size of Lake Chad 90 Percent in 40 Years,' Agence France-Presse. 9 March, 1999.

33 Keen, David. 'Incentives and disincentives for violence,' in M. Berdal and D. Malone, Greed and grievance: Economic agendas and civil war. Boulder: Lynne Reinner, 2000, pp19–42; Goodhand, Jonathan, 'Enduring disorder and persistent poverty: A review of linkages between war and chronic poverty,' World Development, 31, 2003, pp629–646.

34 Intergovernmental Panel on Climate Change. Climate Change 2001: Impacts, Adaptation, and Vulnerability. Contribution of Working Group II to the Third Assessment Report of the Intergovernmental Panel on Climate Change. Cambridge University Press: Cambridge, 2001.

35 Stewart, F. 'Crisis Prevention: Tackling Horizontal Inequalities,' Oxford Development Studies, 28, 2000, pp245–263.

36 Ohlsson, Leif. Livelihood conflicts: Linking poverty and environment as causes of conflict. Stockholm: Environmental Policy Unit, Swedish International Development Cooperation Agency, 2000; Goodhand, Jonathan, 'Enduring disorder and persistent poverty: A review of linkages between war and chronic poverty,' World Development, 31, 2003, pp629–646.

37 Strategic Survey 2007. London: International Institute for Strategic Studies, 2007.

38 Oxfam. Adapting to climate change, Washington: Oxfam International, 2007; IPCC (Intergovernmental Panel on Climate Change), 2007, Climate Change 2007: Impacts, Adaptation, and Vulnerability. Contribution of Working Group II to the Fourth Assessment Report of the Intergovernmental Panel on Climate Change, Cambridge University Press, Cambridge.

39   Intergovernmental Panel on Climate Change. Climate Change 2007: Impacts, Adaptation, and Vulnerability. Contribution of Working Group II to the Fourth Assessment Report of the Intergovernmental Panel on Climate Change, Cambridge University Press: Cambridge, 2007, Chapter 19 on Africa.

40   Collier, Paul. 'Doing Well Out of War: An Economic Perspective,' in Mats Berdal and David M. Malone, eds, Greed & Grievance: Economic Agendas in Civil Wars, Boulder, CO and London: Lynne Rienner, 2000, pp91–111; Keen, David. 'Incentives and disincentives for violence,' in M. Berdal and D. Malone, Greed and grievance: Economic agendas and civil war, Boulder: Lynne Reinner, 2000, pp19–42; Goodhand, Jonathan. 'Enduring disorder and persistent poverty: A review of linkages between war and chronic poverty,' World Development, 31, 2003, pp629–646.

41   Simonett, Otto. 'Potential Impacts of Global Warming,' GRID-Geneva, Case Studies on Climatic Change, Geneva, 1989.

42   Paavola, Jouni & W. Neil Adger. 'Fair adaptation to climate change,' Ecological Economics, 56, 2006, pp 594–609, p.595. Available at: www.elsevier.com/locate/ecolecon.

43   The World Bank GEF. Available at: www.worldbank.org/gef.

44   World Bank: http://siteresources.worldbank.org/ESSDNETWORK/Resources/DialogueonClimateChange_KathySierra.pdf; UNFCCC Factsheet 'Investment and financial flows for a strengthened response to climate change.' Available at: http://unfccc.int/files/meetings/intersessional/awg_4_and_dialogue_4/application/pdf/070828_fin_flow_factsheet.pdf

45   OECD DAC CPDC on Environment, Conflict and Peace. Available at: http://www.oecd.org/document/44/0,2340,en_2649_34567_3552 7980_1_1_1_1,00.html.

46   EU Global Monitoring for Environment and Security (GMES) webpage. Available at: http://europa.eu/scadplus/leg/en /lvb/l28170.htm. Accessed on 22 October 2007.

47   These 22 states are: Bangladesh, Bhutan, Burundi, Cambodia, Comoros, Djibouti, Eritrea, Haiti, Kiribati, Lesotho, Madagascar, Malawi, Mauritania, Niger, Democratic Republic of Congo, Rwanda, Samoa, Senegal, Sudan, Tanzania and Tuvalu (UNFCCC, 2007).

48   UN Statistics Division/CDIAC, carbon dioxide emissions per capita, MDG indicator 28. Available at: http://unstats. un.org/unsd/mi/mi_series_results.asp?rowId=75. Accessed July 2007.

49   UN Statistics Division/CDIAC, carbon dioxide emissions per capita, MDG indicator 28. Available at: http://unstats.un.org/unsd/mi/mi_series_results.asp?rowId=75. Accessed July 2007.

50   World Bank Sri Lanka website: http://web.worldbank.org/nlk. Accessed on 6 June, 2007.

51   Chávez, Franz. 'Bolivia: Cochabamba's "Water War", Six Years On,' IPS News, 8 November, 2006. Available at: http://ipsnews.net/news.asp?idnews=35418. Accessed on 7 September, 2007.

52   Wara, Michael. 'Is it working? Is the global carbon market working?', Nature, 8 February, 2007, pp. 595–596; and Larry Lohmann (editor). Carbon Trading: A Critical Conversation on Climate Change, Privatisation and Power, Dag Hammarskjold Foundation, 2006.

53   Davies, Nick. 'Abuse and incompetence in fight against global warming,' The Guardian, 2 June, 2007. Available at: http://www.guardian.co.uk/environment/2007/jun/02/energy.bu siness; John Vida. 'Guilt-free sins of emission,' The Guardian, 3 February, 2005. Available at: http://www.guardian.co.uk/environment/2005/feb/03/emissionstrading.environment; www.cdmwatch.org

54   'The World Bank and the Carbon Market,' CDM Watch, April 2005. Available at: www.cdmwatch.org

55   CDM Watch Website: www.cdmwatch.org. Accessed on 23 July, 2007.

56   Intergovernmental Panel on Climate Change. Climate Change 2007: Impacts, Adaptation, and Vulnerability. Contribution of Working Group II to the Fourth Assessment Report of the Intergovernmental Panel on Climate Change, Cambridge University Press: Cambridge, 2007, Chapter 17: Assessment of adaptation practices, options, constraints and capacity.

57   Adapted from Twigg, John. 'Characteristics of a Disaster-resilient Community,' 2007.

58   Our analysis of Nepal draws strongly on the research of Dr Fiona Rotberg of the Silk Road Studies Institute at Uppsala University. She generously contributed to our understanding with her current research and we are very grateful for her contribution. While we have benefited from this assistance, our conclusions and any errors of interpretation are our own responsibility and any fault should not

be laid at the door of Dr Rotberg.

59 CIA Factbook 2007. Available at: https://www.cia.gov/library/publications/the-world-factbook/geos/np.html. Accessed on 16 August, 2007.

60 Gregson, Jonathan. Blood Against the Snows, Fourth Estate: London, 2003.

61 Intergovernmental Panel on Climate Change. Climate Change 2007: Impacts, Adaptation, and Vulnerability. Contribution of Working Group II to the Fourth Assessment Report of the Intergovernmental Panel on Climate Change, Cambridge University Press: Cambridge, 2007.

62 Shrestha, A. B. 'Climate Change in Nepal,' in draft proceedings of the Consultative Workshop on Climate Change Impacts and Adaptation Options in Nepal's Hydropower Sector with a Focus on Hydrological Regime Changes Including Glacial Lake Outburst Flooding, Department of Hydrology and Meteorology and Asian Disaster Preparedness Centre: Kathmandu, 2003.

63 OECD DAC. Harmonising Donor Practices for Effective Aid Delivery, OECD: Paris, 2003.

64 World Bank Nepal website: http://web.worldbank.org/np. Accessed on 6 June, 2007.

65 Human Development Report 2006. Available online at: http://hdr.undp.org/hdr2006/statistics/countries/data_sheets/ct y_ds_NPL.html. Accessed on 22 October, 2007.

66 ICIMOD. Climatic and Hydrological Atlas of Nepal, International Centre for Integrated Mountain Development: Kathmandu, 1996; UNEP. Nepal: State of the Environment, United Nations Environment Programme: Nairobi, 2001; Liu, X. and B. Chen. 'Climate Warming in the Tibetan Plateau during Recent Decades,' International Journal of Climatology, 20, 2002, pp1729–1742.

67 OECD DAC. 'Development and Climate Change in Nepal: Focus on Water Resources and Hydropower,' 2003. Available at: http://www.oecd.org/dataoecd/ 6/51/19742202.pdf.

68 OECD DAC. 'Development and Climate Change in Nepal: Focus on Water Resources and Hydropower,' 2003. Available at: http://www.oecd.org/dataoecd/6 /51/19742202.pdf.

69 Examples taken from: Intergovernmental Panel on Climate Change. Climate Change 2007: Impacts, Adaptation, and Vulnerability. Contribution of Working Group II to the Fourth Assessment Report of the Intergovernmental Panel on Climate Change, Cambridge University Press, Cambridge, 2007, Chapter 17: Assessment of adaptation practices, options, constraints and capacity.

70 Intergovernmental Panel on Climate Change. Climate Change 2007: Impacts, Adaptation, and Vulnerability. Contribution of Working Group II to the Fourth Assessment Report of the Intergovernmental Panel on Climate Change, Cambridge University Press: Cambridge, 2007, Chapter 17.

Chapter 7

Nic Maclellan

LOSING PARADISE

Leaders from Small Island Developing States (SIDS) around the world gathered in the Maldives in November 2007, and issued the Malé Declaration on the Human Dimensions of Climate Change. Calling for urgent action by developed nations, they 'committed to an inclusive process that puts people, their prosperity, homes, survival and rights at the centre of the climate change debate'. As Australian politicians debate the technicalities of the CPRS Emissions Trading Scheme and how much compensation to provide the coal industry, it's important we come back to this human dimension.

Over the past year, I've been visiting communities in the Pacific islands, to ask people about their concerns on climate change and to find out what they're doing to respond to the adverse effects of global warming.

From renewable energy initiatives and community-based vulnerability training to advocacy at international meetings, islanders are actively engaged in responding to the climate emergency. But the enormity of the environmental impacts already locked into the ecosystem means that some people are debating whether they'll need to leave their homelands.

You can't help but focus on the human impacts when visiting low-lying islands in the Pacific. The potential hazards are obvious on atolls like South Tarawa in Kiribati, a narrow strip of land 40 kilometres long but only 50–100 metres wide. With land areas just metres above sea-level, there is no retreat to higher ground from the ravages of storm surges and more intense cyclones. Facing salt water inundation of agricultural land and fresh water supplies, these threats to coastal villages tend to concentrate the mind about the powers of the elements. For low-lying atoll nations in Polynesia and Micronesia, the potential failure of the Copenhagen negotiations and delays in the reduction of greenhouse gas emissions will lead to forced displacement. However, the current intergovernmental Pacific Islands Framework for Action on Climate Change (PIFACC), developed by the Forum member countries, makes no mention of displacement or migration.

In spite of this, some Pacific island governments like Kiribati, Tokelau and Niue are openly discussing issues of relocation and resettlement due to climate change. In July 2007, a joint statement from Pacific environment ministers to the Forum Economic Ministers Meeting (FEMM) noted: 'The potential for some Pacific islands to become uninhabitable due to climate change is a very real one. Consequently some in our region have raised the issue of their citizens becoming environment refugees. Potential evacuation of island populations raises grave concerns over sovereign rights as well as the unthinkable possibility of entire cultures being damaged or obliterated'.

In August 2009, the outgoing chair of the Pacific Islands Forum, Premier of Niue Toke Talagi, says it may be time for the regional organisation to formally consider the issue of resettlement of people affected by climate change. Speaking at the official opening of the 2009 Forum leaders meeting in Cairns, Talagi stated, 'While all of us are affected, the situation for small island states is quite worrisome. For them, choices such as resettlement must be considered seriously and I wonder whether the Forum is ready to commence formal discussion on the matter'. Across the Pacific, there are a number of examples where people from low lying islands are considering relocation after being affected by extreme weather events, tectonic land shifts or climatic change that damages food security and water supply.

The case of the Carteret Islands in Bougainville is well known, where Ursula Rakova and the local NGO Tulele Peisa are assisting families to resettle on church-donated land on the main island of Buka. There are similar problems looming in other outlying atoll communities, such as the Duke of York atolls (a number of small low-lying islands in St. George's Channel near Rabaul in Papua New Guinea) or the Mortlock Islands in Chuuk State, Federated States of Micronesia. In the Solomon Islands, tectonic plate movement and sea-level rise may lead to the displacement of people in outlying atolls like Ongtong Java (Lord Howe) or artificial islands like Walande in Malaita Province.

But what will resettlement involve? To hear about the experience of people who've already been forced from their homes, I visited the islands of Western Province in the Solomon Islands, which were hit by a tsunami in April 2007. More than two years after the tsunami, many people on the main island of Gizo are reluctant to return to coastal villages, and are still living in improvised housing up in the hills and mountain ridges. At Titiana, one of the coastal villages on Gizo that bore the brunt of the tsunami, you can see the damage to community infrastructure. Villager Orau Mote shows us where the school, church and pastor's house were swept away—all that remains is a pile of concrete and steel rods. Children in Titiana have been using large tents as their school rooms, provided by UNICEF and the Ministry of Education. Titiana's United Church pastor Motu Tarakabu told me that many residents are still traumatised by the disaster. 'Only about 20 per cent of residents have come back to the village after the destruction of the tsunami. Many others have decided to stay away and remain up in the hills—they have fear in their heart. People are still strong that they won't come back to the village.'

Driving up the mountain ridges, you meet people from the coast who are refusing to resettle in their former homes and are building new houses to replace the tents and tarpaulins supplied after the disaster. Some villagers are rebuilding on land provided by clan relatives, but many are squatting on government land alongside roads and logging tracks. Orau Mote explains that a number of people of Micronesian heritage were relocated from the Gilbert Islands to the British Solomon Islands Protectorate during the era of British colonial rule. For these migrant communities, displaced again by the tsunami, there are new problems—people of Melanesian heritage often have clan and community links that can assist with resettlement. For non-Indigenous communities, even those who have lived in the Solomon Islands for decades, it is harder to find access to land and resources.

The villagers have sought support from Oxfam and the Solomon Islands Red Cross for provision of water tanks, corrugated iron for water catchment and housing, and other support services. But conditions remain difficult for the displaced communities, in spite of their resilience. Sale Sam, who lives in Tiroduke camp up on the ridges above Gizo town, said, 'Until Oxfam provided water tanks, we had to cart water for miles. The hill tops are very exposed—the wind blows from all directions, unlike the village which was sheltered'. Children from lower grades are attending classes up in the hills, but for senior grades the children need to trek down to the coastal villages each day, travelling kilometres to school. On the coast, women used to go out on the reef at low tide to collect crabs, shellfish and other seafood—an important source of protein to add to food grown in village gardens. But now it's harder to easily access this vital food supply. 'Our diet is changing now that we live on the higher ground', said one camp resident. 'The men still go down to the coast to go fishing, but we don't go out so much on the reef.'

Although in his sixties, Sale Sam still works to support his daughter Jocelyn, who relies on a wheelchair for mobility as they make a new home. For me, the resilience of this young woman, living in a wheel chair on a mountainside in the Solomon Islands, symbolises the larger challenge—what will displacement mean for the many thousands of people who face relocation in coming decades because of climate change?

Refugee or migrant?

In recent years, there is a growing academic literature on climate change, forced migration and conflict, but a mixed response to the concept of 'climate refugees.' The United Nations High Commission for Refugees (UNHCR) argues that the term 'environmental refugee' is not appropriate, as the definition of refugee under the 1951 Refugee Convention and international humanitarian law has particular limits, covering people who are seeking protection because of a well-founded belief of persecution related to their religion, ethnicity, political beliefs etc. Signatories to the 1951 Convention have specific legal responsibilities to people who reach their territory and claim asylum and protection, and refugee advocates are reluctant to see these state obligations watered down.

As noted in an October 2008 UNHCR briefing paper 'Climate change, natural disasters and human displacement', UNHCR has serious reservations with respect to the terminology and notion of 'environmental refugees' or 'climate refugees'. These terms have no basis in international refugee law. Furthermore, the majority of those who are commonly described as environmental refugees have not crossed an international border. Use of the terminology could potentially undermine the international legal regime for the protection of refugees and create confusion regarding the link between climate change, environmental degradation and migration.

Part of the problem is that key UN agencies responsible for displaced people have no formal mandate to address the climate issue. UNHCR does not cover people who are displaced internally or seek refuge overseas because of environmental causes. However, because of its practical experience in dealing with large scale forced movement of people, UNHCR staff and resources have increasingly been allocated to support operations in the aftermath of major natural disasters (such as the 2004 Asian tsunami, 2005 South Asian earthquake, 2006 floods in Somalia and 2008 floods in Burma, amongst others).

UNHCR is worried that its existing responsibility for refugees, asylum seekers and internally displaced people will be overwhelmed by the tens of millions of people potentially displaced by climate change. However the numbers of people who meet the definition of 'environmental refugee' are also contested. Studies have cited global figures ranging from 200 million (researcher Norman Myers) to over 1 billion potential refugees (a 2007 Christian Aid report). But migration specialists have questioned these numbers, arguing that people affected by environmental impacts will not necessarily cross international borders to seek refuge.

An important 2008 study on forced migration and climate change from the Norwegian Refugee Council, 'Future floods of refugees', raises crucial qualifications on the term refugees: There seems to be some fear in the developed countries that they, if not flooded literally, will most certainly be flooded by 'climate refugees'. From a forced migration perspective, the term is flawed for several reasons. The term 'climate refugees' implies a mono-causality that one rarely finds in human reality. No one factor, event or process, inevitably results in forced migration or conflict. It is very likely that climate change impacts will contribute to an increase in forced migration. Because one cannot completely isolate climate change as a cause however, it is difficult, if not impossible, to stipulate any numbers. Importantly, the impacts depend not only on natural exposure, but also on the vulnerability and resilience of the areas and people, including capacities to adapt. At best, we have 'guesstimates' about the possible form and scope of forced migration related to climate change.

Agency and choice

When they look at international rather than domestic impacts, climate advocacy groups in Australia and New Zealand have highlighted the issue of 'Pacific climate refugees' in their campaigning. Many have argued that Australia and New Zealand, as the largest members of the Pacific Islands Forum, have particular responsibilities to their island neighbours. But do people debating the issue ever ask those most affected what they really want? It may seem trite to see people in developing countries as actors rather than victims in this global emergency, yet much of the climate literature presents the Pacific's only contribution to the climate debate as a loud 'glug, glug, glug' as the islands sink beneath the waves.

The issue of displacement raises a number of practical, emotional and political responses. In interviews with people around the Pacific, different opinions came from the elderly compared to younger people who have more flexible skills for migration. As one old man in the Solomon Islands told me, 'They talk about us moving. But we are tied to this land. Will we take our cemeteries with us? For we are nothing without our land and our ancestors'. Community activist Annie Homasi from Tuvalu says the slow pace of action by large industrialised countries has the potential to cause uncertainty and even division in the local community, for people who are fearful they may have to relocate from their homes.

'There's quite a debate at home, maybe even a division, between the older generation and the young people. Because they go overseas for school, the young ones say, "Yeah, we have to move". But the older ones say, "This is me, my identity and my heritage—I don't want to go".'

There are also complex cultural responses in the Pacific, with many religious people stating that God will not forsake them. Some old people deny any long term threat from floods and rain, citing Biblical injunctions like God's promise to Noah after the Flood: 'neither will I ever again smite everything living as I have done' (Genesis 8:21).

Most Pacific governments are still reluctant to focus resources on displacement issues, because they feel this will acknowledge defeat and undermine negotiating positions at the international level, as they press for stronger targets in the Copenhagen negotiations.

Government leaders from Kiribati and Tuvalu continue to stress that increased mitigation efforts by industrialised nations should be the focus of activity. Speaking to the UN General Assembly in September 2008, Tuvalu Prime Minister Apisai Ielemia stated: 'We strongly believe that it is the political and moral responsibility of the world, particularly those who caused the problem, to save small islands and countries like Tuvalu from climate change, and ensure that we continue to live in our home islands with long-term security, cultural identity and fundamental human dignity. Forcing us to leave our islands due to the inaction of those responsible is immoral, and cannot be used as quick fix solutions to the problem.'

Open borders

Most of the discussion of climate displacement in Australia focuses on the need for Pacific Rim countries to change their migration policies. But the language of the debate revives past fears about being 'swamped' by immigrants or asylum seekers. Concerned activist groups stress Australia's moral obligations to open its doors while conservatives respond with refrains that echo John Howard's infamous dog whistle, 'We will decide who comes to this country and the circumstances in which they come'.

Environmental groups have argued that Australia's existing humanitarian immigration quotas should not be allocated to climate-related refugees and that an additional category is required. The Migration (Climate Refugees) Amendment Bill 2007 advanced by the Australian Greens proposed the creation of a new visa class to formally recognise climate refugees, but lapsed without support from the major parties. Other options could involve an expanded system of free migration as already exists between Australia and New Zealand, which enjoy shared migration rights of free access and permanent residence. New Zealand's Pacific Access Category, which provides migration quotas for citizens from Tuvalu, Kiribati, Fiji and Tonga, provides a de facto window for migration from climate affected countries, even though the New Zealand government has not explicitly recognised this as an option dedicated to people affected by environmental impacts.

In contrast, some Pacific leaders have suggested that it may be more appropriate to call for Australian and New Zealand financial support for the resettlement of people to other Pacific islands, to provide agricultural land and a suitable cultural context for displaced rural communities. A key feature of environmental displacement in the Pacific is that much of the movement is internal, rather than across international boundaries, which places extra burdens on national government budgets as well as host communities who accept people from other areas.

But money is not enough. A worrying feature of the debate about 'climate refugees' is that the bald predictions of forced relocation give little agency or choice to the affected communities. Compared to a rapid natural disaster like an earthquake or tsunami, the ravages of climate change will mount over time, so people can be engaged in discussing the options. We must learn from the failure of past resettlement projects in developing countries, which comes not just from inadequate inputs of resources but from the inherent complexity of this as a social process involving human beings with hopes, dreams, aspirations and especially memories. Relocation and resettlement is not simply a material infrastructure process—it is also a social process and there are a number of issues of co-operation, voice and justice that need to be addressed. How do you promote resettlement with respect for equality and equity?

Moving to a new location within a country or across international borders is just the first step, and there are a host of political as well as technical dilemmas for communities on the move:

- Do displaced people have a say in the design and construction of new communities (for example, site selection that can provide water, arable land and other resources; culturally appropriate housing in terms of size, design, spacing and materials; settlement design to allow social and cultural interaction)?
- Are people being compensated for need or loss (that is what they need for survival or for what they feel they've lost)?
- Can you be compensated for intangibles, such as the grief of losing a home, or loss of political and cultural identity?
- Will the wealthy leave early and leave behind those with fewer resources?
- Will displaced people be better serviced by donors than existing members of the new host community, causing inter-communal tensions?
- Should old power relations and systems of chiefly rule be recreated, or are they tied to past relationships with the lost land?

This raises the core question of whether funding for adaptation and relocation will be allocated without the engagement and consent of affected communities. Is planning for relocation being done with people or for people? The potential for displacement because of climate change needs extensive community participation and debate, as noted by Betarim Rimon of the Kiribati Ministry of Environment: 'In Kiribati, we are talking about relocation over time rather than forced displacement. We think about relocation as a long, thought out, planned process.' Kiribati President Anote Tong stressed this in his address to the opening session of the 2008 UN General Assembly:

The relocation of 100,000 people of Kiribati, for example, cannot be done overnight. It requires long term foward planning and the sooner we act, the less stressful and less painful it will be for all concerned.

This is why my Government has developed a long-term merit-based relocation strategy as an option for our people. As leaders, it is our duty to the people we serve to prepare them for the worst-case scenario.

Australia refuses to plan ahead

In 'Engaging our Pacific Neighbours on Climate Change'—Australia's latest climate policy statement issued in August 2009—the Rudd government notes: 'The potential for climate change to displace people is increasingly gaining international attention. Australians are aware of and concerned about this issue.' But we need more than awareness and concern.

Successive Australian governments have failed to engage in foward planning involving communities and governments around the region, to address the issues of displacement from a rights-based approach.

For many years, Pacific Rim governments have been reluctant to publicly address this issue. In October 2006, the then Immigration Minister Amanda Vanstone stated that her Department had not made any plans to deal with people displaced by environmental or climate change, arguing, 'There's no such thing as a climate refugee'. In November 2006, Secretary of the Department of Immigration Andrew Metcalfe told a Senate estimates hearing that the Australian Government had done no planning on how people movement caused by climate change in the Asia-Pacific region might affect Australia. Since then, however, the debate has been flourishing amongst security analysts and strategic think tanks, which have focused on border protection and the potential for conflict overland and resources. In 2007, the then Australian Federal Police Commissioner Mick Keelty sparked a political debate when he argued that climate change will turn border security in Australia's biggest policing issue this century. He stated that climate change could increase displacement and migration in our region. 'In their millions, people could begin to look for new land and they will cross oceans and borders to do it. Existing cultural tensions may be exacerbated as large numbers of people undertake forced migration. The potential security issues are enormous and should not be understated.'

The securitisation of the debate has also been highlighted in Force 2030—the May 2009 Defence White Paper issued by the Rudd government. This is the first the climate issue has been discussed in a Defence White Paper, but it does not really reflect a shift in focus from 'national security' to 'human security'. In the paper, action on climate changes is reframed through the prism of border security: The main effort against such developments will of course need to be undertaken through co-ordinated international climate change mitigation and economic assistance strategies should these and other strategies fail to mitigate the strains relating to climate change and they exacerbate existing precursors for conflict, the Government would probably have to use the ADF as an instrument to deal with any threats inimical to our interests.

Will people displaced by global warming be redefined as 'threats inimical to our interests'? Social justice activists need to reframe the debate, to highlight the right to development for affected communities wherever they are, rather than just focusing on the need for mitigation rights.

Nic Maclellan, 'Losing Paradise', *Arena*, November 2009.

PART 2

International legal and institutional framework

Chapter 8

Walter Kälin

DISPLACEMENT CAUSED BY THE EFFECTS OF CLIMATE CHANGE
Who Will Be Affected and What Are the Gaps in the Normative Framework for Their Protection?

[…]

I. The Climate Change – Displacement Nexus

Key findings relevant for the issue of displacement by the Intergovernmental Panel on Climate Change include:[1]

i.   A key impact of climate change will be reduced availability of water, particularly in parts of the tropics, the Mediterranean and Middle Eastern regions and the Southern tips of Africa and Latin America. In contrast, water availability will increase in parts of Eastern Africa, the Indian sub-continent, China, and the Northern Latitudes. Hundreds of millions of people will be exposed to water stress (i.e. droughts and lack of water or flooding, mudslides etc).

ii. A decrease in crop yields is projected which increases the likelihood that additional tens of millions of people will be at risk of hunger. The most affected region is likely to be Africa.

iii. Due to rising sea-levels, the densely populated "mega-deltas" especially in Asia and Africa and small islands are most at risk from floods, storms and coastal flooding and eventual submerging, with a potential impact on tens of millions of people.

iv. The overall impact on health will be negative, especially for the poor, elderly, young and other marginalized sectors of society.

v.   Overall, the areas that will be most affected by climate change are Africa, Asian mega-deltas and small islands.

Climate change is usually referred to as "global warming" and as such does not displace people. Rather climate change produces environmental effects which may make it difficult or even impossible for people to survive where they are. Most causes of displacement triggered by climate change, such as flooding, hurricanes, desertification or even the "sinking" of stretches of land,[2] are not new. However, their frequency and magnitude are likely to increase.

The challenge is to better analyze these causes of displacement, to identify the areas where the effects of climate change are most likely to occur and to examine the character of forced displacement and other population movements they could trigger. In this context, it might be required to review existing international legal instruments and to explore ways of filling potential protection gaps.

In this regard, the following (tentative and hypothetical) typology may be helpful.[3]

(i) The increase of *hydro-meteorological disasters*, such as flooding, hurricanes/typhoons/cyclones or mudslides, will occur in most regions, but the African and Asian mega deltas are likely to be most affected. Such disasters can cause large-scale displacement and huge economic costs, but depending on recovery efforts the ensuing displacement need not be long-term and return in principle remains possible as durable solution.

One should not forget that many hydro-meteorological disasters will occur regardless of climate change and other disasters such as volcanoes or earthquakes presumably have no linkage to such change. Nevertheless, they too may cause movement of persons and such persons should not be treated differently from those affected by the effects of climate change.

(ii) *Environmental degradation and slow onset disasters* (e.g. reduction of water availability, desertification, long-term effects of recurrent flooding, sinking costal zones, increased salinisation of ground-water and soil etc.): With the dramatic decrease of water availability in some regions and recurrent flooding in others, economic opportunities and conditions of life will deteriorate in affected areas. Such deterioration may not necessarily cause forced displacement in the strict sense of the word but, among other reasons, will incite people to move to regions with better income opportunities and living conditions. However, if the areas become uninhabitable because of complete desertification or sinking coastal zones, then population movements amount to forced displacement and become permanent.

(iii) *The case of "sinking" small island states* caused by rising sea levels constitutes a particular challenge. As a consequence, such areas become uninhabitable and in extreme cases the remaining territory of affected states can no longer accommodate the whole population or such states disappear as a whole. When this happens, the population cannot return and becomes permanently displaced to other countries.

(iv) Disasters will increase the need for governments to *designate areas as high-risk zones* too dangerous for human habitation. This means that people may have to be (forcibly) evacuated and displaced from their homes and prohibited from returning there and relocated to safe areas. This could occur, for example, because of increased risk of flooding or mudslides due to the thaw of the permafrost in mountain regions, but also along rivers and coastal plains prone to flooding. The difference between this situation and the previous typology of disaster-induced displacement is that return may not be possible, thus becoming a permanent form of displacement until other durable solutions are found for those affected.

(v) A decrease in essential resources due to climate change (water; food production) most likely will trigger *armed conflict and violence*: This is most likely to affect regions that have reduced water availability and that cannot easily adapt (e.g. by switching to economic activities requiring less water) due to poverty. These armed conflicts may last for as long as resource scarcity continues. This in turn would impede the chances of reaching peace agreements which provide for the equitable sharing of the limited resources and thus prolong the conflicts, leading to more situations of protracted displacement.

II. The Nature of Movements, Affected Persons and Protection Frameworks

These five scenarios can help to identify the character of the movement, i.e. whether it is forced or voluntary, to qualify those who move (migrants[4]; IDPs,[5] refugees,[6] stateless persons, other categories?) and to assess whether and to what extent present international law is equipped and provides adequate normative frameworks to address the protection and assistance needs of such persons.

(i) *Hydro-meteorological disasters* can trigger forced displacement. Two situations should be distinguished:

- Most of the displaced remain inside their country and as internally displaced persons receive protection and assistance under human rights law and in accordance with the 1998 Guiding Principles on Internal Displacement. For these internally displaced persons, the existing normative framework is sufficient.
- Some of the displaced may cross an internationally recognized state border, e.g. because the only escape route leads there, because the protection and assistance capacities of their country are exhausted or because they hope for better protection and assistance outside their own country. They have no particular protected (legal) status, as they do not qualify as refugees, nor are they migrants. In the past, host governments have in some cases allowed such persons to stay temporarily for humanitarian reasons until they could return to their countries in safety and dignity,[7] but practice has not been uniform. The status of these persons remains unclear and despite the applicability of human rights law, including in particular provisions applicable to migrant workers, there is a risk that these persons end up in a legal and operational limbo.

(ii) *Situations of environmental degradation and slow onset disasters* create several types of movements of persons:

- General deterioration of conditions of life and economic opportunities as a consequence of climate change may motivate persons looking for better opportunities and living conditions to move to other parts of the country or abroad before the areas they live in become uninhabitable. These persons are protected by human rights law, including, if they move to a foreign country, guarantees specifically protecting migrant workers.
- If areas start to become uninhabitable, because of complete desertification, salination of soil and ground-water or sinking of coastal zones, people may, during a first phase, leave voluntarily to find better (economic) opportunities elsewhere within or outside their country, but later movements may amount to forced displacement and become permanent as inhabitants of such regions no longer have a choice but to leave permanently. If the people stay within their country, they are internally displaced persons and fall under the ambit of the Guiding Principles on Internal Displacement. If they go abroad, they have no protection other than

that afforded by international human rights law including provisions on economic migrants; in particular, they have no right under international law to enter another country and stay there and thus depend on the generosity of other countries. This scenario poses two particular challenges: 1. there is a lack of criteria to determine where to draw the line between voluntary movement and forced displacement; 2. those forcibly displaced to other countries remain without specific protection as they do not qualify for refugee status and as their movement is forced they cannot be qualified as migrants either.

(iii) The *"sinking" of small island states will be gradual*:

- In the initial phases, this slow-onset disaster will incite persons to migrate to other islands belonging to the same country or abroad in search of better opportunities. If they migrate to another country, these persons are protected by human rights law including guarantees specifically protecting economic migrants.

- Later, such movements can turn into forced displacement because areas of origin could become uninhabitable and in extreme cases the remaining territory of affected states could no longer accommodate the whole population or would disappear entirely, rendering return impossible. When this occurs, the population would become permanently displaced to other parts of their country or to other countries. In this case, besides human rights law in general, the Guiding Principles become applicable in the case of internal displacement. However, there are normative gaps for those who move abroad, leaving them in a legal limbo as they are neither migrants nor refugees. It is also unclear as to whether provisions on statelessness would apply as it remains to be seen whether those affected become stateless persons under international law. These persons do not become stateless as long as there is some remaining part of the territory of their State, and even where a whole country disappears it is not certain that they become stateless in the legal sense. Statelessness means to be without nationality, not without state. It cannot be excluded that such small island states will continue to exist as a legal entity at least for some time even if their territory has disappeared as nobody will be ready to formally terminate statehood. Even if these persons end up without a nationality, international law on statelessness does not provide adequate protection for them. Obviously, such persons will be in need of some form of international protection. Their rights need to be identified and it remains to be determined whether these people require a specific legal status. The question of the responsibility of the international community, in particular regarding relocation, needs clarification as well.

(iv) *The designation of high risk zones too dangerous for human habitation* may trigger (forced) evacuations and displacement:

- Affected persons are internally displaced persons. In terms of durable solutions they cannot return but must be relocated to safe areas or locally integrated in the evacuation area. Sustainability of the solution chosen is important to avoid permanent and protracted displacement situations or even return to high risk zones exposing the lives of returnees to a high risk incompatible with human rights standards. International human rights law, the Guiding Principles and the analogous application of norms and guidelines on relocation in the context of development projects provide a sufficient normative framework for addressing these situations.[8]

- Should people decide to leave their country because they reject relocation sites offered to them or because they are not offered sustainable solutions in accordance with relevant human rights standards by their own government, protection is limited to that offered by general human rights law, including in particular provisions applicable to migrant workers but their status remains unclear and they may not have a right to enter and remain in the country of refuge.

(v) *"Climate change-induced" armed conflict and violence* trigger forced displacement. Those fleeing abroad may qualify as refugees protected by the 1951 Convention on the Status of Refugees and similar instruments or are persons in need of subsidiary forms of protection or temporary protection available for persons fleeing armed conflict; those remaining inside their own country are internally displaced persons. The available normative frameworks are international humanitarian law, human rights law, refugee law and the Guiding Principles. They provide a sufficient normative framework for addressing these situations since affected persons are fleeing armed conflict, rather than the changes brought about by climate change.

This analysis allows drawing the following conclusions:

First: Existing human rights norms and the Guiding Principles on Internal Displacement provide sufficient protection for those *forcibly displaced inside their own country by sudden-onset disasters* (scenario i) or because their *place of origin has become uninhabitable* or been declared too dangerous for human habitation (scenario iv).

Second: Existing international law (international humanitarian law, human rights norms, Guiding Principles on Internal Displacement, refugee law) is sufficient to protect *persons displaced by armed conflict triggered by the effects of climate change* whether or not they cross an internationally recognized state border (scenario v).

Third: There is a need to clarify or even develop the normative framework applicable to other situations, namely

a.   Persons moving across internationally recognized state borders in the wake of sudden-onset disasters (scenario i);
b.   Persons moving inside or outside their country as a consequence of slow-onset disasters (scenario ii); and
c.   Persons leaving "sinking island States" and moving across international recognized state borders (scenario iii).
d.   Persons moving across internationally recognized state borders in the wake of designation of their place of origin as high risk zone too dangerous for human habitation (scenario iv).

Questions to be addressed include three sets of issues:

1.   Should those moving voluntarily and those being forcibly displaced be treated differently not only as regards assistance and protection while away from their homes but also as regards their possibility to be admitted to other countries and remain their temporarily?

2.   If yes, what should be the criteria to distinguish between those who voluntarily leave their homes or places of habitual residence because of the effects of climate change and those who are forced to leave by the effects of climate change?

3.   What would be the respective entitlements to assistance and protection of those leaving voluntarily and those forcibly displaced?

Present international law, while recognizing that all human beings are entitled to the full enjoyment of human rights, does differentiate between persons who move voluntarily and those forcibly displaced for whom special normative regimes (refugee law; Guiding Principles on Internal Displacement) have been developed at least in some cases. The remainder of this paper focuses on criteria to distinguish between forced and voluntary movements.

III. Filling the Normative Gaps: Criteria for Defining the Different Categories of Affected Persons

There are different ways to develop criteria to determine when a movement is no longer voluntary, but happens under compulsion. One option would be based on a vulnerability analysis to assess when vulnerabilities have reached a degree that a person was forced to leave his or her home. It is obviously extremely complex to develop generic criteria on this basis and to apply them individually, in particular in situations of slow onset disasters.

This paper suggests a different approach. The point of departure should not be the subjective motives of individuals or communities for their decision to move, but rather the question *as to whether in light of the prevailing circumstances and the particular vulnerabilities of the persons concerned it would be appropriate to require them to go back to their original homes.*

This should be analyzed on the basis of three elements: permissibility, factual possibility and reasonableness of return.

Permissibility

Human rights law (and refugee law by analogy) sets out cases in which return is *impermissible* per se in all scenarios. This is, first, the prohibition to send people back to a situation where their life or limb is at risk,[9] e.g if the conditions at the only place they can return to is exposing them to high risks; and, second, the prohibition of collective expulsion, i.e. of decisions to collectively send affected persons back, without assessing their individual situation.[10]

Possibility

Return may temporarily not be possible due to technical or administrative impediments, e.g. roads are cut off or destroyed or loss of documentation. Return is also impossible if the country of origin refuses readmission for technical or legal reasons. For example during an emergency a country might not have the capacity to absorb large return flows or it might prevent readmission of persons who cannot prove that they are its citizens because their documents were destroyed, lost or simply left behind when they left.

Citizens of sinking small island states may experience a very special factual situation that makes return impossible, if the territory has shrunk in a way or resources necessary for survival (e.g. drinking water) are destroyed to a degree that doesn't allow the return of persons or if the island as such has disappeared.

Reasonableness

Return cannot be reasonably expected from the persons concerned, e.g. if the country of origin does not provide any assistance or protection at all or far below international standards as long as the displacement lasts. The same is true where it does not provide any kind of durable solutions according to international standards, in particular when zones have become or were declared uninhabitable and return to their homes therefore is no longer an option for the displaced.

If the answer to one of the following question: Is return *permissible*? Is return *possible*? Can return *reasonably be required*? is "no", then individuals concerned should be regarded as victims of forced displacement in need of specific protection and assistance either within their own country (internal displacement) or in another State (external displacement). In the latter case, they should be granted at least a temporary stay in the country where they have found refuge until the conditions for their return are fulfilled. For citizens of sinking island states permanent solutions on the territory of other states must be found.

A next step would consist of identifying for each of the four categories of persons outlined above more detailed criteria to determine under what circumstances return to the country of origin (or in the case of internally displaced persons to their place of former habitual residence) would be *impossible* or could *not be reasonably expected* from them and to develop proposals for temporary protection regimes applicable to those who were forced to cross an international border due to the effects of climate change, which regulates their status. What is also needed is to further determine their entitlements to assistance and protection as well as their obligations.

IV. Conclusion

1. Persons affected by the effects of natural disasters and other effects of climate change should be considered as being forcibly displaced and thus, in principle, of special concern of the international community if:

- As persons displaced inside their country, they cannot return to their homes for factual or legal reasons or cannot reasonably be expected to do so because of a lack of security or sustainable livelihoods there. They are persons internally displaced as a consequence of the effects of climate change or other environmentally related reasons. They should be assisted and protected within the framework of the Guiding Principles on Internal Displacement.
- As persons displaced across internationally recognized state borders, they cannot return to their country of origin for factual or legal reasons or cannot reasonably be expected to do so because of a lack of security or sustainable livelihoods there. They are persons displaced across internationally recognized borders as a consequence of the effects of climate change or other environmentally related reasons. They should be granted admission to and tempo- rary – and in some cases (e.g. sinking island states) even permanent – stay in the country

concerned. The exact scope of their entitlements to assistance and protection needs to be further determined.

2. Persons affected by the effects of natural disasters and other effects of climate change but not falling into any of these categories are not victims of forced movement and thus should not, in principle, be of concern to the international community as a *special category* even if their movement is triggered by such effects. Nevertheless, they may profit from protection and assistance available under general human rights law, including in particular provisions addressing the specific needs of migrants, and the exact scope of these entitlements needs to be further determined.

3. Persons displaced inside or across internationally recognized state borders by armed conflicts over resources becoming scarce as a consequence of climate change, are internally displaced persons, refugees or persons under temporary protection regimes, and their entitlements to assistance and protection are determined by the respective normative regimes (Guiding Principles on Internal Displacement; refugee law; etc.).

Walter Kälin, Representative of the Secretary General on the Human Rights of Internally Displaced Persons, 'Displacement Caused by the Effects of Climate Change: Who Will Be Affected and What Are the Gaps in the Normative Framework for Their Protection?', Background Paper, 10 October 2008.

Notes

1   The following is based on Martin Parry, Humanitarian Implications of Climate Change, ppt-presentation to the IASC Principals Meeting (30 April 2008).
2   See e.g. the submerged ancient Roman cities in the Mediterranean Sea.
3   These scenarios are a typology. In reality, they may coincide and overlap.
4   For the purpose of this paper, the term 'migrant' refers to the definition of migrant worker in Art. 2(1) of the International Convention on the Protection of the Rights of All Migrant Workers and Members of Their Families, i.e., "a person who is to be engaged, is engaged or has been engaged in a remunerated activity in a State of which he or she is not a national."
5   The term 'internally displaced persons' refers to persons covered by the Guiding Principles on Internal Displacement: "Internally displaced persons are persons or groups of persons who have been forced or obliged to flee or to leave their homes or places of habitual residence, in particular as a result of or in order to avoid the effects of armed conflict, situations of generalized violence, violations of human rights or natural or human-made disasters, and who have not crossed an internationally recognized State border."
6   The term 'refugee' refers to the legal definition of the 1951 Convention on the Status of Refugees, the 1969 African Convention governing the specific aspects of refugee problems in Africa as well as the 1984 Cartagena Declaration on Refugees. The 1951 Refugee Convention defines the term 'refugee' in art. 1A(2) as a person who "owing to well-founded fear of being persecuted for reasons of race, religion, nationality, membership of a particular social group or political opinion, is outside the country of his nationality and is unable or, owing to such fear, is unwilling to avail himself of the protection of that country; or who, not having a nationality and being outside the country of his former habitual residence as a result of such events, is unable or, owing to such fear, is unwilling to return to it." The African Convention expands this notion to include "every person who, owing

to external aggression, occupation, foreign domination or events seriously disturbing public order in either part or the whole of his country of origin or nationality, is compelled to leave his place of habitual residence in order to seek refuge in another place outside his country of origin or nationality". The Cartagena Declaration on Refugees adds the criterion "massive violation of human rights".

7   This has been the practice e.g. for persons affected by flooding in different parts of the SADC region or for victims of Hurricane Mitch in the USA who were granted "temporary protection status" in accordance with US migration law that provides for temporary protection not only for persons fleeing armed conflict and situations of generalized violence but also for those who cannot return to their country of origin in the aftermath of a natural disaster. The notion of "temporary protection" has to be distinguished from the concept of temporary protection as used particularly in Europe to handle a situation of mass influx of people fleeing armed conflict or generalized violence (see European Community, Council Directive 2001/55/EC of 20 July 2001 on minimum standards for giving temporary protection in the event of a mass influx of displaced persons).

8   See for example, World Bank Operational Policy 4.12 of 2001.

9   This principle was derived by the European Court of Human Rights and the UN Human Rights Committee from the prohibition of torture, cruel and inhuman treatment (Art. 3 ECHR; Art. 7 ICCPR). The principle of non-refoulement is also a corner-stone principle of international refugee law (Art. 33 1951 Convention) that has gained the quality of international customary law and arguably even us cogens.

10 Art. 4 Protocol 4/ECHR: Prohibition of collective expulsion of aliens.

Chapter 9

UN High Commissioner for Refugees

CLIMATE CHANGE, NATURAL DISASTERS AND HUMAN DISPLACEMENT A UNHCR PERSPECTIVE[1]

> Although there is a growing awareness of the perils of climate change, its likely impact on human displacement and mobility has received too little attention.[2]
>
> (António Guterres, UN High Commissioner for Refugees)

Introduction

The process of climate change – and the multiple natural disasters it will engender – will in all certainty add to the scale and complexity of human mobility and displacement.[3] Thus far the international community has focused primarily on the scientific aspects of climate change, with the aim of understanding the processes at play and mitigating the impact of human activity. Yet climate change is likely to pose humanitarian problems and challenges. As such it is of direct interest to humanitarian agencies,[4] including the Office of the United Nations High Commissioner for Refugees (UNHCR).

UNHCR is a leading agency of the United Nations responsible for and possessing the expertise in the area of forced displacement. It is projected that climate change will over time trigger larger and more complex movements of population, both within and across borders, and has the potential to render some people stateless. Since climate change is certain to have a major impact on future patterns of human mobility, approaches which address environmental issues in isolation from other variables and processes will not be sufficient to solve the problem.

In tandem with deeper understanding of the scientific processes at play, UNHCR would encourage more reflection on the humanitarian and displacement challenges that climate change will generate. It is likely that most of the displacement provoked by climate change manifested, for example, through natural disasters, could remain internal in nature. Great strides have been made in developing the legal framework for the protection of internally displaced persons (IDPs).[5] As part of United Nations humanitarian reform, a consensual division of labor for their assistance has also been established known as the Cluster Approach. It is likely that the multiplication of natural disaster scenarios will test the capacities of humanitarian actors, and may call for a new distribution of roles and/or new models of cooperation.

Some cross-border movement scenarios may be dealt with within the existing international refugee framework, which has proven to be flexible over the past decades, but others may require new approaches, premised upon new forms of inter-State cooperation, international solidarity and responsibility-sharing. Another angle requiring reflection is the phenomenon of statelessness.

This note contains UNHCR's preliminary perspectives on these questions as a contribution to the ongoing debate on climate change. In the interests of brevity, it does not present detailed empirical evidence relating to the issue of climate change, which can be found in successive reports of the Intergovernmental Panel on Climate Change (IPCC).[6]

Instead, the following sections examine:

1.   Foreseeable displacement scenarios,
2.   Their implications for UNHCR,
3.   Terminology and the 1951 Refugee Convention,
4.   Suggestions for the way forward.
5.   Information on UNHCR activities in relation to environmental conservation and the impact of climate change on UNHCR operations is included in the Annex.

Displacement Scenarios

While climate change has been the subject of intense debate and speculation within the scientific community, insufficient attention has been given to the humanitarian consequences it will generate. Just as the causes of climate change are being analyzed and its consequences projected, it is equally vital to anticipate foreseeable movement scenarios and strengthen the responses to the humanitarian consequences.

Climate change is already undermining the livelihoods and security of many people, exacerbating income differentials and deepening inequalities. Over the last two decades the number of recorded natural disasters has doubled from some 200 to over 400 per year. Nine out of every ten natural disasters today are climate-related.[7] The Norwegian Refugee Council recently indicated that as many as 20 million people may have been displaced by climate-induced sudden-onset natural disasters in 2008 alone.[8]

As temperatures rise and land becomes less productive, the process of urbanization will accelerate, generating additional competition for scarce resources and public services in cities across the globe. The incidence of vector-borne diseases will also increase as a result of climate change, as will the cost of food and energy. Increased social tension and political conflict is thus likely, though it may remain difficult to trace the origins of such tensions to climate change.

Just as human movements which are induced or strongly influenced by the process of climate change will vary in character, so will a range of responses and, potentially, new approaches. As in the past, populations will implement adaptation strategies to avert or deal with displacement as a result of unpredictable disasters such as cyclones, floods and mudslides.

In regions affected by the longer-term consequences of climate change, people will also move in large numbers, but will do so over longer periods of time and in more diverse directions. Some will move to more hospitable areas in home countries while others will seek to leave their own country and enter other States. Since new forms and patterns of movement are emerging, the concepts traditionally used to categorize different types of movement are becoming increasingly blurred.[9] New legal frameworks may need to be negotiated.

The Representative of the Secretary-General on the Human Rights of Internally Displaced Persons, Walter Kälin, has identified five climate change-related scenarios that may directly or indirectly cause human displacement. They provide a useful starting point for analyzing the character of displacement and assessing the protection and assistance needs of those affected:[10]

- hydro-meteorological disasters (flooding, hurricanes/typhoons/cyclones, mudslides, etc.);
- zones designated by Governments as being too high-risk and dangerous for human habitation;
- environmental degradation and slow onset disaster (e.g. reduction of water availability, desertification, recurrent flooding, salinization of coastal zones, etc.);
- the case of 'sinking' small island states; and,
- violent conflict triggered by a decrease in essential resources (e.g. water, land, food) owing to climate change.

In each of these scenarios, people may become displaced either within their own countries or across international borders. While the latter type of movement is less likely, at least in the initial phases of displacement, regard must also be had to the situation of migrants who find themselves outside their country of nationality as disaster strikes there, and are thus unable and/or unwilling to return home.

Where hydro-meteorological disasters or environmental degradation cause internal displacement, as States have primary responsibility for their citizens, national and local authorities have a vital role to play in responding to such scenarios. IDPs should receive protection and assistance in accordance with the 1998 Guiding Principles on Internal Displacement. As a result of recurring disasters, some States may exercise the sovereign obligation to protect their citizens by designating areas as high-risk zones, too dangerous for human habitation, owing to their location, for example, in flood-prone or landslide-prone areas. People may have to be forcibly evacuated and displaced from their lands, prohibited from returning to them, and relocated to safe areas. It is likely that the affected persons would qualify as IDPs and, once again, be protected by the 1998 Guiding Principles on Internal Displacement. The most likely durable solutions would be integration in the places of displacement or relocation to new areas inside the country, since return will normally not be possible.

Where people affected by such disasters cross an international border, for example because the only escape routes lead them there, they would not normally qualify as refugees who are entitled to international protection within the existing international refugee framework, nor would they necessarily be classified as migrants. While benefiting from the applicability of human rights norms, "their status remains unclear."[11]

Determining whether cross-border movements are forced or voluntary may not be practicable in the vast majority of cases, but this is not the most relevant element under international law. The crux of the issue will be whether persons have a need for international protection; and, if so, on what grounds this need may be turned into an entitlement.

Another challenge will be posed to low-lying small island States by rising sea levels. The phenomenon may prompt internal relocation as well as migration abroad, until such time as the territory is no longer able to sustain human life. As in the previous scenario, prevailing international refugee law would not automatically apply. The question of statelessness is, however, more directly implicated.[12]

With the disappearance of territory, one of the key constituting elements of statehood, it is not clear that these States would continue to exist as such. The same would apply if the territory would be uninhabitable to such an extent that the entire population and the Government would be forced to relocate to other States. In the event that statehood is deemed to have ceased in such a scenario, the populations concerned would be left stateless unless they acquired other nationalities.

Even where the States continued to exist in legal terms and their Governments attempted to function from the territory of other States, it is unclear that they would be able to ensure the rights which flow from citizenship. If they were unable to ensure such basic rights as the right to return to one's own country or to obtain a passport, statelessness considerations would also arise. In view of the fact that statelessness has not yet arisen, however, the international law principle of prevention of statelessness would be applicable.

The fifth and final scenario of human displacement is a decrease in vital resources (water, land, food production) attributable to climate change, which triggers armed conflict and violence. Regardless of the underlying causes, those displaced by armed conflict inside their country are IDPs in the sense of the 1998 Guiding Principles on Internal Displacement. Those fleeing to other countries could qualify as refugees or be afforded protection under regional refugee law instruments[13] or 'complementary forms of protection' under the relevant international law instruments or under the national law of receiving States.[14]

Persons who find themselves abroad when a natural disaster affects their home country will need protection against forcible return – for a shorter or longer period of time, depending on the circumstances. There may be instances in which the disaster-affected State itself seeks this form of relief on behalf of its citizens abroad. However, there may also be instances where the home State is simply unable to advocate for the protection needs of its own citizens due to the very nature of the disaster.

Implications for UNHCR

It is clear from the above analysis that some movements likely to be prompted by climate change could indeed fall within the traditional refugee law framework, bringing them within the ambit of international or regional refugee instruments, or complementary forms of protection, as well as within UNHCR's mandate.[15]

The most obvious example is that of refugee movements provoked by armed conflict rooted in environmental factors. Such conflicts and displacements have occurred in a number of settings. Already today, some commentators have argued that conflict over energy sources, fertile land and fresh water are among the factors fuelling the crisis in the Darfur region of Sudan.[16] Climate-related issues are projected to become an even more direct and common driver of conflicts. Should more conflicts of this nature manifest themselves in future, the demand for protection and assistance under the refugee framework will grow. This will, in turn, place a potentially unbearable strain on current standards and practices. One significant problem with the existing framework already exists in the context of persons fleeing indiscriminate violence, where the regime is both fragile and prone to be inconsistently applied. An analysis of this particular aspect of the protection regime, and a search for ways to strengthen it, will therefore feature in UNHCR's climate change strategy.

There may also be situations where the victims of natural disasters flee from their homeland because their Government has consciously withheld or obstructed assistance in order to punish or marginalize them on one of the five grounds set out in the refugee definition. In such scenarios, the persons concerned could legitimately be refugees in the traditional sense of the term.

A second scenario with implications for UNHCR relates to the potentially most dramatic manifestation of climate change, that of the 'sinking island' scenario, whereby the inhabitants of island states such as the Maldives and Tuvalu may eventually be obliged to leave their own country as a result of rising sea levels and the flooding of low-lying areas. Some form of UNHCR role regarding those obliged to seek safety abroad may be called for, certainly inasmuch as statelessness would be a concern.

UNHCR's role with conflict-induced internal displacement would be triggered as a result of the Cluster Approach. Under the division of labor introduced by the Cluster Approach, UNHCR has assumed global leadership of the Protection Cluster,[17] and co-leads the global Camp Coordination and Camp Management Cluster[18] (CCCM) with the International Organization for Migration (IOM) and the Emergency Shelter Cluster[19] with the International Federation of Red Cross and Red Crescent Societies (IFRC).

UNHCR's involvement with people who have been displaced within their own country as a result of natural disasters has traditionally been determined on a case-by-case basis. Generally speaking, when UNHCR had an established presence and programme in a country that was struck by such a disaster, the Office offered its support to the authorities as a sign of solidarity and as a contribution to broader international and UN relief efforts. However, the Cluster Approach is also relevant in natural disaster scenarios, since it has been agreed that, at country level, the leadership role for protection in natural disaster situations is decided upon by UNHCR, the Office of the UN High Commissioner for Human Rights and UNICEF, on a case-by-case basis. This formula has, however, come under criticism from, among others, the UN Emergency Relief Coordinator, as it does not bring about the necessary predictability or rapidity of response. All three protection agencies concerned, including UNHCR, are currently engaged in a thorough review of this arrangement.

Any new approach must be rights-based, since experience during the 2004 Indian Ocean tsunami and other recent disasters have confirmed that such emergencies generate new threats to the human rights of affected populations. In terms of preventing and responding to such threats, UNHCR considers the Inter Agency Standing Committee Operational Guidelines on the Protection of Persons Affected by Natural Disasters and the related Pilot Manual to be particularly valuable resources to address the special needs and vulnerabilities of persons forcibly displaced by the effects of natural disasters.[20]

Even where UNHCR's mandate is clear and uncontested, such as in longstanding refugee situations, climate change will affect the delivery of operations, in view of its implications in the areas of water, sanitation, agriculture, environmental protection and health.[21] The risk of flood water contaminating fresh water supplies, for example, will add another layer of complexity to the task of providing clean water and minimizing the spread of disease. For information on how environmental factors, including climate change, have impacted on, and are addressed by, UNHCR at the operational level please refer to the Annex.

In order to respond to these challenges, there is likely to be an increasing dependency on pre-existing partnerships with climate change-affected countries and a need to strengthen, where possible, dialogues between relevant Governments and UNHCR. Improving relationships with Governments and partner organizations will feature prominently in UNHCR's efforts to minimize and adapt to the impacts of climate change in refugee, returnee and IDP operations.

Terminology and the 1951 UN Refugee Convention

In recent times, a growing number of organizations and commentators have employed the notion of 'environmental refugees' or 'climate refugees,' a concept used to refer to people who are obliged to leave their usual place of residence as a result of long-term climate change or sudden natural disasters. UNHCR has serious reservations with respect to the terminology and notion of environmental refugees or climate refugees. These terms have no basis in international refugee law.

The phrase 'refugee' is a legal term. A person who has been determined a refugee will have satisfied the criteria under the 1951 Refugee Convention, the 1969 OAU Convention, or UNHCR's mandate. For this reason, just as a reference to an 'economic refugee' is not a reference to a recognized term under international law, neither are 'climate refugee' nor 'environmental refugee'. While often used, particularly in the media, it would be incorrect to give the words a legal meaning that has not been endorsed by the legal community.

UNHCR is actually of the opinion that the use of such terminology could potentially undermine the international legal regime for the protection of refugees whose rights and obligations are quite clearly defined and understood. It would also not be helpful to appear to imply a link and thus create confusion regarding the impact of climate change, environmental degradation and migration and persecution that is at the root of a refugee fleeing a country of origin and seeking international protection. While environmental factors can contribute to prompting cross-border movements, they are not grounds, in and of themselves, for the grant of refugee status under international refugee law. However, UNHCR does recognise that there are indeed certain groups of migrants, currently falling outside of the scope of international protection, who are in need of humanitarian and/or other forms of assistance.[22]

Some states and NGOs have suggested that the 1951 Refugee Convention should simply be amended and expressly extended to include people who have been displaced across borders as a result of long-term climate change or sudden natural disasters. UNHCR considers that any initiative to modify this definition would risk a renegotiation of the 1951 Refugee Convention, which would not be justified by actual needs. Moreover, in the current political environment, it could result in a lowering of protection standards for refugees and even undermine the international refugee protection regime altogether.

The Way Ahead

The United Nations Secretary-General has committed the UN system to be responsive to the evolution of the inter-governmental discussions on climate change, while at the same time offering proactive leadership in key emerging areas.[23] Addressing the humanitarian consequences is an obvious priority.

The world is currently confronted with an accumulation of negative trends: climate change, an increased incidence of natural disasters, rising food and energy prices, as well as turbulence in financial markets and a global economic slump. While it is impossible to predict the exact outcome of these phenomena, it is evident that they create the conditions in which significant numbers of people may be displaced or feel obliged to migrate. In responding to these circumstances, and for the reasons discussed above, the Office would encourage the international community to adopt an approach based on respect for human rights and international cooperation.[24] This is particularly important in the context of the international protection regime, as it may take some time to reach agreement on the appropriate way forward, but this should not prevent States from recognizing their existing obligations under international law.

Understanding the issues

Climate change is not just a scientific issue. More work is needed to analyze the likely human displacement scenarios which climate change will cause, and to identify and fill any legal and operational gaps. It will also be necessary to assess the potential consequences of climate change on populations who are already of concern to UNHCR and to support appropriate preparedness, adaptation and coping mechanisms.

There is presently an increasing amount of analysis of the relationship between climate change, environmental degradation, armed conflict, displacement and migration, but there is a pressing need for accurate empirical data to be collected and aggregated in order to keep the legal aspects of the impacts of climate change in line with the effects which are already being felt in operations at the field level. Common understandings need to be developed in relation to these issues. The use of innovative technology will play a central part in environmental protection. Significant progress on the mapping of certain hot spots has already been made,[25] additional knowledge is also required with respect to the environmental hot spots where displacement is most likely to be a feature; the coping mechanisms (including migration) employed by people who are most adversely affected by climate change; and the extent and ways in which environmental degradation is acting as a driver of social and political conflict.

Prevention, mitigation and adaptation

The UN Framework Convention on Climate Change provides an important means of pursuing and attaining the objective of combating the root causes of climate change, in particular greenhouse gas emissions. It is undisputed, however, that the devastating consequences of climate change cannot be reversed in the short term – this is why long-term cooperative action within the UNFCCC framework is concerned with both mitigation ('avoiding the unmanageable') and adaptation ('managing the unavoidable').

Migration is often the survival strategy employed by populations whose human security is threatened. UNHCR is convinced that additional international funding will not only be needed to help States mitigate the impact of climate change, but also to bolster adaptation, disaster preparedness and risk reduction, and humanitarian response at the national level. To avoid situations where people are compelled to migrate or become displaced, the resilience of communities must be better understood and reinforced, both in terms of their physical security and their ability to sustain adequate livelihoods.

It is evident that prevention and adaptation activities at the local level should be supported by both the affected States and the broader international community, including relevant components of the UN system and the international financial institutions. UNHCR will assist Governments, where possible, with the implementation of their National Adaptation Programmes of Action, but inter-State cooperation will lie at the heart of effective mitigation and adaptation planning, which will have to be closely linked to development. All UNHCR staff involved at the country level with refugee and IDP settlements, both rural and urban, will need to be equipped with strategies to combat and cope with the effects of climate change, impacting not just on persons of concern to UNHCR but also the broader host communities. Within the context of the United Nations Development Group, UNHCR has assisted in developing guidance to UN country teams on environmental sustainability and disaster risk reduction strategies.[26]

Moving forward: key work sites for UNHCR

UNHCR's overarching policy to tackle the effects of climate change will be reflected in three distinct areas: operations management; protection strategies; and advocacy. There are already some emerging issues within operations management, to which the Office has become more responsive and sensitive. One such issue is that of UNHCR's ability to offer assistance and protection in urban environments. The increase in urban populations as a result of climate change-induced migration will contribute to the 'urban heat island effect'[27] and will therefore lead to increases in temperatures in addition to those brought about by climate change. This will make people particularly vulnerable to health risks and will make the delivery of assistance and protection by UNHCR especially important. Work by the Office on this aspect of climate change will dovetail with UNHCR's broader urban policy work.

The need to look closely at the treatment of persons fleeing armed conflict or indiscriminate violence, sometimes called 'war refugees', will become an issue of increasing importance if predictions of climate change leading to armed conflict materialize. The current regime, which is inconsistently applied by States, places a heavy reliance on subsidiary protection, due in part to the fact that the collective threat faced by persons involved in armed conflict makes the classification of individuals as refugees under the 1951 Refugee Convention difficult. The net result is that persons fleeing armed conflict are often afforded less protection than Convention refugees. Reviewing the legal mechanics of this element of the protection regime and advocating for an appropriate strengthening of the regime, in appropriate fora, will feature as a component of UNHCR's wider policy on climate change.

UNHCR considers advocacy as an important tool in ensuring the realization of the protection needs of persons of concern falling within its mandate. While 2009 is a critical year in negotiating a more effective response to the challenge of climate change, the outcome of the UNFCCC Conference in Copenhagen ('COP 15') in December 2009 needs to be awaited. But 'climate change' will not be solved completely even by the best of outcomes, and genuine progress may or may not be achieved. Much will indeed depend on whether human displacement and humanitarian aspects of climate change, and the protection needs of persons of concern to UNHCR, will be properly referenced as issues that the international community must factor in. UNHCR believes that the need for advocacy on climate change issues will remain in various fora into 2010 and beyond.

[...]

UN High Commissioner for Refugees, '*Climate Change, Natural Disasters and Human Displacement: A UNHCR Perspective*', 14 August 2009

Notes

1   This policy paper updates and supersedes the paper issued by UNHCR in October 2008 and will be updated as necessary, in line with relevant developments within the climate change debate. Further UNHCR publications on climate change are available online at: www.unhcr.org/climate.

2   Foreign Affairs, September/October 2008, available at http://www.foreignaffairs.org/20080901faessay87506/antonio-guterres/millions-uprooted.htmls.

3   The United Nations Framework Convention on Climate Change (UNFCCC, available at http://www.unhcr.org/refworld/docid/3b00f2770.html) defines climate change as "a change of climate which is attributed directly or indirectly to human activity that alters the composition of the global atmosphere and which is in addition to other natural climate variability that has been observed over comparable time periods". Climate change is a process which is manifested in a number of ways, including: a rise in average temperatures, often referred to as 'global warming'; changes in rainfall patterns leading to floods, droughts and, in some areas, desertification; extreme and unpredictable weather patterns leading to more numerous and intense natural disasters; and, the melting of glaciers and the polar ice-caps resulting in rising sea-levels and coastal erosion, rendering low-lying areas uninhabitable.

4   Several members of the Inter-Agency Standing Committee's Informal Task Force on Climate Change have produced papers detailing the humanitarian implications of climate change. These are available at: http://www.humanitarianinfo.org/iasc/pageloader.aspx?page=content-news-newsdetails&newsid=134.

5   See 1998 Guiding Principles on Internal Displacement, available at http://www2.ohchr.org/english/ issues/idp/standards.htm.

6   Available at: http://www.ipcc.ch/publications_and_data/publications_and_data_reports.htm.

7   Sir John Holmes, Under-Secretary-General for Humanitarian Affairs and Emergency Relief Coordinator, Opening Remarks at the Dubai International Humanitarian Aid and Development Conference and Exhibition 'DIHAD 2008 Conference', 8 April 2008, available at: http://www.reliefweb.int/rw/rwb.nsf/db900sid/YSAR–7DHL88?OpenDocument.

8   Elverland, S., '20 million Climate Displaced in 2008', Norwegian Refugee Council, 8 June 2009, available at: http://www.nrc.no/?did=9407544.

9   See 'The Climate Change – Displacement Nexus', presented by Prof. Walter Kälin, Representative of the Secretary-General on the Human Rights of Internally Displaced Persons, Panel on disaster risk reduction and preparedness: addressing the humanitarian consequences of natural disasters, ECOSOC Humanitarian Affairs Segment, 16 July 2008, available at: http://www.brookings.edu/speeches/2008/ 0716_climate_change_kalin.aspx.

10   Kälin, op cit.

11   Kälin, op cit.

12   An analysis of the legal aspects of forced displacement occurring in the context of climate change has recently been submitted by UNHCR to the 6th session of the Ad Hoc Working Group on Long-

Term Cooperative Action (AWG-LCA 6) under the United Nations Framework Convention on Climate Change (UNFCCC); an additional submission was made on the subject of climate change and statelessness. The AWG-LCA met in June 2009 in Bonn, for talks which represented an important opportunity for States Parties to the UNFCCC to negotiate on a draft text in the run up to the United Nations Climate Change Conference in Copenhagen, 7–18 December 2009. Both UNHCR submission papers to the UNFCCC are available on UNHCR's climate change web page at: http://www.unhcr.org/ cgi-bin/texis/vtx/search/?page=&comid=4a2d26df6&cid=49aea9390&keywords=UNFCCC.

13   See for example the broader definitions of 'refugee' contained in the OAU Convention Governing the Specific Aspects of Refugee Problems in Africa 1969, at Article 1 (2), available at http://www.unhcr.org/refworld/docid/3ae6b36018.html, and the Cartagena Declaration on Refugees 1984, at conclusion 3, available at http://www.unhcr.org/refworld/docid/3ae6b36ec.html.

14   The rights under the 1966 International Covenant on Economic, Social and Cultural Rights (ICESCR), for example, apply not just to citizens, but to all persons within the jurisdiction of the State, including refugees, asylum-seekers and 'illegal' migrants. See General Comment No. 3 on the nature of States Parties' obligations, available at http://www.unhcr.org/refworld/docid/4538838e10.html. The EU 'Qualification Directive' (Council Directive 2004/83/EC, available at http://www.unhcr.org/refworld/docid/4157e75e4.htm) widens the European protection regime to encompass certain persons not classified as refugees, but nonetheless in need of international protection.

15   In addition to the 1951 Refugee Convention definition, the mandate includes: victims of manmade disasters and persons of concern to the High Commissioner. See ECOSOC Resolution 2011 (LXI) of 2 August 1976, available at http://www.unhcr.org/refworld/docid/3ae69ef418.html, General Assembly Resolution 31/35 of 30 November 1976, available at http://www.unhcr.org/refworld/docid/3b00f0375c.html, and General Assembly Resolution 48/118 of 20 December 1993, available at http://www.unhcr.org/refworld/docid/3b00f2641c.html.

16   Boano, C., Zetter, R. and Morris, T., Environmentally displaced people: Understanding the linkages between environmental change, livelihoods and forced migration, Refugee Studies Centre, University of Oxford, November 2008, available at http://repository.forcedmigration.org/pdf/?pid=fmo:4960.

17   See http://www.humanitarianreform.org/humanitarianreform/Default.aspx?tabid=79.

18   See http://www.humanitarianreform.org/humanitarianreform/Default.aspx?tabid=672.

19   See http://www.humanitarianreform.org/humanitarianreform/Default.aspx?tabid=77.

20   IASC Operational Guidelines on the Protection of Persons Affected by Natural Disasters and the related Pilot Manual, Brookings–Bern Project on Internal Displacement, March 2008, available at http://www.unhcr.org/refworld/docid/49a2b8f72.html.

21   See, for example, the UNFCCC submission paper 'Protecting the health of vulnerable people from the humanitarian consequences of climate change and climate related disasters', submitted to the 6th session of the Ad Hoc Working Group on Long-Term Cooperative Action, 1–12 June 2009, by the World Health Organization in collaboration with UNHCR and other agencies, available at http://www.unhcr.org/refworld/docid/4a2d189e1a.html. See also: 'Climate Change and AIDS: A Joint Working Paper', UNEP and UNAIDS, June 2009, available at http://data.unaids.org/pub/BaseDocument/2008/20081223_unep_unaids_joint_working_paper_on_cca_en.pdf.

22   Mixed migration and associated gaps in the international protection regime featured prominently within the High Commissioner's Dialogue on Protection Challenges, held on 11 and 12 December 2007. The Chairman's summary of the dialogue is available at the following link: http://www.unhcr.org/refworld/ docid/479744c42.html.

23   The Secretary-General will host a Climate Change Summit in New York on 22 September 2009 as part of a strategy to assist countries with their negotiations in the run up to the UNFCCC in Copenhagen, to be held on 7–18 December 2009.

24   See report of the Office of the High Commissioner for Human Rights on the relationship between climate change and human rights, 15 January 2009 (document A/HRC/10/61, available at http://www.unhcr.org/refworld/docid/498811532.html), which contains a useful analysis of the issue of

international legal obligations and inter-State cooperation in the context of climate change.

25   For example, the United Nations University (in collaboration with CARE International and UNHCR) recently presented a side event at the June 2009 Bonn Climate Change Talks entitled 'In Search of Shelter: Mapping the Effects of Climate Change'.

26   See UNDG Working Group on Programming Issues, Integrating Disaster Risk Reduction into the Common Country Assessment and United Nations Development Assistance Framework, April 2009, available at http://www.undg.org/docs/9866/UNDG-DRR-Guidance-Note-2009_DUP_08-07-2009_11- 43-02-734_AM.PDF.

27   See 'Protecting the health of vulnerable people from the humanitarian consequences of climate change and climate related disasters', World Health Organization (in collaboration with UNHCR), above at footnote 21.

Chapter 10

UN High Commissioner for Refugees

CLIMATE CHANGE AND STATELESSNESS
An Overview

[…]

'Sinking island states' present one of the most dramatic scenarios of the impact of climate change. The entire populations of low-lying States such as the Maldives, Tuvalu, Kiribati and the Marshall Islands may in future be obliged to leave their own country as a result of climate change. Moreover, the existence of their State as such may be threatened. Entire populations of affected states could thus become stateless.

Article 1 of the 1954 Convention Relating to the Status of Stateless Persons defines a stateless person as "a person who is not considered a national by any state under the operation of its law". Should a state cease to exist, citizenship of that state would cease, as there would no longer be a state of which a person could be a national.[1] The question is then the extent to which climate change could affect statehood.

Although there is no internationally agreed definition of what constitutes a State, there is agreement in the existing doctrine that there must be territory inhabited by a permanent population under the control of an effective government. Additionally, independence has been cited by publicists as a central criterion of statehood. The criteria are not absolute and are, moreover, applied less strictly once a State is established. The temporary exile of a government even for an extended period would thus not necessarily lead to extinction of the State. There is no precedent for loss of the entire territory or the exile of the entire population; presumably statehood would similarly not cease if such loss or exile were temporary. The implication, however, is that where such a situation would be permanent, statehood could be questioned.[2]

Should, the entire territory of a State be permanently submerged, inevitably there could be no permanent population attached to it or a government in control of it. The loss of all territory has been cited most frequently as a possible ground for loss of statehood. It appears, however, unlikely to occur before the end of the century, even with the upwardly revised rates in rising sea-levels announced by scientists recently.[3]

A threat to statehood may nonetheless arise far earlier. It is projected that the number and severity of extreme events such as storms and flooding will increase considerably. Extensive loss of fresh water and arable land due to contamination and seepage is expected. As well, destruction of the economic base is additionally expected due to erosion, as well as damage to corals and fishing grounds due to rising sea levels and global warming. The Intergovernmental

Panel on Climate Change (IPCC) has thus indicated that "rapid sea-level rise that inundates islands and coastal settlements is likely to limit adaptation possibilities, with potential options being limited to migration".[4] It has also confirmed that rising sea-levels are unavoidable.[5]

Low-lying island States are thus very likely to be entirely uninhabitable long before their full submersion, causing entire populations and the governments to be externally displaced. Unless territory could be protected or territory was ceded by another State, the exile of the population and the government would presumably be permanent. The population and the government would be entirely dependent on the status the host State would be willing to grant it. The government's independence could thus also be questioned. The IPCC itself has noted the threat to the sovereignty of low-lying island States likely to be affected.[6]

Should statehood cease, the population would be rendered stateless. Disappearance of a State due to loss of territory or the permanent exile of the population or the government is without precedent. The international community could agree that the affected States would continue to exist nonetheless. Even in such a case, however, governments of affected States would face many constraints in practice[7] and their populations would be likely to find themselves largely in a situation that would be similar to if not the same as if statehood had ceased. The population could thus be considered de facto stateless.[8]

To the extent that statelessness is foreseeable, efforts should focus on preventing it from arising. The principle of prevention of statelessness is a general one recognized in international law as a corollary to the right to a nationality. Both have been iterated in numerous international and regional human rights instruments.[9] As well, specific instruments address prevention and reduction of statelessness, including *inter alia* the 1961 Convention on the Reduction of Statelessness (hereafter the 1961 Convention).[10] In addition, the General Assembly in 1996 entrusted UNHCR with a global mandate to engage in preventing and reducing statelessness as well as to protect stateless persons.[11]

To prevent statelessness in the context of low-lying island States, one option would be that territory elsewhere would be ceded to the affected State to ensure its continued existence. If other States agreed that this was the same State, statelessness would not arise. Union with another State would be another option. In such a case, the 1961 Convention and the Draft Articles on the Nationality of Natural Persons in Relation to Succession of States provide for specific safeguards to prevent statelessness.[12] Otherwise, the acquisition of the nationality of another State would need to be foreseen. As existing instruments do not provide specific guidance for such a case, situation-specific arrangements would be required.

Ideally, multilateral comprehensive agreements would provide where, and on what legal basis such populations would be permitted to move elsewhere and their status. To prevent temporary statelessness, acquisition of an effective nationality should be foreseen prior to the dissolution of the affected State. Dual nationality may therefore need to be permitted at least for a transitional period. As well, a waiver may be required of formal requirements for renunciation or acquisition of nationality which might be difficult to fulfil for affected populations. Such arrangements would need to provide *inter alia* for the right of residence, military obligations, health care, pensions and other social security benefits. Citizens of affected States that might have been displaced earlier, possibly to third States not party to the agreement, may also need to be considered. Provision should also be made to ensure that any resettlement be environmentally, socially and economically viable and sustainable over the longer term. Further, the principle of family unity should be considered and due regard given to permitting the affected populations to protect their identity as a people, including their language, culture, history and traditions.

The early introduction of educational and other measures to prepare for such displacement, such as labour migration schemes, could serve not only to increase the resilience and ability to adapt in the host country, but also provide further resources and reinforce the resilience of the population remaining on the islands. Although complete relocation of the entire population would be a measure of last resort, early preparedness could also help avert a humanitarian catastrophe by promoting orderly movements of affected populations and increasing the viability of the move.

Such arrangements would ideally be elaborated on the basis of participatory involvement of the population as well as the Government of affected island States. Other interested States, and relevant organizations and agencies could be involved as partners, with due account being taken of existing links with other States.

To this end, recognition would be required *inter alia* in the United Nations Framework Convention on Climate Change (UNFCCC) that external displacement will be inevitable and that statelessness may arise as a consequence of climate change, particularly in the case of low-lying island States. Funding should be made available at an early stage to prepare for prevention of statelessness and for displacement and/or migration as a possible adaptation measure.

In view of its mandate to engage in preventive actions related to statelessness, UNHCR would be pleased to support efforts by States to devise appropriate solutions for potentially affected populations, in partnership with other actors.

UNHCR, '*Climate Change and Statelessness: An Overview*', Submission to the 6th session of the Ad Hoc Working Group on Long-Term Cooperative Action under the UNFCCC, 1 to 12 June 2009, 15 May 2009

Notes

1   This was also confirmed by the International Law Commission: "[w]hen a state disappears by dissolution, its nationality also disappears", International Law Commission, Draft Articles on Nationality of Natural Persons in Relation to the Succession of States (With Commentaries), 3 April 1999. Supplement No. 10 (A/54/10), UNHCR Refworld, Commentary (1) to Article 23, http://www.unhcr.org/refworld/docid/4512b6dd4.html (accessed 16 September 2008).

2   See inter alia Ian Brownlie, Principles of public international law, Sixth edition, Oxford: Oxford University Press, 2003, pp.64, 70–79, 86–88, 105–107, 117–118; James Crawford, The Creation of States in International Law, Second edition, Oxford: Oxford University Press, 2006, pp.3–74; Malcolm N. Shaw, International Law, Sixth edition, Cambridge: Cambridge University Press, 2008, pp.211–214; Peter Malanczuk, Akehurst's Modern Introduction to International Law, Seventh revised edition, London, New York: Routledge, 1997, reprinted 1998, pp. 75–79, 88, 152–154. In the absence of an internationally agreed definition under treaty or customary law, resort is had here to the writings of publicists as a subsidiary source of international law is in accordance with Article 38 of the Statute of the International Court of Justice.

3   In the context of the International Scientific Congress on Climate Change held in March 2009 in Copenhagen, scientists highlighted that likely sea-level rise by 2100 could be up to one meter or more, and was unlikely to be less than 0.5m, see inter alia Climate Secretariat, University of

Copenhagen, "Rising sea levels set to have major impacts around the world", 10 March 2009, http://climatecongress.ku.dk/newsroom/rising_sealevels/ (accessed 18 March 2009).

4   Intergovernmental Panel on Climate Change (IPCC), Climate Change 2007, Fourth Assessment Report, "Report of the International Working Group II Report "Impacts, Adaptation and Vulnerability", p.73 http://www.ipcc.ch/ipccreports/ar4-wg2.htm (accessed 18 September 2008). The IPCC is a scientific inter- governmental body established by the World Meteorological Organization (WMO) and the United Nations Environment Programme (UNEP) to provide decision-makers and others interested in climate change with an objective source of information about climate change.

5   Ibid, p.317.

6   Ibid, p.736 .

7   See Brownlie, supra, pp.64, 86–88, Crawford, supra, pp.26–28, 93, Malanczuk, supra, p.84.

8   There is no universally accepted definition of de facto statelessness. It has been referred to in different instruments as well as by publicists. The Final Act of the 1961 Convention on Reduction of Statelessness indicates that "persons who are stateless de facto should as far as possible be treated as stateless de jure to enable them to acquire an effective nationality", thus indicating that lack of effective nationality would be considered as a form of de facto statelessness. See also Council of Europe Convention on the Avoidance of Statelessness in Relation to State Succession, Strasbourg, 19.V.2006, Council of Europe Treaty Series, No. 200, Explanatory Report, Article 3 Prevention of statelessness, para. 16. "State succession may well create situations of de facto statelessness where persons do have the nationality of one of the States concerned but are unable to benefit from the protection of that State".

9   Article 15 of the 1948 Universal Declaration of Human Rights provides that "[e]veryone has a right to a nationality. No one shall be arbitrarily deprived of his nationality nor denied the right to change his nationality". The right to a nationality is iterated as well inter alia in the 1966 International Covenant on Civil and Political Rights, and the 1989 Convention on the Rights of the Child, as well as the 1990 International Convention on the Protection of the Rights of All Migrant Workers and Members of their Families, although the formulation varies. The 1979 Convention on the Elimination of All Forms of Discrimination Against Women, the 1965 International Convention on the Elimination of All Forms of Racial Discrimination and the 1957 Convention on the Nationality of Married Women also contain relevant provisions. At a regional level, see the 1990 African Charter on the Rights and Welfare of the Child, the 2005 Covenant on the Rights of the Child in Islam, the 1948 American Declaration on the Rights and Duties of Man, the 1969 American Convention on Human Rights and the 1997 European Convention on Nationality. The principle of prevention of statelessness is iterated in many of the same instruments. Additionally, both principles have been iterated numerous times by the General Assembly, the Human Rights Council and the International Law Commission.

10 Other instruments include inter alia the Draft Articles on the Nationality of Natural Persons in Relation to Succession of States and, at a regional level, the 2006 Council of Europe Convention on the avoidance of statelessness in relation to State succession.

11 General Assembly Resolution GA/RES/50/152, 9 February 1996, paras.14–15. This mandate is additional to its earlier mandate to undertake the functions foreseen under Article 11 of the 1961 Convention first given in 1974 and then extended through Resolutions GA/RES/3274 (XXIX), 10 December 1974, and GA/RES/31/36, 30 November 1976.

12 The 1961 Convention in Article 10 provides that in the absence of any treaty ensuring that no persons are left stateless, any contracting State to which territory is transferred shall grant its nationality on persons who would otherwise be stateless. The Draft Articles on Nationality of Natural Persons in Relation to the Succession of States are also relevant. Article 21 prescribes that citizenship should be automatically extended to all citizens of the predecessor island State. As well, habitual residents who may have left owing to climatic change, should have right of option to obtain same status in successor state including acquired rights; see commentary (3) to above Article. The International Law Commission noted that Article 21 in its view embodied a rule of customary international law; see commentary (6) to above article.

Chapter 11

Bruce Burson

SECURING MEANINGFUL INTERNATIONAL AGREEMENT ON CLIMATE CHANGE RELATED DISPLACEMENT AND MIGRATION
The Refugee Convention as a Window on International Burden (Responsibility) – Sharing in an Involuntary Movement Context[1]

Introduction

Despite one of the earliest warning of the Intergovernmental Panel on Climate Change[2] being that migration may be one of the single greatest impacts of climate change, neither the UNFCCC nor the Kyoto Protocol make any reference to the human mobility consequences of climate change. The failure of COP 15 to reach a binding global agreement to address climate change mitigation and adaptation and the furore surrounding the Copenhagen Accord demonstrates just how problematic securing international cooperation to deal with climate change related issues can be. The multi-causal nature of migration and the difficulty in identifying a uniquely environmental push factor in many cases will only increase the complexity of the task ahead.

And yet, there seems little doubt that securing international cooperation to deal with climate change related[3] migration will be vital. After all, migration has been an important mechanism for responding to climate stress for millennia.[4] However, predictions as to the numbers of people at risk of being displaced or migrating as a result of climate change have tended to be 'guesstimates' and not scientifically credible, producing huge variation in numbers of displaced persons from the tens to hundreds of millions up to, according to one study, as many as 1 billion. The most commonly cited figure is that of around 200 million persons displaced by climate change by 2050.[5] To put this in perspective, this figure equates to what the IOM currently estimates to be the total number of migrants worldwide.[6] The lack of accurate information on the possible migratory consequences of climate change impedes our ability to adequately prepare for and comprehensively respond to the humanitarian and protection needs of environmental migrants. Understanding more clearly the potential scale and patterns of climate change related migration is therefore *the* most pressing policy issue.[7] Recent studies[8] which combine sophisticated geophysical analysis with demographic data to highlight at risk populations and suggest likely migration pathways represent a welcome progression in the body evidence available to policy makers. While more detailed data is required in terms of shaping the contours of policy, a sufficiently solid evidential base already exists to make clear that the possible human mobility and humanitarian consequences of climate change constitute not only a *major* policy problem, but also constitute a *near-term* policy problem requiring immediate attention. It is no longer a question of 'if', but 'where, when and how'.

Certainly, the latest IPCC assessment report contains observations which suggest the need for a precautionary approach to be adopted. The IPCC report that even under favourable scenarios substantial numbers of persons are likely to be negatively affected by climate change over the coming decades. In particular, densely populated regions of the Mediterranean coastline and central, western and southern Africa are expected to be adversely affected. Specifically, the IPCC concluded with high confidence[9] that, by 2020, between 75 and 250 million people in Africa may be exposed to increased water stress. Coupled with increased demand, this will adversely affect livelihoods and exacerbate water-related problems.[10] Yields from rain-fed agriculture could be reduced by up to 50% in some countries.[11] As for Asia, the IPCC state[12] that more than 1 billion people could be adversely affected by projected decreases in freshwater availability in central, south, east and south-east Asia by 2050. The heavily populated mega-deltas in the region will be at greatest risk of increased flooding. While different parts of Asia will experience increases and decreases in crop yield, factoring in rapid population growth and urbanisation, the IPCC concluded, risk of hunger is projected to rise in several developing countries.

Plainly it would be a mistake to conclude that all or even most people adversely affected by these events and processes will respond by migrating, or that those who do will want, or be able, to migrate externally. These environmental factors will interact with numerous other conditions in places of origin and destination as well as factors at the individual and household level (e.g. age, gender) to determine when and if migration occurs, the type of migration engaged in and by whom.[13] By the same token, however, the reality is that some of the people adversely affected by climate change will engage in cross-border migration. Climate change, therefore, has at least the potential to become a significant driver of international migration on a large scale over coming decades. Consistent with the precautionary principle that underpins much of the efforts in international environmental governance, it seems only sensible that the international community starts thinking about and planning for the movement of many millions of people, some internationally, over the coming decades as a result of climate change.

This chapter argues that examination of the operation of the Refugee Convention is useful in this context. The Convention is relevant because it comprises an international regime predicated on the need for responsibility-sharing in an involuntary movement context. Furthermore, it is premised on recognition that international co-operation is necessary to alleviate the "unduly heavy burden that refugee flows may place on certain countries".[14] Regional Refugee instruments and texts also refer to the need for international solidarity and burdens sharing.[15] Scholarly debate of its design and operation, successes and failures, provides a window through which to view the issues that will arise in designing arrangements to effectively respond to international climate migration.

The need for a rights-centred approach

A key feature of the Refugee Convention regime that it is rights-centred. It guarantees minimum standards of rights to social goods such as housing and employment and did so prior to the adoption of the two International human rights Covenants of 1966. While it is axiomatic that the content of any international policy response should also be rights centred, there is a burgeoning literature on this and this chapter does not propose to enter into examination of how human rights norms can aid this response. Rather, the focus in this chapter is on lessons in regime design that can be learned from examination of the refuge protection regime.

A word on nomenclature – the "Environmental Refugee" debate

Controversy surrounds how to describe persons who migrate because of environmental degradation and, in particular, whether they can legitimately be described as "environmental refugees". This chapter will not enter into the debate surrounding usage of the term "environmental refugee" although it is, I believe, a term best avoided save in a narrow set of circumstances.[16] Instead, I propose use for the purpose of this chapter, the term "international climate migrant"[17] which seems to me to capture the essential characteristics of those persons in the category I am considering. In particular:

- *International* – captures the notion that movement of is into the territory of a state other than their own.
- *Climate* – limits the field of concern to cases where loss of ecosystem and ecosystem services as a result of climate change related events and processes constitutes a significant, but not exclusive, factor in driving the population movement.[18]
- *Migrant* – migration typically describes movement from one administrative unit to another, whether internal or international[19] and captures a range of population movement such as circular, temporary, seasonal and permanent migration, which may all be involved.

The Convention as a window – what lessons?

An ad hoc or managed response?

Examination of the refugee protection regime sheds light on the fundamental question as to whether it is desirable that there is an agreed and managed response in which the burden is shared by states or an ad-hoc response in which states are left to deal with climate change migrants on a case-by-case, scenario-by-scenario basis. The Convention's regime provides examples of both managed and *ad hoc* processes at work. Analysis of the strengths and weaknesses of each will usefully inform development of a similar regime in respect of international climate migration.

The managed process is composed of the UNHCR engaging in a program of refugee status determination (RSD) as an aspect of its core protection function.[20] This is considered by UNHCR to be an integral aspect of the Convention's responsibility-sharing arrangements.[21] This managed process results in third country states, typically a few states of the 'global north', agreeing to admit to their territory, persons from the 'global south' found to be in need of resettlement by UNHCR and for whom other durable solutions such as repatriation or local integration are not viable or feasible. This process can be called managed in that there is a greater element of coordination and control between the state and institutional actors involved in the resettlement process. The key point here is that states of the global north are, to some extent, *already* funding[22] and taking part in a managed population movement regime which results in persons[23] at risk of serious harm being resettled into other states. A similar regime of a kind in respect of international climate migration would not, therefore, be novel. However, the managed process suffers from major weaknesses insofar as resettlement is concerned. First, there is no binding obligation on States parties to accept UNHCR recognised refugees for resettlement. Second, the limited number of states participating in the managed process means only a small fraction of those found to be in need of resettlement are actually resettled. This forces asylum seekers to rely on *ad hoc* processes.

The *ad hoc* process involves claims for recognition as a refugee being made by persons arriving within the jurisdiction of States Parties to the Convention. These spontaneous claims are determined within the domestic RSD systems of the state concerned. While there is nothing in the Convention which requires permanent resettlement under this process, the norm is that those recognised as refugees are given some form of settled immigration status in the country of refuge. Given the projected numbers, continuing to rely on an ad hoc process over the coming decades to deal with the challenges of international climate migration has the potential to impose heavy costs on a number of levels:

- *Human costs* – asylum seekers are often forced to rely on people smugglers to access ad hoc arrangements. Transported in overcrowded and unsafe lorries and boats, deaths are all too common. Receiving states are often ill-equipped to cope with the influx[24] leaving refugees to live marginal and insecure existences – in many cases for years. The human costs of the *ad hoc* process are large.
- *Structural costs* – *ad hoc* responses will perpetuate regime imbalance. Many of the areas at risk of climate change induced water or food shortages resulting from climate change are located in conflict-prone states of the global south. Those affected are likely to rely on migration pathways established in relation to conflict displacement. This will tend to increase the numbers moving to neighbouring states in the global south as the country of first asylum.[25]
- *Fiscal costs* – for refugee receiving states of the global north, one of the lessons of the last 20 years has been that many people fleeing poverty and economic marginalisation have been forced to use the ad hoc route in the absence of alternative employment-centred, managed migration polices. The number of claimants accessing the managed process is already dwarfed by those relying on the *ad hoc* process.[26] Unsurprisingly, the costs of administering the *ad hoc* process far dwarf the contributions made to UNHCR to fund its activities. According to one account, approximately US$10 billion was spent by OECD governments on RSD determination procedures in 1997 compared to the US$ 1 billion given to UNHCR to fund its programmes.[27] The ratio does not appear to have changed notably over time.[28] In the absence of managed policy in place to deal with international climate migration, countless persons whose livelihoods are sufficiently affected by climate change may feel they have little choice but to enter the spontaneous *ad hoc* process with all the fiscal costs this entails. Once lodged, the claim must be examined and considered in some way. Even with truncated processes, there will be costs that will be borne entirely by the tax-payer of the state concerned.

There are further powerful reasons why the policy response to international climate migration should be a shared and managed response. First, it provides states with something akin to an insurance scheme that they will not face a migration emergency alone and engender more predictable responses and lower costs.[29] Second, for migrants, as long as the scheme is based on international human rights norms, it can provide certainty and better protection and assistance. For all these reasons it seems desirable to strive towards some agreed framework for dealing with international climate migration.

If a managed response is considered desirable, examination of the Convention provides insights into a range of other relevant issues such as:

- What barriers to agreement might arise;
- The form of agreement;
- The arena of management – global or regional?
- What it is that is being managed/shared.

Identification of potential barriers to effective burden-sharing

That only a small fraction of those found to be in need of international protection by UNHCR are actually resettled or otherwise offered some from of temporary international protection points to the reality that states will need to be persuaded that a managed response within an agreed framework of rules is in their interests. The academic literature on the Convention provides a rich source of analysis of state reluctance to fully embrace the concept of burden-sharing in its implementation. Three issues emerge from this literature of critical importance in enhancing the prospect of burden-sharing in respect of international climate migration:

Appeal to state interest

Some scholars[30] argue a '*demise of interest-conversion*' has led to an increasing marginalisation of the Convention by states as response to displacement. Hathaway identifies two contributing factors. First, the Convention's arbitrary assignment of full legal responsibility with only limited acknowledgement of the receiving state's interests and, second, the tendency away from the envisaged temporary protection of refugees as important contributors to this demise. Drawing on examples where states of the north and south have come together to solve displacement crises, Betts argues that creating substantive linkages to pre-existing state interests so as to allow cross issue persuasion by the UNHCR has been key to securing international cooperation.[31] Betts argues:[32]

> The concept of cross-issue persuasion sheds light on the conditions under which UNHCR has been able to appeal to the interests of states to induce them to contribute to refugee protection. Northern states have generally had few incentives to engage in burden-sharing, but they have done so insofar as they have been persuaded that there is a relationship between refugee protection in the south and their wider interests in other areas.

The importance of seeking interest conversion or linkage cannot be underestimated – states will not agree to a regime not thought to be in their interest. Betts identifies a number of areas of national interest – migration, security, development, trade and peace-building – within which linkages can be made to enhance refugee protection. These areas also constitute areas of engagement in the context of international climate migration. In the development context, there is an increasing body of evidence highlighting the positive role migration can play in furthering development both in the country of origin and host state. The Global Forum on Migration and Development, a state-led dialogue designed to enhance synergy between migration policy and development outcomes in both the countries of origin and destination, provides a useful mechanism through which states may come to reach agreement on facilitating international climate migration where necessary. In relation to security, it has been noted.[33]

While the most immediate victims of climate change and failed adaptation will be the world's poor, the fall-out will not respect the neat divides of national borders. Climate change has the potential to create humanitarian disasters, ecological collapse and economic dislocation on a far greater scale than we see today. Rich countries will not be immune to the consequences. Mass environmental displacement, the loss of livelihoods, rising hunger and water shortages have the potential to unleash national, regional and global security threats. Already fragile states could collapse under the weight of growing poverty and social tensions. Pressures to migrate will intensify. Conflicts over water could become more severe and widespread. In an interdependent world, climate change impacts will inevitably flow across national borders. Meanwhile, if the countries that carry primary responsibility for the problem are perceived to turn a blind eye to the consequences, the resentment and anger that will surely follow could foster the conditions for political extremism.

It is, however, critical that if seeking to create issue linkage via 'security', the positive role migration can play does not become subordinate to the 'high politics of security'[34] and regarded only as a problem.

Confront the disincentives

Disincentives are said to exist for states in the refugee context that are absent in other international contexts such as security or environmental harm.[35] The argument is that states are reluctant to pay up-front for something they have no control over and in respect of which the future benefits are unclear. Also, there exists a residual state capacity to ward-off the refugee 'threat' by taking unilateral action such as closing or tightening border control. Finally unlike environmental agreements which deal with the *causes* of harm, the Convention deals with *consequences*.

However, important contextual differences between international climate migration and classic refugee flows exist which can be emphasised to incentivise participation in any new regime. First, while uncertainties in the climate system together with variations in exposure to risk, vulnerability to natural hazards and adaptive capacity mean that it is impossible to obtain a totally precise picture climate change related migration flows, in general, the environmental context provides great predictive opportunity than the political. The fact that changes in the planetary climate and its effects on ecosystem and ecosystem services are able to be observed, measured and monitored means that should still be easier to predict more accurately the likely climate change related environmental conditions which *might* produce cross-border population movement over the coming decades in a particular state or region than predict the socio–political conditions which might do likewise. An important development in this regard is the notion of 'hotspots' – areas in which natural hazards such as drought, flooding or cyclones are likely to intensify in the coming decades.[36] Recent studies have combined sophisticated geophysical analysis with detailed case studies of the migration responses of affected populations in a number of locations across the globe to produce "present plausible future developments that provide decision makers a basis for focusing their discussions on the role of human mobility in adaptation."[37] With this and other research into how affected communities have coped with natural disasters and other environmental shocks and stressors, it should prove far easier to quantify both the numbers of people at risk of displacement and their geographic distribution in the context of climate change than is true in relation to existing refugee flows. This will afford greater opportunity to plan and allocate resources and generally manage the situation.

Second, states do in fact have the ability to exercise a greater degree of control over the causes of climate change and not just the consequences. Via agreement on *their own* mitigation efforts, states can exert far more control over displacement in the climate change arena than they in the political arena where control over causes of displacement is heavily constrained by the sovereignty of *other* states.[38] Moreover, the positive relationship between migration and development means arguments around future benefits lose their force.

Compensate for the accident of geography[39]

Overwhelmingly, the majority of refugees do not enter the ad hoc processes operated by the states of the 'north'. Rather, geographical proximity to displacement is the primary determinant of which state hosts current refugee populations. This means that many millions of refugees end up in refugee camps in countries of first asylum – often for many years – countries that are often developing countries with much less financial ability to absorb the costs associated with hosting refugee populations. Against this background, it is unsurprising that issues of equitable burden-sharing have arisen. The debate centres around the issue of financial transfers to assist those states bearing these disproportionate burdens and, in particular, the *additionality* of such funding vis-à-vis development assistance generally. Attempts to introduce more equitable burden-sharing arrangements in respect of refugee movements in Africa foundered because of failure to reach a north–south consensus on this issue.[40]

Achieving consensus around the basis for and extent of financial transfers to assist with the financial costs associated with the hosting of displaced populations will be critical to the success of any international agreement dealing with international climate migration. Such funding should, in my view, be under the umbrella of the UNFCCC and not as an aspect of development aid for two reasons. First the UNFCCC will likely set the international policy agenda for years to come. Second, the UNFCCC already has built into it ideas that states of the global north bear the historical responsibility for the climatic change that is causing the relevant environmental stresses. This gives rise to a moral obligation to address its consequences. Third, the UNFCCC *already* contemplates north–south financial transfers to deal with costs of adapting to climate change providing an existing framework. Article 4(4) obliges developed country parties to assist developing country parties that are particularly vulnerable to the adverse effects of climate change "in meeting their adaptation costs". As the notion that migration is an identifiable and has been a historically verifiable response for communities within developing country parties in adapting to changes in their natural environment, there is no reason in principle why adaptation costs associated with relocating vulnerable communities on either a temporary or permanent basis cannot be captured by this. Indeed, such north–south financial assistance has happened on a bilateral basis in the Pacific already.[41] However, Article 4(4) will be of more contestable application in context of international climate migration than in the cases of internal relocation and resettlement. Where the receiving state is also a UNFCCC developing country party, it is possible to argue that Article 4(4) covers costs associated with hosting non-national displaced populations although this is likely to be contested by developed country parties. The call in the Bali Action Plan for enhanced action on "ways to strengthen the catalytic role of the Convention....as a means to support adaptation in a coherent and integrated manner"[42] at least provides some impetus for discussion of this issue. The planning for adaptation required by UNFCCC Article 4(1)(b) should now explicitly encompass the issue of migration. UNHCR and other agencies are already engaging with the Bali Action Plan and have made submissions to the Ad Hoc Working Group on Long-Term Cooperative Action under the UNFCCC.[43]

Achieving agreement around financial transfers will not be easy. The fraught nature of negotiations surrounding north-south financial transfers in relation to mitigation and *physical* adaptation costs in the UNFCCC context points to highly difficult negotiations on transfers relating to costs linked to *human* adaptation via migration. There are well-known problems around the chronic underfunding and fragmentation of existing UNFCCC adaptation funding mechanisms. It will also in many cases be difficult to establish direct linkage between climate change related events and processes and migration decisions. But, given the body of evidence available suggesting that the impacts of climate change related events and processes will become an increasingly important factor in many people's decisions to migrate, *as a matter of principle*, a migration assistance fund should be established and administered under the broad umbrella of the UNFCCC.

The form of agreement

The issue here is whether burdens should be shared through a rigid legal treaty or through some form of framework agreement or other legal instrument which prescribes an agreed procedure for dealing with emergencies as and when they arrive. Each approach has risks. The former may result in states pegging their commitments at "a low level not responding to actual protection needs" whereas the latter "entails a comparatively higher risk of defection in situations of crisis."[44]

There have been explicit calls for a managed migration response of a treaty-based nature. In 2008, the German Advisory Council on Global Change, an independent, scientific advisory body to the German Federal Government recommended the "creation of a quantitative distribution formula which would involve the whole international community admitting environmental migrants" within a new negotiated international agreement.[45] Others have also argued for a specific global protocol to protect "climate refugees".[46]

A formal treaty dealing with international climate migration, while ideal in terms of the binding nature of the obligations that would flow from it, may not be politically feasible. States "have been especially wary of institutionalised schemes that would commit them to take a certain proportion of refugee flows of unknown frequency and magnitude".[47] While there is to some extent a greater degree of predictive certainty in this context, there remains a residual level of uncertainty over the extent and timing of climate related environmental events and processes as well as their humanitarian consequences. Furthermore, the migratory impacts of climate change will be felt differently between states at different times. States will differ in their adaptive and absorptive capacities over time. While binding commitments are ideal, it seems better, and more politically feasible, that there is a degree of flexibility in the system, albeit with sufficient rigidity to minimise the potential for 'defection'. Initially, it may be politically more feasible for the agreement to take the form of an agreed statement of general principles, or a memorandum of understanding provided that such soft law options. Relying on soft-law mechanisms may prove the most practicable way to both galvanise and harmonise state response over time. Certainly, in Latin America, the 1984 Cartagena Declaration, technically a piece of "academic reflection", has been instrumental in shaping regional asylum policy.[48] Similarly, the Guiding Principles on Internally Displaced Persons, nominally a collection of provisions from treaties, have had an important and identifiable normative effect. They have been expressly acknowledged as being an important inspiration behind the African Union's decision to invite African states to adopt a regional Convention Of The Protection And Assistance Of Internally Placed Persons In Africa.[49]

The breadth of international co-operation: global, regional or bilateral?

Closely linked to the issue of the form of the agreement is the question of who shares the burden or responsibility.[50] The Convention is relevant to this issue because it provides examples of regional inter-state cooperation to deal with refugee flows occurring in specific places or contexts. Suhrke argues that the two successful cases of inter-state burden-sharing involved *regional* arrangements.[51] Others[52] also argue that a regional response may be the most effective way to secure burden-sharing including in the context of international climate migration.

While established regional coalitions of states are more likely to have sensitivity to and understanding of the local conditions and causes of the displacement, a hybrid model involving global and regional and even bi-lateral features seems better suited. Bi-lateral arrangements comprise an important component of current international migration management and will continue to do so. However, the fact that all states will, to some extent, be affected by climate change means that, ideally, the agreement should also contain an element of global management via some linkage to the UNFCCC and/or any successor to the Kyoto protocol for two reasons. First, as Betts' case studies highlight, successful north–south cooperation has only occurred when the refugee protection agreement was sufficiently embedded with a wider process or related to issues of sufficient foreign policy concern to the stakeholder states.[53] The UNFCCC represents a wider process designed to respond to the prospect of catastrophic climate change.

Second, linkage to the UNFCCC is apposite because at its heart[54] lies the notion of common but differentiated state responsibility (CBDR).[55] Hathaway's[56] prescription for a 'reformulation' of the Convention's regime to re-legitimatise it in the eyes draws on the need to be a move towards CBDR. The concept of CBDR points to the contours of a hybrid model of this kind with both hard and soft law components. By disaggregating the issues of 'who goes where?' and 'who pays?', it is possible to envisage a regime with variable but broadly balanced commitments and responsibilities.[57] Building upon these notions in the international climate migration context, one can contemplate an architecture whereby preferences by displaced populations to resettle in regionally proximate states with shared cultural ties or in states with colonial or community links can be accommodated by the different state actors have differentiated responsibilities to deal with the common problem. The states of the settling and host populations would enter into bi-lateral arrangements and bear a responsibility to engage in proper planning and implementation with a view to achieving the sustainable integration and development of the migrating communities. Other states and, in particular, the states of the global north not otherwise engaged at a bilateral level, would bear the responsibility of providing financial and technical assistance as required.

What is to be shared?

Financial costs, however, are not the only thing that can be shared. The German Advisory Council on Global Change proposal advocates for a people-sharing arrangement. The Convention's implementation by states provides examples of attempts to share both people and norms. A good example of a functioning people-sharing arrangement and the problems which can arise from them is the EU's *Convention Determining the State Responsible For Examining Applications for Asylum Lodged in One of the Member States of the European Community 1990* (the Dublin Convention). As the title suggests, the Dublin Convention seeks to allocate responsibility for determining refugee status amongst member states according to the place where the claimant first entered the territory of the EU. A successful claimant would then be entitled to whatever immigration status that responsible state conferred according to its domestic immigration policy. The system is cumbersome. It requires a state receiving a refugee claim by a person it believes to have passed through another EU member state to formally notify that member state that it considers it to be the responsible state under the Dublin Convention for processing of the claim for asylum. If accepted by the other state, the claimant is then removed to the state of first entry.

The Dublin Convention is also controversial. UNHCR[58] has been critical of the operation of the Dublin Convention because its basic underlying premise, that asylum seekers are able to "enjoy generally equivalent levels of procedural and substantive protection, pursuant to harmonized laws and practices" is not fulfilled. It encountered particular legal difficulties in the United Kingdom, a destination often only reached after claimants had prior entered states on mainland Europe. At one point in time, member states and their domestic courts took divergent views on fundamental issues of refugee law such as whether a non-state agent could be the agent of persecution. As a result, the United Kingdom courts held that removals from the United Kingdom which did consider non-state agents to be agents of persecution to states such as France and Germany which, at the time,[59] did not, were unlawful.[60] In other words, the introduction

of a people-sharing arrangement without ensuring a harmonisation or sharing of norms created much anxiety and trauma for an already vulnerable group, increased the costs to the state as public money funded the legal challenges, and the lengthy delays that ensued – sometimes for years when legal challenges were mounted – reduced the overall effectiveness of the scheme.

Conscious of these problems, the EU has for some time attempted to secure the harmonisation of norms of refugee-related procedure and rights among member states. This constitutes a form of indirect people-sharing as the idea is to minimise 'asylum-shopping' by minimising the perceived comparative advantage for an asylum seeker in lodging a claim in one State as opposed to another. The introduction of the *Council Directive on the Minimum standards for the qualification and status of third country nationals or stateless persons as refugees or as persons who otherwise need international protection and the content of the protection granted* Council Directive 2004/83/EC (29 April 2004) ("the Qualification Directive") is the latest attempt to equalise approaches to interpretation of the Convention and protection standards. However, the introduction of the Qualification Directive has not led to the intended harmonisation. A study by UNHCR concluded that, while there had been come harmonisation on some issues across member states, divergent practice on fundamental issues was still in evidence.[61]

Analysis of the application of the Convention reveals the difficulties and limitations in using 'people sharing', an effective burden-sharing mechanism. Issues of human and social capital grounded in family and community relationships are critical in the establishment of migration networks.[62] Such relationships affect migration preferences and decisions which may not align with the state-level agreed distribution scheme. As Suhrke[63] remarks, "refugees may not want to be shared with countries where they have few cultural and social ties." From a more statist perspective, it has been observed[64] that while people sharing is attractive for States because it redistributes the source of all costs, for the individual concerned, this has costs which may affect their ability to integrate.[65] Sharing people might therefore increase the total cost of protection in individual states. The EU experience in respect of the Dublin Convention illustrates how dysfunctional a people-sharing regime can be without true equalisation of norms.

Who manages?

The central role of UNHCR in the managed refugee protection process points towards the issue of governance. Which institution, if a managed approach is to be adopted, should be responsible for managing the process? Migration in general, and climate change related migration even more so, is a truly cross-cutting phenomenon.[66] It raises difficult policy issues related to not only migration but also development, the environment and humanitarian assistance. Unsurprisingly, a range of UN and other international organisations have been involved in workshops, conferences and preparing reports in climate change informed by these particular policy perspectives. The movement towards what has become known as the 'cluster approach' to humanitarian responses to natural disasters recognises that the needs of persons affected by natural disasters are best met by a coordinated and predicable response in which agencies work together with each bringing a particular institutional expertise to bear on a particular aspect of the problem.[67] Recognising this, the Inter-Agency Standing Committee has allocated a particular agency or agencies to be the global cluster leader and focal point for coordination of humanitarian efforts. It may be that climate change related migration can be accommodated within this framework. Yet climate change related migration introduces an additional layer of complexity. In terms of design of the system, allocating, for example, a point in time at which persons must move across international borders to avoid the worst effects of slow-onset events involves a significant scientific component, the expertise for which is channelled largely through the IPCC. This raises the issue of whether a new agency, under the auspices of the UNFCCC but drawing on the expertise of the existing institutional actors, should be established.

Conclusion

Migration and displacement resulting from climate change has the potential to become one of the most significant humanitarian challenges of the 21st century. Responding to it will comprise an equally significant political challenge in a highly fractured and divided world. Yet there seems little doubt that some form of international cooperation will be required if the challenge is to be successfully met. To be meaningful, there will need to be some agreed framework in place by which the various states assume known but not necessarily identical obligations in particular circumstances of displacement and migration. The Refugee Convention, as the cornerstone of a functioning protection and resettlement regime, provides a valuable frame of reference for policy makers thinking about the design of an international regime to respond to climate change related migration. It has had to respond to a variety of involuntary population movements on both large and small scales, in all corners of the world. While it is far from a perfect regime, its successes and failures provide a useful window through which to identify and reflect upon critical issues and thereby enhance the prospects for reaching international cooperation based on an agreed framework.

Bruce Burson 'Securing Meaningful International Agreement on Climate Change Related Displacement and Migration: The Refugee Convention as a Window on International Burden (Responsibility) – Sharing in an Involuntary Movement Context 2010

Notes

1   This chapter represents an amalgam of papers presented at two conferences – Australia New Zealand Society of International Law, 17th Annual Conference Wellington, NZ July 2009 and Waikato University Population Studies Centre, Pathways Conference Wellington, NZ November 2009.
2   IPCC (1990:4).
3   I acknowledge that climate change will not, in itself, lead to population displacement and migration but rather will do so though related events and processes that have a negative effect on local ecosystems and ecosystem services.
4   Brown (2008:21), Warner et al (2009:v).
5   Brown (2008: 11), citing Myers (1993).
6   See: http://www.iom.int/jahia/Jahia/about-migration/facts-and-figures/global-estimates-and-trends.
7   Boncour and Burson (2009).
8   Hugo (2009) Warner et al (2009).
9   Defined as "about an 8 out of 10 chance" – see IPCC (2007:21).
10 IPCC 2007: 13.

11   IPCC 2007: 13 Looking at longer term horizons, toward the end of the 21st century projected
      sea level rise will affect low lying coastal areas with large populations and it is likely that wheat
      production will disappear from Africa by 2080 (Boko et al 2007, 448).

12   (2007: 13).

13   Boncour and Burson (2009:4).

14   See Preambular Paragraph 4 of the 1951 Refugee Convention. At the outset I acknowledge Noll's
      (2003: 238) argument that burden sharing is a negative term and that 'responsibility' or 'risk' sharing
      may be more suitable. However, for purposes of this paper, the term burden sharing will be used
      because this is the terminology of the Refugee Convention.

15   See, for Example Article II(4) of the 1969 Organisation of African Unity (OAU) Convention
      Governing the Specific Aspects of Refugees in Africa.

16   Burson (2010). The potential for issues of environmental degradation to interact with issues of
      poverty, inequality and discriminatory modes of government means it is an oversimplification to
      argue that there is an insurmountable barrier between environmentally induced migration and
      recognition as a Convention refugee, many of those displaced for environmental reasons will simply
      be unable to come within the Convention definition. Moreover, the term 'environmental refugee'
      is, at best, counter-productive in that it may lead refugee-receiving states to deny current levels of
      protection afforded to persons existing and future Convention refugees. See here, Black (2001: 11-
      12); Castles (2002:5); UNHCR (2008A).

17   I acknowledge that the term displacement may be more appropriate in cases where movement takes
      places for sheer survival in cases of sudden onset natural disasters.

18   Plainly, in the context of climate change, environmental factors cannot be separated from the non-
      environmental factors such as industrial, economic and social policy choices over the last 250 years
      which have caused the build up of the current level of atmospheric greenhouse gas emissions.

19   See IOM (2004:19) Displacement: a forced removal of a person from his/her home or country, often
      due to armed conflict or natural disasters. See also Kliot (2004: 76).

20   Paragraph 8, Statute of the Office of the United Nations High Commissioner For Refugees, Annex to
      UNGA Res. 428(V) , 14 December 1950.

21   UNHCR (2004: I/7).

22   Overwhelmingly the contributions to the managed process came from the global north. Of the
      US$1,116, 884, 686 received in programme contributions, some US$932,125,884 originated in
      states of the global north (including here the European Commission) (UNHCR 2008c). The global
      programmes aimed at promotion of refugee law and resettlement project comprised only a small
      faction of these overall costs (UNHCR 2008c: 92).

23   In 2007, some 75,300 persons were admitted by 14 resettlement states. Of these, the overwhelming
      majority were taken by developed states such as the USA (48,300), Canada (11,200), Australia
      (9,600), Sweden (1,800), Norway (1,100) and New Zealand (740) (UNHCR 2008b:11).

24   This affects no only the states of the global south but also increasingly some EU states, particularly
      Greece which has been unable to cope with a substantial rise in numbers of persons entering its
      island territories illegally – see UNHCR Press Release Greece's infrastructure struggles to cope with
      mixed migration flows (19 January 2009). For a sobering overview of increases in the human toll, see
      UNHCR Briefing Notes Boat Peoples Arrivals in the Mediterranean, Gulf of Aden already top 2007
      (4 November 2008).

25   Although what constitutes a safe neighbouring country in this context may be different to the political
      context of classic refugee flows. Environmental events, unlike the repressive policies of government,
      are not contained to recognise international boundaries. Rather, the same climatic event might affect
      a swathe of territory across a number of states.

26   In 2007, over 540,000 spontaneous claims were lodged with individual States Parties to the Convention
      (UNHCR 2008b: 13). In contrast, during 2007, UNHCR offices globally received nearly 80,000 new
      applications for refugee status. In 2008 the total number of persons lodging individual applications
      for refugee status rose to 839, 000 of which UNHCR received 9% – see UNHCR (2009a).

27   Acharya and Dewit (1997: 117).

28   Betts (2005: 5).
29   Noll (2003: 241); Suhrke (1998: 398).
30   Betts (2005: 44); Hathaway (1997: xviii).
31   Betts (2005, 2009).
32   2009 at 175–176.
33   UNDP (2007: 185–186). See also Barnett and Adger (2007).
34   Brown et al. (2007: 1154).
35   Suhrke (1998:400–402).
36   For a detailed examination of hotspots see Thow and de Blois (2008). See also Raleigh et al (2008: 13).
37   Warner et al (2009: 1). See also Hugo (2009).
38   Witness here the debate surrounding the 'Responsibility to Protect' (R2P) doctrine.
39   The phraseology is from Betts (2005: 4).
40   Betts (2005: 23–30).
41   In 2005, SPREP, with financial assistance from the Canadian Government, relocated some 100 families to higher ground after frequent flooding and erosion made site of original settlement uninhabitable – see SPREP Press Release: Pacific villagers become first climate change refugees (7 December 2005) http://www.sprep.org/article/news_print.asp?id=247m.
42   Paragraph 1c(v).
43   Copy on file with author.
44   Noll (2003: 247).
45   Schubert et al (2008: 206).
46   See Biermann and Boas (2008).
47   Suhrke (1998: 397).
48   See (Cuellar et al 1991: 487).
49   See speech by Macrine Mayanja, Head of Humanitarian Affairs, Refugees and Internally Displaced Persons Division of the Political Affairs department of the AU to Conference of Ten Years of Guiding Principles on Internal Displacement: Achievements and Future Challenges, Oslo, 16–17 October 2008 – see http://www.internal-displacement.org/8025708F004BE3B1/(httpInfoFiles)/2E5A75E E5FE1678DC12574F300339FD7/$file/GP10_speech_Macrine%20Mayanja.pdf.
50   Noll (2003: 241).
51   (1998: 413). Citing the resettlement of European refugees after World War Two and the resettlement of Vietnamese people in the 1970s. She argues that these arrangements worked because the participating states shared a sense of values with, and obligations towards, the displaced (the instrumental-communitarian model) or where a major actor was successful in pressuring other states into collective action (the hegemonic scheme). This is contrasted with the situation that prevailed in the 1990s in relation to the displacement associated with the disintegration of the former Yugoslavia where a more restricted approach emerged because of an absence of these factors.
52   Acharya and Dewitt (1997: 132) Hans and Suhrke (1997: 108). Each was writing as part of the 'reformulation project' led by Hathaway which was conceived as a mechanism for increasing the Convention's relevance to contemporary responses to displacement. Williams (2008: 518) explicitly argues for a regional approach in the context of forced climate migration.
53   2009 at 94, 142. In the case of the refugees in Latin America, the agreement was part of a 'wider regional peace process and its related post conflict reconstruction and development initiative' under the auspices of the Esquipulas II declaration which expressly referred to displaced persons and the Special Programme for Economic Cooperation for Central America.
54   Article 3(1).
55   Rajamani (2001:130) aptly describes CBDR as 'the ethical anchor' of the developmental process behind the current global climate change architecture.
56   1997: xxiv.
57   Boncour and Burson (2009).
58   2009b: 2.

59   The position now is different following the introduction of the Council Directive on the Minimum standards for the qualification and status of third-country nationals or stateless persons as refugees or as persons who otherwise need international protection and the content of the protection granted Council Directive 2004/83/EC (29 April 2004) (The Qualification Directive). Binding on all EU member states it provides at Article 6 that "actors of persecution" can include both state and non-state entities.

60   See Adan and Aitsegeur v Secretary of State for the Home Department [2001] 2 WLR 143 HL).

61   See UNHCR (2007).

62   Castles and Miller (2003: 27).

63   1998: 400.

64   Noll (2003: 243).

65   For example, presence of family in a host state is a classic pull factor and one that determines where many refugees lodge their application for asylum. If these links are absent in the place to which they are subsequently relocated under any people-sharing scheme, this will inevitably impact upon their integration into the society of the host state.

66   Morton et al. (2008: 5).

67   See Inter-Agency Standing Committee (IASC) Guidance Note on Using the Cluster Approach to Strengthen Humanitarian Response (24 November 2006). The note records 9 different sectors or areas of activity assigned an institution(s) as follows:

   1. Nutrition, UNICEF
   2. Health, WHO
   3. Water/Sanitation, UNICEF
   4. Emergency Shelter: IDPs (from conflict), UNHCR
       Disaster situations IFRC (Convener)
   5. Camp Coordination/Management: IDPs (from conflict), UNHCR
       Disaster situations, IOM
   6. Protection: IDPs (from conflict), UNHCR
       Disasters/civilians affected by conflict (other than IDPs), UNHCR/OHCHR/UNICEF
   7. Early Recovery, UNDP.

[References omitted]

Chapter 12

Aurelie Lopez

THE PROTECTION OF ENVIRONMENTALLY DISPLACED PERSONS IN INTERNATIONAL LAW

[...]

III. The Maelstrom Stemming from the Notion of "Environmental Refugee"

According to the Refugee Convention,

> the term "refugee" shall apply to any person who . . . owing to well-founded fear of being persecuted for reasons of race, religion, nationality, membership of a particular social group or political opinion, is outside the country of his nationality and is unable or, owing to such fear, is unwilling to avail himself of the protection of that country; or who, not having a nationality and being outside the country of his former habitual residence as a result of such events, is unable or, owing to such fear, is unwilling to return to it.[75]

This Article discusses each of the elements of the refugee definition with respect to environmentally displaced persons. An analysis of the present legal definition of "refugee" highlights the legal hurdles presented when attempting to adapt the situation of environmentally displaced persons to fit the refugee definition.

A. The Blunt, Well-Founded Fear of Persecution of Environmentally Displaced Persons

Most authors concur that in the vast majority of situations neither persecution nor concerted state action is involved, so that "unlike victims of persecution, those fleeing environmental disaster can, in most cases, turn to their own government for help and support."[76] The Office of the United Nations High Commissioner on Refugees (UNHCR) makes this distinction by stating that:

> [r]efugees are distinguished by the fact that they lack the protection of their state and therefore look to the international community to provide them with security. Environmentally displaced people, on the other hand, can usually count upon the protection

of their State, even if it is limited in its capacity to provide them with emergency relief or longer-term reconstruction assistance.[77]

Nevertheless, Christopher Kozoll rightly points out that "[n]othing in either international or national standards explicitly disavows the idea that one may be persecuted through environmental harm."[78] Therefore, at least some of the world's environmentally displaced persons probably already fall within the settled meanings of refugee. Such people are entitled to recognition of their status as refugees and should be able to claim any international protections available to recognized refugees.

The Handbook on Procedures and Criteria for Determining Refugee Status (Handbook), which is undoubtedly the most authoritative interpretation of the 1951 Refugee Convention and the 1967 Refugee Protocol, affirms that "[t]here is no universally accepted definition of 'persecution,' and various attempts to formulate such a definition have [been] met with little success."[79] As B.S. Chimni further explains, "[i]t is widely accepted that the drafters of the Convention deliberately left the meaning of 'persecution' undefined as it was an impossible task to enumerate in advance the myriad forms it might assume."[80] Certainly, therefore, environmental harm may be considered as persecutory according to the provisions of the Handbook.[81]

Persecution is "an act of government against individuals."[82] In other words, in order to be recognized as a refugee because of environmental impairment, one must establish that the adverse consequences on the environment are due to governmental actions. Environmental disruption is not, however, always caused by human activities. In such circumstances, it is impossible to affirm that people are fleeing persecution. In this regard, the Handbook explicitly rules out victims of famine or natural disaster, unless they also have a well-founded fear of persecution for one of the reasons stated.[83] In addition, the Handbook does not prevent victims of human-made environmental impairment from receiving refugee status. To the contrary, "the fact that the direct harm inflicted is on the environment rather than on the individual should not change the fact that such harm is persecution."[84] Indeed, a serious environmental crisis will certainly threaten the health, life, or freedom of persons and accordingly amount to persecution.[85] Christopher Kozoll affirms therefore that "[e]nvironmental harm is as capable of being a means of persecution as any other form of harm."[86]

Jessica Cooper emphasizes that governments are involved in most cases of environmental disaster[87] and argues that, whether the environment is depredated because of negligent decision making or because of decisions intentionally sacrificing the environment of a region for the benefit of national economic interests, government-induced environmental degradation is a form of persecution.[88] She concludes that "[w]ith governments playing so pertinent a role in the occurrence of environmental crises, refugees seeking refuge from the resulting environmental degradation are effectively seeking refuge from their governments as well."[89] Nonetheless, this argument is too simplistic since the refugee definition requires further qualitative elements for the governmental action or inaction to be persecutory under the terms of the 1951 Refugee Convention.[90]

First, as Christopher Kozoll stresses, "[t]o establish that flight based on environmental harm is indeed flight based on well-founded fear of persecution, an individual will have to show more than a generalised environmental degradation."[91] Thus, an individual has to bring evidence of a severe environmental harm that either threatens his life or freedom, or is of such nature or extent that it would reasonably induce fear. In addition, for the persecution to be individualized, the environmental harm must affect the individual in his capacity as a member of a protected category to a greater degree than other persons.[92]

Secondly, in order to establish persecution, a person has to demonstrate both the persecutory impact and persecutory intent on the part of the governmental entity. The nature of the intent required is more than volition or awareness of consequences. Thus, the governmental entity must have been negligent or inactive "because of," and not merely "in spite of" its adverse effects upon an identifiable group.[93] As a consequence, Jessica Cooper's argument of state persecution in cases of environmental degradation, as illustrated by examples of governmental negligence or inaction in the African Sahel and Chernobyl, is unconvincing.[94]

James Hathaway states, however, that "persecution may be defined as the sustained or systemic violation of basic human rights demonstrative of a failure of State protection" and further affirms that

[a] well-founded fear of persecution exists when one reasonably anticipates that remaining in the country may result in a form of serious harm which the government cannot or will not prevent, including either "specific hostile acts or an accumulation of adverse circumstances such as discrimination existing in an atmosphere of insecurity and fear."[95]

Similarly, Jeanhee Hong contends that the substantial element in the refugee definition is the absence of state protection.[96] She further argues that the language of the Convention and the Protocol suggests that refugeehood can result even in the absence of persecution, for instance from circumstances that simply render a government unable to extend effective protection.[97]

Her argument is based on the Handbook's recognition of "grave circumstances" rendering a government's protection "ineffective," which suggests that the refugee definition should be revised to consider environmentally-hazardous circumstances.[98] Moreover, she argues that the Handbook's statement that "whether unable or unwilling to avail himself of the protection of his Government, a refugee is always a person who does not enjoy such protection"[99] emphasizes the significance of the absence of state protection in the refugee definition, regardless of its cause.[100] Nevertheless, she admits that reinterpreting or revising the refugee definition to include all environmentally displaced persons who lack the protection of their States would open the door to a flood of refugees far beyond what the international community is able to manage. Such an interpretation, therefore, would have to be limited by specific requirements, such as the occurrence of certain threshold levels of environmental destruction in the country of origin, and the existence of specific circumstances rendering the applicants unable to avail themselves of their government's protection within a designated period of time.[101]

The 1951 Refugee Convention considers the lack of state protection as persecutory if, and only if, the persecutory intent on the part of the governmental entity may be established. According to the provisions of the Handbook, corroborated by the Refugee Convention, persecution may take the form of environmental damage.[102] Persecution may be inflicted directly by the governmental entity or indirectly by the lack of protection from these governmental entities. In any case, the harm must be inflicted on cognizable groups with the particular intent to harm these groups because of a valuation of the lives and cultures of the people harmed.[103] As will be explained in the following subpart, it is paramount to establish that the persecution is based on one of the five grounds specified in the refugee definition, namely "on account of race, religion, nationality, membership in a particular social group, or political opinion."[104]

B. The Blurred Grounds of the Persecution Suffered by Environmentally Displaced Persons

At first glance, natural disasters, which have widespread impacts and affect people indiscriminately without regard to race, religion, nationality, political opinion, or membership in particular social groups, do not give rise to refugee status. Similarly, in most cases, human-caused environmental degradation is not carried out in order to hurt people for any of the reasons enumerated in the refugee definition. Yet, Christopher Kozoll argues that victims of natural disasters and human-caused environmental degradation meet the traditional definition of refugee, as enshrined in international law, in at least two circumstances. First, when "a government systematically imposes the risks and burdens of decisions impacting environmental quality on members of a particular race, religion, nationality, social group or political opinion on account of one or more of these protected factors," and second, "where the relevant authority refuses to mitigate or mitigates inadequately environmental disasters, whether of human origin or not, and in so doing 'targets' a group based on one of the listed factors."[105] More precisely, Jessica Cooper posits that these victims of environmental damage, identified in the environmental justice literature as persons discriminated against and adversely affected by environmental decisions, belong to a social group of persons who are politically powerless to protect their environment.[106]

The fifth category, "membership in a particular social group," was added to the refugee definition without discussion or dissent in order to fill gaps left by the four more specific grounds of persecution.[107] It is therefore the most flexible ground of persecution. T. Alexander Aleinikoff observes that "the history of the Convention provides no support for a narrow reading of the grounds of persecution. Rather, it displays an intent to write a definition of refugee sufficiently broad to cover existing victims of persecution."[108]

Though subject to extensive interpretation, a social group must nevertheless be characterized by different elements. In particular, it must be defined independently from the persecution at issue. For this reason, a social group composed of persons lacking political power to protect its environment seems to be defined by nothing more than the harm sought to be remedied. To the contrary, Jessica Cooper argues that these persons form a group characterized by a "common experience" which is a component required to form a "social group" under the refugee definition: "[u]nder international refugee law, 'social group' has been interpreted to mean 'a recognisable or cognisable group within . . . society that shares some . . . experience in common.'"[109] Indeed, to reassert the wording of the Handbook, these persons are "of similar background, habits or social status" and for this reason are subject to environmental discrimination.[110]

Dana Zartner Falstrom, however, argues that "political powerlessness is not an immutable characteristic that will make a person or group of persons members of a particular social group."[111]Unless environmental victims are also associated by other factors, such as religion or culture, they do not constitute a social group. Eventually, some authors observe not only that political powerlessness is not a common characteristic to all victims of environmental damage,[112] but also that the argument of political powerlessness as a reason for persecution is itself questionable.

The theory that people suffer the risks and burdens of environmental degradation for reasons of race, religion, nationality, social group, or political opinion, which are the elements usually underlying political powerlessness, is controversial. Studies on environmental justice have focused on the socio-economic status of the victims of environmental degradation and the related issue of equity in burden-sharing of pollution.[113] Although a large number of authors affirm that some persons experience violence through environmental discrimination,[114] Roliff Purrington and Michael Wynne acknowledged that, at present, social science research on the topic of environmental racism is new and immature.[115] Nothing approaching a consensus has been reached in the literature in regard to the standards or methodologies that should be employed to make a judgement about a claim of environmental racism. Thus, to argue that research on environmental racism has reached a conclusion is a serious mischaracterization. An appropriated methodology to objectively assess, measure, or even define environmental racism has yet to be established.[116]

Moreover, it is noteworthy that although the burden of environmentally-polluting facilities and practices may indeed fall most heavily on lower income and minority groups, the use of the concept of environmental racism as a basis for legal relief presents several procedural and evidentiary problems. Environmental racism implies that industrial or governmental actors have intentionally placed polluting facilities in poor and minority neighbourhoods with the particular intent of imposing the adverse effects upon these persons because of their "race" or "ethnicity" or "membership in a particular social group," while in fact these decisions may have been made for economic reasons.[117] Therefore, Roliff Purrington and Michael Wynne conclude that "[t]o the extent minorities are disproportionately affected, the problem may be that some minorities unfairly find themselves at the lower end of the economic spectrum, suggesting a deeper societal problem of which the environmental pollution component is but a symptom."[118]

Though difficult, demonstrating that environmentally displaced persons meet the criteria of the refugee definition is not impossible. In particular, victims of environmental degradation taking the form of an "environmental cleansing," which can be defined as the "deliberate manipulation and misuse of the environment so as to subordinate groups based on characteristics such as race, ethnicity, nationality, religion and so forth,"[119] may be granted refugee status.

The most illustrative example of persecution through environmental damage for reasons of one or more of the five grounds enumerated in the refugee definition is the drainage of the marshes in southern Iraq. For millennia, the marshes covered an area of about 20,000 square kilometres in southern Iraq, where the Euphrates River meets the Tigris River.[120] They were inhabited by various tribes of Shi'a Muslims, collectively known as the Ma'dan or Marsh Arabs, who had been living in the marshes for over 5,000 years and had based their livelihood on the marshes through fishing, hunting, manufacturing handicraft from reeds and cane, buffalo breeding, and agriculture.[121]

Scholars posit that the Iraqi government has committed genocide of the Marsh Arabs, "deliberately inflicting on the group conditions of life calculated to bring about its physical destruction in whole or in part."[122] Systematic draining of the marshes was an environmental attack that contributed to the genocide. Between 1991 and 1997, Iraq developed a water diversion project, which involved constructing giant canals to "dry the land and expel the inhabitants."[123] The Iraqi government seriously affected the course of the rivers, preventing water from reaching two-thirds of the marshlands.[124] As a result, "[b]y 2000, the Iraqi portion of the Fertile Crescent was dry and the surrounding land crusted with salt."[125] The subsequent loss of livelihood and forced deportations resulted in the death of thousands of Marsh Arabs.[126]

The specific intent to destroy the Marsh Arabs may be established by, among other circumstantial evidence, the "Plan of Action for the Marshes" approved by Saddam Hussein.[127] "Though Iraqi officials may argue that the drainage served developmental purposes, researchers have found that 'many of the canals and other engineering structures serve no agricultural, economic, or developmental purpose.'"[128]

Eventually, the Iraqi popular perception of the Marsh Arabs came to demonstrate the intent to destroy the group because of their religious and ethnic lines. As Tara Weinstein explains:

> [t]he former Iraqi government targeted the Marsh Arabs not only as Shi'a but also as a specific ethnic group connected with Iran. Marsh Arabs were singled out in the media "among the Shi'a for their alleged poverty, backwardness and immorality; they are disparagingly described as 'monkey-faced' people who are not 'real Iraqis,' but rather the descendants of black slaves."[129]

Armed conflicts are notorious for using the environment as a tool of war. However, customary principles of international law prohibit acts having adverse effects on the environment when they cause unnecessary damage or excessive destruction.[130] Environmental damage has not, however, been the main concern of international tribunals and courts. Asylum claims are more likely to focus on other issues rather than the environmental aspects of the conflict.[131] Thus, one may indeed suffer environmental persecution for any of the reasons enumerated in the 1951 Refugee Convention. A survey of the legal literature highlights, however, the limited number of cases illustrating this hypothesis, effectively excluding most environmentally displaced persons from refugee protection.[132]

C. The Problematic Dichotomy among Environmentally Displaced Persons

The 1951 Refugee Convention requires that a refugee be outside his country of origin, and accordingly does not encompass situations of internal displacement.[133] In international law, there is a dichotomy between the protection afforded to "environmental refugees" and persons internally-displaced for environmental reasons. Yet reference to the term "persons internally-displaced for environmental reasons" is in itself problematic. Indeed, there is no authoritative definition of "internally-displaced persons" in international law. Since 1975, however, displaced persons have been included in the mandate of UNHCR, which considers "internally-displaced persons" as any person or group of persons who, if they had breached an international border, would be refugees.[134] The definition appears to exclude most of the persons internally-displaced for environmental reasons since environmentally displaced persons are often escaping environmental pressures rather than the enumerated persecutions.[135]

In 1998, the UN Secretary-General's representative for displaced persons, Francis M. Deng, proposed the following definition of internally-displaced persons:

> internally displaced persons are persons or groups of persons who have been forced or obliged to flee or to leave their homes or places of habitual residence, in particular as a result of or in order to avoid the effects of armed conflict, situations of generalized violence, violations of human rights or natural or human-made disasters, and who have not crossed an internationally recognized State border.[136]

This definition would encompass persons internally-displaced for environmental reasons, but it is not legally binding. It serves only as a base for discussion on the actual content of international law's protection of internally-displaced persons.[137]

In any case, there is little benefit in according people affected by environmental degradation the status of displaced persons since it is only a descriptive term, not a status that confers obligations on states.[138] The Guiding Principles on Internal Displacement are not in any way binding on states, nor are they part of customary law.[139]

D. Conclusions Concerning the Environmental Refugee

Although the 1951 Refugee Convention cannot be reasonably interpreted to include environmentally displaced persons, and some authors affirm that it was not drafted with those persons in mind,[140] nothing actually prevents a country from granting refugee status to a person persecuted through environmental damage for one of the reasons enumerated in the 1951 Refugee Convention. Environmental damage or degradation is not recognized by Article 1 of the 1951 Refugee Convention as a valid ground for seeking asylum, yet it may certainly be a tool of persecution.

Thus the reasons to grant refugee status in circumstances of environmental degradation may be grounded in the letter of the 1951 Refugee Convention, namely for reasons of race, religion, nationality, membership of a particular social group, or political opinion. In other words, the political or social reasons underlying the environmental degradation appear to legitimate refugee protection. Nonetheless, the actual state of refugee law presents some drawbacks. It is noteworthy that people displaced forcibly because of environmental impairment do not receive equal consideration and protection. Depending on the nature of the incident provoking environmental degradation, the migrants appear to be more or less likely to fit the international legal definition of refugee. For instance, Gregory McCue observes that if the environmental destruction is caused by war, then migrants are more likely to be recognized as refugees and, accordingly, to receive the attendant protection.[141] In contrast, he asserts that people for whom environmental degradation is a primary cause of forced migration do not deserve refugee protection.[142]

Thus, though environmentally displaced persons may be granted refugee status, the term "environmental refugees" appears to be a legal misnomer.[143] The term is legally meaningless and confusing since it does not refer to a consistent category of displaced persons. Environmental degradation is not in itself a ground of persecution; rather it is one tool of persecution. For this reason, UNHCR, the principal international body entrusted with the task of ensuring the proper treatment of refugees and finding enduring solutions for their plight,[144] as well as the International Organization for Migration (IOM) and the Refugee Policy Group, have all opted not to use the term "environmental refugee," and instead to use the term "environmentally displaced persons."[145] Nevertheless, Gregory McCue reports that "[i]n environmentally-driven, transboundary migrations where countries or regional organizations have extended their refugee definition to people fleeing 'events seriously disturbing the public order,'[146] the UNHCR can and does administer this broader statutory language."[147]

In conclusion, a substantial number of people have fled or are likely to flee across national borders for environmental circumstances that fall outside of the scope of the 1951 Refugee Convention, and are not accordingly afforded any legal protection. Likewise, the refugee protection excludes environmentally displaced persons who are still within their country of origin. Legal scholars have therefore made propositions in relation to the protection of environmentally displaced persons that will be discussed in the next Part.

IV. Propositions and Arguments in Relation to the Protection of Environmentally Displaced Persons

A. The Conundrums of Revising the Traditional Refugee Definition to Encompass Environmentally Displaced Persons

In response to the problems of revising the traditional definition of refugee to encompass the environmentally displaced person, several propositions have emerged. At the regional level, the Organization of African Unity (OAU) adopted the Convention Governing the Specific Aspects of Refugee Problems in Africa in 1969.[148] Reasserting the definition of the 1951 Refugee Convention, it further states:

> the term refugee shall also apply to every person who, owing to external aggression, occupation, foreign domination or events seriously disturbing the public order in either part or the whole of his country of origin or nationality, is compelled to leave his place of habitual residence in order to seek refuge in another place outside his country of origin or nationality.[149]

The OAU Convention was not primarily designed with environmentally displaced persons in mind, but unequivocally includes victims of environmental crises since such events seriously disturb the public order. Accordingly, these "environmental refugees" are included in UNHCR's mandate.[150]

Nevertheless, persons meeting only the additional definitional requirements, but not the requirements congruent to the 1951 Refugee Convention, are only entitled to temporary protection. These refugees are therefore protected from forcibly being returned to their home states but are not allowed to resettle in the receiving state.[151] At the time the 1951 Refugee Convention was elaborated, temporary protection was justified by the fact that the 1951 Refugee Convention was primarily aimed at assisting the mass influx of people displaced by wars of independence throughout Africa, who would ultimately return to their home states.[152] In contrast, environmentally displaced persons often leave behind an environment that no longer sustains a living and may remain devastated for a long period of time. Consequently, a mere provisional protection improves the situation of environmental refugees in Africa. It does not, however, provide an adequate remedy to their vulnerable situation.

Similarly, in 1984 the Organization of American States (OAS) adopted the Cartagena Declaration on Refugees, which provides in its article 3 that in addition to containing the elements of the 1951 Convention . . . and the 1967 Protocol . . . [the definition] includes among refugees persons who have fled their country because their lives, safety or freedom have been threatened by generalized violence, foreign aggression, internal conflicts, massive violations of human rights or other circumstances which have seriously disturbed the public order.[153]

Like the OAU Convention, the Cartagena Declaration on Refugees was not primarily designed with environmentally displaced persons in mind, but arguably includes victims of environmental crisis since such events are seriously disturbing the public order.[154] The International Conference on Central American Refugees (CIREFCA) report interpreting the Cartagena Declaration distinguishes, however, between "victims of natural disasters" and other events "seriously disturbing the public order," and observes that the former do not qualify as refugees.[155] In contrast, human-made events, such as accidents, would qualify since they are not "natural" disasters. Therefore, determining the causal role of human activities in slow-onset disruptions, although difficult, remains critical to qualify as a refugee under the Cartagena Declaration. Yet, the instrument is not legally binding on states that are parties to the OAS, so its practical effects are doubtful.

B. Proposed Definition of "Environmental Refugee" at the International Level

Contemporary causes of mass migration have undoubtedly changed since the formulation of the 1951 Refugee Convention. Environmental factors now play a critical role in migration phenomena. For this reason, scholars advocate a constructive and innovative interpretation of the 1951 Refugee Convention. In particular, James Hathaway highlights that "there has, for far too long, been an anachronistic fixation with literalism, with insufficient attention paid to the duty to read text in line with the context, object and purpose of a treaty."[156] Thus, there has been a failure to adequately develop the potential for treaty law to play a genuinely transformative role in the international system. As Hathaway states, this approach "misreads the authentic rules of treaty interpretation, and bespeaks a lack of creativity within the bounds expressly sanctioned by States."[157]

In 1985, Essam El-Hinnawi used the term "environmental refugees" to describe "those people who have been forced to leave their traditional habitat, temporarily or permanently, because of a marked environmental disruption (natural and/or triggered by people) that jeopardized their existence and/or seriously affected the quality of their life."[158] By this definition "environmental disruption" means any physical, chemical, or biological changes in the ecosystem (or the resource base) that render it, temporarily or permanently, unsuitable to support human life.[159]

Norman Myers further described the political, social, and economic components of the term "environmental refugees" in a comprehensive study on the issue.[160] Eventually, Jessica

Cooper suggested expressing the proposed definition in legal terms, advocating an expansion of the traditional refugee definition in the following terms:

> any person who owing (1) to well-founded fear of being persecuted for reasons of race, religion, nationality, membership of a particular social group, or political opinion, or (2) to degraded environmental conditions threatening his life, health, means of subsistence, or use of natural resources, is outside the country of his nationality and is unable or, owing to such fear, is unwilling to avail himself of the protection of that country.[161]

Nevertheless, a definition that has survived without modifications for more than five decades is not realistically likely to be modified, especially when considering the attitudes and priorities of governments. Moreover, some authors challenge an expansion of the refugee definition encompassing environmental degradation as a valid ground for seeking asylum. Since in most cases environmental change cannot be meaningfully separated from political, economic, and social changes, it appears meaningless to recognize environmental refugees.[162] Eventually, some authors posit that expanding the refugee definition presents limited assets for environmentally displaced persons since the refugee definition requires the crossing of internationally recognized borders and therefore excludes most environmentally displaced persons from international protection.[163]

Therefore, there appears to be a certain impetus in citing refugees as a reason for protecting the environment, and in citing the environment as a reason for protecting refugees. Yet, it is not practical to advocate an expansion of the refugee definition to include environmental degradation as a ground to seek asylum. What is of more practical import is to understand the relationship between people and the environment as part of an analysis of the causes and consequences of movement, rather than as the sole cause or consequence.[164] This is the focal point of the alternative propositions made in order to deal specifically with the issue of environmentally displaced persons, as discussed in the following subpart.

C. The Limits of Applying the Complementary Forms of International Protection to Environmentally Displaced Persons

Although the refugee definition may vary according to the different legal systems,[165] refugee status and its related rights remain circumscribed to a purposefully narrow proportion of the persons in need of international protection. This Article has highlighted the major impediments to recognizing refugee status of environmentally displaced persons. Yet, UNHCR has acknowledged the particular vulnerability of environmentally displaced persons and, accordingly, has emphasized the need to focus international attention on the increasing number of persons affected by environmental crisis by setting up an international protection regime.[166] Until now, states have recognized (on an ad hoc basis) complementary forms of protection to categories of persons, who do not fit neatly in the refugee definition, but nonetheless deserve international protection.[167]

These forms of protection are in all respects based on human rights treaties or on more general humanitarian principles.[168] The regimes of complementary protection are not, however, regulated by international law. To the contrary, they are subject to national legislation, which specifies the eligibility criteria as well as the rights and entitlements of complementary protection beneficiaries. Such legislation usually provides a lower level of protection and depends on the political and economic interests of the host country.[169] It is noteworthy to emphasise that complementary protection broadens the basis for international protection, particularly limited under the 1951 Refugee Convention and the 1967 Refugee Protocol. Therefore, such a source of alternative protection should not be considered to be of less importance. Complementary protection has traditionally been discretionary and based on compassion, although frameworks predetermining the criteria for complementary protection, in a similar fashion to the 1951 Refugee Convention, have flourished in recent years.[170]

This Article reviews the different means of protection elaborated in the European Union and the United States. Describing the purpose as well as the main point of these instruments' legal provisions, this Article analyzes whether environmentally displaced persons may be eligible for complementary protection in any, if not all, of these regimes. Later, it discusses the need to further elaborate new instruments, in particular an international instrument dealing specifically with the issue of environmentally displaced persons.

1. Complementary Protection in Europe: The Forgotten Category of Environmentally Displaced Persons

At the European level, several instruments on complementary protection have been adopted. Some examples are the Directive on Minimum Standards for Giving Temporary Protection in the Event of a Mass Influx of Displaced Persons and on Measures Promoting a Balance of Efforts Between Member States in Receiving Such Persons and Bearing the Consequences Thereof (Directive on Temporary Protection),[171] and the Directive on Minimum Standards for the Qualification and Status of Third Country Nationals or Stateless Persons as Refugees or as Persons Who Otherwise Need International Protection and the Content of the Protection Granted (Directive on Subsidiary Protection).[172] This Article analyzes the conundrum related to the present regime of complementary protection in Europe, and further attempts to demonstrate that, as a matter of course, the instruments do not provide a satisfactory outcome for environmentally displaced persons. Environmentally displaced persons may qualify for temporary protection, yet it is obviously not an appropriate protection considering the peculiarity of the human rights violations. Conversely, environmentally displaced persons have not been considered at the drafting nor at the adoption of the Directive on Subsidiary Protection, and may not be eligible for protection under its provisions.

A. ELIGIBILITY CRITERIA FOR INTERNATIONAL PROTECTION UNDER THE DIRECTIVE ON TEMPORARY PROTECTION

Under the terms of the Directive on Temporary Protection, temporary protection is defined as:

> a procedure of exceptional character to provide, in the event of a mass influx or imminent mass influx of displaced persons from third countries who are unable to return to their country of origin, immediate and temporary protection to such persons, in particular if there is also a risk that the asylum system will be unable to process this influx without adverse effects for its efficient operation, in the interests of the persons concerned and other persons requesting protection.[173]

The provisions further read:

> "displaced persons" means third-country nationals or stateless persons who have had to leave their country or region of origin, or have been evacuated, in particular in response to an appeal by international organisations, and are unable to return in safe and durable conditions because of the situation prevailing in that country, who may fall within the scope of Article 1A of the Geneva Convention or other international or national instruments giving international protection, in particular: (i) persons who have fled areas of armed conflict or endemic violence; (ii) persons at serious risk of, or who have been the victims of, systematic or generalised violations of their human rights.[174]

The use of the words "in particular" implies that the drafters had in mind situations that may lead to mass influx. However, other situations that prevent a person from returning in safe and durable conditions may qualify for temporary protection. The provisions are therefore flexible enough to address new situations arising under international law.

The Council of Europe (Council) has established the temporary protection regime and is in charge of determining who may qualify as a mass influx of displaced persons entitled to temporary protection. The decision, binding on all member states, was adopted by a qualified majority on a proposal from the European Commission (Commission).[175] Arguably, environmentally displaced persons may qualify for temporary protection. There is no need to further demonstrate the relationship between degradation of the environment and human rights violations, such as the right to food, water, housing, and so on, extensively described and discussed in the legal literature. Nevertheless, it is paramount to determine whether temporary protection is appropriate considering the peculiarity of the situation of environmentally displaced persons.

The Directive on Temporary Protection specifies that unless terminated by another Council decision, the normal duration of temporary protection is one year, with an automatic extension of two six-month periods.[176] The maximum possible duration of temporary protection is a total of three years because if the reasons for temporary protection persist, the Council may decide (again, by qualified majority and on a proposal by the Commission) to extend the protection for another year.[177] Although Article 3.2 provides that "Member States shall apply temporary protection with due respect for human rights and fundamental freedoms and their obligations regarding non-refoulement," after a period of three years the regime of temporary protection expires.[178]

Temporary protection is based on the conviction that it "can represent a reasonable administrative policy in an emergency situation only where individual refugee status determination is not immediately practicable and where its application will enhance admission to the territory."[179] Thus, temporary protection may relieve persons affected by sudden environmental disaster who may return to their country of origin once the crisis is over. Yet, for those persons affected by severe and durable environmental degradation, return to the country of origin in the near future remains impossible, so temporary protection is of limited assistance.

Notwithstanding the controversial dichotomy in international law between the level of protection afforded to refugees and the level of protection for other persons in need of international protection, which has been severely criticized for being inconsistent with international obligation of nondiscrimination,[180] the main concern in relation to environmentally displaced persons is the type of protection afforded and its ability to remedy the difficult situation.

B. Eligibility Criteria for International Protection Under the Directive on Subsidiary Protection

The Qualification Directive is a supranational instrument harmonizing the regime of complementary protection in Europe. From the outset, however, the ability of the instrument to alleviate the problem of forced migrants that do not qualify for refugee status appears dubious. Indeed, Jane McAdam points out that "[t]hough it has shifted complementary protection beyond the realm of ad hoc and discretionary national practices to a codified regime, it does not reflect best practice and unjustifiably entrenches a protection hierarchy."[181]

The Directive on Subsidiary Protection sets forth the legal status as well as the rights associated with the subsidiary protection regime. The directive reads:

> "person eligible for subsidiary protection" means a third country national or a stateless person who does not qualify as a refugee but in respect of whom substantial grounds have been shown for believing that the person concerned, if returned to his or her country of origin, or in the case of a stateless person, to his or her country of former habitual residence, would face a real risk of suffering serious harm as defined in Article 15, and to whom Article 17(1) and (2) do not apply, and is unable, or, owing to such risk, unwilling to avail himself or herself of the protection of that country.[182]

Some scholars have expressed concern about the definition given to the terms of the Directive on Subsidiary Protection.[183] This Article will only examine the points of interest relevant to the issue of the protection of environmentally displaced persons. Setting out the constitutive elements of subsidiary protection, Article 15 is certainly the critical provision of the Directive on Subsidiary Protection. Article 15 states that:

> serious harm consists of: (a) death penalty or execution; or (b) torture or inhuman or degrading treatment or punishment of an applicant in the country of origin; or (c) serious and individual threat to a civilian's life or person by reason of indiscriminate violence in situations of international or internal armed conflict.[184]

The provision is very disappointing since it is based on "the least contestable human rights-based protections which already form part of most Member States' protection policies."[185]

The definition of subsidiary protection employed in the Directive on Subsidiary Protection is based largely on international human rights instruments relevant to subsidiary protection. The most pertinent of these being Article 3 of the European Convention on Human Rights and Fundamental Freedoms (ECHR), Article 3 of the UN Convention Against Torture and Other Cruel, Inhuman or Degrading Treatment or Punishment, and Article 7 of the International Covenant on Civil and Political Rights.[186]

Jane McAdam argues that Article 15(b) "would apply to persons who are unable to demonstrate a link to a Convention ground, which may amount to cases where perpetrators resort to torture based on purely criminal motivation."[187] Similarly, Article 15(c) would apply to persons who do not qualify for refugee status, in particular in times of civil wars or internal armed conflicts when the well-founded fear of persecution based on one of the five grounds is difficult to determine.[188] Thus, McAdam posits that "[t]he provision reflects the existence of consistent, albeit varied, State practice of granting some form of complementary protection to persons fleeing the indiscriminate effects of armed conflict or generalised violence without a specific link to Convention grounds."[189]

At first glance, the Directive on Subsidiary Protection's purpose suggests that the instrument may provide protection to environmentally displaced persons allegedly in need of international protection. Nevertheless, the directive further qualifies the types of harm that may trigger subsidiary protection. It then becomes difficult to argue that environmental crises fit within one of the enumerated categories of serious harm as defined in the Directive on Subsidiary Protection under Article 15. Moreover, the wording of the provision does not leave any room for an additional type of "serious harm" in which environmental degradation could fit and any argument to the contrary would challenge both the letter and the spirit of the directive as expressed during the drafting of the instrument.

During the drafting process, the content of Article 15 was extensively discussed.[190] A human rights paragraph, applying specifically to acts outside the scope of subparagraphs (a) to (c), was drafted, affirming that serious harm could consist of a "violation of a human right, sufficiently severe to engage the Member State's international obligations."[191] The provision was nonetheless deleted, confining the provision to three types of serious harm. Jane McAdam emphasises the drawback of the provision as it was finally adopted, stressing that:

> [t]he deletion of this broader human rights provision has significantly reduced the scope of the directive. It was the provision which allowed for the greatest development of the human rights-refugee law nexus, providing flexibility for addressing new situations arising in international law and relevant developments in the jurisprudence of the European Court of Human Rights. As Article 15 stands now, there is little room for interpretation and it may be that "inhuman or degrading treatment or punishment" becomes the focal point for seeking to broaden the directive's scope, functioning in a similar fashion to the Convention's "membership of a particular social group" category.[192]

Although the Directive on Subsidiary Protection was discussed and elaborated at a time when the international community seemed to be particularly aware of the substantial problem of environmental degradation and, as a consequence, of the vulnerability of large groups of persons in search of international protection, the issue was completely and deliberately ignored. The concerns related to environmental causes of forced migration were mentioned during the discussion on the Directive on Subsidiary Protection.[193] Eventually, however, it was decided that the instrument would simply harmonize existing concepts and methods of subsidiary protection in the European Union, drawing on the "best" elements of the member states' national systems, and would not create a new system of protection per se.[194] It is therefore an instrument of compromise and not a comprehensive and systematic analysis of all protection possibilities within international law.[195]

Discussions on the "environmental refugee" issue originated in the mid-1980s, and have become increasingly frequent. Thus, the development of such a directive acknowledging the failure of the 1951 Refugee Convention to cope with every situation of forced migration would have provided a good opportunity to consider the issue of environmentally displaced persons.

C. Conclusion Concerning Complementary Protection in Europe

Rather than creating new obligations incumbent on member states, the Directive on Subsidiary Protection clarifies and codifies existing international and community obligations and practices. Thus, the directive appears to be of limited interest for environmentally displaced persons, until now overlooked in present instruments. Nonetheless, the expectation of a forthcoming European policy specifically dealing with the issue is not senseless. Firstly, the Directive on Subsidiary Protection provides for minimum standards on complementary protection but does not prevent states from according more favorable conditions, leaving certain points entirely to the discretion of member states.[196] Thus, the European Council on Refugees and Exiles advocates for lobbying national decision makers to implement the directive in a way which is conducive to higher standards or provides for necessary legal and other safeguards.[197] As far as environmentally displaced persons are concerned, however, an extensive interpretation of the directive may not reasonably include this category of displaced persons.

The Directive on Subsidiary Protection remains subject to amendments deemed "necessary."[198] The issue of environmentally displaced persons is increasingly the object of discussion, and could therefore become a central element of forthcoming instruments. Whether incorporated within the framework of the Directive on Subsidiary Protection or addressed through the creation of a separate instrument, there is a legal basis for the protection of environmentally displaced persons. Incorporating the protection of environmentally displaced persons in the existing directive or in a separate document would find a legal basis in the European instruments and fulfil the objective of a common asylum system based on minimum protection standards. The existing forms of complementary protection are based on Article 63 of the treaty establishing the European Community, which requires the European Union Council of Ministers to adopt minimum standards for granting temporary protection to displaced persons who need international protection.[199]

2. Complementary Protection in the United States: The Uncertain Protection of Environmentally Displaced Persons

In the recent past, natural disasters of unparalleled proportions have triggered the need to reevaluate and reexamine immigration policies not adequately dealing with the issue. As a result, the United States enacted the Immigration Act of 1990 (IMMACT),[200] which sets forth a regime of temporary protection. Temporary Protected Status (TPS) is granted to eligible nationals of a country where:

1) There is an ongoing armed conflict, and requiring return would pose a serious threat to personal safety;

2) There has been an earthquake, flood, drought, epidemic, or other environmental disaster resulting in a substantial, but temporary, disruption of living conditions; the foreign State is unable, temporarily, to handle adequately the return of its nationals; and the foreign State officially has requested temporary protection for its nationals in the United States; or

3) There exist extraordinary and temporary conditions that prevent nationals from returning in safety, unless the Attorney General finds that permitting the aliens to remain temporarily is contrary to the national interest.[201]

TPS designation is a purely discretionary decision, but once established, the status applies to all the residents of that country (or a region of that country that is so designated) who arrive in the United States before a date specified by the Attorney General.[202] At the present time, the United States has granted TPS to inhabitants of Montserrat,[203] Honduras, and Nicaragua[204] because of environmental disasters that have substantially disrupted living conditions, as a result of which those nations were unable, temporarily, to handle adequately the return of their nationals.

Certainly, "by allowing them to remain longer, the United States gives the affected nations time to cope with destabilizing conditions and rebuild, rather than overwhelm them with more people to care for."[205] Yet, to be eligible for TPS, a person needs to be already in the United States on the date of designation.[206] Thus, TPS is not used to provide protection to persons directly affected by an environmental catastrophe who as a result try to migrate, and in any case, it is not meant to facilitate the admission of persons from outside the United States.[207] Moreover, TPS is valid for a maximum of six months, although it may be extended for up to eighteen months if the secretary determines that sufficient conditions exist to trigger TPS.[208] Thus, the protection accorded does not lead to permanent residence. Once it expires, beneficiaries resume their prior immigration status (or any other status granted while a beneficiary of TPS).[209] In practice, however, permanent residence may be accorded, yet the law foresees "strict procedures for allowing those protected to become permanent residents."[210] Accordingly, the rights and entitlements recognized are limited. The persons protected are allowed to work during the period the status is in effect,[211] but they may be deemed ineligible for public assistance by states and localities.[212] Moreover, individuals granted TPS cannot apply for the admission of their spouses or children.[213]

TPS reflects the international obligation of non-refoulement, protecting persons who have entered the United States and who would face life-threatening circumstances if they returned home. Nonetheless, it "raises concerns since it permits the deportation of individuals who entered after the cutoff date, even though they would face substantially similar circumstances in the home country as would be faced by those granted protection."[214]

D. Proposed International Regime of Complementary Protection Specifically Dealing with Environmentally Displaced Persons

The existing refugee structure and current refugee norms are not a panacea to protect environmentally displaced persons in a systematic fashion. First and foremost, environmentally displaced persons do not fit neatly within the refugee definition. Likewise, the existing forms of complementary protection do not properly address the plea of environmentally displaced persons. Although in theory some persons who are forced to leave their country of origin, partly because of environmental reasons, may receive refugee status or complementary protection, some authors express concern about such an outcome. Dana Zartner Falstrom, for instance, points out that an appropriate remedy to the issue must address both the cause of the problem (environmental issues) and the result (environmental refugees).[215] Yet the 1951 Refugee Convention and the complementary forms of protection presently existing, while addressing the result, do not consider the cause of the problem.[216] As a matter of course, "[merely allowing environmentally displaced individuals to move does not solve the problem. Not only is their homeland continually decimated, but also the massive influx of environmental refugees to other areas creates a vicious cycle of environmental problems in these new areas."[217]

Nevertheless, Falstrom does not criticise the arguments advanced by her fellow scholars without making her own contribution in pursuing a remedy to the problem of environmentally displaced persons. Following the framework of the Convention Against Torture and Other Cruel, Inhuman and Degrading Treatment or Punishment (Convention Against Torture),[218] she advocates the elaboration of a new document that would focus not only on protecting those individuals who are forced to leave their homes due to environmental displacement, but also would require specific obligations from state parties to prevent the root causes from occurring.[219] As with the Convention Against Torture, she suggests that states "offer temporary protection to those fleeing from environmental problems, and also assume obligations and duties in order to solve these problems within their own jurisdictions, thus preventing the creation of environmental refugees from the start."[220] She affirms that sufficient evidence of support for a new convention governing environmentally displaced persons already exists in international treaty law and customary international law, and can provide the necessary sense of state obligation for a new treaty to succeed.[221]

This Article has explained that complementary protection reflects international obligations such as the obligation of non-refoulement, as expressed in Article 3 of the Convention Against Torture. Those provisions are applied to involuntarily-displaced persons who do not meet the refugee definition but are in need of international protection for the reasons enumerated in the different instruments on complementary protection. Likewise, the moral and legal obligation to protect and not to expose people to inhuman treatments could provide the base for a model of protection dealing specifically with the issue of environmentally displaced persons. After analyzing the Convention Against Torture, the Article assesses Falstrom's argument to elaborate a convention addressing the specific issue of environmentally displaced persons in the same fashion.

The Convention Against Torture sets forth both rights for individuals and affirmative obligations for signatory states.[222] Interestingly, the convention applies to persons who fear torture, regardless of whether the person has committed a crime or entered a country illegally. Moreover, it does not require that torture, cruel, inhuman, or degrading punishments be based on race, religion, nationality, membership in a particular social group, or political opinion. The obligation for the state not to repress those persons is, however, temporary and lasts only as long as the threat of torture exists.[223] Falstrom exposes the assets of the Convention Against Torture, an instrument with extensive provisions protecting individuals and requiring positive acts from states, supervised by the convention body entitled to require reports and to investigate.[224] She emphasises that "the Convention Against Torture is one of the most widely ratified and widely implemented treaties in international human rights law,"[225] and stresses that "the explicit purpose of the Convention Against Torture, coupled with its clear provisions enumerating specific obligations States must satisfy and protections they must provide, as well as the support the prohibition enjoys from customary international law and the law of nations make the Convention very effective."[226] Eventually she concludes that "the positive features of the Convention Against Torture can be emulated in a new document protecting environmentally displaced persons."[227]

For the purpose of a convention addressing the specific issue of environmentally displaced persons, "an environmentally displaced person is an individual forced to leave his or her home due to environmental reasons."[228] States would have the obligation, on a temporary basis, to take legislative, administrative, judicial, or any other necessary action to protect these people who arrive in their territory because of any of the listed environmental problems.[229] Falstrom does not advocate for permanent residency. Instead, she states that "once the basis for the protection has ended, the State may reexamine the case and return the person to his or her home if it is deemed safe."[230] First, she argues that "[t]his solution avoids one of the problems posed by the proponents for including environmentally displaced persons under existing refugee protections: States are more likely to assist victims of environmental degradation and disaster if it is seen as a temporary protection, rather than a permanent resettlement."[231] More importantly, the proposed convention establishes a temporary protection regime that would only come to an end once the reasons that compelled the person to flee his or her country of origin have ceased to exist, thereby circumventing the concerns posed at the present time by the limits established in months.[232]

Furthermore, "[t]his proposed Convention on the Protection of Environmentally Displaced Persons would address the root cause of the migration."[233] Indeed, "[a]s in the Convention Against Torture, the Convention on the Protection of Environmentally Displaced Persons should incorporate extensive provisions outlining State responsibility to find, correct, and prevent occurrences of the environmental degradation and destruction that force people to migrate."[234] Eventually, she proposes that "the new Convention . . . establish an oversight body, reporting mechanisms, dispute resolution procedures, and sanction provisions to encourage active compliance by all State parties."[235]

In order to support her proposition, Falstrom refers to sources of international conventional and customary law underlying protection of environmentally displaced persons.[236] She enumerates the various instruments adopted over the past decades "for the purpose of protecting the environment, reducing environmental damage, and protecting the rights of persons living within the environment."[237] Like many authors, including Lynn Berat,[238] Falstrom argues that rights related to the environment are customary international law, lending credence to the need to protect environmentally displaced persons.

Protection of environmentally displaced persons may be implied from the provisions enshrined in several instruments of paramount importance setting the basic human rights. In particular, Article 3 of the Universal Declaration of Human Rights proclaims that "[e]veryone has the right to life, liberty and the security of person."[239] Whereas Article 25 of the Universal Declaration of Human Rights provides that:

> [e]veryone has the right to a standard of living adequate for the health and well-being of himself and of his family, including food, clothing, housing and medical care and necessary social services, and the right to security in the event of unemployment, sickness . . . or other lack of livelihood in circumstances beyond his control.[240]

Although some authors advance that from these provisions may be inferred an obligation for states to protect environmentally displaced persons,[241] Falstrom argues that these instruments reflect the international awareness of environmental issues and accordingly the willingness to remedy them, but do not in themselves affirm such positive obligation for states.[242] Indeed, she posits that these provisions "can be used to test international support, [for the obligation of states to protect environmentally displaced persons] and lead to the formulation of a separate, cohesive document based on the belief that protecting the environment to prevent persons from being displaced from their homes has become a principle of customary international law."[243]

Falstrom acknowledges that the elaboration of such an instrument, recognizing the rights of environmentally displaced persons, will require time and energy.[244] Her proposition is nonetheless more than praiseworthy as it contributes to a vivid dialogue on the issue and provides constructive remedy. The hurdles related to the cost of implementation and compliance may be problematic. Yet, UN Agencies, in particular the UNHCR, nongovernmental organizations, and even states may provide humanitarian assistance to environmentally displaced persons. At the present time, UNHCR has developed a programme to provide assistance to states who take in environmentally displaced persons.[245] Of particular concern for UNHCR is the issue of environmental degradation in the host country as a consequence of mass displacement. Hence, UNHCR has elaborated guidelines to safeguard the environment around refugee operations and to encourage management of natural resources with a view to long-term sustainability.[246]

In addition to working towards confining the impact of environmentally displaced persons in the receiving country, UNHCR assists states with rehabilitation and cleanup operations.[247] Some authors recommend that UNHCR further assist countries with the costs of supporting these refugees, and preventing environmental disasters that cause the displacement. Examples of solidarity funds already exist, but should be more widely acknowledged in order to inspire other similar initiatives. For instance, in February 1997, following a conference on population movements in the Commonwealth of Independent States, organizations such as UNHCR and the International Organization for Migration issued an appeal for funds for the purpose of the "resettlement and integration of environmental migrants in the Central Asian States."[248] In 1997, UNHCR stated that "[d]onor states and international organisations [had] also taken a particular interest in the Aral Sea basin, although the environmental problems in that area [were] so extreme that remedial and preventive activities [would] be required for many years to come."[249] In 2006, however, a newspaper article was titling the renaissance of the Aral Sea thanks to the construction of a dam.[250] The article explains that the Aral Sea lost seventy-five percent of its volume due to poorly designed and badly managed projects diverting the two main rivers pouring into the sea in order to irrigate cotton plantations in

Central Asia.[251] Yet the World Bank and the government of Kazakhstan set up a project of dam construction, fostering the rise of the small Aral Basin and as a consequence the return of inhabitants previously living from the activities of the sea.[252] The project illustrates how international cooperation may remedy or reduce the adverse consequences of environmental degradation and therefore enable people to return to the area in which they were previously living. The Aral Sea was declared dead, forcing thousands of people to flee in search of an environment that could sustain their life; today the area is repopulated.

V. Conclusion

Although the international community acknowledged the correlation between environmental degradation and human rights violations years ago, the present set of international norms related to the environment do not address the situation of environmentally displaced persons. Considering the blatant vulnerability of these persons, the failure of international law to address the issue is of great concern. The propositions advanced in pursuing a remedy to the problem of environmentally-induced migrations are rare, but some of them are nonetheless invaluable because they present practical and feasible means of protection.

Reflecting the purpose of the instruments on complementary protection, namely to provide protection to persons in need of international protection who do not meet the criteria of the 1951 Refugee Convention, Falstrom proposes a convention addressing the specific issue of environmentally-induced migration.[253] The convention would be elaborated in a similar fashion to the Convention Against Torture. Interestingly, not only does the Convention Against Torture reassert the obligation of *non-refoulement*, the essential element of instruments on international protection, but also it provides an extensive set of rights and obligations, combined with an elaborated mechanism of implementation, that has proven to be effective to protect individuals against torture.[254]

The proposition to adapt these provisions to the case of environmentally displaced persons is particularly praiseworthy. Indeed, Falstrom is one of the few authors to propose a concrete framework to deal with the issue of environmentally displaced persons.[255] The literature on environmental refugees usually discusses whether environmentally displaced persons fit the refugee definition. As a matter of course, the discussion has lasted for many years and has not proven constructive. Moreover, the proposition to broaden the refugee definition does not address the root causes of the problem, and thereby neglects an important aspect to remedy comprehensively the issue. Although it is not the purpose of the Refugee Convention to make any judgement on the political, economic, and social attributes of foreign countries when receiving nationals from those countries, in the particular case of mass displacement for environmental reasons, addressing the root causes is paramount. Working on projects of sustainable use of the environment may prevent the multiplication of further, and in some circumstances irremediable, mass displacement. Eventually, considering the present dichotomy in international law between refugees and internally-displaced persons, authors supporting the idea that environmental refugees fall under the 1951 Refugee Convention advocate for protection only of those environmental refugees who cross an internationally recognized state border, as this is one of the essential requirements of the refugee definition.

Disregarding internally-displaced persons leaves many environmentally displaced persons without international protection.

This Article has reviewed the problems encountered in protecting environmentally displaced persons. The particular issue of protecting those environmentally displaced does not, however, pose any major legal or political difficulties. Although it is a pressing issue, it is not a controversial issue. To the contrary, dealing adequately with the problems related to environmental migration simply requires more rigorously structured and coordinated international action with regard to environmental issues in general, and the protection of environmentally displaced persons in particular. Moreover, the variety of sources relating to the problem of environmental migration reveals the increased interest in the issue, and engenders the belief that the international community may confront the problem in the near future. In this regard, an instrument like the proposed convention provides the assets for a comprehensive remedy of the issues related to environmentally-induced migration, and would be worth further discussion.

[…]

Aurelie Lopez, 'Protection of Environmentally Displaced Persons in International Law', *Environmental Law Review*, Volume 37, Issue 2, 2007, pp. 365–409.

Notes

75   1951 Refugee Convention, supra note 8, art. 1.

76   Christopher Kozoll, Poisoning the Well: Persecution, the Environment, and Refugee Status, 15 Colo. J. Int'l Envtl. L. & Pol'y 271, 272 (2004); see also Masters, supra note 12, at 866 (noting that many argue for a more expansive definition of "refugee" – one which will include "environmental refugee[s]"); Falstrom, supra note 9, at 13 (stating that "even if the government did not regulate a nuclear plant or did not prevent soil erosion from occurring, these are not actions that rise to the level of persecution").

77   U.N. High Comm'r for Refugees, The State of the World's Refugees: A Humanitarian Agenda, box 1.2 (1997), available at http://www.unhcr.org/publ/3eef1d896.html.

78   Kozoll, supra note 69, at 274.

79   United Nations High Comm'r for Refugees, Handbook on Procedures and Criteria for Determining Refugee Status Under the 1951 Convention and the 1967 Protocol Relating to the Status of Refugees, ch. II.B.(2)(b), U.N. Doc. HCR/1P/4/Eng. Rev. 1 (1979) (reedited Jan. 1992) [hereinafter Handbook on Procedures and Criteria for Determining Refugee Status], available at http://www.unhcr.org/publ/PUBL/3d58e13b4.pdf; see Jeanhee Hong, Refugees of the 21st Century: Environmental Injustice, 10 Cornell J.L. & Pub. Pol'y 323, 338 (2001).

80   B.S. Chimni, International Refugee Law 5 (2000).

81   See Handbook on Procedures and Criteria for Determining Refugee Status, supra note 79, ch. II.B.(2)(b) (noting that there is no single definition of "persecution" and that attempting to define the term has been unsuccessful).

82   Suhrke, supra note 7, at 157–59.

83   Handbook on Procedures and Criteria for Determining Refugee Status, supra note 72, at 9; see also Hong, supra note 72, at 331 (asserting that "a leading study of the time, conducted at the request of the UNHCR, explicitly stated that the Convention's definition excludes victims of natural disasters from acquiring refugee status. According to this source, the events that cause displacement must 'derive from the relations between the State and its nationals'"). She further explains that Jacques Vernant, invited in 1951 by the UNHCR to conduct an independent and scientific survey of the refugee situation at the time, described the definition of refugee in international law as consisting of

two elements: 1) persons qualifying for refugee status [that] must have left the territory of the State of which they were nationals, and 2) the root-causes of a person's displacement must be of a political nature and "accompanied by persecution or the threat of persecution against himself or at least against a section of the population with which he identifies himself." Id. at 331 n.54. Vernant concluded that the second condition "excludes victims of natural disasters from the definition of the refugee known to international law." Jacques Vernant, The Refugee in the Post-War World 4–7 (1953).

84   Kozoll, supra note 69, at 273.

85   See Handbook on Procedures and Criteria for Determining Refugee Status, supra note 79, at 51 (stating that serious violations of human rights based on "a threat to life or freedom on account of race, religion, nationality, political opinion or membership of a particular social group" would constitute persecution).

86   Kozoll, supra note 69, at 297. He refers to Sarei v. Rio Tinto, 221 F. Supp. 2d 1116 (C.D. Cal. 2002), and observes that the allegations present an example of how individuals might suffer persecution through harm inflicted by environmental damage, although the plaintiffs did not raise the issue. Kozoll, supra note 69, at 299.

87   This argument is corroborated by different studies that have exposed the role of government in both causing disasters and in causing populations to be more vulnerable to them. Cooper, supra note 12, at 502; see Wijkman & Timberlake, supra note 17, at 11–17; Leon Gordenker, Refugees in International Politics 13 (1987); see also Myers & Kent, supra note 19, at 169 (listing "governmental shortcomings" as a factor that could generate environmental refugees).

88   Cooper, supra note 12, at 486–87.

89   Cooper, supra note 12, at 502; see also Hong, supra note 72, at 323 (affirming that "developing countries increasingly confront dangerous environmental conditions due to industrial activity and exploitation, often at the request or with the approval of their governments. When such government actions create life-threatening circumstances, the most seriously affected victims of those situations should be entitled to refugee status").

90   E.g., the refugee must have a fear of being persecuted because of "race, religion, nationality, [or] membership of a particular social group or political opinion." 1951 Refugee Convention, supra note 8, art. 1.

91   Kozoll, supra note 69, at 284; see also Handbook on Procedures and Criteria for Determining Refugee Status, supra note 72, at 39.

92   Kozoll, supra note 69, at 284.

93   Roliff Purrington & Michael Wynne, Environmental Racism: Is a Nascent Social Science Concept a Sound Basis for Legal Relief?, 35 Hous. Law. 34, 35 (Mar.–Apr. 1998).

94   Cooper, supra note 12, at 505–06.

95   James Hathaway, The Law of Refugee Status 105 (1991); see also Handbook on Procedures and Criteria for Determining Refugee Status, supra note 72, at 12 ("Where serious discriminatory or other offensive acts are committed by the local populace, they can be considered as persecution if they are knowingly tolerated by the authorities, or if the authorities refuse, or prove unable, to offer effective protection.").

96   Hong, supra note 72, at 339.

97   Id.

98   Id.

99   Handbook on Procedures and Criteria for Determining Refugee Status, supra note 72, at 97.

100   Hong, supra note 72, at 339.

101   Hong, supra note 72, at 339–40.

102   See supra text accompanying note 74.

103   Kozoll, supra note 69, at 306–07.

104   8 U.S.C. § 1101(a)(42) (2000); Alexander Aleinikoff observes that the U.S. Board of Immigration Appeals focuses primarily on whether the persecution likely to be suffered by the applicant is based on one of the five grounds specified in the refugee definition. Moreover, the adjudicators appear to adopt a narrow and technical reading of the specified grounds for persecution. T. Alexander Aleinikoff, The

Meaning of "Persecution" in U.S. Asylum Law, in Refugee Policy: Canada and the United States 292, 296 (Howard Adelman ed., 1990).

105 Kozoll, supra note 69, at 273–74.

106 Cooper, supra note 12, at 523–26.

107 Conference of Plenipotentiaries on the Status of Refugees and Stateless Persons, Jan. 1, 1970, Summary Record of the Twenty-third Meeting, 8, U.N. Doc. A/CONF.2/SR.23, available at http://www.unhcr.org/protect/PROTECTION/3ae68cda10.html; see also 1 Atle Grahl-Madsen, The Status of Refugees in International Law 219 (1966) (noting that the social group category covers many of the same cases as the other categories but is also broader and was designed to prevent gaps).

108 Aleinikoff, supra note 97, at 298.

109 Cooper, supra note 12, at 522 n.233 (quoting Denissenko v. Haskett FED No. 404/96 20 (Mar. 22, 1996, Austl.), available at http://www.austlii.edu.au//cgi-bin/disp.pl/au/cases/cth/ federal_ct/ unrep8379.html?query=denissenko%20v%20haskett).

110 Handbook on Procedures and Criteria for Determining Refugee Status, supra note 72, 77.

111 Falstrom, supra note 9, at 13; see also Ira J. Kurzban, Immigration Law Sourcebook 368–72 (10th ed. 2006) (discussing the meaning of membership in a particular social group). The U.S. Board of Immigration Appeals, which was struggling with the concept of "particular social group," finally stated that it is a group of persons all of whom share a common, immutable characteristic. The shared characteristic might be an innate one such as sex, color, or kinship ties, or in some circumstances it might be a shared past experience such as a former military leadership [position] or land ownership. The particular kind of group characteristic that will qualify under this construction remains to be determined on a case-by-case basis. However, whatever the common characteristic that defines the group, it must be one that the members of the group either cannot change, or should not be required to change because it is fundamental to their individual identities or consciences. In re Acosta, 19 B.I.A. 211, 233 (1985), available at www.usdoj.gov/eoir/vll/ intdec/vol19/2986.pdf.

112 Falstrom, supra note 9, at 15.

113 See Robert W. Collin, Review of the Legal Literature on Environmental Racism, Environmental Equity, and Environmental Justice, 9 J. Envtl. L. & Litig. 121, 121–69 (1994) (reviewing the legal literature on environmental racism, equity, and justice); Deeohn Ferris & David Hahn-Baker, Environmentalists and Environmental Justice Policy, in Environmental Justice-Issues, Policies, and Solutions 66, 66–67 (Bunyan Bryant ed., 1995) (advocating expanding the environmental movement to include environmental justice).

114 Balakrishnan Rajagopal, The Violence of Development, Wash. Post, Aug. 9, 2002, at A19; Gunther Baechler, Violence Through Environmental Discrimination 1–3 (2005).

115 Purrington & Wynne, supra note 86, at 34.

116 Id.

117 Id.

118 Id. at 35.

119 Mark A. Drumbl, The International Responses to the Environmental Impacts of War, 17 Geo. Int'l Envtl. L. Rev. 565, 627 (2005).

120 Human Rights Watch, The Iraqi Government Assault on the Marsh Arabs (5 Jan., 2003), available at http://www.hrw.org/backgrounder/mena/marsharabs1.pdf.

121 Alexander Tkachenko, The Economy of the Iraq Marshes in the 1990s, in The Iraqi Marshlands 36, 41 (Emma Nicholson & Peter Clark eds., 2002).

122 Convention on the Prevention and Punishment of the Crime of Genocide, art. 2(c), Dec. 9, 1948, 78 U.N.T.S. 277, 280 (entered into force Jan. 12, 1951) [hereinafter Genocide Convention]; see also Kriangsak Kittichaisaree, International Criminal Law 71 (2001) (defining the act of genocide); Steven R. Ratner & Jason S. Abrams, Accountability for Human Rights Atrocities in International Law 29 (2001) (citing the Genocide Convention for definition of genocide).

123 Weinstein, supra note 16, at 715.

124 Tkachenko, supra note 114, at 47.

125   Weinstein, supra note 16, at 715 (citing Human Rights Watch, supra note 113, at 4–5 (describing "that the marshland ecosystem had collapsed by 2000," resulting in the presence of large salt crusts)).

126   See Ernestina Coast, Demography of the Marsh Arabs, in The Iraqi Marshlands 19, 19–21 (Emma Nicholson & Peter Clark eds., 2002); Christopher Mitchell, Assault on the Marshlands, in The Iraqi Marshlands 64, 66 (Emma Nicholson & Peter Clark eds., 2002).

127   See Max van der Stoel, Report on the Situation of Human Rights in Iraq 94–98 (1993), available at http://www.unhchr.ch/Huridocda/Huridoca.nsf/0/dae4fa61e79b30c6c1256e580058 f523/$FILE/G9310695.pdf (translating and reprinting the "Plan of Action for the Marshes"); see also Mitchell, supra note 119, at 64 (describing Saddam Hussein's plans for the Marsh Dwellers); Weinstein, supra note 16, at 718 (reporting that "the plan called for the destruction of homes, poisonings, assassinations, and the resettlement of Marsh Arabs on dry land, and specifically mentioned 'poisoning the environment and burning homes'").

128   Weinstein, supra note 16, at 718–19; see also Mitchell, supra note 119, at 64, 67–68 (pointing out that along with the drainage, the Iraqi forces allegedly filled the waters themselves with toxic chemicals).

129   Weinstein, supra note 16, at 719.

130   See Carl E. Bruch, All's not Fair in (Civil) War: Criminal Liability for Environmental Damage in Internal Armed Conflict, 25 Vt. L. Rev. 695, 710 (2001) (noting that principles of international customary law are "a source of norms to prevent, minimize, and punish environmental damage" during armed conflicts).

131   See Myers & Kent, supra note 19, 17–19 (1995) (explaining that environmental factors cause the problems that lead refugees to migrate).

132   See id. at 154 (arguing that the current regime "ignore[s] environmental refugees simply because there is no established mode of dealing with the problem they represent").

133   David A. Korn, Exodus Within Borders: An Introduction to the Crisis of Internal Displacement 2 (1999).

134   United Nations High Comm'r for Refugees, Protecting Refugees, http://www.unhcr.org/protect/3b84c7e23.html (last visited Apr. 15, 2007).

135   Int'l Org. for Migration, supra note 12, at 23.

136   Francis Deng, Guiding Principles on Internal Displacement: Report of the Representative of the Secretary-General, 2, U.N. Doc. E/CN.4/1998/53/ADD.2 (Feb. 11, 1998), available at http://www.unhchr.ch/html/menu2/7/b/principles.htm.

137   But see Francis M. Deng, Foreword to The Brookings Inst., Handbook for Applying the Guiding Principles on Internal Displacement at i (1999) (noting that the Guiding Principles "are based on and consistent with international human rights law, humanitarian law, and refugee law by analogy").

138   Keane, supra note 6, at 217.

139   See Marco Simons, The Emergence of a Norm Against Arbitrary Forced Relocation, 34 Colum. Hum. Rts. L. Rev. 95, 128 (2002) (stating that, while the Guiding Principles are not yet customary in international law, "they may soon reach that status").

140   See, e.g., McCue, supra note 9, at 173 (stating that "political pressures keep the system from expanding to address environmentally displaced migrants"); Falstrom, supra note 9, at 2 (commenting that the 1951 Refugee Convention has been implemented in ways that "make it practically impossible for an environmentally displaced person to gain asylum"); Keane, supra note 6, at 215 (noting that "the term 'environmental refugees' is a misnomer, as environmentally displaced persons are not recognized as refugees").

141   McCue, supra note 9, at 156. Persons for whom environmental degradation is the primary cause of movement may be defined as people who would not have moved if it had not been for one of a few general kinds of degradation of their natural environment. Id. at 157.

142   McCue, supra note 9, at 177.

143   JoAnn McGregor, Refugees and the Environment, in Geography and Refugees: Patterns and Processes of Change 157, 157–70 (Richard Black & Vaughn Robinson eds., 1993). The term "environmental refugee" is poorly defined and legally meaningless and confusing. See also Kibreab Gaim, Migration,

Environment and Refugeehood, in Environment and Population Change 115, 115–29 (Basia Zaba & John Clarke eds., 1994) (discussing the difficulty of determining the exact role played by environmental change in contributing to mass population displacements); Richard Black, supra note 63, at 1–2 (questioning the value of the very notion of "environmental refugees"). Black sees the concept as a myth or, in other words, as a misleading, highly politicized, and potentially damaging concept. Black argues that there are no environmental refugees as such. While environmental factors do play a part in forced migration, they are always closely linked to a range of other political and economic factors, so that focusing on the environmental factors in isolation does not help in understanding specific situations of population displacement. Id. at 2–3.

144   See Staff of S. Comm. on the Judiciary, 96th Cong., 1st Sess., World Refugee Crisis: The International Community's Response 253–320 (Comm. Print 1979) (identifying UNHCR as the primary international agency for protecting refugees and coordinating action on their behalf); see also McCue, supra note 9, at 151, 170–73 (explaining the historical development of UNHCR).

145   See Norman Myers, Environmentally-Induced Displacements: The State of the Art, in Environmentally-Induced Population Displacements and Environmental Impacts Resulting From Mass Migrations 21, 21 (U.N. High Comm'r for Refugees et al. eds., 1996) (defining environmentally displaced persons as "persons who are displaced within their own country of habitual residence or who have crossed an international border and for whom environmental degradation, deterioration or destruction is a major cause of their displacement, although not necessarily the sole one").

146   As McCue states, this language has only been adopted by the Organization of African Unity (OAU) and the Organization of American States (OAS). McCue, supra note 9, at 153 n.17; see Convention Governing the Specific Aspects of the Refugee Problems in Africa, art. 1, Sept. 10, 1969, 1001 U.N.T.S. 45, available at http://www.africa-union.org/root/au/Documents/Treaties/Text/Refugee_Convention.pdf [hereinafter OAU Convention] (defining the term refugee as applying "to every person who, owing to external aggression, occupation, foreign domination or events seriously disturbing public order . . . is compelled to leave his place of habitual residence in order to seek refuge in another place outside his country of origin or nationality"); Cartagena Declaration on Refugees, Annual Report of the Inter-American Commission on Human Rights, OAS Doc. OEA/Ser.L/V/II.66/doc.10, rev. 1, at 190–93 (Nov. 19–22, 1984) [hereinafter Cartagena Declaration], available at http://www.unhcr.org/home/RSDLEGAL/3ae6b36ec.html (explaining that the definition of refugee should be expanded to include those displaced by "other circumstances which have seriously disturbed public order").

147   McCue, supra note 9, at 153–54 (citing an Interview by Gregory McCue with Ruven Menikdiwela, UNHCR Associate Legal Officer, in Washington D.C. (Mar. 25, 1993)).

148   OAU Convention, supra note 139, art. 1.

149   Id.

150   McCue, supra note 9, at 153–54.

151   Cooper, supra note 12, at 480, 498.

152   Id.

153   Cartagena Declaration, supra note 139, art. 3.

154   Cooper, supra note 12, at 499; Keane, supra note 6, at 216.

155   United Nations High Comm'r on Refugees, Document CIREFCA/89/9 (1989) (English version); see also Cancado Trindade, The Contribution of International Human Rights Law to International Change, in Environmental Change and International Law: New Challenges and Dimensions 244, 299 (Edith Brown Weiss ed., 1992).

156   James Hathaway, The Rights of Refugees Under International Law 49 (2005).

157   Id.

158   El-Hinnawi, supra note 6, at 4.

159   McCue, supra note 9, at 158.

160   Myers, supra note 19, at 14–18.

161   Cooper, supra note 12, at 480, 494.

162   McGregor, supra note 136, at 158.

163   Keane, supra note 6, at 217.

164   Id. at 223.

165   See Guy Goodwin-Gill, The Refugee in International Law 3–29 (1996) (noting the different legal
      definitions given to the term refugee).

166   United Nations High Comm'r for Refugees, UNHCR's Partners, http://www.unhcr.org/ protect/
      PROTECTION/3b03da604.html (last visited Apr. 15, 2007).

167   Jane McAdam, Complementary Protection and Beyond: How States Deal with Human Rights
      Protection 1 (United Nations High Comm'r of Refugees, Working Paper No. 118, 2005).

168   Id. at 3. See generally D. Perluss & J.F. Hartman, Temporary Refuge: Emergence of a Customary
      Norm, 26 Va. J. Int'l L. 551 (1986) (discussing the 1951 UN Convention pertaining to refugees); G.S.
      Goodwin-Gill, Commentary, Non-Refoulement and the New Asylum Seekers, 26 Va. J. Int'l L. 897
      (1986) (discussing the Geneva Conventions); K. Hailbronner, Non-Refoulement and "Humanitarian"
      Refugees: Customary International Law or Wishful Legal Thinking?, 26 Va. J. Int'l L. 857 (1986)
      (discussing UN Convention among other protections or lack thereof for refugees).

169   See European Council on Refugees & Exiles, European Asylum Systems: Legal and Social Conditions
      for Asylum Seekers and Refugees in Western Europe, 2003, http://www.ecre.org/ conditions/2003/
      austria2003.pdf (last visited Apr. 15, 2007) (listing of Austria's legislative approach). For a discussion
      of European Union subsidiary protection practices, see generally Jane McAdam, The European
      Qualification Directive: The Creation of a Subsidiary Protection Regime, 17 Int'l J. Refugee L. 461
      (2005).

170   McAdam, supra note 169, at 465 n.21.

171   Council Directive 2001/55, 2001 O.J. (L212) 12 (EC) [hereinafter Directive on Temporary
      Protection]. The Directive on Temporary Protection is the first piece of legislation flowing from the
      asylum agenda of the Amsterdam Treaty. The Directive on Temporary Protection "entered into force
      on 7 August 2001 (date of publication in the Official Journal of the EC) and applies to all European
      Union Member States except for Denmark and Ireland. According to article 32(1) of the directive,
      the 13 Member States bound by the Directive have to ensure the necessary domestic legislation is
      in place by Dec. 31 2002." European Council on Refugees & Exiles, ECRE Information Note on
      the Council Directive 2001/55/EC of 20 July 2001 on Minimum Standards for Giving Temporary
      Protection in the Event of a Mass Influx of Displaced Persons and on Measures Promoting a Balance
      of Efforts Between Member States in Receiving Such Persons and Bearing the Consequence Thereof,
      http://www.ecre.org/statements/tpsumm.shtml (last visited Apr. 15, 2007).

172   Council Directive 2004/83, 2004 O.J. (L 304) 12 (EC) [hereinafter Directive on Subsidiary
      Protection]. Other Member-States of the European Union must bring into force laws, regulations
      and administrative provisions to comply with the Directive of Subsidiary Protection by October 10,
      2006. See European Council on Refugees & Exiles, Information Note on the Directive on Subsidiary
      Protection, IN1/10/2004/ext/CN, http://www.ecre.org/statements/ qualpro.pdf (last visited
      Apr. 15, 2007) (analyzing the key provisions of the Directive on Subsidiary Protection).

173   Directive on Temporary Protection, supra note 164, art. 2(a).

174   Id. art. 2(c).

175   Id. art. 5.

176   Id. art. 4.1.

177   Id. art. 4.2.

178   Directive on Temporary Protection, supra note 164, art. 3.2 (emphasis added).

179   European Council on Refugees and Exiles, supra note 165.

180   The Directive on Temporary Protection reminds that "[w]ith respect to the treatment of persons
      enjoying temporary protection under this Directive, the Member States are bound by obligations
      under instruments of international law to which they are party and which prohibit discrimination."
      Directive on Temporary Protection, supra note 164, art. 16. In practice, however, the rights and
      entitlements accorded to refugees and beneficiaries of temporary protection differ. See McAdam,
      supra note 167, at 5 (discussing the problem in relation to the Directive on Subsidiary Protection).
      McAdam stresses that "[l]egally, there is no reason why the source of protection should require

differentiation in the rights and status accorded to a beneficiary. UNHCR has stated that rights and benefits should be based on need rather than the grounds on which a person has been granted protection, and that there is accordingly no valid reason to treat beneficiaries of subsidiary protection differently from Convention refugees." Id. at 5–6.

181 McAdam, supra note 169, at 462.

182 Directive on Subsidiary Protection, supra note 165, art. 2(e).

183 See generally McAdam, supra note 169, at 461 (analyzing each term of the definition, as well as the practical implication for the recognition of international protection and the ensuing entitlement to specific rights).

184 Directive on Subsidiary Protection, supra note 165, art. 15.

185 McAdam, supra note 169, at 474.

186 European Convention on Human Rights and Fundamental Freedoms, art. 3, Nov. 4, 1950, 213 U.N.T.S. 222; Convention Against Torture and Other Cruel, Inhumane or Degrading Treatment or Punishment, art. 3, G.A. Res. 39/46, U.N. Doc. A/RES/39 (Dec. 10 1984); International Covenant on Civil and Political Rights, art. 7, G.A. Res. 2200A (XXI), U.N. GAOR, 21st Sess., Supp. No. 16, U.N. Doc. A/6316 (Dec. 16, 1966).

187 McAdam, supra note 169, at 479.

188 Id. at 480.

189 Id. at 479–80.

190 Id. at 474.

191 Id. at 477.

192 McAdam, supra note 167, at 3.

193 McAdam, supra note 169, at 464.

194 Id. at 464–65. As stated by McAdam, the purpose of the Directive on Subsidiary Protection, as set forth in Article 1, is to "lay down minimum standards for the qualification of third country nationals or stateless persons as refugees or as persons who otherwise need international protection and the content of the protection granted." Id. 466–67. The directive is "based on existing international and EC obligations and current Member State practice." Id. at 464–65.

195 Id. at 465.

196 Directive on Temporary Protection, supra note 164, art. 3(5) ("The directive shall not affect the prerogative of the Member States to adopt or retain more favourable conditions for persons covered by temporary protection.").

197 See European Council on Refugees & Exiles, supra note 164 (providing the ECRE's position on the Directive on Subsidiary Protection provisions).

198 Directive on Subsidiary Protection, supra note 165, art. 37.
    1. By April 10 2008, the Commission shall report to the European Parliament and the Council on the application of this Directive and shall propose any amendments that are necessary. These proposals for amendments shall be made by way of priority in relation to articles 15, 26 and 33. Member States shall send the Commission all the information that is appropriate for drawing up that report by Oct. 10, 2007. 2. After presenting the report, the Commission shall report to the European Parliament and the Council on the application of this Directive at least every five years. Id.

199 Consolidated Version of the Treaty Establishing the European Community, 2002 O.J. (C 325) 58, 59, available at http://europa.eu.int/eur-lex/en/treaties/dat/C_2002325EN.003301.html.

200 Immigration Act of 1990, Pub. L. No. 101–649, 104 Stat. 4978 (codified as amended in scattered sections of 8 U.S.C.).

201 Immigration and Nationality Act, 8 U.S.C. 1254a(b)(1) (2000).

202 Id. 1254a(b)(1), (c)(1)(A).

203 Montserrat was touched by a volcanic eruption in August 1997. Between August 22, 1997 and August 22, 1999, 300 habitants from Montserrat were granted TPS. See Susan Martin, Andy Schoenholtz & Deborah Waller Meyers, Temporary Protection: Towards a New Regional and Domestic Framework, 12 Geo. Immigr. L.J. 543, 550 (1998) (discussing TPS and arguing for a more expansive use of that

designation); see also Shazneen Rabadi, Developments in the Executive Branch, 13 Geo. Immigr. L.J. 137, 138 (1998) (outlining a Federal Register entry extending Montserrat's designation under the TPS program). Rabadi notes that the Attorney General has extended Montserrat's designation under the TPS program. Extension of Designation of Montserrat Under Temporary Protected Status Program, 63 Fed. Reg. 45,864, 45,864–65 (Aug. 27, 1998) (stating that there is ongoing difficulty related to the volcanic eruption and extending TPS designation by 12 months).

204 Honduras and Nicaragua have been particularly affected by Hurricane Mitch. Andrew I. Schoenholtz & Thomas F. Muther, Immigration and Nationality, 33 Int'l Law. 517, 521 (1999). In response to the devastating impact of Hurricane Mitch and to requests from Central American governments, the Clinton Administration granted temporary protection to some 150,000 Hondurans and Nicaraguans. Id.

205 Id. at 521.

206 Id. at 549.

207 Id. at 522.

208 Id.

209 Jane McAdam, supra note 167, at 18.

210 Martin, Schoenholtz, & Meyers, supra note 196, at 549; 8 U.S.C. 1254a(h)(2) (2000).

211 8 U.S.C. § 1254a(a)(1)(B), (a)(2) (2000).

212 Martin, Schoenholtz & Meyers, supra note 196, at 549; 8 U.S.C. § 1254a(f)(2) (2000).

213 See Temporary Protected Status for Nationals of Designated States, 8 C.F.R. 244.2(b) (2006) (requiring the applicant to be "continuously physically present in the United States since the effective date of the most recent designation of that foreign state").

214 Martin, Schoenholtz & Meyers, supra note 196, at 549.

215 Falstrom, supra note 9, at 2.

216 Id.

217 Id.

218 Convention Against Torture and Other Cruel, Inhuman or Degrading Treatment or Punishment, Dec. 10, 1984, 1456 U.N.T.S. 85 [hereinafter Convention Against Torture].

219 Falstrom, supra note 9, at 21.

220 Id. at 2.

221 Id. at 2–3.

222 Id. at 20. Not only does the Convention Against Torture acknowledge that torture, and other cruel, inhuman, and degrading punishments are a human rights violation, but also it urges states to take affirmative action in order to prevent and remedy those human rights violations. Id. Article 2 requires states to take "legislative, administrative, judicial, or other measures to prevent acts of torture within its territory"; Article 3 prohibits a state from returning any individual to a state where it is likely (s) he will suffer those treatments; Article 4 requires states to make all acts of torture offences under the state's domestic criminal law; Article 12 requires a "prompt and impartial investigation" of any possible acts of torture; and Article 14 requires states to ensure that victims of torture have adequate means of redress. Id. (citing Convention Against Torture, supra note 211, arts. 2, 3, 4, 12, 14). Eventually, the convention requires states to undertake educational and training initiatives to ensure that tortuous acts are not being committed by individuals within its territory. Falstrom, supra note 8, at 20 (citing Convention Against Torture, supra note 211, arts. 10–12).

223 Falstrom, supra note 9, at 19.

224 Convention Against Torture, supra note 211, arts. 17–24; Falstrom, supra note 9, at 20.

225 Falstrom, supra note 9, at 20.

226 Id. at 21.

227 Id.

228 Falstrom, supra note 9, at 22.

229 Falstrom proposes a provision that would read as follows: "No State Party shall expel, return or extradite an environmentally displaced person to any State where there are substantial grounds for believing that he or she would be in danger due to one of the environmental problems listed in this Convention." Id.

230 Id.

231 Id.

232 See infra note 199 and accompanying text (discussing the TPS limits).

233 Falstrom, supra note 9, at 23.

234 Id. at 23. Falstrom further provides practical examples of how to act responsibly in order to prevent environmental degradation:

> For example, the new Convention could include provisions requiring all State parties to ensure that acts of environmental sabotage (as in the case of the oil companies in Nigeria) are made illegal under domestic law. Each State party could be required to provide education and information to rural farmers regarding sustainable agriculture and conservation of water. States would be required to closely regulate those who sell pesticides and other toxic materials to ensure that the products are being used correctly and safely. States could also be required to set up strict regulations and guidelines for hazardous industries such as nuclear plants, which can create widespread environmental harms if operated improperly. Education, oversight, and inspection would be required for such industries. Moreover, individuals living within a certain radius of these types of operations should be educated about the dangers and the proper response should a problem arise. The list could go on, but the general idea would be to require and encourage States to strengthen their existing mechanism or create a mechanism to inform and educate their populations to prevent environmental disasters and degradation before they start. Id.

235 Id.

236 Id. at 24.

237 Id.; see also id. at n.88 (providing a list of such instruments).

238 Lynn Berat, Defending the Right to a Healthy Environment: Toward a Crime of Genocide in International Law, 11 B.U. Int'l L.J. 327, 329 (1993).

239 Universal Declaration of Human Rights, G.A. Res. 217A, art. 3, U.N. GAOR, 3d Sess., 1st plen. mtg., U.N. Doc. A/810 (Dec. 12, 1948); see also Michelle Leighton Schwartz, International Legal Protection for Victims of Environmental Abuse, 18 Yale J. Int'l L. 355, 361–64 (1993) (discussing the right to life as one of the human rights threatened by environmental problems).

240 Universal Declaration of Human Rights, supra note 239, art. 25.

241 Cooper, supra note 11, at 489–93. She argues that "[t]he comprehensive language of these provisions can be interpreted as setting broad environmental standards and creating an implicit human right to freedom from life-threatening and otherwise intolerable environmental conditions." Id. (citing Schwartz, supra note 232, at 361–64, 367, 367 n.68 (listing articles 12, 17, 23, and 27 of the Universal Declaration of Human Rights as embodying human rights which may be violated by environmental threats)). Hence, Cooper posits that "this human right to freedom from intolerable environmental degradation ought to command protection" under the Refuge Convention. Id. at 492.

242 Falstrom, supra note 9, at 23–26.

243 Id. at 25. See generally Maria Stavropoulou, The Right Not to Be Displaced, 9 Am. U. J. Int'l L. & Pol'y 689, 706–35 (1994) (discussing the current state of international law with respect to displacement of persons, and identifying the human rights violated by displacement).

244 Falstrom, supra note 9, at 27.

245 Id.

246 Id. at 28 n. 101.

247 Id.

248 United Nations High Comm'r on Refugees, Humanitarian Codes of Conduct, in The State of the World's Refugees 1997: A Humanitarian Agenda, ch. 1 (1997), available at http://www.unhcr.org/publ/PUBL/3eb789912.pdf.

249 Id. ch. 1 box 1.2. See generally Philip P. Micklin, Desiccation of the Aral Sea: A Water Management Disaster in the Soviet Union, 241 Sci. 1170 (1988) (describing the environmental deterioration and economic impact due to the desiccation of the Aral Sea, as well as some possible steps to restore water inflows to the Aral).

250 Ilan Greenberg, As a Sea Rises, So Do Hopes for Fish, Jobs and Riches, N.Y. Times, Apr. 6, 2006, at A4.

251 Id.
252 Id.
253 Falstrom, supra note 9, at 18–28.
254 Id.
255 Similarly, McCue advocates for the adoption of a convention based on international environmental law principles to cope with environmental migration. McCue, supra note 9, 189–90. For a comprehensive catalogue of which instruments include what norms, see generally, Edith Brown Weiss, Environmental Disasters in International Law, 1986 Anuario Jurídico Interamericano 141 (1987) (discussing ways to minimize damage and provide emergency assistance with international environmental disasters). Professor Weiss has exhaustively researched these principles and noted their use in numerous conventions, treaties and so-called "soft-law" sources from bodies such as the International Legal Association and the World Commission on Environment and Development. Id. at 145, 147. Interestingly, the proposed convention would add the migratory effects of environmental events to the responsibility of States under these principles. Id. at 146. It would also enhance the development of financial and technical assistance for the particular concern of environmental and population pressures. Id. at 145–50.

Chapter 13

UN Security Council

SECURITY COUNCIL HOLDS FIRST-EVER DEBATE ON IMPACT OF CLIMATE CHANGE ON PEACE, SECURITY, HEARING OVER 50 SPEAKERS

Some delegations raise doubts regarding council's role on issue, while others, particularly small island states, welcome council's consideration

With scientists predicting that land and water resources will gradually become more scarce in the coming years, and that global warming may irreversibly alter the face of the planet, the United Nations Security Council today held its first-ever debate on the impact of climate change on security, as some delegates raised doubts over whether the Council was the proper forum to discuss the issue.

The day-long meeting, called by the United Kingdom, aimed to examine the relationship between energy, security and climate, and featured interventions from more than 50 delegations, representing imperiled island nations and industrialized greenhouse gas emitters alike. While some speakers praised the initiative, there were reservations from developing countries, which saw climate change as a socio-economic development issue to be dealt with by the more widely representative General Assembly. Many delegations also called for the United Nations to urgently consider convening a global summit on the issue.

The session was chaired by British Foreign Secretary, Margaret Beckett, whose country holds the presidency of the 15-nation Council for April. She said that recent scientific evidence reinforced, or even exceeded, the worst fears about climate change, as she warned of migration on an unprecedented scale because of flooding, disease and famine. She also said that drought and crop failure could cause intensified competition for food, water and energy.

She said that climate change was a security issue, but it was not a matter of narrow national security — it was about "our collective security in a fragile and increasingly interdependent world". By holding today's debate, the Council was not seeking to pre-empt the authority of other bodies, including the General Assembly and the Economic and Social Council. The decisions that they came to, and action taken, in all those bodies required the fullest possible understanding of the issues involved. "[So] climate change can bring us together, if we have the wisdom to prevent it from driving us apart," she declared.

Calling for a "long-term global response" to deal with climate change, along with unified efforts involving the Security Council, Member States and other international bodies,

Secretary-General Ban Ki-moon said that projected climate changes could not only have serious environmental, social and economic implications, but implications for peace and security, as well.

"This is especially true in vulnerable regions that face multiple stresses at the same time – pre-existing conflict, poverty and unequal access to resources, weak institutions, food insecurity and incidence of diseases such as HIV/AIDS," he said. The Secretary-General outlined several "alarming, though not alarmist" scenarios, including limited or threatened access to energy increasing the risk of conflict, a scarcity of food and water transforming peaceful competition into violence and floods and droughts sparking massive human migrations, polarizing societies and weakening the ability of countries to resolve conflicts peacefully.

China's representative was among those who argued that the Council was not the proper forum for a debate on climate change. "The developing countries believe that the Security Council has neither the professional competence in handling climate change – nor is it the right decision-making place for extensive participation leading up to widely acceptable proposals," he said.

The issue could have certain security implications, but, generally speaking, it was, in essence, an issue of sustainable development. The United Nations Framework Convention on Climate Change had laid down the fundamental principles for the international community's response to climate change. The Kyoto Protocol had set up targets for developed countries – limited, but measurable – for reducing greenhouse gas emissions. To effectively respond to climate change, he said it was necessary to follow the principle of "common, but differentiated, responsibilities" set forth in the Convention, respect existing arrangements, strengthen cooperation and encourage more action.

The representative of Pakistan, speaking on behalf of the "Group of 77" developing countries and China, agreed, saying that the Council's primary duty was to maintain international peace and security. Other issues, including those related to economic and social development, were assigned to the Economic and Social Council and the General Assembly. The ever-increasing encroachment of the Security Council on the roles and responsibilities of the other main organs of the United Nations represented a "distortion" of the principles and purposes of the Charter, infringed on the authority of the other bodies and compromised the rights of the Organization's wider membership.

But Papua New Guinea's representative, who spoke on behalf of the Pacific Islands Forum, said that the impact of climate change on small islands was no less threatening than the dangers guns and bombs posed to large nations. Pacific island countries were likely to face massive dislocations of people, similar to population flows sparked by conflict. The impact on identity and social cohesion were likely to cause as much resentment, hatred and alienation as any refugee crisis.

"The Security Council, charged with protecting human rights and the integrity and security of States, is the paramount international forum available to us," he said. The Forum did not expect the Council to get involved in Climate Change Convention negotiations, but it did expect the 15-member body to keep the issue of climate change under continuous review, to ensure that all countries contributed to solving the problem and that those efforts were commensurate with their resources and capacities. It also expected the Council to review sensitive issues, such as implications for sovereignty and international legal rights from the loss of land, resources and people.

Singapore's speaker said that, while it was obvious that there was some discomfort about the venue and nature of today's debate, it was equally obvious that climate change was "the" global environmental challenge. Given their paucity of resources, developing countries would be the hardest hit, and some had their survival at stake. But it was not only the poor that would suffer. There was broad consensus that it was necessary to act to arrest what "we ourselves are responsible for". Many of the problems caused by climate change could only be tackled if nations worked together.

"Let us view our procedural disagreements against this backdrop," he said. While it might be difficult to quantify the relationship between climate change and international peace and security, there should be no doubt that climate change was an immediate global challenge, whose effects were transboundary and multifaceted. He was not advocating that the Security Council play a key role on climate change, but neither could he deny that body "some sort of a role, because it seems obvious to all but the wilfully blind that climate change must, if not now, then eventually have some impact on international peace and security.

Also participating in today's debate were the Minister for Foreign Affairs of Slovakia, the Under-Secretary of State for Foreign Affairs of Italy, the Federal Minister for Economic Cooperation and Development of Germany (on behalf of the European Union), the Minister for Development and Cooperation of the Netherlands and the Minister for State and Foreign Affairs of the Maldives.

Others taking part in the meeting were the representatives of Belgium, Ghana, Congo, Qatar, United States, France, Indonesia, Panama, South Africa, Russian Federation, Peru, Switzerland, Japan, Namibia, Barbados, Ukraine, Egypt, Australia, New Zealand, Tuvalu, Bangladesh, Venezuela, Sudan (on behalf of the African Group), Solomon Islands, Palau, Denmark, Iceland, Marshall Islands, Philippines, Mexico, Brazil, India, Republic of Korea, Norway, Federated States of Micronesia, Argentina, Cuba (on behalf of the Non-Aligned Movement), Liechtenstein, Bolivia, Cape Verde, Costa Rica, Israel, Canada, Mauritius and Comoros.

The meeting began at 10:20 a.m. and suspended at 1:20 p.m. The Council resumed its debate at 3:15 p.m. and wrapped up at 6:35 p.m.

[...]

UN Security Council 'Security Council Holds First-ever Debate on Impact of Climate Change on Peace, Security, Hearing Over 50 Speakers', Department of Public Information, News and Media Division, New York, Security Council, 5663rd Meeting, 17 April 2007.

Chapter 14

United Nations General Assembly

UNITED NATIONS GENERAL ASSEMBLY RESOLUTION 63/281
Climate Change and its Possible Security Implications

The General Assembly,

Recalling its resolution 63/32 of 26 November 2008 and other resolutions relating to the protection of the global climate for present and future generations of mankind,

Recalling also Article 1 of the Charter of the United Nations, which established the purposes of the United Nations,

Recognizing the respective responsibilities of the principal organs of the United Nations, including the primary responsibility for the maintenance of international peace and security conferred upon the Security Council and the responsibility for sustainable development issues, including climate change, conferred upon the General Assembly and the Economic and Social Council,

Noting the open debate in the Security Council on "Energy, security and climate" held on 17 April 2007,[1]

Reaffirming that the United Nations Framework Convention on Climate Change[2] is the key instrument for addressing climate change,

Recalling the provisions of the United Nations Framework Convention on Climate Change, including the acknowledgement that the global nature of climate change calls for the widest possible cooperation by all countries and their participation in an effective and appropriate international response, in accordance with their common but differentiated responsibilities and respective capabilities and their social and economic conditions,

Reaffirming the Programme of Action for the Sustainable Development of Small Island Developing States,[3] the Mauritius Declaration[4] and the Mauritius Strategy for the Further Implementation of the Programme of Action for the Sustainable Development of Small Island Developing States,[5]

Recalling the 2005 World Summit Outcome,[6]

Deeply concerned that the adverse impacts of climate change, including sea-level rise, could have possible security implications,

1. *Invites* the relevant organs of the United Nations, as appropriate and within their respective mandates, to intensify their efforts in considering and addressing climate change, including its possible security implications;

2. *Requests* the Secretary-General to submit a comprehensive report to the General Assembly at its sixty-fourth session on the possible security implications of climate change, based on the views of the Member States and relevant regional and international organizations.

85th plenary meeting 3 June 2009

UN General Assembly Resolution 63/281, '*Climate Change and its Possible Security Implications*', 11 June 2009, UN Doc. A/RES/63/281.

Notes

1　Including the letter dated 5 April 2007 from the Permanent Representative of the United Kingdom of Great Britain and Northern Ireland to the United Nations addressed to the President of the Security Council (S/2007/186), the letter dated 12 April 2007 from the Chargé daffier a.i. of the Permanent Mission of Cuba to the United Nations on behalf of the Movement of Non-Aligned Countries, addressed to the President of the Security Council (S/2007/203) and the letter dated 16 April 2007 from the Permanent Representative of Pakistan to the United Nations, on behalf of the Group of 77 and China, addressed to the President of the Security Council (S/2007/211). See S/PV.5663.

2　United Nations, Treaty Series, vol. 1771, No. 30822.

3　Report of the Global Conference on the Sustainable Development of Small Island Developing States, Bridgetown, Barbados, 25 April–6 May 1994 (United Nations publication, Sales No.E.94.I.18 and corrigenda), chap. I, resolution 1, annex II.

4　Report of the International Meeting to Review the Implementation of the Programme of Action for the Sustainable Development of Small Island Developing States, Port Louis, Mauritius, 10–14 January 2005 (United Nations publication, Sales No. E.05.II.A.4 and corrigendum), chap. I, resolution 1, annex I.

5　Ibid., annex II.

6　See resolution 60/1.

PART 2.1

Human rights law

Chapter 15

Human Rights Council

HUMAN RIGHTS COUNCIL RESOLUTION 7/23
Human Rights and Climate Change

The Human Rights Council,

Concerned that climate change poses an immediate and far-reaching threat to people and communities around the world and has implications for the full enjoyment of human rights,

Recognizing that climate change is a global problem and that it requires a global solution,

Reaffirming the Charter of the United Nations, the Universal Declaration of Human Rights, the International Covenant on Economic, Social and Cultural Rights, the International Covenant on Civil and Political Rights and the Vienna Declaration and Programme of Action,

Noting the findings of the fourth assessment report of the Intergovernmental Panel on Climate Change, including that the warming of the climate system is unequivocal and that most of the observed increase in global average temperatures since the mid-twentieth century is very likely human-induced,

Recognizing that the United Nations Framework Convention on Climate Change remains the comprehensive global framework to deal with climate change issues, reaffirming the principles of the Framework Convention as contained in article 3 thereof, and welcoming the decisions of the United Nations Climate Change Conference held in Bali, Indonesia, in December 2007, and in particular the adoption of the Bali Action Plan,

Recalling that the Vienna Declaration and Programme of Action reaffirmed the right to development, as established in the Declaration on the Right to Development, as a universal and inalienable right and as an integral part of fundamental human rights,

Recognizing that human beings are at the centre of concerns for sustainable development and that the right to development must be fulfilled so as to equitably meet the development and environmental needs of present and future generations,

Recognizing also that the world's poor are especially vulnerable to the effects of climate change, in particular those concentrated in high-risk areas, and also tend to have more limited adaptation capacities,

Recognizing further that low-lying and other small island countries, countries with low-lying coastal, arid and semi-arid areas or areas liable to floods, drought and desertification, and developing countries with fragile mountainous ecosystems are particularly vulnerable to the adverse effects of climate change,

Recalling the relevant provisions of declarations, resolutions and programmes of action adopted by major United Nations conferences, summits and special sessions and their follow-up meetings, in particular Agenda 21 and the Rio Declaration on Environment and Development, and the Johannesburg Declaration on Sustainable Development and the Johannesburg Plan of Implementation,

Recalling also Commission on Human Rights resolution 2005/60 of 20 April 2005 on human rights and the environment as part of sustainable development,

Recalling further Council resolution 6/27 of 14 December 2007 on adequate housing as a component of the right to an adequate standard of living and in particular paragraph 3 thereof, and Council decision 2/104 of 27 November 2006 on human rights and access to water,

Taking note of the contribution provided by special procedures of the Council in examining and advancing the understanding of the link between the enjoyment of human rights and the protection of environment,

Taking note also of the conclusions and recommendations contained in the report of the Special Rapporteur on the right of everyone to the enjoyment of the highest attainable standard of physical and mental health to the General Assembly (A/62/214), which include a call for the Council to study the impact of climate change on human rights,

1. *Decides* to request the Office of the United Nations High Commissioner for Human Rights, in consultation with and taking into account the views of States, other relevant international organizations and intergovernmental bodies including the Intergovernmental Panel on Climate Change and the secretariat of the United Nations Framework Convention on Climate Change, and other stakeholders, to conduct, within existing resources, a detailed analytical study on the relationship between climate change and human rights, to be submitted to the Council prior to its tenth session;

2. *Encourages* States to contribute to the study conducted by the Office of the High Commissioner;

3. *Decides* to consider the issue at its tenth session under agenda item 3, and thereafter to make available the study, together with a summary of the debate held during its tenth session, to the Conference of Parties to the United Nations Framework Convention on Climate Change for its consideration.

41st meeting 28 March 2008. Adopted without a vote.

Human Rights Council Resolution 7/23, '*Human Rights and Climate Change*', 41st Meeting Human Rights Council, 28 March 2008.

Chapter 16

UN High Commissioner for Human Rights

THE RELATIONSHIP BETWEEN CLIMATE CHANGE AND HUMAN RIGHTS

[...]

Introduction

1. The present report is submitted pursuant to Human Rights Council resolution 7/23 in which the Office of the United Nations High Commissioner for Human Rights (OHCHR) was requested to conduct a detailed analytical study of the relationship between climate change and human rights, taking into account the views of States and other stakeholders.

2. Written submissions were received from States, intergovernmental organizations, national human rights institutions, non-governmental organizations, and individual experts. OHCHR also organized a one-day open-ended consultation on the relationship between climate change and human rights, held on 22 October 2008 in Geneva. The inputs received during the consultation process have informed the preparation of this report.[1]

3. This report seeks to outline main aspects of the relationship between climate change and human rights. Climate change debates have traditionally focused on scientific, environmental and economic aspects. As scientific understanding of the causes and consequences of climate change has evolved and impacts on human lives and living conditions have become more evident, the focus of debates has progressively broadened with increasing attention being given to human and social dimensions of climate change. Human Rights Council resolution 7/23 on human rights and climate change exemplifies this broadening of the debate.

4. Special procedures of the Human Rights Council have also addressed the human rights implications of climate change in recent statements and reports,[2] while the Organization of American States and the Alliance of Small Island States have recently drawn attention to the relationship between climate change and human rights.[3] In addition, a growing volume of reports and studies address the interface between climate change and human rights.[4]

I. Climate change: an overview global warming and its causes

5. The United Nations Framework Convention on Climate Change, which has near universal membership, provides the common international framework to address the causes and consequences of climate change, also referred to as "global warming". The Convention defines

climate change as "a change of climate which is attributed directly or indirectly to human activity that alters the composition of the global atmosphere and which is in addition to natural climate variability observed over comparable time periods".[5]

6.   The Intergovernmental Panel on Climate Change (IPCC) has greatly contributed to improving understanding about and raising awareness of climate change risks.[6] Since the publication of its First Assessment Report (IPCC AR1) in 1990, climate science has rapidly evolved, enabling the IPCC to make increasingly definitive statements about the reality, causes and consequences of climate change. Its Fourth Assessment Report (IPCC AR4), issued in 2007, presents a clear scientific consensus that global warming "is unequivocal" and that, with more than 90 per cent certainty, most of the warming observed over the past 50 years is caused by manmade greenhouse gas emissions.[7] Current levels of greenhouse gas concentrations far exceed pre-industrial levels as recorded in polar ice cores dating back 650,000 years, and the predominant source of this increase is the combustion of fossil fuels.[8]

7.   The IPCC AR4 presents the current scientific consensus on climate change. It is based on the contributions of three working groups focusing on: the physical science basis (Working Group I); impacts, adaptation and vulnerability (Working Group II); and mitigation of climate change (Working Group III). The Synthesis Report and Summaries for Policymakers have been adopted and approved by member States at an IPCC plenary session. The findings provide the main scientific resource for this study in exploring the relationship between climate change and human rights.

Observed and projected impacts

8.   Amongst the main observed and projected changes in weather patterns related to global warming are:[9]

- Contraction of snow-covered areas and shrinking of sea ice
- Sea level rise and higher water temperatures
- Increased frequency of hot extremes and heatwaves
- Heavy precipitation events and increase in areas affected by drought
- Increased intensity of tropical cyclones (typhoons and hurricanes).

9.   The IPCC assessments and a growing volume of studies provide an increasingly detailed picture of how these changes in the physical climate will impact on human lives. IPCC AR4 outlines impacts in six main areas: ecosystems; food; water; health; coasts; and industry, settlement and society,[10] some of which are described further below in relation to their implications for specific human rights.

Unequal burden and the equity principle

10.   Industrialized countries, defined as annex I countries under the United Nations Framework Convention on Climate Change, have historically contributed most to manmade greenhouse

gas emissions. At the same time, the impacts of climate change are distributed very unevenly, disproportionally affecting poorer regions and countries, that is, those who have generally contributed the least to human-induced climate change.

11.   The unequal burden of the effects of climate change is reflected in article 3 of the Convention (referred to as "the equity article"). It stipulates that parties should protect the climate system "on the basis of equity and in accordance with their common but differentiated responsibilities and respective capabilities"; that developed countries "should take the lead in combating climate change and the adverse effects thereof" and that full consideration should be given to the needs of developing countries, especially "those that are particularly vulnerable to the adverse effects of climate change" and "that would have to bear a disproportionate or abnormal burden under the Convention".[11] Giving operational meaning to the "equity principle" is a key challenge in ongoing climate change negotiations.

Response measures: mitigation and adaptation

12.   Mitigation and adaptation are the two main strategies to address climate change. Mitigation aims to minimize the extent of global warming by reducing emission levels and stabilizing greenhouse gas concentrations in the atmosphere. Adaptation aims to strengthen the capacity of societies and ecosystems to cope with and adapt to climate change risks and impacts.

13.   Reaching an agreement on required global mitigation measures lies at the heart of international climate change negotiations. Article 2 defines the "ultimate objective" of the Convention and associated instruments as "the stabilization of greenhouse gas concentrations in the atmosphere at a level that would prevent dangerous anthropogenic interference with the climate system". A key question is to operationally define the term "dangerous".[12]

14.   Over the past decades, scientific studies and policy considerations have converged towards a threshold for dangerous climate change of a maximum rise in global average temperature of 2°C above the pre-industrial level.[13] Staying below this threshold will significantly reduce the adverse impacts on ecosystems and human lives. It will require that global greenhouse gas emissions peak within the next decade and be reduced to less than 50 per cent of the current level by 2050. Yet, even this stabilization scenario would lead to a "best estimate" global average temperature increase of 2°C–2.4°C above pre-industrial levels.[14] Moreover, the possibility of containing the temperature rise to around 20°C becomes increasingly unlikely if emission reductions are postponed beyond the next 15 years.

15.   Adaptation and the financing of adaptation measures are also central in international climate change negotiations. Irrespective of the scale of mitigation measures taken today and over the next decades, global warming will continue due to the inertia of the climate system and the long-term effects of previous greenhouse gas emissions. Consequently, adaptation measures are required to enable societies to cope with the effects of now unavoidable global warming. Climate change adaptation covers a wide range of actions and strategies, such as building sea defences, relocating populations from flood-prone areas, improved water

management, and early warning systems. Equally, adaptation requires strengthening the capacities and coping mechanisms of individuals and communities.

II. Implications for the enjoyment of human rights

A. climate change, environmental harm and human rights

16.   An increase in global average temperatures of approximately 2°C will have major, and predominantly negative, effects on ecosystems across the globe, on the goods and services they provide. Already today, climate change is among the most important drivers of ecosystem changes, along with overexploitation of resources and pollution.[15] Moreover, global warming will exacerbate the harmful effects of environmental pollution, including higher levels of ground-level ozone in urban areas. In view of such effects, which have implications for a wide range of human rights, it is relevant to discuss the relationship between human rights and the environment.

17.   Principle 1 of the 1972 Declaration of the United Nations Conference on the Human Environment (the Stockholm Declaration) states that there is "a fundamental right to freedom, equality and adequate conditions of life, in an environment of a quality that permits a life of dignity and well-being". The Stockholm Declaration reflects a general recognition of the interdependence and interrelatedness of human rights and the environment.[16]

18.   While the universal human rights treaties do not refer to a specific right to a safe and healthy environment, the United Nations human rights treaty bodies all recognize the intrinsic link between the environment and the realization of a range of human rights, such as the right to life, to health, to food, to water, and to housing.[17] The Convention on the Rights of the Child provides that States parties shall take appropriate measures to combat disease and malnutrition "through the provision of adequate nutritious foods and clean drinking water, taking into consideration the dangers and risks of environmental pollution".[18]

19.   Equally, the Committee on Economic, Social and Cultural Rights (CESCR) has clarified that the right to adequate food requires the adoption of "appropriate economic, environmental and social policies" and that the right to health extends to its underlying determinants, including a healthy environment.[19]

B. Effects on specific rights

20.   Whereas global warming will potentially have implications for the full range of human rights, the following subsections provide examples of rights which seem to relate most directly to climate change-related impacts identified by IPCC.

1. The right to life

21.   The right to life is explicitly protected under the International Covenant on Civil and Political Rights and the Convention on the Rights of the Child.[20] The Human Rights Committee has described the right to life as the "supreme right", "basic to all human rights", and it is a

right from which no derogation is permitted even in time of public emergency.[21] Moreover, the Committee has clarified that the right to life imposes an obligation on States to take positive measures for its protection, including taking measures to reduce infant mortality, malnutrition and epidemics.[22] The Convention on the Rights of the Child explicitly links the right to life to the obligation of States "to ensure to the maximum extent possible the survival and development of the child".[23] According to the Committee on the Rights of the Child, the right to survival and development must be implemented in a holistic manner, "through the enforcement of all the other provisions of the Convention, including rights to health, adequate nutrition, social security, an adequate standard of living, a healthy and safe environment...".[24]

22.   A number of observed and projected effects of climate change will pose direct and indirect threats to human lives. IPCC AR4 projects with high confidence an increase in people suffering from death, disease and injury from heatwaves, floods, storms, fires and droughts. Equally, climate change will affect the right to life through an increase in hunger and malnutrition and related disorders impacting on child growth and development; cardio respiratory morbidity and mortality related to ground-level ozone.[25]

23.   Climate change will exacerbate weather-related disasters which already have devastating effects on people and their enjoyment of the right to life, particularly in the developing world. For example, an estimated 262 million people were affected by climate disasters annually from 2000 to 2004, of whom over 98 per cent live in developing countries.[26] Tropical cyclone hazards, affecting approximately 120 million people annually, killed an estimated 250,000 people from 1980 to 2000.[27]

24.   Protection of the right to life, generally and in the context of climate change, is closely related to measures for the fulfilment of other rights, such as those related to food, water, health and housing. With regard to weather-related natural disasters, this close interconnectedness of rights is reflected in the Inter-Agency Standing Committee (IASC) operational guidelines on human rights and natural disasters.[28]

2. The right to adequate food

25.   The right to food is explicitly mentioned under the International Covenant on Economic, Social and Cultural Rights, the Convention on the Rights of the Child and the Convention on the Rights of Persons with Disabilities and implied in general provisions on an adequate standard of living of the Convention on the Elimination of All Forms of Discrimination against Women and the International Convention on the Elimination of All Forms of Racial Discrimination.[29]

In addition to a right to adequate food, the International Covenant on Economic, Social and Cultural Rights also enshrines "the fundamental right of everyone to be free from hunger".[30] Elements of the right to food include the availability of adequate food (including through the possibility of feeding oneself from natural resources) and accessible to all individuals under the jurisdiction of a State. Equally, States must ensure freedom from hunger and take necessary action to alleviate hunger, even in times of natural or other disasters.[31]

26.   As a consequence of climate change, the potential for food production is projected initially to increase at mid to high latitudes with an increase in global average temperature in the range of 1–3°C. However, at lower latitudes crop productivity is projected to decrease, increasing the risk of hunger and food insecurity in the poorer regions of the world.[32] According to one estimate, an additional 600 million people will face malnutrition due to climate change,[33] with a particularly negative effect on sub-Saharan Africa.[34]

Poor people living in developing countries are particularly vulnerable given their disproportionate dependency on climate-sensitive resources for their food and livelihoods.[35]

27.   The Special Rapporteur on the right to food has documented how extreme climate events are increasingly threatening livelihoods and food security.[36] In responding to this threat, the realization of the right to adequate food requires that special attention be given to vulnerable and disadvantaged groups, including people living in disaster prone areas and indigenous peoples whose livelihood may be threatened.[37]

3. The right to water

28.   CESCR has defined the right to water as the right of everyone to sufficient, safe, acceptable, physically accessible and affordable water for personal and domestic uses, such as drinking, food preparation and personal and household hygiene.[38] The Convention on the Elimination of All Forms of Discrimination against Women and the Convention on the Rights of Persons with Disabilities explicitly refer to access to water services in provisions on an adequate standard of living, while the Convention on the Rights of the Child refers to the provision of "clean drinking water" as part of the measures States shall take to combat disease and malnutrition.[39]

29.   Loss of glaciers and reductions in snow cover are projected to increase and to negatively affect water availability for more than one-sixth of the world's population supplied by melt water from mountain ranges. Weather extremes, such as drought and flooding, will also impact on water supplies.[40]

Climate change will thus exacerbate existing stresses on water resources and compound the problem of access to safe drinking water, currently denied to an estimated 1.1 billion people globally and a major cause of morbidity and disease.[41] In this regard, climate change interacts with a range of other causes of water stress, such as population growth, environmental degradation, poor water management, poverty and inequality.[42]

30.   As various studies document, the negative effects of climate change on water supply and on the effective enjoyment of the right to water can be mitigated through the adoption of appropriate measures and policies.[43]

4. The right to health

31.   The right to the highest attainable standard of physical and mental health (the right to health) is most comprehensively addressed in article 12 of the International Covenant on Economic, Social and Cultural Rights and referred to in five other core international human

rights treaties.[44] This right implies the enjoyment of, and equal access to, appropriate health care and, more broadly, to goods, services and conditions which enable a person to live a healthy life. Underlying determinants of health include adequate food and nutrition, housing, safe drinking water and adequate sanitation, and a healthy environment.[45] Other key elements are the availability, accessibility (both physically and economically), and quality of health and health-care facilities, goods and services.[46]

32.   Climate change is projected to affect the health status of millions of people, including through increases in malnutrition, increased diseases and injury due to extreme weather events, and an increased burden of diarrhoeal, cardio respiratory and infectious diseases.[47] Global warming may also affect the spread of malaria and other vector borne diseases in some parts of the world.[48] Overall, the negative health effects will disproportionately be felt in sub-Saharan Africa, South Asia and the Middle East. Poor health and malnutrition increases vulnerability and reduces the capacity of individuals and groups to adapt to climate change.

33.   Climate change constitutes a severe additional stress to health systems worldwide, prompting the Special Rapporteur on the right to health to warn that a failure of the international community to confront the health threats posed by global warming will endanger the lives of millions of people.[49] Most at risk are those individuals and communities with a low adaptive capacity. Conversely, addressing poor health is one central aspect of reducing vulnerability to the effects of climate change.

34.   Non-climate related factors, such as education, health care, public health initiatives, are critical in determining how global warming will affect the health of populations.[50] Protecting the right to health in the face of climate change will require comprehensive measures, including mitigating the adverse impacts of global warming on underlying determinants of health and giving priority to protecting vulnerable individuals and communities.

5. The right to adequate housing

35.   The right to adequate housing is enshrined in several core international human rights instruments and most comprehensively under the International Covenant on Economic, Social and Cultural Rights as an element of the right to an adequate standard of living.[51] The right to adequate housing has been defined as "the right to live somewhere in security, peace and dignity".[52] Core elements of this right include security of tenure, protection against forced evictions,[53] availability of services, materials, facilities and infrastructure, affordability, habitability, accessibility, location and cultural adequacy.[54]

36.   Observed and projected climate change will affect the right to adequate housing in several ways. Sea level rise and storm surges will have a direct impact on many coastal settlements.[55] In the Arctic region and in low-lying island States such impacts have already led to the relocation of peoples and communities.[56] Settlements in low-lying mega-deltas are also particularly at risk, as evidenced by the millions of people and homes affected by flooding in recent years.

37.   The erosion of livelihoods, partly caused by climate change, is a main "push" factor for increasing rural to urban migration. Many will move to urban slums and informal settlements where they are often forced to build shelters in hazardous areas.[57] Already today, an estimated 1 billion people live in urban slums on fragile hillsides or flood-prone riverbanks and face acute vulnerability to extreme climate events.[58]

38.   Human rights guarantees in the context of climate change include: (a) adequate protection of housing from weather hazards (habitability of housing); (b) access to housing away from hazardous zones; (c) access to shelter and disaster preparedness in cases of displacement caused by extreme weather events; (d) protection of communities that are relocated away from hazardous zones, including protection against forced evictions without appropriate forms of legal or other protection, including adequate consultation with affected persons.[59]

6. The right to self-determination

39.   The right to self-determination is a fundamental principle of international law. Common article 1, paragraph 1, of the International Covenant on Economic, Social and Cultural Rights and the International Covenant on Civil and Political Rights establishes that "all peoples have the right of self-determination", by virtue of which "they freely determine their political status and freely pursue their economic, social and cultural development".[60] Important aspects of the right to self-determination include the right of a people not to be deprived of its own means of subsistence and the obligation of a State party to promote the realization of the right to self-determination, including for people living outside its territory.[61] While the right to self-determination is a collective right held by peoples rather than individuals, its realization is an essential condition for the effective enjoyment of individual human rights.

40.   Sea level rise and extreme weather events related to climate change are threatening the habitability and, in the longer term, the territorial existence of a number of low-lying island States. Equally, changes in the climate threaten to deprive indigenous peoples of their traditional territories and sources of livelihood. Either of these impacts would have implications for the right to self-determination.

41.   The inundation and disappearance of small island States would have implications for the right to self-determination, as well as for the full range of human rights for which individuals depend on the State for their protection. The disappearance of a State for climate change-related reasons would give rise to a range of legal questions, including concerning the status of people inhabiting such disappearing territories and the protection afforded to them under international law (discussed further below). While there is no clear precedence to follow, it is clear that insofar as climate change poses a threat to the right of peoples to self-determination, States have a duty to take positive action, individually and jointly, to address and avert this threat. Equally, States have an obligation to take action to avert climate change impacts which threaten the cultural and social identity of indigenous peoples.

C. Effects on specific groups

42.   The effects of climate change will be felt most acutely by those segments of the population who are already in vulnerable situations due to factors such as poverty, gender, age, minority status, and disability.[62] Under international human rights law, States are legally bound to address such vulnerabilities in accordance with the principle of equality and non-discrimination.

43.   Vulnerability and impact assessments in the context of climate change largely focus on impacts on economic sectors, such as health and water, rather than on the vulnerabilities of specific segments of the population.[63] Submissions to this report and other studies indicate awareness of the need for more detailed assessments at the country level and point to some of the factors which affect individuals and communities.

44.   The present section focuses on factors determining vulnerability to climate change for women, children and indigenous peoples.

1. Women

45.   Women are especially exposed to climate change-related risks due to existing gender discrimination, inequality and inhibiting gender roles. It is established that women, particularly elderly women and girls, are affected more severely and are more at risk during all phases of weather-related disasters: risk preparedness, warning communication and response, social and economic impacts, recovery and reconstruction.[64] The death rate of women is markedly higher than that of men during natural disasters (often linked to reasons such as: women are more likely to be looking after children, to be wearing clothes which inhibit movement and are less likely to be able to swim). This is particularly the case in disaster-affected societies in which the socio-economic status of women is low.[65] Women are susceptible to gender-based violence during natural disasters and during migration, and girls are more likely to drop out of school when households come under additional stress. Rural women are particularly affected by effects on agriculture and deteriorating living conditions in rural areas. Vulnerability is exacerbated by factors such as unequal rights to property, exclusion from decision-making and difficulties in accessing information and financial services.[66]

46.   Studies document how crucial for successful climate change adaptation the knowledge and capacities of women are. For example, there are numerous examples of how measures to empower women and to address discriminatory practices have increased the capacity of communities to cope with extreme weather events.[67]

47.   International human rights standards and principles underline the need to adequately assess and address the gender-differentiated impacts of climate change. In the context of negotiations on the United Nations Framework Convention on Climate Change, States have highlighted gender-specific vulnerability assessments as important elements in determining adaptation options.[68] Yet, there is a general lack of accurate data disaggregated by gender data in this area.

2. Children

48.   Studies show that climate change will exacerbate existing health risks and undermine support structures that protect children from harm.[69] Overall, the health burden of climate change will primarily be borne by children in the developing world.[70] For example, extreme weather events and increased water stress already constitute leading causes of malnutrition and infant and child mortality and morbidity. Likewise, increased stress on livelihoods will make it more difficult for children to attend school. Girls will be particularly affected as traditional household chores, such as collecting firewood and water, require more time and energy when supplies are scarce. Moreover, like women, children have a higher mortality rate as a result of weather-related disasters.

49.   As today's children and young persons will shape the world of tomorrow, children are central actors in promoting behaviour change required to mitigate the effects of global warming. Children's knowledge and awareness of climate change also influence wider households and community actions.[71] Education on environmental matters among children is crucial and various initiatives at national and international levels seek to engage children and young people as actors in the climate change agenda.[72]

50.   The Convention on the Rights of the Child, which enjoys near universal ratification, obliges States to take action to ensure the realization of all rights in the Convention for all children in their jurisdiction, including measures to safeguard children's right to life, survival and development through, inter alia, addressing problems of environmental pollution and degradation. Importantly, children must be recognized as active participants and stewards of natural resources in the promotion and protection of a safe and healthy environment.[73]

3. Indigenous peoples

51.   Climate change, together with pollution and environmental degradation, poses a serious threat to indigenous peoples, who often live in marginal lands and fragile ecosystems which are particularly sensitive to alterations in the physical environment.[74] Climate change-related impacts have already led to the relocation of Inuit communities in polar regions and affected their traditional livelihoods. Indigenous peoples inhabiting low-lying island States face similar pressures, threatening their cultural identity which is closely linked to their traditional lands and livelihoods.[75]

52.   Indigenous peoples have been voicing their concern about the impacts of climate change on their collective human rights and their rights as distinct peoples.[76] In particular, indigenous peoples have stressed the importance of giving them a voice in policymaking on climate change at both national and international levels and of taking into account and building upon their traditional knowledge.[77] As a study cited by the IPCC in its Fourth Assessment Report observes, "Incorporating indigenous knowledge into climate change policies can lead to the development of effective adaptation strategies that are cost-effective, participatory and sustainable".[78]

53.   The United Nations Declaration on the Rights of Indigenous Peoples sets out several rights and principles of relevance to threats posed by climate change.[79] Core international human rights treaties also provide for protection of indigenous peoples, in particular with regard to the right to self-determination and rights related to culture.[80] The rights of indigenous peoples are also enshrined in ILO Convention No. 169 (1989) concerning Indigenous and Tribal Peoples in Independent Countries.

54.   Indigenous peoples have brought several cases before national courts and regional and international human rights bodies claiming violations of human rights related to environmental issues. In 2005, a group of Inuit in the Canadian and Alaskan Arctic presented a case before the Inter-American Commission on Human Rights seeking compensation for alleged violations of their human rights resulting from climate change caused by greenhouse gas emissions from the United States of America.[81] While the Inter-American Commission deemed the case inadmissible, it drew international attention to the threats posed by climate change to indigenous peoples.

D. Displacement

55.   The First Assessment Report of the IPCC (1990) noted that the greatest single impact of climate change might be on human migration. The report estimated that by 2050, 150 million people could be displaced by climate change-related phenomena, such as desertification, increasing water scarcity, and floods and storms.[82] It is estimated that climate change-related displacement will primarily occur within countries and that it will affect primarily poorer regions and countries.[83]

56.   It is possible to distinguish between four main climate change-related displacement scenarios,[84] where displacement is caused by:

- Weather-related disasters, such as hurricanes and flooding
- Gradual environmental deterioration and slow onset disasters, such as desertification, sinking of coastal zones and possible total submersion of low-lying island States
- Increased disaster risks resulting in relocation of people from high-risk zones
- Social upheaval and violence attributable to climate change-related factors.

57.   Persons affected by displacement within national borders are entitled to the full range of human rights guarantees by a given State,[85] including protection against arbitrary or forced displacement and rights related to housing and property restitution for displaced persons.[86] To the extent that movement has been forced, persons would also qualify for increased assistance and protection as a vulnerable group in accordance with the Guiding Principles on Internal Displacement.[87] However, with regard to slow-onset disasters and environmental degradation it remains challenging to distinguish between voluntary and forced population movements.

58.   Persons moving voluntarily or forcibly across an international border due to environmental factors would be entitled to general human rights guarantees in a receiving State, but would

often not have a right of entry to that State. Persons forcibly displaced across borders for environmental reasons have been referred to as "climate refugees" or "environmental refugees". The Office of the United Nations High Commissioner for Refugees, the International Organization for Migration and other humanitarian organizations have advised that these terms have no legal basis in international refugee law and should be avoided in order not to undermine the international legal regime for the protection of refugees.[88]

59.   The Representative of the Secretary-General on human rights of internally displaced persons has suggested that a person who cannot be reasonably expected to return (e.g. if assistance and protection provided by the country of origin is far below international standards) should be considered a victim of forced displacement and be granted at least a temporary stay.[89]

60.   One possible scenario of forcible displacement across national borders is the eventual total submergence of small island States.[90] Two working papers of the Sub-Commission on the Promotion and Protection of Human Rights point to some of the human rights issues such situations would raise, such as the rights of affected populations vis-à-vis receiving States and possible entitlement to live in community.[91] Human rights law does not provide clear answers as to the status of populations who have been displaced from sinking island States. Arguably, dealing with such possible disasters and protecting the human rights of the people affected will first and foremost require adequate long-term political solutions, rather than new legal instruments.[92]

E. Conflict and security risks

61.   Recent reports and studies identify climate change as a key challenge to global peace and stability.[93] This was also recognized by the Norwegian Nobel Committee when, in 2007, it awarded the Nobel Peace Prize jointly to the IPCC and Al Gore for raising awareness of man-made climate change.[94] Equally, in 2007, the Security Council held a day-long debate on the impact of climate change on peace and security.

62.   According to one study, the effects of climate change interacting with economic, social and political problems will create a high risk of violent conflict in 46 countries – home to 2.7 billion people.[95] These countries, mainly in sub-Saharan Africa, Asia and Latin America, are also the countries which are particularly exposed to projected negative impacts of climate change.

63.   Climate change-related conflicts could be one driver of forced displacement. In such cases, in addition to the general human rights protection framework, other international standards would be applicable, including the Guiding Principles on Internal Displacement, international humanitarian law, international refugee law and subsidiary and temporary protection regimes for persons fleeing from armed conflict. Violent conflict, irrespective of its causes, has direct implications for the protection and enjoyment of human rights.

64.   It should be noted, however, that knowledge remains limited as to the causal linkages between environmental factors and conflict and there is little empirical evidence to substantiate the projected impacts of environmental factors on armed conflict.[96]

F. Human rights implications of response measures

65.   The United Nations Framework Convention on Climate Change and its Kyoto Protocol commit States parties to minimize adverse economic, social and environmental impacts resulting from the implementation of measures taken to mitigate or adapt to climate change impacts ("response measures").[97] With regard to measures to reduce the concentration of greenhouse gases in the atmosphere (mitigation), agro-fuel production is one example of how mitigation measures may have adverse secondary effects on human rights, especially the right to food.[98]

66.   Whereas agro-fuel production could bring positive benefits for climate change and for farmers in developing countries, agro-fuels have also contributed to increasing the price of food commodities "because of the competition between food, feed and fuel for scarce arable land".[99] CESCR has urged States to implement strategies to combat global climate change that do not negatively affect the right to adequate food and freedom from hunger, but rather promote sustainable agriculture, as required by article 2 of the United Nations Framework Convention on Climate Change.[100]

67.   Apart from the impact on the right to food, concerns have also been raised that demand for befouls could encroach on the rights of indigenous peoples to their traditional lands and culture.[101]

68.   Concerns have also been raised about possible adverse effects of reduced emissions from deforestation and degradation (REDD) programmes. These programmes provide compensation for retaining forest cover and could potentially benefit indigenous peoples who depend on those forest resources. However, indigenous communities fear expropriation of their lands and displacement and have concerns about the current framework for REDD. The Permanent Forum on Indigenous Issues stated that new proposals for reduced emissions from deforestation "must address the need for global and national policy reforms…respecting rights to land, territories and resources, and the rights of self-determination and the free, prior and informed consent of the indigenous peoples concerned".[102]

III. Relevant human rights obligations

69.   There exists broad agreement that climate change has generally negative effects on the realization of human rights. This section seeks to outline how the empirical reality and projections of the adverse effects of climate change on the effective enjoyment of human rights relate to obligations assumed by States under the international human rights treaties.

70.   While climate change has obvious implications for the enjoyment of human rights, it is less obvious whether, and to what extent, such effects can be qualified as human rights violations in a strict legal sense.[103]

Qualifying the effects of climate change as human rights violations poses a series of difficulties. First, it is virtually impossible to disentangle the complex causal relationships linking historical greenhouse gas emissions of a particular country with a specific climate change-related effect, let alone with the range of direct and indirect implications for human rights. Second, global warming is often one of several contributing factors to climate change-related effects, such as hurricanes, environmental degradation and water stress. Accordingly, it is often impossible to establish the extent to which a concrete climate change-related event with implications for human rights is attributable to global warming. Third, adverse effects of global warming are often projections about future impacts, whereas human rights violations are normally established after the harm has occurred.[104]

71.   Irrespective of whether or not climate change effects can be construed as human rights violations, human rights obligations provide important protection to the individuals whose rights are affected by climate change or by measures taken to respond to climate change.

A. National level obligations

72.   Under international human rights law, individuals rely first and foremost on their own States for the protection of their human rights. In the face of climate change, however, it is doubtful, for the reasons mentioned above, that an individual would be able to hold a particular State responsible for harm caused by climate change. Human rights law provides more effective protection with regard to measures taken by States to address climate change and their impact on human rights.

73.   For example, if individuals have to move away from a high-risk zone, the State must ensure adequate safeguards and take measures to avoid forced evictions. Equally, several claims about environmental harm have been considered by national, regional and international judicial and quasi-judicial bodies, including the Human Rights Committee, regarding the impact on human rights, such as the right to life, to health, to privacy and family life and to information.[105] Similar cases in which an environmental harm is linked to climate change could also be considered by courts and quasi-judicial human rights treaty bodies. In such cases, it would appear that the matter of the case would rest on whether the State through its acts or omissions had failed to protect an individual against a harm affecting the enjoyment of human rights.

74.   In some cases, States may have an obligation to protect individuals against foreseeable threats to human rights related to climate change, such as an increased risk of flooding in certain areas. In that regard, the jurisprudence of the European Court of Human Rights gives some indication of how a failure to take measures against foreseeable risks could possibly amount to a violation of human rights. The Court found a violation of the right to life in a case where State authorities had failed to implement land-planning and emergency relief policies while they were aware of an increasing risk of a large-scale mudslide. The Court also noted that the population had not been adequately informed about the risk.[106]

1. Progressive realization of economic, social and cultural rights.

75.   As discussed in chapter II, climate change will have implications for a number of economic, social and cultural rights. As specified in the relevant treaty provisions, States are obliged to take measures towards the full realization of economic, social and cultural rights to the maximum extent of their available resources.[107] As climate change will place an additional burden on the resources available to States, economic and social rights are likely to suffer.

76.   While international human rights treaties recognize that some aspects of economic, social and cultural rights may only be realized progressively over time, they also impose obligations which require immediate implementation. First, States parties must take deliberate, concrete and targeted measures, making the most efficient use of available resources, to move as expeditiously and effectively as possible towards the full realization of rights.[108] Second, irrespective of resource limitations, States must guarantee non-discrimination in access to economic, social and cultural rights. Third, States have a core obligation to ensure, at the very least, minimum essential levels of each right enshrined in the Covenant. For example, a State party in which "any significant number of individuals is deprived of essential foodstuffs, of essential primary health care, of basic shelter and housing, or of the most basic forms of education" would be failing to meet its minimum core obligations and, prima facie, be in violation of the Covenant.[109]

77.   In sum, irrespective of the additional strain climate change-related events may place on available resources, States remain under an obligation to ensure the widest possible enjoyment of economic, social and cultural rights under any given circumstances. Importantly, States must, as a matter of priority, seek to satisfy core obligations and protect groups in society who are in a particularly vulnerable situation.[110]

2. Access to information and participation in decision-making

78.   Awareness-raising and access to information are critical to efforts to address climate change. For example, it is critically important that early-warning information be provided in a manner accessible to all sectors of society. Under the United Nations Framework Convention on Climate Change, the parties commit to promote and facilitate public access to information on climate change.[111] Under international human rights law, access to information is implied in the rights to freedom of opinion and expression.[112] Jurisprudence of regional human rights courts has also underlined the importance of access to information in relation to environmental risks.[113]

79.   Participation in decision-making is of key importance in efforts to tackle climate change. For example, adequate and meaningful consultation with affected persons should precede decisions to relocate people away from hazardous zones.[114] Under the Convention, States parties shall promote and facilitate "public participation in addressing climate change and its effects and developing adequate responses".[115] The right to participation in decision-making is implied in article 25 of the International Covenant on Civil and Political Rights which guarantees the right to "take part in the conduct of public affairs". Equally, the United Nations

Declaration on the Rights of Indigenous Peoples states that States shall consult and cooperate with indigenous peoples "to obtain their free, prior and informed consent" before adopting measures that may affect them.[116] The Convention on the Rights of the Child in article 12 enshrines the right of children to express their views freely in all matters affecting them.

3. Guiding principles for policymaking

80.   Human rights standards and principles should inform and strengthen policymaking in the area of climate change, promoting policy coherence and sustainable outcomes. The human rights framework draws attention to the importance of aligning climate change policies and measures with overall human rights objectives, including through assessing possible effects of such policies and measures on human rights.

81.   Moreover, looking at climate change vulnerability and adaptive capacity in human rights terms highlights the importance of analysing power relationships, addressing underlying causes of inequality and discrimination, and gives particular attention to marginalized and vulnerable members of society. The human rights framework seeks to empower individuals and underlines the critical importance of effective participation of individuals and communities in decision-making processes affecting their lives.

82.   Equally, human rights standards underline the need to prioritize access of all persons to at least basic levels of economic, social and cultural rights, such as access to basic medical care, essential drugs and to compulsory primary education free of charge.

83.   The human rights framework also stresses the importance of accountability mechanisms in the implementation of measures and policies in the area of climate change and requires access to administrative and judicial remedies in cases of human rights violations.[117]

B. Obligations of international cooperation

84.   Climate change can only be effectively addressed through cooperation of all members of the international community.[118] Moreover, international cooperation is important because the effects and risk of climate change are significantly higher in low-income countries.

85.   International cooperation to promote and protect human rights lies at the heart of the Charter of the United Nations.[119] The importance of such cooperation is explicitly stated in provisions of the International Covenant on Economic, Social and Cultural Rights, the Convention on the Rights of the Child, the Convention on the Rights of People with Disabilities and in the Declaration on the Right to Development.[120] According to CESCR and the Committee on the Rights of the Child, the obligation to take steps to the maximum of available resources to implement economic, social and cultural rights includes an obligation of States, where necessary, to seek international cooperation.[121] States have also committed themselves not only to implement the treaties within their jurisdiction, but also to contribute, through international cooperation, to global implementation.[122] Developed States have a particular responsibility and interest to assist the poorer developing States.[123]

86.   The Committee on Economic, Social and Cultural Rights identifies four types of extraterritorial obligations to promote and protect economic, social and cultural rights. Accordingly, States have legal obligations to:

- Refrain from interfering with the enjoyment of human rights in other countries
- Take measures to prevent third parties (e.g. private companies) over which they hold influence from interfering with the enjoyment of human rights in other countries
- Take steps through international assistance and cooperation, depending on the availability of resources, to facilitate fulfilment of human rights in other countries, including disaster relief, emergency assistance, and assistance to refugees and displaced persons
- Ensure that human rights are given due attention in international agreements and that such agreements do not adversely impact upon human rights.[124]

87.   Human rights standards and principles are consistent with and further emphasize "the principle of common but differentiated responsibilities" contained in the United Nations Framework Convention on Climate Change. According to this principle, developed country Parties (annex I) commit to assisting developing country Parties (non-annex I) in meeting the costs of adaptation to the adverse effects of climate change and to take full account of the specific needs of least developed countries in funding and transfer of technology.[125] The human rights framework complements the Convention by underlining that "the human person is the central subject of development",[126] and that international cooperation is not merely a matter of the obligations of a State towards other States, but also of the obligations towards individuals.

88.   Human rights standards and principles, underpinned by universally recognized moral values, can usefully inform debates on equity and fair distribution of mitigation and adaptation burdens. Above all, human rights principles and standards focus attention on how a given distribution of burden affects the enjoyment of human rights.

Intergenerational equity and the precautionary principle

89.   The United Nations Framework Convention on Climate Change stresses principles of particular importance in the context of climate change which are less well developed in human rights law. Notably, these include the notion of "intergenerational equity and justice" and "the precautionary principle", both of which are well-established in international environmental law.

90.   Human rights treaty bodies have alluded to the notion of intergenerational equity.[127] However, the human rights principles of equality and non-discrimination generally focus on situations in the present, even if it is understood that the value of these core human rights principles would not diminish over time and be equally applicable to future generations.[128]

91.   The precautionary principle reflected in article 3 of the United Nations Framework Convention on Climate Change, states that lack of full scientific certainty should not be used

as a reason for postponing precautionary measures to anticipate, prevent or minimize the causes of climate change and mitigate its adverse effects. As discussed above, human rights litigation is not well-suited to promote precautionary measures based on risk assessments, unless such risks pose an imminent threat to the human rights of specific individuals. Yet, by drawing attention to the broader human rights implications of climate change risks, the human rights perspective, in line with the precautionary principle, emphasizes the need to avoid unnecessary delay in taking action to contain the threat of global warming.

IV. Conclusions

92.   Climate change-related impacts, as set out in the assessment reports of the Intergovernmental Panel on Climate Change, have a range of implications for the effective enjoyment of human rights. The effects on human rights can be of a direct nature, such as the threat extreme weather events may pose to the right to life, but will often have an indirect and gradual effect on human rights, such as increasing stress on health systems and vulnerabilities related to climate change-induced migration.

93.   The effects of climate change are already being felt by individuals and communities around the world. Particularly vulnerable are those living on the "front line" of climate change, in places where even small climatic changes can have catastrophic consequences for lives and livelihoods. Vulnerability due to geography is often compounded by a low capacity to adapt, rendering many of the poorest countries and communities particularly vulnerable to the effects of climate change.

94.   Within countries, existing vulnerabilities are exacerbated by the effects of climate change. Groups such as children, women, the elderly and persons with disabilities are often particularly vulnerable to the adverse effects of climate change on the enjoyment of their human rights. The application of a human rights approach in preventing and responding to the effects of climate change serves to empower individuals and groups, who should be perceived as active agents of change and not as passive victims.

95.   Often the effects of climate change on human rights are determined by non-climatic factors, including discrimination and unequal power relationships. This underlines the importance of addressing human rights threats posed by climate change through adequate policies and measures which are coherent with overall human rights objectives. Human rights standards and principles should inform and strengthen policy measures in the area of climate change.

96.   The physical impacts of global warming cannot easily be classified as human rights violations, not least because climate change-related harm often cannot clearly be attributed to acts or omissions of specific States. Yet, addressing that harm remains a critical human rights concern and obligation under international law. Hence, legal protection remains relevant as a safeguard against climate change-related risks and infringements of human rights resulting from policies and measures taken at the national level to address climate change.

97.   There is a need for more detailed studies and data collection at country level in order to assess the human rights impact of climate change-related phenomena and of policies and measures adopted to address climate change. In this regard, States could usefully provide information on measures to assess and address vulnerabilities and impacts related to climate change as they affect individuals and groups, in reporting to the United Nations human rights treaty monitoring bodies and the United Nations Framework Convention on Climate Change.

98.   Further study is also needed of protection mechanisms for persons who may be considered to have been displaced within or across national borders due to climate change-related events and for those populations which may be permanently displaced as a consequence of inundation of low-lying areas and island States.

99.   Global warming can only be dealt with through cooperation by all members of the international community. Equally, international assistance is required to ensure sustainable development pathways in developing countries and enable them to adapt to now unavoidable climate change. International human rights law complements the United Nations Framework Convention on Climate Change by underlining that international cooperation is not only expedient but also a human rights obligation and that its central objective is the realization of human rights.

[Annex omitted]

Annual Report of the United Nations High Commissioner for Human Rights and Reports of the Office of the High Commissioner and the Secretary-General, Human Rights Council, 10th Session, UN Doc. A/HRC/10/61, 15 January 2009.

Notes

1   Most of the submissions made and a summary of discussions of the consultation meeting containing various recommendations made by participants are available at http://www2.ohchr.org/english/issues/climatechange/study.htm.

2   For example, in a joint statement on International Human Rights Day, 10 December 2008, the special procedures mandate holders of the Human Rights Council emphasized that climate change has "potentially massive human rights and development implications".

3   AG/RES.2429 (XXXVIII-O/08), Human rights and climate change in the Americas; Male' Declaration on the Human Dimension of Global Climate Change, 2007.

4   Many of these studies and reports have been submitted to the Office of the United Nations High Commissioner for Human Rights (OHCHR) and are available at http://www2.ohchr.org/english/issues/climatechange/submissions.htm.

5   United Nations Framework Convention on Climate Change (UNFCCC), art. 1, para. 2. The Intergovernmental Panel on Climate Change (IPCC) uses a similar definition, the main difference being that IPCC covers all aspects of climate change and does not make a distinction between climate change attributable to human activity and climate change and variability attributable to natural causes.

6   IPCC was set up jointly by the World Meteorological Organization (WMO) and the United Nations Environment Programme (UNEP) in 1988 to provide authoritative assessments, based on the best available scientific literature, on climate change causes, impacts and possible response strategies.

7   Climate Change 2007 Synthesis Report, adopted at IPCC Plenary XXVII, Valencia, Spain, 12–17

November 2007 (IPCC AR4 Synthesis Report), p. 72.

8   See IPCC AR4 Working Group I (WGI) Report, pp. 23–25.

9   With the exception of impacts on tropical cyclones, the IPCC AR4 considers these impacts very likely (more than 90 per cent certainty). Projections on increased intensity of tropical cyclones are considered likely (more than 66 per cent certainty).

10 See IPCC AR4 Synthesis Report, pp. 48–53. 11 UNFCCC, art. 3, paras. 1 and 2.

11 UNFCCC, art. 3, paras. 1 and 2.

12 While UNFCCC does not include specific greenhouse gas reduction targets, its Kyoto Protocol assigns legally binding caps on greenhouse gas emissions for industrialized countries and emerging economies for the period 2008–2012. The Protocol entered into force in 2005 and has to date been ratified by 183 parties to UNFCCC.

13 See IPCC AR4 Working Group III (WGIII) Report, pp. 99–100.

14 Four other scenarios of higher stabilization levels estimate the likely temperature increases in the range of 2.8°C to 6.1°C, IPCC AR4 WGIII Report, pp. 227–228. A/HRC/10/61, p. 7.

15 See Millennium Ecosystems Assessment 2005, Ecosystems and Human Well-being, Synthesis, pp. 67 and 79.

16 A joint seminar on human rights and the environment organized by OHCHR and UNEP in 2002 also documented a growing recognition of the connection between human rights, environmental protection and sustainable development (see E/CN.4/2002/WP.7).

17 ILO Convention No. 169 (1989) concerning Indigenous and Tribal Peoples in Independent Countries provides for special protection of the environment of the areas which indigenous peoples occupy or otherwise use. At the regional level, the African Charter on Human and Peoples' Rights and the San Salvador Protocol to the American Convention on Human Rights recognize the right to live in a healthy or satisfactory environment. Moreover, many national constitutions refer to a right to an environment of a certain quality.

18 Convention on the Rights of the Child (CRC), art. 24, para. 2 (c).

19 Committee on Economic, Social and Cultural Rights (CESCR), general comments No. 12 (1999) on the right to adequate food (art. 11), para. 4, and No. 14 (2000) on the right to the highest attainable standard of health (art. 12), para. 4.

20 International Covenant on Civil and Political Rights (ICCPR), art. 6; CRC, art. 6.

21 Human Rights Committee, general comments No. 6 (1982) on art. 6 (Right to life), para. 1, and No. 14 (1984) on art. 6 (Right to life), para. 1.

22 Human Rights Committee, general comment No. 6, para. 5. Likewise, the Committee has asked States to provide data on pregnancy and childbirth-related deaths and gender-disaggregated data on infant mortality rates when reporting on the status of implementation of the right to life (general comment No. 28 (2000) on art. 3 (The equality of rights between men and women), para. 10).

23 CRC, art. 6, para. 2.

24 Committee on the Rights of the Child, general comment No. 7 (2006) on implementing rights in early childhood, para. 10.

25 IPCC AR4 Working Group II (WGII) Report, p. 393.

26 United Nations Development Programme (UNDP), Human Development Report 2007/2008, Fighting climate change: Human solidarity in a divided world, p. 8.

27 IPCC AR4 Working Group II Report, p. 317.

28 Inter-Agency Standing Committee, Protecting Persons Affected by Natural Disasters – IASC Operational Guidelines on Human Rights and Natural Disasters, Brooking-Bern Project on Internal Displacement, 2006.

29 International Covenant on Economic, Social and Cultural Rights (ICESCR), art. 11; CRC, art. 24 (c); Convention on the Rights of Persons with Disabilities (CRPD), art. 25 (f) and art. 28, para. 1; Convention on the Elimination of All Forms of Discrimination against Women (CEDAW), art. 14, para. 2 (h); International Convention on the Elimination of All Forms of Racial Discrimination (ICERD), art. 5 (e).

30 ICESCR, art. 11, para. 2.

31 CESCR general comment No. 12 (1999) on the right to adequate food (art. 11), para. 6.
32 IPCC AR4 Synthesis Report, p. 48.
33 UNDP Human Development Report 2006, Beyond scarcity: Power, poverty and the global water crisis.
34 IPCC AR4 WGII Report, p. 275.
35 IPCC AR4 WGII, p. 359. United Nations Millennium Project 2005, Halving Hunger: It Can Be Done, Task Force on Hunger, p. 66. Furthermore, according to the Human Rights Council Special Rapporteur on the right to food, "half of the world's hungry people … depend for their survival on lands which are inherently poor and which may be becoming less fertile and less productive as a result of the impacts of repeated droughts, climate change and unsustainable land use" (A/HRC/7/5, para. 51).
36 See e.g. A/HRC/7/5, para. 51; A/HRC/7/5/Add.2, paras. 11 and 15.
37 See e.g. CESCR general comment No. 12 (1999) on the right to adequate food (art. 11), para. 28.
38 CESCR general comment No. 15 (2002) on the right to water (arts. 11 and 12), para. 2. While not explicitly mentioned in ICESCR, the right is seen to be implicit in arts. 11 (adequate standard of living) and 12 (health). General comment No. 15 provides further guidance on the normative contents of the right to water and related obligations of States.
39 See CEDAW, art. 14, para. 2 (h); CRPD, art. 28, para. 2 (a); CRC, art. 24, para. 2 (c).
40 IPCC AR4 Synthesis Report, pp. 48–49.
41 Millennium Ecosystems Assessment 2005, Ecosystems and Human Well-being, Synthesis, p. 52.
42 According to the UNDP Human Development Report 2006, the root causes of the current water crisis lie in poor water management, poverty and inequality, rather than in an absolute shortage of physical supply.
43 IPCC AR4 WGII Report, p. 191. UNDP Human Development Report 2006.
44 CEDAW, arts. 12 and 14, para. 2 (b); ICERD, art. 5 (e) (iv); CRC, art. 24; CRPD, arts. 16, para. 4, 22, para. 2, and 25; International Convention on the Protection of the Rights of All Migrant Workers and Members of Their Families (ICRMW), arts. 43, para. 1 (e), 45, para. 1 (c), and 70. See also ICESCR arts. 7 (b) and 10.
45 CESCR general comment No. 12, para. 8.
46 See CESCR general comment No. 12, CEDAW general recommendation No. 24 (1999) on art. 12 of the Convention (women and health); CRC general comment No. 4 (2003) on Adolescent health and development in the context of the Convention on the Rights of the Child.
47 IPCC AR4 Synthesis, p. 48.
48 Uncertainty remains about the potential impact of climate change on malaria at local and global scales because of a lack of data and the interplay of other contributing non-climatic factors such as socio-economic development, immunity and drug resistance (see IPCC WGII Report, p. 404).
49 A/62/214, para. 102. A/HRC/10/61 page 13
50 IPCC AR4 WGII Report, p. 12.
51 ICESCR, art. 11. See also Universal Declaration of Human Rights, art. 25, para. 1; ICERD, art. 5 (e) (iii); CEDAW, art. 14, para. 2; CRC, art. 27, para. 3; ICRMW, art. 43, para. 1 (d); CRPD, arts. 9, para. 1 (a), and 28, paras. 1 and 2 (d).
52 CESCR general comment No. 12, para. 6.
53 See CESCR general comment No. 7 (1997) on the right to adequate housing (art. 11 (1) of the Covenant): Forced evictions.
54 CESCR general comment No. 12, para. 8.
55 IPCC AR4 WGII Report, p. 333.
56 IPCC AR4 WGII Report, p. 672.
57 A/63/275, paras. 31–38.
58 UNDP Human Development Report 2007/2008, Fighting climate change: Human solidarity in a divided world, p. 9.
59 In this regard the Guiding Principles on Internal Displacement (E/CN.4/1998/53/Add.2, annex) provide that "at the minimum, regardless of the circumstances, and without discrimination, competent

authorities shall provide internally displaced persons with and ensure safe access to: ... basic shelter and housing" (principle 18).

60   The right to self-determination is enshrined in Articles 1 and 55 of the Charter of the United Nations and also contained in the Declaration on the Right to Development, art. 1, para. 2, and the United Nations Declaration on the Rights of Indigenous Peoples, arts. 3 and 4.

61   Human Rights Committee, general comment No. 12 (1984) on art. 1 (Right to self-determination), para. 6. See also Committee on the Elimination of Racial Discrimination (CERD), general recommendation 21 (1996) on the right to self-determination.

62   See e.g. IPCC AR4 WGII Report, p. 374.

63   National communications, submitted according to arts. 4 and 12 of UNFCCC, make frequent references to the human impacts of climate change, but generally do so in an aggregate and general manner, mentioning for example that people living in poverty are particularly vulnerable.

64   IPCC AR4 WGII, p. 398. See also submission by the United Nations Development Fund for Women available at: http://www2.ohchr.org/english/issues/climatechange/index.htm.

65   E. Neumayer and T. Plümper, The Gendered Nature of Natural Disasters: The Impact of Catastrophic Events on the Gender Gap in Life Expectancy, 1981–2002, available at http://ssrn.com/abstract=874965. As the authors conclude, based on the study of disasters in 141 countries, "[a] systematic effect on the gender gap in life expectancy is only plausible if natural disasters exacerbate previously existing patterns of discrimination that render females more vulnerable to the fatal impact of disasters" (p. 27).

66   Y. Lambrou and R. Laub, "Gender perspectives on the conventions on biodiversity, climate change and desertification", Food and Agriculture Organization of the United Nations (FAO), Gender and Population Division, pp. 7–8.

67   See e.g. IPCC AR4 WGII Report, p. 398; International Strategy for Disaster Reduction, Gender Perspectives: Integrating Disaster Risk Reduction into Climate Change Adaptation. Good Practices and Lessons Learned, UN/ISDR 2008.

68   UNFCCC, Climate Change: Impacts, Vulnerabilities and Adaptation in Developing Countries, 2007, p. 16.

69   UNICEF Innocenti Research Centre, Climate Change and Children: A Human Security Challenge, New York and Florence, 2008; UNICEF UK, Our Climate, Our Children, Our Responsibility: The Implications of Climate Change for the World's Children, London, 2008.

70   World Bank, Global Monitoring Report 2008 – MDGs and the Environment: Agenda for Inclusive and Sustainable Development, p. 211.

71   UNICEF UK (see footnote 69 above), p. 29.

72   For example, UNEP and UNICEF have developed an environmental resource pack for child-friendly schools designed to empower children (see footnote 69 above, UNICEF Innocenti Research Centre, p. 28).

73   See e.g. CRC, general comment No. 4 (2003) on adolescent health and development in the context of the Convention on the Rights of the Child.

74   M. Macchi and others, Indigenous and Traditional Peoples and Climate Change, International Union for Conservation of Nature, 2008.

75   See e.g. report of the Special Rapporteur on the situation of human rights and fundamental freedoms of indigenous peoples, A/HRC/4/32, para. 49.

76   In April 2008, the Permanent Forum for Indigenous Issues stated that climate change "is an urgent and immediate threat to human rights" (E/C.19/2008/13, para. 23).

77   E/C.19/2008/13, para. 4. The Permanent Forum also recommended that a mechanism be put in place for the participation of indigenous peoples in climate change negotiations under UNFCCC (ibid., para. 30).

78   IPCC AR4 WGII Report, p. 865 (citing Robinson and Herbert, 2001).

79   Key provisions include the right to effective mechanisms for prevention of, and redress for, actions which have the aim or effect of dispossessing them of their lands, territories or resources (art. 8); the principle of free, prior and informed consent (art. 19), the right to the conservation and protection of

the environment and indigenous lands and territories (art. 29), the right to maintain, control, protect and develop their cultural heritage and traditional knowledge and cultural expressions (art. 31).

80 See the provisions on cultural rights in ICCPR, art. 27, and ICESCR, art. 15.

81 Available at: http://inuitcircumpolar.com/files/uploads/icc-files/FINALPetitionICC.pdf.

82 More recent studies refer to estimates for the same period of 200 million (Stern Review on the Economics of Climate Change, 2006, available at http://www.hm-treasury.gov.uk/sternreview_index.htm) and 250 million (Human tide: the real migration crisis, Christian Aid 2007). See also IPCC AR4 WGII Report, p. 365 and the Norwegian Refugee Council, Future floods of refugees: A comment on climate change, conflict and forced migration, 2008.

83 See e.g. contributions to Forced Migration Review, vol. 1, No. 31, October 2008.

84 Adapted from typology proposed by the Representative of the Secretary-General on human rights of internally displaced persons and also used in the working paper submitted by the IASC informal group on migration/displacement and climate change, "Climate Change, Migration and Displacement: who will be affected", 31 October 2008.

85 Guiding Principles on Internal Displacement (E/CN.4/1998/53/Add.2, annex), principles 1, para. 1, and 6, para. 1 .

86 Principle 8.2, Principles on Housing and Property Restitution for Refugees and Displaced Persons (endorsed by the Sub-Commission on the Promotion and Protection of Human Rights in resolution 2005/2); FAO/IDMC/NRC/OCHA/OHCHR/UN-Habitat/UNHCR: Housing and Property Restitution for Refugees and Displaced Persons: Implementing the "Pinheiro Principles", 2007.

87 The Guiding Principles have gained wide acceptance and were recognized by the General Assembly in the 2005 World Summit Outcome (A/RES/60/1) "as an important international framework for the protection of internally displaced persons".

88 See IASC working paper referred to in footnote 84 above.

89 Representative of the Secretary-General on human rights of internally displaced persons, Displacement Caused by the Effects of Climate Change: Who will be affected and what are the gaps in the normative framework for their protection? background paper, 2008, available at: http://www2.ohchr.org/english/issues/climatechange/submissions.htm.

90 In the face of rising sea levels, migration is one adaptation strategy which is already being implemented in low-lying island States, such as Kiribati, the Maldives, and Tuvalu. So far this population movement has mainly taken the form of in-country resettlement schemes (IPCC AR4 WGII Report, p. 708).

91 The papers (E/CN.4/Sub.2/AC.4/2004/CRP.1; E/CN.4/Sub.2/2005/28) were prepared by Françoise Hampson pursuant to a request from the Commission on Human Rights (decision 2004/122) to prepare a report on the legal implications of the disappearance of States for environmental reasons. A questionnaire was prepared in 2006 (E/CN.4/Sub.2/AC.4/2006/CRP.2) with a view to obtaining more accurate data on the nature, scale and imminence of the problem, but as yet no follow-up has been given to this initiative.

92 This point was made by Ms Hampson and other panellists at the consultation meeting organized by OHCHR on 22 October 2008, summary of discussions available at: http://www2.ohchr.org/english/issues/climatechange/docs/SummaryofDiscussions.doc.

93 See e.g. Government of the United Kingdom of Great Britain and Northern Ireland, The National Security Strategy of the United Kingdom: Security in an interdependent world, 2008 and German Advisory Council on Global Change, World in Transition – Climate Change as a Security Risk, 2008.

94 As the Chairman of the Nobel Committee stated: "The chief threats may be direct violence, but deaths may also have less direct sources in starvation, disease, or natural disasters" (Presentation speech 10 December 2007).

95 International Alert and Swedish International Development Cooperation Agency (SIDA), A Climate of Conflict, 2008, p. 7. In the same vein, the Special Rapporteur on the right to food observes that conflicts in Africa, including in the Darfur region, are linked to land degradation and related fights over resources (A/HRC/7/5, para. 51).

96 See e.g. H. Buhaug, N.P. Gleditsch and O.M. Theisen, Implications of Climate Change for Armed Conflict, 2008. As the IPCC AR4 WGII Report points out (citing Fairhead, 2004) there are many

other intervening and contributing causes of conflict and many environmentally-influenced conflicts in Africa are related to abundance of natural resources (e.g. oil and diamonds) rather than scarcity, suggesting "caution in the prediction of such conflicts as a result of climate change" (p. 365).

97   UNFCCC, art. 4, para. 8, and Kyoto Protocol, arts. 2, para. 3, and 3, para. 14.

98   For a discussion of the human rights dimensions of mitigation and adaptation policies see International Council on Human Rights Policy, Climate Change and Human Rights: A Rough Guide, 2008, chapter II.

99   Statement of the Special Rapporteur on the right to food, 22 May 2008, at the special session of the Human Rights Council on the global food crisis.

100   E/C.12/2008/1, para. 13.

101   See e.g. M. Macchi and others, Indigenous and Traditional Peoples and Climate Change, International Union for Conservation of Nature, 2008. CERD expressed concern about plans to establish a large-scale biofuel plantation and the threat it constituted to the rights of indigenous peoples to own their lands and enjoy their culture (CERD/C/IDN/CO/3, para. 17).

102   E/C.19/2008/13, para. 45.

103   In recent years, several lawsuits related to greenhouse gas emissions and their contribution to climate change have been filed at national level against State authorities and private actors. However, the Inuit petition to the Inter-American Commission on Human Rights (see footnote 81 above) remains the only case to have invoked human rights law. For an overview of recent climate change-related lawsuits, see e.g. International Council for Human Rights Policy, Climate Change and Human Rights: A Rough Guide, 2008.

104   The Human Rights Committee has clarified that for a person to claim to be a victim of a violation of a right, "he or she must show either that an act or an omission of a State party has already adversely affected his or her enjoyment of such a right, or that such an effect is imminent …" Aalbersberg v. The Netherlands (No. 1440/2005). In several cases concerning environmental harms, the Committee has found that the author(s) did not meet these criteria for a victim of a human rights violation.

105   For a review of relevant jurisprudence, see Asia Pacific Forum of National Human Rights Institutions, Human Rights and the Environment, 12th Annual Meeting, Sydney, 2007; D. Shelton, "Human rights and the environment: jurisprudence of human rights bodies", background paper No. 2, Joint UNEP-OHCHR Expert Seminar on Human Rights and the Environment, January 2002, available at http://www.unhchr.ch/environment/bp2.html.

106   Budayeva and Others v. Russia, European Court of Human Rights (ECHR), No. 15339/02.

107   See CESCR general comment No. 3 (1990) on the nature of States parties' obligations (art. 2, para. 1, of the Covenant). For a discussion of the concept of progressive realization under the international human rights treaties, see report of the United Nations High Commissioner for Human Rights to the Economic and Social Council (E/2007/82).

108   See e.g. CESCR general comments No. 3, paras. 2 and 9, and No. 14 (2000) on the right to the highest attainable standard of health (art. 12), para. 31.

109   CESCR general comment No. 3, para. 10.

110   See Statement by CESCR (E/C.12/2007/1, paras. 4 and 6).

111   UNFCCC, art. 6.

112   Universal Declaration of Human Rights, art. 19, and ICCPR, art. 19.

113   See e.g. Guerra and Others v. Italy, ECHR 14967/89; Inter-American Court of Human Rights, Case of Claude Reyes et al. v. Chile. Merits, Reparations and Costs, Series C, No. 151.

114   See A/63/275, para. 38.

115   Article 6. The amended New Delhi work programme on article 6 elaborates on and reinforces this point (FCCC/CP/2007/6/Add.1, decision 9/CP.13, annex, para. 17 (k)).

116   United Nations Declaration on the Rights of Indigenous Peoples, art. 19.

117   Useful guidance on how human rights standards and principles can be incorporated into policy measures are found in various guidance tools, including Frequently Asked Questions on a Human Rights-Based Approach to Development Cooperation; OHCHR (2006), Principles and Guidelines for a Human Rights Approach to Poverty Reduction Strategies, available at http://www.ohchr.org/

EN/PublicationsResources/Pages/SpecialIssues.aspx.

118   In the words of the special procedures mandate holders of the Human Rights Council, in a joint statement on International Human Rights Day, 10 December 2008: "Today the interests of States, and the impacts of actions by States, are ever more interconnected. New challenges include ensuring global access to food, and those presented by climate change and financial crisis have potentially massive human rights and development implications. If we are to confront them effectively we must do so collectively."

119   See articles 1, paragraph 3, 55 and 56.

120   ICESCR, arts. 2, para. 1, 11, para. 2, 15, para. 4, 22 and 23; Convention on the Rights of the Child, arts. 4 and 24, para. 4; CRPD, art. 32; Declaration on the Right to Development, arts. 3, 4 and 6.

121   CESCR, general comment No. 3, para. 11; Committee on the Rights of the Child, general comment No. 5 (2003) on general measures of implementation of the Convention on the Rights of the Child (arts. 4, 42 and 44, para. 6), para. 7.

122   See e.g. CRC, general comment No. 5, para. 7.

123   See CESCR general comment No. 3, para. 14.

124   See e.g. CESCR general comments No. 12 (1999) on the right to adequate food (art. 11); No. 13 (1999) on the right to education (art. 13); No. 14 (2000) on the right to the highest attainable standard of health (art. 12); and No. 15 (2002) on the right to water (arts. 11 and 12 of the Covenant).

125   UNFCCC, art. 4, paras. 4 and 9.

126   Declaration on the Right to Development, art. 2, para. 1.

127   See CESCR general comments No. 12, para. 7, and No. 15, para. 11. Equally the concern for how current needs and rights affect the future health and development of the child is central to the Convention on the Rights of the Child (see e.g. Committee on the Rights of the Child general comment No. 4 (2003) on adolescent health and development in the context of the Convention on the Rights of the Child, para. 13).

128   For a discussion on the relationship between intergenerational equity and human rights in the context of climate change, see S. Caney, "Human rights, climate change, and discounting", Environmental Politics, vol. 17, No. 4, August 2008, p. 536.

Chapter 17

Françoise Hampson

THE HUMAN RIGHTS SITUATION OF INDIGENOUS PEOPLES IN STATES AND OTHER TERRITORIES THREATENED WITH EXTINCTION FOR ENVIRONMENTAL REASONS

I. Introduction

1.   Following discussion of the issue at the twenty-first session of the Working Group on Indigenous Populations (see E/CN.4/Sub.2/2003/22) and during the fifty-fourth session of the Sub-Commission, the Sub-Commission adopted resolution 2003/24.

2.   The Commission, in decision 2004/122 decided "urgently to call upon the Sub Commission on the Promotion and Protection of Human Rights to prepare a report on the legal implications of the disappearance of States for environmental reasons, including the implications for the human rights of their residents, with particular reference to the rights of indigenous people".

3.   Françoise Hampson produced a working paper (E/CN.4/Sub.2/AC.4/2004/CRP.1) which was discussed at the twenty-second session of the Working Group on Indigenous Populations and the fifty-fifth session of the Sub-Commission. In resolution E/CN.4/Sub.2/2004/10, the Sub-Commission requested the Commission, inter alia, for the authority to send out a questionnaire. In decision 2005/112, the Commission approved the Sub-Commission's request. Ms. Hampson was also requested to submit an expanded working paper to the Sub-Commission at its fifty-seventh session. The present document is submitted in accordance with that request.

4.   In the time available since the adoption of the Commission's decision, it has not been possible to formulate and distribute the questionnaire. It is hoped that that will be done before the end of August 2005.

II. The issue

5.   Certain States face the likelihood of the disappearance of the whole or a significant part of their surface area for environmental reasons. This report does not inquire into the cause of such disappearance but takes it as a fact. There will come a point at which life is not sustainable in those States. The need for the evacuation of such States gives rise to a variety of human rights issues. Whilst the previous discussions have focused principally on rising sea levels and salt water

entering freshwater aquifers, it is clear that that is not the only environmental threat facing these vulnerable populations. The tsunami in December 2004 exposed the vulnerability to tidal waves. The questionnaire will seek to establish what type of environmental threats face different populations. At first sight, they could include volcanoes, earthquakes, tidal waves and rising sea levels. Whilst the actual occurrence of the first three is unpredictable, the fact of vulnerability already exists. In the case of rising sea levels, the danger is not from an event but from a process. That process is already under way in the case of certain States. The eventual need for the population to move is not merely predictable but inevitable, unless the process is halted or reversed. What may be unpredictable is the time frame within which decisions have to be taken.

6.   Three different kinds of situation, all of which relate to environmental degradation, need to be distinguished. There is first the problem of environmental damage caused to the land of indigenous peoples, with an impact on the enjoyment of their land rights and with implications for a wide variety of their personal rights. As defined, that issue concerns exclusively indigenous peoples and could be addressed by the Permanent Forum on Indigenous Issues and the Special Rapporteur on the situation of human rights and fundamental freedoms of indigenous people. Indeed, in his report to the sixty-first session of the Commission (E/CN.4/2005/88), Mr. Stavenhagen addressed this issue and recommended that "participatory scientific research be promoted in this area (with special attention paid to vulnerable environments such as the Arctic, the forests of the far North, tropical forests and high mountain areas)" (para. 92).

7.   The second situation is where life is no longer sustainable on the land of a particular indigenous group, requiring that they move and/or be given different or additional land within the same State. If such a situation currently exists or is likely to exist soon, the situation could form the basis of a joint visit and/or study by the Special Rapporteur and the Representative of the Secretary General on the human rights of internally displaced persons.

8.   The third situation, the one with which this report is concerned, regards the population of sovereign States. It is not a question of their being able to move within the State. They will not become internally displaced persons. Nor will they become refugees, in the sense of the 1951 Convention on the Status of Refugees. That treaty does not recognize as refugees persons who have to leave their country for reasons relating to the environment. In certain circumstances, they may be regarded as indigenous peoples but, following their displacement, they will not be indigenous in relation to receiving countries. It is not clear that they come within the mandate of any existing special procedure and yet they face the prospect of complete disruption to their lives and livelihood.

9.   Affected States fall into three categories:

   a.   States which will totally disappear;
   b.   States a significant proportion of whose territory will disappear, leaving only such territory as will be unable to support the existing population. This may arise where what will remain will not be of a size to support the existing population and/or where what

remains is not of a nature to support the existing population;

c.   States a significant proportion of whose territory will disappear, with serious implications for the existing population.

One of the goals of the questionnaire will be to establish which are the affected States.

10.   Certain territories whose populations may be similarly affected appear not to be sovereign States. They may be colonies, some other form of non-self-governing territory or States in free association with another State. Insofar as the inhabitants of the territories in question are citizens of a sovereign State independent of the territory, that State would appear to have the responsibility to secure the protection of the rights of that population. A second goal of the questionnaire will be to establish which States have such a relationship with one or more territories threatened with disappearance and whether they acknowledge a responsibility to secure the future of the population of those territories. Prima facie, such an acknowledgement would take the population group outside the scope of this study. If, however, it emerges that the people concerned are to be regarded as an indigenous group, it will be necessary to determine whether the State acknowledges that they have that identity and how it would propose to deal with that dimension of the situation.

III. Legal implications of the situation

11.   Whilst States Members of the United Nations are used to addressing issues of State succession, it would appear that the extinction of a State, without there being a successor is unprecedented. In the case of State succession, one State may become several or several States may become one. In addition, part of the territory may be detached from one State and joined to another State. Any form of State succession gives rise to a variety of legal questions. In some cases, populations have been forcibly transferred. In others, the population remains in place but experiences a change of citizenship or runs the risk of becoming stateless. In certain recent cases, people previously belonging to a majority population have found themselves new minorities in the newly created State. In recent cases of State succession, citizenship and other status issues have been addressed by putting pressure on the successor State to make it possible, in practice, for long-term residents to acquire citizenship or equivalent rights.

12.   In the case of those States which are likely to disappear for environmental reasons, there would appear to be no successor State on whom obligations can be imposed and eventually the predecessor State will no longer be in existence.

13.   The disappearance of States for environmental reasons will give rise to a variety of legal questions, not all of which relate specifically to human rights. The first difficulty is when determining a State counts as having disappeared: at the point when the population can only survive by leaving, even if parts of the territory remain above water, or only when the entire territory is submerged? Will the State simply cease to exist or will some form of recognition of non-existence, at least on the part of the United Nations, be required? Will

the same principle apply in relation to all international organizations of which the State in question is a member? Who determines whether a territory should be considered as having disappeared for environmental reasons?

14.   The following examples merely illustrate the issues which may arise:

- Who, if anyone, will assume responsibility for any national debt?
- What will be the status of what were internal or territorial waters or an exclusive economic zone, particularly in the light of article 121.3 of the United Nations Convention on the Law of the Sea?
- Who will assume responsibility for marking hazards to navigation in a barely submerged State?
- What will be the status of citizens and of legal persons registered or incorporated in such territories?
- What will be the status of diplomatic representatives of the State outside the territory of the State at the time of its disappearance?

15.   A study by the Sub-Commission can only address the human rights implications of the situation for the populations concerned. The other issues will need to be examined elsewhere.

IV. Human rights of the affected populations

16.   There are a variety of concerns affecting the populations of such States, including citizenship rights, rights relating to forced relocation, rights in the admitting State and, currently, rights in relation to their State of origin.

A. Citizenship rights

17.   It is difficult to see how citizenship can retain any meaning when the State itself has ceased to exist. Is citizenship simply an issue of individual right, or does it have a collective dimension? Does it include, for example, a right to live together with other citizens? Does the population of a State have a right to remain a collectivity? If so, what are the implications for a State willing to admit at least some members of the group? If the citizens of a State which has ceased to exist are relocated to one State, do they become a national minority? Is it possible to belong to a nation which has ceased to exist? Where they formed an indigenous group in the State of origin, can they still be regarded as an indigenous group in the receiving State, even though they are anything but indigenous to it? Can they remain citizens of their State of origin? Are they entitled to some form of government-in-exile? Could such governments exist indefinitely, or only for so long as there were citizens whom they could, in some sense, represent? Would they have to be prohibited from allowing transmission of such "citizenship"? Does such a population group have any claim to a territory of its own to replace the territory it has lost? Presumably any such claim would be against the international community as a whole, rather than against a particular State. States at risk of disappearing for environmental reasons tend to be geographically closest to States which cannot reasonably be held solely responsible

for the environmental factors giving rise to the disappearance of the State. There would appear to be no moral or legal reason for assuming that the closest State should assume the entirety of the burden of resettlement. How, if at all, are any of these questions affected by the possibly indigenous character of some of the affected populations?

18.   If the possibly indigenous character of some of the affected populations is thought to make a difference to their entitlements, it may become necessary to define "indigenous". In certain contexts, the term has historically been used to identify populations which were there first, in contrast to those who came later. If such a limited definition were used, many of the affected populations would probably not be regarded as indigenous. If the term can be used to describe those who were there first, where none have come later, some of the people affected would appear to be indigenous. They may be able to rely on self-identification as indigenous peoples.

19.   In view of the significance of self-identification, the questionnaire will need to seek to establish whether or not the affected populations regard themselves as indigenous.

B. Rights relating to forced relocation

20.   What rights, if any, do the affected populations have vis-à-vis other States? A status analogous to refugee status would not appear to be adequate. The people affected need more than a right to claim environmental asylum. They need a right to be granted it. Any State that might be willing to admit hundreds or even a few thousand may well be reluctant to accept entry entitlement as of right or to accept the entitlement of some, if that is thought to imply the entitlement of all.

C. Rights in the receiving State

21.   What rights will such people have within the receiving State? Will they be entitled to insist on living in community? Do they have any claim to land on which to live in community? Can those individuals who do not wish to live in community claim freedom of movement within the receiving State, even whilst others claim the right to live in community? Are they entitled to receive the same support, in kind and in services, as is provided to refugees? Are they entitled to claim citizenship? There may be a need for a fast-track procedure to claim citizenship. If not, is the receiving State at least required to issue such people with documentation, including travel documents? Some of the issues identified at paragraph 17 supra involve claims against receiving States.

D. Rights in relation to their own States

22.   At first sight, the easiest question to address is the issue of the rights of the affected populations vis-à-vis their own States. It is not clear, however, what that entails in this situation. Where there is action which their States could take to mitigate or delay the harm, there may be a human rights issue. The previous lack of a tsunami early-warning system in the Indian Ocean

may be an example of such an issue. Such a system already exists for States in or bordering on the Pacific Ocean. So far as forced relocation is concerned, the people are not yet victims and, when they do become victims, it will not be on account of an act or omission on the part of their own State. The current issue is the need for international representations to be made, solutions to be found and plans put in place, preferably in consultation with the people who will be affected. Those concerns may be understood as being included within the right to development, but they extend beyond the traditional scope of that notion.

E. Which are the affected States?

23.   There does not appear to be any agreed list of the affected States. An examination of sources such as The World Factbook does not resolve the question. Whilst the entry for each territory gives the maximum elevation, that cannot determine the question. First, just because the maximum elevation is low does not necessarily mean that the territory is at risk of being submerged by rising water levels. The experience of rising water levels does not appear to be being experienced evenly, all over the globe. Second, just because the territory contains high ground does not mean that it will be capable of sustaining its population, should it lose what is at present a coastal area in which the population lives. This is one of the reasons why the questionnaire is so important. It will need to seek to identify vulnerability by reference to the need for population movement outside national territory. In some cases, an island or group of islands may be able to absorb the likely displacement of population. In other cases, such displacement might make life unsustainable for the population as a whole but sustainable for the current population. In that case, the displaced population would need to leave the territory but the current population could continue to live in what remains of the territory. In other cases, the entire population may need to move. It will therefore be necessary to obtain information with regard to the number who will need to be displaced and the degree of vulnerability.

24.   A certain number of territories with a maximum elevation of below 100 metres appear to be territories with no indigenous population, on which there are small military garrisons and/or meteorological observers and/or scientists. A larger number of such territories appear to be in free association with a sovereign State or non-self-governing territories. Paragraph 10 above explains how the questionnaire will be designed to address the problems in such territories. Whilst the primary responsibility may be thought to attach to that State, at least in the case of non-self-governing territories, particular issues may arise which may be distinct from individual citizenship rights.

25.   The States most likely to be affected would seem to be:

- Tuvalu: highest point 5 m; population 11,468;
- Nauru: highest point 61 m; population 12,809;
- Kiribati: highest point 81 m; population 100,798;
- Maldives: highest point 2.4 m; population 339,330;
- Bahamas: highest point 63 m; population 299,697.

26.   The States most likely to be affected are principally in the Pacific Ocean but also include States in the Indian Ocean and the Caribbean. The States listed in paragraph 25 are all islands or groups of islands. One State which will not disappear but which risks losing a significant proportion of its surface area for environmental reasons is Bangladesh.

27.   The total population of the States identified in paragraph 25 is under half a million. For reasons indicated in paragraph 17, there would seem to be no reason, in law or morality, why the closest States should shoulder the whole of the burden. It should also be borne in mind that one such State, New Zealand, already has responsibilities for the populations of territories in free association with it, many of whose territories are likely to disappear for environmental reasons.

F. What is currently being done about the problem?

28.   The issue of sustainable development is being addressed. The Declaration of Barbados was adopted in 1994, in the context of the Global Conference on the Sustainable Development of Small Island Developing States. The Conference does not, however, appear to be addressing the problem of States which will completely disappear for environmental reasons and does not seem to be addressing the rights of affected populations. There is a reference to the possibility of small island developing States (SIDS) potentially becoming uninhabitable, but the provisions of the Declaration are all directed at promoting sustainability.

29.   The United Nations Environment Programme has developed Assistance in the Implementation of the Barbados Programme of Action for the Sustainable Development of Small Island Developing States (SIDS). Again, the focus is on sustainability and not on what happens when it is determined that the situation of a particular State is unsustainable. In January 2005, there occurred the follow-up meeting to the Barbados Declaration (Barbados + 10) in Mauritius. It appears that there were no discussions in official sessions of the need to provide for the displacement of the population of certain States.

30.   The otherwise impressive Mauritius Strategy for the Further Implementation of the Programme of Action for the Sustainable Development of Small Island Developing States (A/ CONF.207/CRP.7, 13 January 2005) acknowledges that climate change and rising sea levels may threaten the very existence of some small island developing States, but its proposals in that regard only address the mitigation of the environmental threats and do not address the need for population displacement. Whilst recognizing the vulnerability of small island developing States, there is nothing in the Mauritius Declaration on the need for population movement.

31.   There are also regional and sub-regional groupings such as the Pacific Islands Forum. That particular body deals with a wider range of issues encompassing good governance and regional cooperation. Its vision statement makes express reference to the quality of people's lives and respect for indigenous and other values, customs and traditions. There is no specific reference, however, to the situation of member States which disappear for environmental reasons.

32.   It is striking that such representations as have already been made appear to have been made bilaterally, rather than through a regional or international framework. The Government of Tuvalu has warned that it may need to evacuate the islands within the coming decades. New Zealand has agreed to admit an annual quota. Australia is reported as having refused to do so. It is reported that the Government of Tuvalu has sought to institute legal proceedings against the United States of America and Australia for failing to control global warming.

33.   Other States are making arrangements within the territory available to them and planning on relocating people from islands which may disappear to other islands. It is not clear that such relocation is sustainable in very vulnerable environments.

34.   Many of the territories and States likely to disappear for environmental reasons are members of the Commonwealth. Over half of the members of the Commonwealth are small States. Whilst the Commonwealth has undertaken a variety of initiatives with regard to small States, including ones relating to their vulnerabilities, nothing on its web site suggests that it has, at the level of the organization, addressed the problem of the relocation of citizens of States which disappear for environmental reasons. Contact has been established with a member of the Commonwealth Secretariat and Ms. Hampson hopes to have discussions with members of that organization, both with regard to identifying what measures are needed and also with regard to the best way of proceeding to maximize the participation of all interested parties in the process.

G. Next steps

35.   Within the next few months, Ms. Hampson, in consultation with NGOs, academic experts and the staff of the Office of the United Nations High Commissioner for Human Rights, will draft and send out a questionnaire covering the ground indicated in this working paper. To that end, she hopes to organize a meeting with interested persons during the twenty-third session of the Working Group on Indigenous Populations and, if necessary, during the fifty-sixth session of the Sub-Commission.

36.   Within the next few months, Ms. Hampson hopes to organize a meeting with members of the Commonwealth Secretariat, as indicated in paragraph 34.

37.   The issue appears to be a serious problem for a very limited number of States. The situation is, at least in the case of Tuvalu, urgent. Little appears to be known about the question outside the affected States. The first task appears to be the obtaining of accurate and specific data on the nature, scale and imminence of the problem. This may be one of those situations in which the concern may be better addressed as a humanitarian issue, rather than as a matter of legal right. It nevertheless has legal ramifications, particularly with regard to the relationship between the displaced population and a receiving State. Previous experience relating to displaced indigenous populations suggests that, if handled badly, the consequences can be disastrous both for the displaced population and the host community. It is to be hoped

that by addressing the issue in advance of the crisis and by seeking to ensure the greatest possible participation of the affected populations, some of those difficulties can be avoided.

38.   In view of the time lag between the request for a study and its authorization by the Commission, it is suggested that the Sub-Commission request the appointment of a Special Rapporteur this session. If approved, the first report of the Special Rapporteur would be made to the fifty-eighth session of the Sub-Commission. By that time, the results of the questionnaire should be available. The first report would analyse the results of the questionnaire. Subject to what emerges from the questionnaire, a second report could report on the views of the affected populations and interested States and international organizations with regard to possible solutions and consider the implications of the current situation. The final report would suggest possible solutions and possible ways of achieving them.

[Footnotes omitted]

Françoise Hampson, Commission on Human Rights, Sub-Commission on the Promotion and Protection of Human Rights, Fifty-Seventh Session, UN Doc. E/CN.4/Sub.2/2005/28, 16 June 2005.

Chapter 18

Raquel Rolnik

MISSION TO MALDIVES
Report of the Special Rapporteur on the Right
to Adequate Housing

I. Introduction

1.   At the invitation of the Government, the Special Rapporteur on adequate housing undertook
an official visit to Maldives from 18 to 25 February 2009. The main purpose of the mission was
to examine the impact of climate change on the right to adequate housing and the achievements
and difficulties encountered in the post-tsunami reconstruction process. Over the course of her
mission, the Special Rapporteur identified a number of additional issues affecting the right to
adequate housing that will be detailed in the present report.

2.   During her visit, the Special Rapporteur met with high-ranking officials and representatives
of the Government in Malé and other islands, including Their Excellencies Dr. Mohamed
Waheed, Vice-President; Mr. Ahmed Naseem, Minister of State for Foreign Affairs; Mr.
Mohamed Aslam, Minister of Housing, Transport and Environment; Mr. Abdulla Shahid,
Minister of State for Housing, Transport and Environment and Mr. Ali Hisham, Minister
of Finance and Treasury. The Special Rapporteur also held meetings with parliamentarians,
Maldivian citizens and members of civil society and non-governmental organizations (NGOs)
and international agencies.

3.   In addition to Malé and the surrounding islands of HulhuMalé and Villingili, the Special
Rapporteur visited the islands of Naifaru, Hinnavaru, Dhuvaafaru, Kandhulhudhoo, and
Madduvari, where she met with local authorities and communities.

4.   She wishes to warmly express her gratitude to the Government of Maldives for the
invitation, the constructive dialogue, its support throughout and after the visit and its
commitment to progressing with the implementation of human rights.

II. General overview

5.   Maldives is an archipelago composed of a chain of coral atolls with 1,192 islands covering
an area of more than 90,000 km^2 and stretching 820 km in length. The islands were traditionally
inhabited for more than 3,000 years by fishermen, sailors and their families. Of the islands,

only 203 (59 per cent of the total land area) are inhabited, and those by approximately 300,000 individuals. Of all islands, 96 per cent are smaller than 1km².[1]

6.   The distinctive geography of Maldives plays an important role in political, economic and social issues. Around one third of the population of Maldives is concentrated in Malé.

The remaining population is dispersed among the islands, with only three islands aside from Malé having a population greater than 5,000. Seventy-four islands have a population of less than 500. This dispersion contributes to difficulties in mobility and the provision of governmental services, as reflected in the wide disparities in income and access to social services and infrastructure between the capital and the outer atolls.[2]

7.   Demographic growth and the rise in consumption and internal migration associated with the development of the country in recent decades has put pressure on small inhabited islands, posing new challenges to the traditional practices of land use and allocation for community expansion.

8.   Adding to these constraints and challenges, the situation of the Maldivian national territory makes the country particularly vulnerable to global phenomena such as potential climate change and its consequences, including the increasing number of natural disasters and the rising sea level. About 80 per cent of the islands are less than 1m above sea level.

These factors have an important impact on both the fragile environment and the communities of the archipelago. The tsunami that hit the country in 2004 was the most manifest illustration of this vulnerability. Pushing many families into even more precarious living conditions, the tsunami contributed to bringing the issue of adequate housing to the forefront of political debates and general concern.

9.   The traditional concept of receiving a plot of land and dividing it amongst new generations has resulted in families inheriting progressively smaller plots of land and houses, which adds tension to an already difficult situation with regard to accessing land and adequate housing. New trends such as high rates of population growth and increasing migration, combined with limited available land for construction, has increased pressure on housing.[3]

10.   The Special Rapporteur notes with satisfaction that housing was declared one of the five priorities of the current Government. She also highly commends the new Constitution for its reference to the right to adequate housing and the legislative commitment to achieve this right for all.

III. Climate change and natural disasters

11.   In her introductory report to the General Assembly, the Special Rapporteur declared her intention to investigate the link between climate change and the right to adequate housing and to take as full advantage as possible of her country visits to gather information and practical examples on the topic.[4]

12.   In its third assessment report, the Intergovernmental Panel on Climate Change stated that climate change would increase the magnitude and frequency of weather extremes, such

as heavy rainstorms, cyclones or hurricanes.[5] Such events would pose specific risks to small islands. The geographic and natural characteristics of Maldives make it particularly vulnerable to environmental changes, especially climate change and related problems such as rising sea levels and other natural disasters associated with changes in temperature and rainfall patterns.

13.   The average height of Maldivian islands is 1.5 m above mean sea level.[6] Being a low-lying, small island archipelago State, a rise in sea level could lead to or exacerbate land loss from beach erosion.

14.   The Special Rapporteur believes that climate change and its consequences impact significantly on the human rights of Maldivians. With regard to the right to adequate housing, such impact can be broadly divided into the following groups:

a.   Threat to life and health of persons and communities: this includes not only physical threats due to increase in temperatures and natural disasters, but also mental effects of insecurity and traumatizing experiences such as the 2004 tsunami;
b.   Loss of houses and property: total or partial destruction of houses and properties is happening in Maldives because of a rise in the sea level and natural disasters such as floods, cyclones, etc;
c.   Loss of livelihood: in Maldives, the economy is concentrated in two sectors — tourism and fishing — both of which are affected by extreme weather and sea events. Tourism and fishing account for around 28 per cent and 8 per cent of gross domestic product, respectively. As many economic activities depend on the coastal ecosystem, climate change would affect communities' livelihoods, including through loss of land, environmental changes affecting fisheries and agriculture and other livelihood activities;[7]
d.   Loss or contamination of freshwater sources;
e.   Threat to security of persons and communities linked to their housing location;
f.   Threat to the very social fabric of Maldives due to displacement; and
g.   Impact on infrastructure and services.

15.   Communities have sometimes expressed their demand for Government intervention through protests and other means.[8]

16.   In some cases, a response to island vulnerability has been heavy engineering projects. The Special Rapporteur is concerned that some engineering projects have been carried out without proper environmental assessment. It seems quite clear that some infrastructure on islands has a significant impact on the island and the atoll environment. In some cases, coastal erosion has increased as the result of improper planning and design of harbours.

17.   Soft engineering solutions, like the rehabilitation of mangroves or other natural barriers to flooding, can, in many cases, be sustainable and make islands more resilient to sea movements, without the need for costly and heavy building works.

18.   The Special Rapporteur commends various initiatives taken by Maldives in relation to climate change and human rights. In this context, she welcomes the recent pledge made by the Government to become the world's first carbon-neutral country in 10 years by fully switching to renewable sources of energy.[9]

19.   The Special Rapporteur is also concerned that Maldives is becoming more and more dependent on carbon; patterns of consumption are increasing and the importation of goods, including the movement of all construction materials, is heavily dependent on transport by boat or by air. In the context of the National Adaptation Strategy Plan, a consolidation programme to adapt Maldives to climate change was considered. The consolidation programme aimed to concentrate population levels on a number of so-called "safer" islands as a means of ensuring adequate housing in the context of climate change. The consolidation programme existed before the climate change agenda was raised both nationally and internationally and was justified as the only way to provide services to small and dispersed communities. After the 2004 tsunami, the policy was renamed as the "Safer Island" programme. This policy has raised tensions and provoked protests in the country and the Special Rapporteur welcomes a Government statement which notes that steps will be taken to review the policy and consider countermeasures for activities already implemented.

[...]

VIII. Conclusions and recommendations

67.   The Special Rapporteur recommends that the authorities carefully consider and implement the recommendations contained in the report of the Human Rights Commission of the Maldives. For this purpose, a land and housing policy council should be established, with the participation of community representatives from the different atolls and members of the different governmental agencies and the private sector, to participate in the formulation of land and housing policies and monitor reform implementation. This would strengthen the commitment of civil society to the reforms and create a system that ensures dialogue and responds to Maldivian needs.

68.   The Special Rapporteur believes that climate change has aggravated and will further amplify some of the problems linked with characteristics of Maldives, including land scarcity and vulnerability of the islands to natural phenomena. The impact of climate change on the acceleration of coastal erosion, frequency of storms and flooding and the rise of the sea level would increasingly affect the housing and livelihood of many Maldivians. This places a responsibility on the international community to support adaptation strategies in Maldives.

69.   Any adaptation programme should give priority to eco-friendly solutions, and the impact of hard engineering solutions should be carefully assessed.

70.   Adaptation strategies need to draw lessons from post-disaster reconstruction processes and in particular the tsunami aftermath.

71.   Disaster prevention and post-disaster reconstruction processes must be designed within a human rights-based approach. These processes must be carried out with consultation and participation of concerned communities, be gender-sensitive, draw upon local knowledge and be culturally appropriate. Special attention must be given to vulnerable groups and those who may face discrimination and exclusion.

72.   Resettlement should be carried out after considering other potential alternatives and in full consultation with concerned communities. In cases where no other option is available, resettlement should be carried out in conformity with international human rights standards.

73.   The Special Rapporteur believed that there is a need to rethink hard engineering projects, including construction of roads between islands and land reclamation, in order to minimize impact on the islands' environment and their innate protection from natural disasters.

74.   Consultation and participation of communities is an essential part of reconstruction planning. This element should be systematically included in all reconstruction processes to ensure they can really achieve long-term positive results.

75.   Internationally funded studies, internal capacity-building and innovative approaches are required to allow adequate housing and infrastructure and climate change adaptation programmes and projects to be designed and innovative local solutions, focused on a human rights-based approach, to the very particular situation of Maldives to be developed.

76.   Various factors including the growth of population and the scarcity in land make a new approach to land distribution and territorial planning unavoidable. Yet, the Special Rapporteur believes that any new approach should keep the very positive aspects of traditional land allocation, which provides access to land for housing purposes to all, regardless of social class and wealth.

77.   The Special Rapporteur strongly recommends that any housing policy should examine various options to cope with the diversity in incomes, livelihood and household arrangements that exist in the country. For instance, rent regulations and subsidies could be combined with microfinance schemes, community loans and different tenure arrangements.

78.   The Special Rapporteur commends the constitutional commitment to the right to adequate housing. This has now to be transcribed into all relevant programmes and policies which should fully take into account the various dimensions of the right to adequate housing.

79.   There is an urgent need to address legislative gaps, such as the lack of tenancy regulation and basic building standards, in full accordance with the right to adequate housing. Adequate complaint mechanisms should be put in place to enforce that and other housing legislation.

80.   Maldives should develop a taxation system in order to better finance its activities, including on housing and social issues, and diminish its reliance on external aid.

81.   Maldives need to put in place a coherent and efficient system of land registration and administration.

82.   While creating an environment that permits individuals to ask for credit to buy houses on the private market can be positive, priority should be placed on providing alternative housing options. In this context, the Special Rapporteur encourages the State authorities to consider her report on the impact of the financial crisis and the recommendations it contains.[33]

83.   The Special Rapporteur commends the joint project of the State and the United Nations Development Programme entitled "Affordable Housing for All" that was signed in March 2009. She hopes that the project will take into full account the various elements of the right to adequate housing and include issues that have been discussed in the present report.[34]

84.   There is a need to increase the number of social housing units and introduce rent regulation. This requires reliable data on income and an efficient monitoring system.

85.   The Special Rapporteur urges the authorities and private contractors to take immediate measures to improve the housing and living conditions of international migrants.

86.   The direct participation of the communities in all stages of housing and urban planning decision-making is crucial for them to be successful. The State must not only provide full access to, and information on, housing and infrastructure planning, it must also build capacity in communities through public awareness and mobilization. The communities themselves should engage and take responsibility for the implementation of their rights and collective decisions.

87.   Local capacity-building should be supported by the international community and international organizations.

Raquel Rolnik, Special Rapporteur on Adequate Housing as a Component of the Right to an Adequate Standard of Living, and on the Right to Non-discrimination in this Context, Human Rights Council, Thirteenth session, UN Doc. A/HRC/13/20/Add.3, 11 January 2010.

Notes

1   Maldives submission under Human Rights Council resolution 7/23, 25 September 2008, p. 15, available at http://www2.ohchr.org/english/issues/climatechange/docs/submissions/Maldives_Submission.pdf.
2   Ibid.
3   Between 1900 and the 1960s, Maldives had a population of 70,000–80,000 inhabitants. From the 1960s to the 1980s, the population growth rate exceeded 3 per cent per year. From the 1980s until 2000, internal migrants represented around 20 per cent of the total population. In 2006, that figure rose to 33 per cent, half of them living in Malé, the capital. See Maldives Population and Housing Census 2006.
4   The Special Rapporteur focused her 2009 thematic report to the General Assembly on the relation between climate change and the right to adequate housing (A/64/255).
5   Intergovernmental Panel on Climate Change, Climate Change 2001 (GRID-Arendal, 2003).
6   Ministry of Home Affairs, Housing and Environment, First National Communication of the Republic of Maldives to the United Nations Framework Convention on Climate Change, 2001, p. 16.
7   See Ministry of Home Affairs, Housing and Environment, First National Communication of the Republic of Maldives to the United Nations Framework Convention on Climate Change, 2001.
8   See e.g. M. Omidi, "Islanders Protest for Beach Erosion Action", Minivan News, 15 December 2008.

9 Through the United Nations Environment Programme initiative which promotes the global transition to low-carbon economies and societies. See United Nations News Centre, "Maldives joins UN emissions scheme in drive to be first carbon neutral country", 5 May 2009.

[…]

33 A/HRC/10/7.

34 The Government and the United Nations Development Programme signed a project on housing on 8 March 2009.

Chapter 19

International Council on Human Rights Policy

CLIMATE CHANGE AND HUMAN RIGHTS
A Rough Guide

[...]

The human rights dimensions of adaptation policies

"Adaptation" refers to actions taken to adjust lives and livelihoods to the new conditions brought about by warming temperatures and associated climate changes.[43] It is commonly used in three distinct ways. It refers first to actions that individuals take at their own initiative. Confronted by warmer weather or more severe storms, for example, people may choose to use new materials in home construction or switch crops or livelihoods. It refers, second, to government measures designed to achieve the same or similar ends (as the Netherlands plans to build sea-walls to protect against rising tides, for example). Third, adaptation has a more technical meaning derived from the UNFCCC and subsequent negotiations. Because the resource imbalance between the perpetrators of climate change and its victims was recognised from the outset, the UNFCCC included a requirement that wealthier countries should provide "new and additional funding" to poorer countries to enable them to address climate change.[44] This funding was to be "additional" to official development assistance (ODA). The practical content of "additionality" (to use the jargon) has remained elusive, however. This is partly because there is no clear baseline, since few wealthy countries have reached the agreed international aid target of 0.7% of GDP (gross domestic product), and partly because very little adaptation funding has ever materialised. In what follows, adaptation is used in this third sense, to refer to the elaboration of an international policy that will deliver adaptation funding to countries that most need it, and to programmes that such funding might support.

Extrapolating from existing "climate sensitive" ODA, the World Bank reckons that adaptation is likely to cost anywhere from US$4 billion to $37 billion each year.[45] Yet at present adaptation funding has not reached even close to the lower end of this scale; and what has been pledged has not been committed or spent. Four adaptation funds exist, all managed by the Global Environmental Facility (GEF), which works through implementing agencies (the World Bank, the United Nations Development Programme (UNDP) and the United Nations Environment Programme (UNEP)) to channel multilateral funding for projects related to the principal multilateral environmental treaties.[46] Climate change is one of six GEF focal areas, but adaptation has consistently been a much lower priority for the GEF than mitigation. Finally, to address long-standing criticism of its lack of an effective adaptation policy, the GEF introduced a Special Priority on Adaptation (SPA) in 2005. The SPA (which never graduated beyond a "pilot" phase), was available to developing countries on application, subject to a complex assessment of their capacity. An original allocation of US$50 million to the SPA had not been spent by the end of the initial pilot period, but no further funds were added for the next "replenishment" period (2007–2010).[47] Expenditure has been and remains excruciatingly slow. According to GEF's latest report, for example, only one of 10 GEF-supported climate change projects in financial year 2006–07 concerned adaptation through the SPA, amounting to just US$1 million of a total US$81 million spent on climate change projects.[48] The rest was geared towards mitigation (developing countries do not have mitigation obligations). Application procedures for the SPA are complex and many developing countries are not aware of what is on offer or how to access these funds.

Three other adaptation funds have been created under international instruments; all are moving at an equally slow pace.[49] Adaptation is one of four programme areas of the Special Climate Change Fund (SCCF) created under the UNFCCC and funded by discretionary pledges of developed countries. Funds may only cover adaptation costs that are "additional" to ordinary ODA.[50] Inactive until recently, seven SCCF projects were finally approved in 2006–07 and involved eight countries (there are 121 developing country parties to the UNFCCC).[51] A Least Developed Country Fund (LDCF), also created under the UNFCCC, is likewise managed by the GEF, and funded through discretionary pledges. It has provided US$200,000 apiece for the preparation of National Adaptation Programmes of Action (NAPAs), designed in-country to address urgent and priority adaptation needs (32 have been finished to date). On the basis of NAPAs existing at the time, the Stern Review projected that US$1.3 billion would be required for the "immediate" adaptation needs of the 47 Least Developed countries (LDCs).[52] So far nothing close to this amount is forthcoming.[53] Finally, an Adaptation Fund was created through the Kyoto Protocol, to be replenished from a 2% levy on Clean Development Mechanism (CDM) projects.[54] Procedures for its management were eventually approved at the 13th Conference of the Parties in Bali and involve a Board with strong developing country representation. The GEF acts as the Secretariat of the Board and is to take direction from the Board and the Parties. This is a compromise hard fought for by developing country representatives in agreeing to allow the GEF a further managerial role in adaptation funding, given its poor track record.[55]

It is widely recognised that adaptation funding cannot be delivered effectively until it is known where assistance will bring the most benefit. Unfortunately, it is just this information that is generally lacking. The reason, as with so much in the climate change debate, is resource related. Because expertise and financing are concentrated in wealthy countries, the latter have much more complete information about the likely impacts of climate change and suitable responses to it, compared with sub-Saharan Africa, for example. The IPCC reports cite countless practical examples of adaptation in rich countries, many of which are already underway; forecasts for poorer countries, by contrast, remain vague and sweeping. The Stern Review makes the point as follows:

> Adaptation will depend on comprehensive climate monitoring networks, and reliable
> scientific information and forecasts on climate change – a key global public good…
> [D]eveloping country governments should provide information to their own citizens but
> currently lack the capacity to do this, demonstrated by the shortage of weather watch
> stations. The international community should therefore support global, regional and
> national research and information systems on risk, including helping developing-country
> governments build adequate monitoring and dissemination programs at the national level.
> Priorities include measuring and forecasting climatic variability, regional and national
> floods, and geophysical hazards.[56]

The list of priority areas identified in the Stern Review demonstrates the scale of the
challenge. Physical science data must necessarily precede, and provide a base for, research on
social and rights impacts. But the latter too are critically important, since the primary purpose
of policy in this area is to reshape the human, social and economic environment. In this context,
human rights thresholds can provide a compass for policy orientation, helping to decide where
research should be directed and what policy should prioritise. So while it is vital to know at
what temperature increase we might expect severe droughts to occur or sea-levels to rise, for
example, it is no less important to learn who these events will affect and where precisely;
what institutional or other support is available; and how this support might be strengthened.

These considerations fit naturally within the agenda outlined in the Bali Action Plan of
December 2007, which calls for:

> Enhanced action on adaptation, including…International cooperation to support urgent
> implementation of adaptation actions, including through vulnerability assessments,
> prioritization of actions, financial needs assessments, capacity-building and response
> strategies, integration of adaptation actions into sectoral and national planning, specific
> projects and programmes, means to incentivize the implementation of adaptation actions,
> and other ways to enable climate-resilient development and reduce vulnerability of all
> Parties, taking into account the urgent and immediate needs of developing countries
> that are particularly vulnerable to the adverse effects of climate change, especially the
> least developed countries and small island developing States, and…countries in Africa
> affected by drought, desertification and floods.[57]

There is already, therefore, a good basis in the emerging climate change regime for the
integration of human rights-focused research into adaptation policy. Human rights organisations
have developed considerable expertise in identifying the risks that vulnerable and less visible
communities face. Combined with more detailed assessments of physical impact, their
methodologies can usefully set social and economic funding priorities for adaptation programmes.

The short-term benefits are evident. Certain climate change impacts are now being felt and
others cannot be halted, because of the extent of historical and current emissions and the timelag
between emissions and their effects on the climate. In the most vulnerable places – Arctic regions,
for example, Saharan Africa, and some small island states such as Tuvalu – a human rights optic
can help make the case for swift, substantial and directed adaptation funding. Who is at risk and
what can be done where crop-based or coastal livelihoods are threatened? What kind of local

and international mechanisms exist to handle the practical and legal complexities of relocating threatened island communities from sinking territories? (Such individuals find themselves in the unprecedented situation of being citizens of a state that no longer has territory, and relocating as de facto refugees, but outside any existing Convention definition of the term.)[58] How might existing mechanisms in each of these contexts be improved? The moral imperative to act in identifying and treating such cases joins neatly with the legal duty to make adaptation funding available.

Long-term adaptation needs are more complex. Considerable information already exists on the expected human impacts of climate change. Adopting a rights focus would help to orient future research, set priorities, assist in evaluation and galvanise support. Excerpts from the IPCC AR4 and the Stern Review, provided in appendices at the end of this report, outline the expected impacts by affected human right (not Stern's term) and by region. These predictions illustrate both the scale of human rights impacts expected in the short- to middle-term, and the extent to which more information will be required in order to locate affected communities and to provide the institutional support they will need.

Both reports further point out that the effect of climate change impacts in developing countries are exacerbated by the relatively greater dependence of their economies on climatic conditions, on one hand, and by the relatively less comprehensive management of natural resources, such as water, on the other. Even where water supplies are predicted to increase, as in South and East Asia, "much of the extra water will come during the wet season and will only be useful for alleviating shortages in the dry season if storage could be created (at a cost)".[59] Furthermore, climate change throws existing development policies off course. In parts of Africa, for example, development scenarios would ordinarily have relied upon future massive irrigation schemes; but as their viability has been undermined by climate change, no obvious alternative strategy has become available.

Exacerbating Factors. Stern Review, pp. 93–97 (references excised)

Already fragile environments: Developing countries are especially vulnerable to the physical impacts of climate change because of their exposure to an already fragile environment, an economic structure that is highly sensitive to an adverse and changing climate, and low incomes that constrain their ability to adapt.

Dependence on agriculture: Developing economies are very sensitive to the direct impacts of climate change given their heavy dependence on agriculture and ecosystems, rapid population growth and concentration of millions of people in slum and squatter settlements, and low health levels. Agriculture and related activities are crucial to many developing countries, in particular for low income or semi-subsistence economies. The rural sector contributes 21% of GDP in India, for example, rising to 39% in a country like Malawi, whilst 61% and 64% of people in South Asia and sub-Saharan Africa are employed in the rural sector. This concentration of economic activities in the rural sector – and in some cases around just a few commodities – is associated with low levels of income. The concentration of activities in one sector also limits flexibility to switch to less climate-sensitive activities such as manufacturing and services.

Dependence on vulnerable ecosystems: All humans depend on the services provided by natural systems. However, environmental assets and the services they provide are especially important for poor people, ranging from the provision of subsistence products and market income, to food security and health services. Poor people are consequently highly sensitive to the degradation and destruction of these natural assets and systems by climate change.

Population growth and urbanisation: Developing countries are also undergoing rapid urbanisation, and the trend is set to continue as populations grow. The number of people living in cities in developing countries is predicted to rise from 43% in 2005 to 56% by 2030. In Africa, for example, the 500km coast between Accra and the Niger delta will likely become a continuous urban megalopolis with more than 50 million people by 2020.

Adaptive capacity: People will adapt to changes in the climate as far as their resources and knowledge allow. But developing countries lack the infrastructure (most notably in the area of water supply and management), financial means, and access to public services that would otherwise help them adapt.

Finally, adequate fulfilment of human rights within vulnerable states would itself provide a solid basis for autonomous adaptation (that is, measures spontaneously initiated by citizens in response to climate changes). Local provision of information, guarantees of public participation in government, and freedom of speech and association all provide affected communities with the voice and capacity to force change in their local settings. Economic and social rights also matter. Education is as important as health: a well-educated population is better equipped to recognise in advance the threats posed by a changing climate and to make preparations. This is one of many areas where ordinary development aid, properly directed, can potentially achieve multiple objectives at once, serving classic development and human rights aims while at the same time contributing to societies' long-term ability to adapt to climate change.

The human rights dimensions of mitigation policies

Perhaps inevitably, the greater part of climate change negotiation is devoted to "mitigation". This term refers to the actions and policies that seek to prevent global warming from causing "dangerous anthropogenic interference" with the climate, as required by the UNFCCC.[60] Although no "dangerous" threshold is mentioned in the treaty, a rise of average global temperatures by no more than 2°C above preindustrial levels was until recently cited in most policy documents (although it now seems increasingly unlikely that it will be achieved). Before investigating the human rights dimensions of mitigation policies, the scientific and policy context is briefly set down in the following two paragraphs.[61]

In the IPCC AR4, greenhouse gas emission levels in the atmosphere were estimated at 455 parts per million of carbon dioxide equivalent (ppm CO_2e),[62] almost double preindustrial levels and rising fast. Current concentrations of greenhouse gases have already warmed the globe and will lead to further warming even if all new emissions were stopped immediately. However, high levels of emissions are certain to continue in the short- to mid-term, and discussion has therefore centred on identifying a point at which emissions concentrations might be stabilised in future to keep warming to a minimum. There is little agreement on the appropriate stabilisation level: different studies reach different conclusions, and all are couched in the language of probability. Recent estimates reckon that if emissions levels are stabilised at 445–490 ppm CO_2e there will be an even chance (50%) that the average global temperature rise will still exceed 2–2.4°C.[63] At 550 ppm CO_2e, the probability of temperatures exceeding 2°C is closer to 80%, and there is an even chance that average global temperatures will rise by 3°C over preindustrial levels.

International Council on Human Rights Policy, 'Climate Change and Human Rights: A Rough Guide', 2008, pp. 21–27

Notes

43 The third IPCC Assessment Report defined adaptation as "adjustments in ecological, social, or economic systems in response to actual or expected climatic stimuli and their effects or impacts. [Adaptation] refers to changes in processes, practices, and structures to moderate potential damages or to benefit from opportunities associated with climate change". Smit and Pilifosova 2001, pp. 877–912.

44 UNFCCC Article 4 (3). This paragraph, and much of the section, relies on Mace, 2005; Müller, 2006 and 2007.

45 Cited in Stern Review, Part V, Chapter 20, p. 442.

46 See for a good overview, Stern Review, Part VI, p. 557. Known as the Rio Conventions because they were all signed in Rio in 1992, these are the UNFCCC, the UN Convention on Biodiversity and the UN Convention to Combat Desertification.

47 See FCCC/CP/2007/3, Report of the Global Environment Facility to the Conference of the Parties, 13th session Bali, 3–14 December 2007 (27 November 2007), para. 8: "Once the remainder of the initial US$50 million of funds devoted to the SPA is committed to projects, an evaluation will be undertaken to draw initial lessons and to assess the potential for mainstreaming adaptation into GEF's focal areas."

48 Ibid., paras. 16–17.

49 Figures are from ibid, paras. 19–27.

50 US$71.5 million has been pledged to date.

51 These amounted to US$24.4 million of SCCF funds. A further US$92.7 million of funding from other sources was leveraged through these projects.

52 Stern Review, p. 442.

53 By late 2007, US$0.6 million (of a pledged total of US$163 million) had been allocated to preparing NAPA projects in four countries. The GEF notes that "approximately US$150m remains to be programmed to meet the urgent and immediate adaptation needs of the LDCs under the LDCF". FCCC/CP/2007/3, para. 27.

54 For a description of the Clean Development Mechanism, see below, pp. 36–40.

55 The Adaptation Fund is set to become operational in 2008. To these four might be added the World Bank's new Pilot Program for Climate Resilience, one of its Climate Investment Funds introduced in 2008, although it was not created under the UNFCCC and lacks official status or widespread support. For more on this new fund, see http://go.worldbank.org/58OVAGT860.

56 Stern Review, Part VI, p. 563.

57 Decision -/CP.13, Bali Action Plan (Advance Unedited Version), Article 1(c)(i).

58 For an informed discussion, see E/CN.4/Sub.2/2005/28, expanded working paper by Françoise Hampson on the human rights situation of indigenous peoples in States and other territories threatened with extinction for environmental reasons (16 June 2005). A total of just under half a million individuals are likely to be affected, from the islands of Tuvalu, Nauru, Kiribati, Maldives and the Bahamas. Ibid, para. 25.

59 Stern Review, p. 63.

60 For a discussion, see the Stern Review, Part III, Chapter 13, p. 289.

61   This account relies on IPCC AR4, WGIII, Technical Summary, and on the Stern Review, Part III, especially Chapters 7–10. More detailed information is provided in IPCC AR4, WGIII, Chapters 1–3.

62   The figure of 455 ppm CO_2e accounts for the intensity of all greenhouse gases in the atmosphere, measured as equivalents of Carbon Dioxide. The amount of Carbon Dioxide itself is estimated at 379 ppm. IPCC AR4, WGIII, Technical Summary, p. 27, which adds: "Incorporating the cooling effect of aerosols, other air pollutants and gases released from land-use change into the equivalent concentration, leads to an effective 311–435 ppm CO_2-eq concentration."

63   See Table TS.2 in IPCC AR4, WGIII, Technical Summary, p. 39. Also UNDP, 2007, p. 46.

Chapter 20

Stand Up for Your Rights

THE HUMAN SIDE OF CLIMATE CHANGE
Human Rights Violations and Climate Refugees

[...]

3 A Human Rights Perspective

In various principles, factors and recent developments in the area of human rights, one could find support for a human rights-based approach to climate change. This chapter will outline such principles and developments and will also show that a great deal of research and reporting, supporting such a human rights approach, has already been done by and on behalf of the UN over the last few years.

3.1 Human Rights Principles and Relevant Developments

Equity

Human rights conventions contain codifications of the principle of equity, more specifically in articles relating to equal treatment/non-discrimination. Human rights approaches can contribute concretely to efforts to tackle inequality. Examples of this are the human rights approach to poverty and the Millennium Development Goals. Climate change impacts are unequal and unfair as the impacts on the poorer parts of the world are expected to be much bigger than in the wealthier parts. Based on this the principle of equity paves the way for affirmative action and change based on human rights.

The normative basis of human rights

A human rights approach anchors the conceptual framework of climate change policies to a firm normative basis in international human rights. It provides clear principles against which to evaluate the policies and it makes visible the people that are affected by climate change and climate change policies.

A human rights-based approach will force to not merely develop policies focused solely on the climate itself, over the heads of the individual people, as it were, but to focus also directly on the individual, and without any form of distinction or discrimination. It will obligate States to protect, respect and fulfil the human rights of their subjects, by making the largest possible efforts to mitigate the effect of climate change in the course of implementing internationally agreed climate change policies. A human rights-based approach will also focus on inclusion of excluded and marginalised populations. It will further encourage participation and democratic processes and it will encourage accountability and transparency in policy decisions.

Human Rights and Business

Business has played its part in causing climate change. Business, however, could also play a key role in the mitigation of global warming, through investing in researching and implementing new energy technologies and energy efficiency measures which would result in reducing GHG emissions.

Traditionally 'business' or corporations were not seen as accountable actors in the Human Rights arena. Although it is still well recognised in international law that States have the primary responsibility to protect, respect and secure the fulfilment of human rights, the international arena is changing and also business have been, and will be, held accountable for human rights violations.

It has become obvious that, although the activities of businesses sometimes results in fulfilling certain human rights (such as work or education), many daily business practices negatively impact human rights. Companies may violate human rights through their employment practices, or through how their production processes impact on workers, communities and the environment.

During the 1980s and 1990s many companies adopted codes of conduct to improve their profile and policies.[19]

In April 2008 Professor John Ruggie, the UN Secretary-General's Special Representative on Human Rights and transnational corporations and other business enterprises, proposed a policy framework for the improvement of the relationship between business and business and human rights: Protect, Respect and Remedy.[20] This framework was accepted by the Human Rights Council unanimously and has won considerable acclaim globally with business and NGOs.

The framework rests on three mutually reinforcing pillars:

1. States have a duty to protect their citizens against any human rights abuses by corporations, through appropriate policies, effective regulation and access to justice.
2. Companies have a responsibility to respect the human rights of all those who have to deal with the corporation, including employees, customers and neighbours.
3. Victims of corporate-related human rights violations should have greater access to effective remedies.

In April 2009 Prof. Ruggie published a follow-up report in which he addressed the question of how this framework can be implemented and further strengthened.[21] In the report he identified several hurdles to be taken if the framework is to work effectively. In a speech, held in Geneva in early October 2009, Prof. Ruggie formulated five key challenges that need to be addressed:

- Almost all human rights are relevant to business, although some rights in specific circumstances may be more relevant than others. The policies of governments and companies should be based on this observation.

- Governments lack a sound set of instruments for managing the complex relationships between business and human rights. In particular such discrepancies are found between policies of government departments dealing with economic affairs or industry and departments involved in human rights.
- With few exceptions, even major multinationals lack a well-developed management system to conduct human rights due diligence. Companies focus so much on legal requirements for their actions that they only gradually discover their failure to meet the broad expectation that their actions will respect all human rights, which is at the basis of their 'social license to operate'.
- Most companies lack grievance mechanisms to which individuals and communities can bring concerns about the companies' activities. Again, a legalistic approach prevails: if we are not legally bound to do it, we will not do it. Companies thus deprive individuals and communities of an opportunity to resolve conflicts in a relatively simple manner, and themselves of an alarm system that can alert them before differences of views escalate into campaigns or lawsuits.
- The incidence of human rights abuses by companies is highest in countries with weak governance institutions, where laws are not enforced or lacking altogether. This applies even more in countries that are in conflict. A systematic solution calls for better enforcement of existing laws, clearer standards and more innovative policy responses from states in which companies operate and where their headquarters are based.[22]

More and more it becomes clear that the human rights framework binds business. And so it should. We agree with former UN High Commissioner for Human Rights, Ms. Mary Robinson's clear statement, that "Human Rights are everyone's business".

As said, business has been and is a major player with regard to GHGs emissions and other climate change causes. Business could, on the other hand, also contribute to cutting back GHG emissions by working more efficiently. It will play a crucial role regarding the introduction and further development of changes in technology, such as new technology for renewable energy.

This report, with many others, shows that climate change is, and progressively will be, a threat to the realisation of human rights and that as the impacts of climate change grow, we know that it will increasingly violate human rights. Based on the responsibility to respect human rights, business should therefore take an active role into combating climate change and mitigating its affects.

Linking Human Rights and Environmental Issues, towards a Right to Environment, the 'climate change and human rights' debate has been preceded by a more general 'human rights and the environment' and 'a healthy environment as a substantive human right' debate over the past decades. International concerns with human rights, health and environmental protection have expanded considerably in the past several decades. The international community has created a vast array of international legal instruments, specialised organs, and agencies at both the global and regional levels to respond to identified concerns in these areas.

Often these have seemed to develop in isolation from one another. For instance, the Right to Health ("Everyone has the right to the enjoyment of the highest attainable standard of physical and mental health") has been recognised for long and was codified in the International Covenant on Economical Social and Cultural Rights[23] (ICESCR) in 1966. The development of this right (for instance through "Special Comments" by the UN ICESCR Committee) has also emphasised the links between health and environment.

Yet, the links between human rights, health and environmental protection were apparent at least as early as the first United Nations Conference on the Human Environment, held in Stockholm in 1972. Indeed, health has seemed to be one of the subjects that bridges the two fields of environmental protection and human rights.

The Stockholm Declaration established a foundation for linking human rights, health and environmental issues, declaring: "Man has the fundamental right to freedom, equality and adequate conditions of life, in an environment of a quality that permits a life of dignity and well-being."

Following the Stockholm Declaration, a number of international human rights instruments progressively included environmental values in their protection systems, for example the Convention on the Elimination of All Forms of Discrimination against Women ('CEDAW', 1979), the Convention on the Rights of the Child ('CRC', 1989) and ILO Convention 169 concerning Indigenous and Tribal Peoples in Independent Countries (1989).

The UN Committee on Economic, Social and Cultural Rights has clearly linked the right to food and the right to health to the environment. The Committee argued that the right to food requires appropriate environmental policies, and that the successful protection of health is determined, among others, by a healthy environment.[24]

Looking back in time, one can clearly distinguish a change in approach from merely linking human rights to environmental disruptions or violations, towards a healthy and clean environment as a separate human right. Professor Dinah Shelton also reaches this conclusion in her Working Paper prepared for the World Health Organisation.[25] This development currently continuous to progress.

Examples of a codification of such a human right to environment are the regional American[26] and African[27] human rights treaties, the 2007 UN General Assembly Declaration on the Rights of Indigenous Peoples[28] and over 100 constitutions globally.

The European Convention on Human Rights (ECHR), however, currently lacks a codified substantive right to environment, but change is happening as we write this report:[29] the Parliamentary Assembly of the Council of Europe (PACE) late September 2009 almost unanimously voted to acknowledge a Right to a healthy Environment in the ECHR. The (nearly non-amended draft) recommendation describes in detail what this right would mean and why we and future generations need for it to be codified.[30]

Stand Up For Your Rights obviously applauds this development and calls on the Committee of Ministers to swiftly accept this recommendation and act accordingly. For more information on this human right, international developments and case law, please see www.RightToEnvironment.org.

Following over a hundred constitutions, two regional human rights treaties and many international declarations having done so, the time has come to acknowledge 'A Healthy and Clean Environment' as a substantive human right at global (UN), regional (e.g. ECHR) and local (domestic constitutional) level.

> Obviously, and as a basis for all other rights and life itself, all people have a right to an environment that is not harmful to their health or well-being and to have nature protected, for the benefit of present and future generations.[31]

The time has come to acknowledge, protect and respect our human right to a healthy environment. Such a substantive human right will without doubt contribute to a rights based and thus a more human approach to climate change.

UN Human Rights Bodies and Climate Change

In early 2008, in resolution 7/23, the UN Human Rights Council asked the Office of the High Commissioner on Human Rights (OHCHR) to undertake a detailed research into the relationship between climate change and human rights. After having received many submissions by a diverse group of actors and stakeholders, a consultation meeting was held in Geneva on 22 October 2008. All on the consultation and outcomes and all of the submissions and documents can be found on the specific OHCHR website on 'Human rights and climate change'.[32]

The outcome, an official OHCHR report[33] on the matter was discussed by the Human Rights Council during its March 2009 meetings and has officially been made available to the COP15 meetings. The report recommends that human rights norms and standards need to be taken into consideration in climate change policies. This will ensure that action on both areas — human rights and climate change — will be harmonised, which in return will lead to sustainable results. It also stresses that the realisation of human rights needs to continue to be a central goal of national as well as international action in the area of climate change.

On 15 June 2009, during the 11th session of the Human Rights Council, a panel discussion took place on 'Climate Change and Human Rights'. What came out of this, is that OHCHR has taken on a more active role towards climate change, as Ms. Kyung-wha Kang, UN Deputy High Commissioner for Human Rights, said in her speech:

> …climate change-related impacts have a range of implications, both direct and indirect, for the effective enjoyment of human rights'. She referred to the coming Copenhagen Climate Change Conference (COP 15) negotiations: 'As you engage in those negotiations, you must bear in mind the grave human rights consequences of a failure to take decisive action now.

In an official COP15 weblog,[34] UN High Commissioner for Human Rights, Ms. Navanethem Pillay adds:

> Climate change is one of the most serious challenges mankind has ever faced and has serious implications for the realization of human rights. A human rights analysis brings into focus how lives of individuals and communities are affected and why human rights safeguards must be integrated into policies and measures to address climate change.

3.2 Individual Protected Rights Affected by Climate Change

The following section will enumerate the specific human rights potentially at risk by climate change. Most of these, but not all, have been addressed in the aforementioned 2009 OHCHR report as well.

The Right to Life

As the Deputy High Commissioner for Human Rights affirmed in her June 2009 Human Rights Council address, climate change can have both a direct and indirect impact on human life. Life might be immediately threatened on account of climate induced extreme weather events, such as heat waves, floods, storms, fires and droughts, or may be threatened gradually, through depletion of food supply, diminishing access to safe drinking water, deterioration in health and susceptibility to disease. The protection of the right to life is therefore closely related to the fulfilment of other rights, such as the right to food, water and sanitation, health and housing.

The right to life, enshrined in Article 6 of the 1966 International Convention on Civil and Political Rights ('ICCPR'), has extensive links to the environment. The most obvious connections manifest themselves in situations where people die due to environmental pollution, such as the Bhopal gas leak and, quite recently, the toxic 'Proba Koala' dumping in the Ivory Coast.[35]

According to CCPR General Comment No 6,[36] the right to life cannot properly be understood in a restrictive manner, and the protection of this right requires that States adopt positive measures. It could be linked to any environmental disruption that directly contributed to the loss of lives – including to air pollution causing 2.4 million deaths per year as mentioned in chapter 2.1. But also to casualties due to rising sea levels and extreme weather events due to climate change. The Committee has further clarified that the right to life imposes an obligation on States to take positive measures for its protection.[37]

That state obligations entailing positive actions when it comes to the right to life and environmental issues, has been confirmed by the European Court of Human Rights in, for example, the 2008 Budayeva case.[38] In this case[39] the European Court of Human Rights further confirmed that, in cases of natural disaster, a positive obligation lies with the State in which territory the disaster occurs, to act with due diligence to prevent threats to the life and property of persons.

The Court based its findings in this case on the recognition that a duty on States exists to protect life against the consequences of disasters by reaffirming that the right to life "does not solely concern deaths resulting from the use of force by agents of the State but also [...] lays down a positive obligation on States to take appropriate steps to safeguard the lives of those within their jurisdiction" emphasizing that "this positive obligation entails above all a primary duty on the State to put in place a legislative and administrative framework designed to provide effective deterrence against threats to the right to life."[40] Such threats would include climate change impacts and environmental disruption.

The Committee on the Rights of the Child clarified that the right to survival and development of the child must be implemented in a holistic manner, "through the enforcement of all the other provisions of the Convention, including rights to health, adequate nutrition, social security, an adequate standard of living, a healthy and safe environment".[41]

The Right to Health

The right to the highest attainable standard of physical and mental health is expressed in the International Covenant on Economic, Social and Cultural Rights ('ICESCR', Art. 12) and is referred to in other core international human rights treaties (e.g. CEDAW Arts. 12 and 14 and CRC Art. 24).

This right, closely linked to the right to life, is often violated in cases of pollution of air, land or water. For example, poorly regulated industries and mines cause severe contamination among local people, resulting in a slew of physical problems and thus impacting health. The CESCR General Comment 14 on the right to the highest attainable standard of health[42] specifically refers, on several occasions, to healthy environmental conditions.

Obviously, climate change impacts that deteriorate health cause a violation of this right as well. Examples of that are: increased diseases and injuries due to extreme weather events, increases in malnutrition, an increased burden of diarrheal, cardio respiratory and infectious diseases, malaria and other vector borne diseases in some parts of the world.

The UNDP 2007/2008 Human Development Report states in addition to this that population displacement undermines the provision of medicare and vaccination programmes, complicating the handling of infectious diseases and making them more deadly. It further reminds that "forced migrants, especially those forced to flee quickly from climate events, are also at greater risk of sexual exploitation, human trafficking and sexual and gender-based violence".

The Right to Adequate Food

The right to adequate food is expressed in a number of international human rights treaties, such as the ICESCR (Art. 11, and additionally the fundamental right to be free from hunger, Art. 11 (2)), CRC (Art. 24) and the Convention on the Rights of Persons with Disabilities (Art. 25(f) and Art. 28, para.1). It is also implied by general provisions on an adequate standard of living of the CEDAW (Art. 14, para.2(h)) and the International Convention on the Elimination of All Forms of Racial Discrimination (Art. 5(e)).

Elements of the right to food include the availability of adequate food (including through the ability to feed oneself from natural resources) and accessible to all individuals under the jurisdiction of a State. Equally, States must ensure freedom from hunger and take necessary action to alleviate hunger, even in times of natural or other disasters.[43]

The IPCC reports set out that food scarcity is likely to rise dramatically owing to extreme drought storms and other events caused by climate change, which in turn result in failing harvests. Consequently, the right to physical and economic access at all times to adequate food or means for its procurement, will progressively be under pressure.

The Right to Water and Sanitation

Although this right is not codified as such in an international treaty, more frequently it is accepted and invoked as such through CESCR General Comment No. 15.[44] It is obviously linked to the rights to life and health. Without access to clean drinking, cooking and bathing water in adequate quantities, individuals and communities worldwide die and suffer serious illnesses. Billions already lack access to clean and safe drinking water.

This number, according to the IPCC reports, is likely to rise dramatically due to glacial melt and other climate change impacts. While human rights treaties do not explicitly contain the right to water as such, international human rights law has recognised the right to water as an integral part of other human rights. Why? Well, water is the most essential ingredient for life. No water = No life.

In the aforementioned CESCR General Comment No. 15, the Committee acknowledged that access to water and sanitation is an independent right. It defines the right as entitling "everyone to sufficient, safe, acceptable, physically accessible and affordable water for personal and domestic uses", such as for consumption, cooking, personal and domestic hygienic requirements.[45] Other international human rights treaties also refer to access to water regarding an adequate standard of living (CEDAW, Art. 14) or to combat disease and malnutrition (CRC, Art. 24).

Water shortages are also predicted to be a main source of (armed) conflict in the future. Water problems and drought are seen as main contributing factors to the Darfur conflict. Water scarcity is most particularly felt in the Middle East, North Africa and sub-Saharan Africa, but also in other regions of the world. The Director-General of UNESCO, Mr. Koïchiro Matsuura, on the occasion of World Water Day 2009, stated in his address, among others, that "...competition over water is [...] leading to serious tensions between different groups of users".

Some have even raised the spectre of future "water wars".[46] These conflicts could potentially cause major refugee flows to move within or across borders.

The Right to Adequate Housing

The right to adequate housing is included in several international human rights treaties as an element of the right to an adequate standard of living (e.g. Art. 11 ICESCR; Art. 14 CEDAW and Art. 27 CRC). The Committee on Economic, Social and Cultural Rights has clarified the right to housing in several of its general comments, such as General Comment No. 4, 7 and 12. It has defined the right as "the right to live somewhere in security, peace and dignity".[47] Basic elements of this right include security of tenure, protection against forced evictions,[48] availability of services, materials, facilities and infrastructure, affordability, habitability, accessibility, location and cultural adequacy.[49]

Many coastal human settlements will be directly impacted by sea level rise and storm surges. In the polar region and in low-lying island States such impacts have already led to the relocation of peoples and communities. Settlements in low-lying mega-deltas are also particularly at risk, as is shown by the fact that millions of people and homes have been affected by flooding in recent years.

Right to Development

This right[50] enunciated in UN GA Declaration on the Right to Development (Resolution 41/128 of 4 December 1986) shares considerable common ground with the right to a healthy environment. A holistic model of sustainable development recognises that environmentally destructive economic progress does not produce long-term societal progress.

Impacts on the poorer parts of the world are expected to be bigger than in the wealthier parts. The less developed nations have done little to cause global warming. However, they are most exposed to its effects and also have fewer resources for coping with the effects of climate change. Whilst they are eager for economic development themselves, they are prognosed to be limited in this due to climate change.

The Right to Self-determination

The right to self-determination is a fundamental principle of international law. Article 1 of the ICCPR and ICESCR establishes that "all peoples have the right of self-determination", by virtue of which "they freely determine their political status and freely pursue their economic, social and cultural development".

Important aspects of the right to self-determination include the right of a people not to be deprived of its own means of subsistence and the obligation of a State party to promote the realisation of the right to self-determination, including for people living outside its territory. Sea-level rise and extreme weather events related to climate change are threatening the habitability and, in the longer term, the territorial existence of a number of low-lying island States. Equally, changes in the climate threaten to deprive indigenous peoples of their traditional territories and sources of livelihood. Either of these impacts would have implications for the right to self-determination.

According to the mentioned recent OHCHR Climate Change and Human Rights report: "…it is clear that insofar as climate change poses a threat to the right of peoples to self-determination, States have a duty to take positive action, individually and jointly, to address and avert this threat."[51]

Right of Access to Information and Right to Participate

The United Nations Framework Convention on Climate Change commits party States to promote and facilitate public access to information on climate change. Further, access to information is implied in the international human rights to freedom of opinion and expression. Jurisprudence of regional human rights courts has also underlined the importance of access to information in relation to environmental risks.

Women and Children's Rights

Although CEDAW and CRC have been referred to in relation to many of the rights mentioned before, women and children's rights deserve specific attention. Why? Well: women and children are even more impacted by environmental disruption than men, because they are even more dependent than men upon primary natural resources for food and water, such as land, forests and water sources. As set out before, 80% of the 1.2 billion people that lack access to clean drinking water are the rural poor. And, more specifically, of the 1.3 billion people living in extreme poverty 70% are women.

Women are especially exposed to climate change-related risks due to existing gender discrimination, inequality and inhibiting gender roles. It is established that women, particularly elderly women and girls, are affected more severely and are more at risk during all phases of weather-related disasters: risk preparedness, warning communication and response, social and economic impacts, recovery and reconstruction. The death rate of women is markedly higher than that of men during natural disasters (often linked to reasons such as: women are more likely to be looking after children, to be wearing clothes which inhibit movement and are less likely to be able to swim).[52]

Children are more vulnerable than adults to diseases because their immune systems and detoxification mechanisms are not fully developed making them also more vulnerable to contaminated water (bacterially or viral) and environmental pollution in general.

The higher vulnerability of women and children is generally recognised in international law, which has resulted in human rights instrument such as the CEDAW and CRC.

Studies show[53] that climate change will exacerbate existing health risks and undermine support structures that protect children from harm. Overall, the health burden of climate change will primarily be borne by children in the developing world. For example, extreme weather events and increased water stress already constitute leading causes of malnutrition and infant and child mortality and morbidity.

The world of tomorrow will consist of today's children. Children are central actors in promoting behaviour change required to mitigate the effects of global warming. Children's knowledge and awareness of climate change also influence wider households and community actions.

The Convention on the Rights of the Child, which enjoys near universal ratification, obliges States to take action to ensure the realization of all rights in CRC for all children in their jurisdiction, including measures to safeguard children's right to life, survival and development through, inter alia, addressing problems of environmental pollution and degradation.

The Rights of Indigenous People

Some of the most obvious examples of climate change impacts involve indigenous peoples, whose lifestyles often depend on their relationship with the natural environment. Impacts of climate change on the Arctic environment and tropical islands and indigenous people having their forests disappear through logging, for example, have disproportionate effects on local culture.

In September 2007, the UN General Assembly adopted the UN Declaration on the Rights of Indigenous Peoples, which – for the first time in a UN Declaration – acknowledged the conservation and protection of the environment and resources as a human right under international law (Art 29).

Also other relevant rights of indigenous peoples are expressed in this declaration and in the core international human rights treaties ICCPR and ICESCR, in particular with regard to the right to self-determination and rights related to culture. The rights of indigenous peoples are enshrined as well in ILO Convention No. 169 (1989) concerning Indigenous and Tribal Peoples in Independent Countries.

The OHCHR emphasises that indigenous peoples, who often live on marginal lands and in fragile ecosystems which are particularly sensitive to changes in the physical environment, face serious threats from climate change, coupled with pollution and environmental degradation.

Climate change-related effects have caused harm to the traditional livelihoods of Inuit communities in polar regions and some communities have already relocated. Indigenous communities of low-lying island States face similar problems.[54]

Apart from the impact on their right to food, concerns have also been raised that demand for bio fuels, to cut down greenhouse gas emissions, could negatively affect the rights of indigenous peoples to their traditional lands and culture.[55]

Stand Up for Your Rights, '*The Human Side of Climate Change: Human Rights Violations and Climate Refugees*', December 2009, pp. 17–23.

Notes

19   Examples of such codes are found on our website under 'Business' and are: Guidelines for Multinational Enterprises (Organisation for Economic Cooperation and Development, OECD); Declaration of Principles concerning Multinational Enterprises and Social Policy (International Labour Organization, ILO); ICC Rules of Conduct and Recommendations (International Chamber of Commerce, ICC); and Global Compact (United Nations Global Compact).

20   Protect, Respect and Remedy: a Framework for Business and Human Rights, United Nations Human Rights Council (A/HRC/8/5), New York, 7 April 2008. Download at http://www.unglobalcompact. org/docs/issues_doc/human_rights/Human_Rights_Working_Group/29Apr08_7_Report_of_ SRSG_to_HRC.pdf.

21   Business and human rights: Towards operationalizing the "protect, respect and remedy" framework, United Nations Human Rights Council (A/HRC/11/13), New York, 22 April 2009. Download at http://www2.ohchr.org/english/bodies/hrcouncil/docs/11session/A.HRC.11.13.pdf.

22   Opening Remarks at the Consultation on operationalising the framework for business and human rights presented by the Special Representative of the Secretary-General on the issue of human rights and transnational corporations and other business enterprises, Professor John G. Ruggie (SRSG), Geneva, 5–6 October 2009.

23   International Covenant on Economic, Social and Cultural Rights (16 Dec 1966), 993 U.N.T.S.

24   Committee on Economic, Social and Cultural Rights (CESCR), general comments No. 12 (1999) on the right to adequate food (Art. 11), para. 4, and No. 14 (2000) on the right to the highest attainable standard of health (Art. 12), para. 4.

25   For a more extensive outline of these approaches, see the working paper by Professor Dinah Shelton: Health and Human Rights Working Paper Series No 1, Human Rights, Health & Environmental Protection: Linkages in Law & Practice, A Background Paper for the World Health Organisation, Prof. Dinah Shelton, London, 2002.

26   Article 11 of the Additional Protocol American Convention on Human Rights in the Area of Economic, Social and Cultural Rights (1988) is titled: "Right to a healthy environment" and proclaims (1) Everyone shall have the right to live in a healthy environment and to have access to basic public services and (2) The States parties shall promote the protection, preservation and improvement of the environment The Protocol provides for both a right to environment and a right to health. Article 10 states that (1) Everyone shall have the right to health, understood to mean the enjoyment of the highest level of physical, mental and social well-being.

27   The African Charter on Human and Peoples' Rights (1981) contains both a right to health and a right to environment. Article 16 of the Charter guarantees the right to enjoy the best attainable state of physical and mental health to every individual. Article 24 declares that all peoples shall have the right to a general satisfactory environment favourable to their development.

28   Article 29 of the Declaration declares that Indigenous peoples have the right to the conservation and protection of the environment and the productive capacity of their lands or territories and resources. States shall establish and implement assistance programmes for indigenous peoples for such conservation and protection, without discrimination.

29   It must be said here that the European Court on Human Rights has developed jurisprudence which accepts for several sorts of environmental disruption to be human rights violations, mostly under

Article 2 (right to life), Article 6 (procedural rights) and Article 8 (family life) of the European Convention on Human Rights (ECHR). The Court has given its most explicit judgments based on Article 8.

30   http://assembly.coe.int/Main.asp?link=/Documents/WorkingDocs/Doc09/EDOC12003.htm, the ink of the final recommendation was still wet when we published this report, but will likely be online in due time.
31   www.RightToEnvironment.org.
32   http://www2.ohchr.org/english/issues/climatechange/index.htm.
33   UN Doc. A/HRC/10/61, 15 January 2009.
34   For the full blog on the COP15 website, see http://en.cop15.dk/blogs/view+blog?year=2009&month=6&blogid=677.
35   For the report of the Special Rapporteur on the adverse effects of the movement and dumping of toxic and dangerous products and wastes on the enjoyment of human rights and his 'Proba Koala' visit to the Netherlands, see http://www.reliefweb.int/rw/RWFiles2009.nsf/FilesByRWDocUnidFilename/VVOS-7VZR9F-full_report.pdf/$File/full_report.pdf.
36   http://www.unhchr.ch/tbs/doc.nsf/0/84ab9690ccd81fc7c12563ed0046fae3.
37   Human Rights Committee, General Comments No. 6 (1982) on Art. 6 CCPR, para. 1 and 5, and No. 14 (1984) on Art. 6 CCPR, para. 1.
38   Budayeva and Others vs. Russia, Application, 15339/02, 21166/02, 20058/02, 11673/02 and 15343/02 [2008] (20 March 2008).
39   The Court was asked to confirm that the Russian government had failed to fulfil its obligations under Article 2 of the European Convention on Human Rights (ECHR), namely to protect the right to life. The case concerned events between 18 to 25 July 2000, when a mudslide led to a catastrophe in the Russian town Tyrnauz: it threatened the applicants' lives and caused eight deaths, among them the husband of one of the applicants. The Court found that the Russian government breached Article 2 ECHR, both in its substance and in its procedural aspects. First, the authorities omitted to implement land-planning and emergency relief policies despite the fact that the area of Tyrnauz was particularly vulnerable for mudslides, thus exposing the residents to "mortal risk". Second, the Court determined that the lack of any state investigation or examination of the accident also constituted a violation of Article 2 ECHR.
40   Quote taken from: Walter Kaelin and Claudine H. Dale, Disaster Risk Mitigation – Why Human Rights Matter (2008).
41   Committee on the Rights of the Child, General Comment No. 7 (2006) on Implementing Rights in Early Childhood, para. 10.
42   http://www.unhchr.ch/tbs/doc.nsf/(symbol)/E.C.12.2000.4.En.
43   CESCR, General Comment No.12 (1999) on the Right to Adequate Food (art.11).
44   http://www.unhchr.ch/tbs/doc.nsf/0/a5458d1d1bbd713fc1256cc400389e94/$FILE/G0340229.pdf.
45   CESCR General Comment No. 15 (2002) on the Right to Water (arts. 11 and 12), para. 2.
46   Address of the Director-General of UNESCO, Mr. Koïchiro Matsuura, on World Water Day, 22 March 2009, at: http://www.unwater.org/worldwaterday/downloads/181210e.pdf.
47   CESCR general comment No. 12, para. 6.
48   CESCR general comment No. 7 (1997) on the right to adequate housing (art. 11 (1) of the Covenant), reg. forced evictions.
49   CESCR general comment No. 12, para. 8.
50   http://www.unhchr.ch/html/menu3/b/74.htm.
51   UN Doc. A/HRC/10/61, 15 January 2009.
52   Idem.
53   Idem.
54   Idem.
55   Idem.

PART 2.2

Selected climate change standards

Chapter 21

Republic of Maldives

MALE' DECLARATION ON THE HUMAN DIMENSION OF GLOBAL CLIMATE CHANGE

We the representatives of the Small Island Developing States having met in Male' from 13 to 14 November 2007.

Aware that the environment provides the infrastructure for human civilization and that life depends on the uninterrupted functioning of natural systems;

Accepting the conclusions of the WMO/UNEP Intergovernmental Panel on Climate Change (IPCC) including, inter alia, that climate change is unequivocal and accelerating, and that mitigation of emissions and adaptation to climate change impacts is physically and economically feasible if urgent action is taken;

Persuaded that the impacts of climate change pose the most immediate, fundamental and far-reaching threat to the environment, individuals and communities around the planet, and that these impacts have been observed to be intensifying in frequency and magnitude;

Emphasizing that small island, low-lying coastal, and atoll states are particularly vulnerable to even small changes to the global climate and are already adversely affected by alterations in ecosystems, changes in precipitation, rising sea-levels and increased incidence of natural disasters;

Convinced that immediate and effective action to mitigate and adapt to climate change presents the greatest opportunity to preserve the prospects for future prosperity, and that further delay risks irreparable harm and jeopardizes sustainable development;

Reaffirming the United Nations Charter and the Universal Declaration of Human Rights;

Recalling the relevant provisions of declarations, resolutions and programmes of action adopted by major United Nations conferences, summits and special sessions and their follow-up meetings, in particular the Declaration of the United Nations Conference on the Human Environment of 1972 (Stockholm Declaration), the 1992 Rio Declaration on Environment and Development and Agenda 21, and the 2002 Johannesburg Declaration on Sustainable Development and Plan of Implementation of the World Summit on Sustainable Development;

Noting that the fundamental right to an environment capable of supporting human society and the full enjoyment of human rights is recognized, in varying formulations, in the constitutions of over one hundred states and directly or indirectly in several international instruments;

Recognizing the leadership of the Alliance of Small Island States in promoting and organizing international responses to climate change for the benefit of their citizens and humanity through inter alia the Male' Declaration on Sea Level Rises, the Barbados Programme of Action, and the Mauritius Strategy;

Acknowledging the United Nations Framework Convention on Climate Change (UNFCCC) and its Kyoto Protocol as important initial multilateral efforts to address climate change through global legal instruments, and the primacy of the United Nations process as the means to address climate change;

Anticipating the publication of the United Nations Development Programme's (UNDP) Human Development Report and the meeting of Commonwealth Heads of Government in Uganda, both of which will emphasise the human aspects of sustainable development;

Concerned that climate change has clear and immediate implications for the full enjoyment of human rights including inter alia the right to life, the right to take part in cultural life, the right to use and enjoy property, the right to an adequate standard of living, the right to food, and the right to the highest attainable standard of physical and mental health;

Do solemnly request:

1.   The international community to commit in Bali to a formal process that will ensure a post-2012 consensus to protect people, planet and prosperity by taking urgent action to stabilize the global climate and ensure that temperature rises fall well below 2°C above pre-industrial averages, and that greenhouse gas concentrations are less than 450ppm, consistent with the principles of common but differentiated responsibilities.

2.   The members of AOSIS in New York to consider including the human dimension of global climate change as one of the agenda items for the meeting of AOSIS Ministers in Bali, and to explore possible alternatives for advancing this initiative in Bali in order to stress the moral and ethical imperatives for action.

3.   The Conference of the Parties of the United Nations Framework Convention on Climate Change, with the help of the Secretariat, under article 7.2(l), to seek the cooperation of the Office of the United Nations High Commissioner for Human Rights and the United Nations Human Rights Council in assessing the human rights implications of climate change.

4.   The Office of the United Nations High Commissioner for Human Rights to conduct a detailed study into the effects of climate change on the full enjoyment of human rights, which includes relevant conclusions and recommendations thereon, to be submitted prior to the tenth session of the Human Rights Council.

5. The United Nations Human Rights Council to convene, in March 2009, a debate on human rights and climate change.

Committed to an inclusive process that puts people, their prosperity, homes, survival and rights at the centre of the climate change debate, other AOSIS members not present in Male' are invited to endorse this Declaration;

Republic of Maldives, 'Male' Declaration on the Human Dimension of Global Climate Change', 14 November 2007.

United Nations Framework Convention on Climate Change

COPENHAGEN ACCORD

1.   We underline that climate change is one of the greatest challenges of our time. We emphasise our strong political will to urgently combat climate change in accordance with the principle of common but differentiated responsibilities and respective capabilities. To achieve the ultimate objective of the Convention to stabilize greenhouse gas concentration in the atmosphere at a level that would prevent dangerous anthropogenic interference with the climate system, we shall, recognizing the scientific view that the increase in global temperature should be below 2 degrees Celsius, on the basis of equity and in the context of sustainable development, enhance our long-term cooperative action to combat climate change. We recognize the critical impacts of climate change and the potential impacts of response measures on countries particularly vulnerable to its adverse effects and stress the need to establish a comprehensive adaptation programme including international support.

[…]

3.   Adaptation to the adverse effects of climate change and the potential impacts of response measures is a challenge faced by all countries. Enhanced action and international cooperation on adaptation is urgently required to ensure the implementation of the Convention by enabling and supporting the implementation of adaptation actions aimed at reducing vulnerability and building resilience in developing countries, especially in those that are particularly vulnerable, especially least developed countries, small island developing States and Africa. We agree that developed countries shall provide adequate, predictable and sustainable financial resources, technology and capacity-building to support the implementation of adaptation action in developing countries.

[…]

United Nations Framework Convention on Climate Change, Copenhagen Accord, 2009, paras. 1 & 3.

Chapter 23

The Pacific Islands Forum

THE NIUE DECLARATION ON CLIMATE CHANGE

We, the Leaders of the Pacific Islands Forum, meeting in Niue:

Deeply concerned by the serious current impacts of and growing threat posed by climate change to the economic, social, cultural and environmental well-being and security of Pacific Island countries; and that current and anticipated changes in the Pacific climate, coupled with the region's vulnerability, are expected to exacerbate existing challenges and lead to significant impacts on Pacific countries' environments, their sustainable development and future survival;

Recalling that despite being amongst the lowest contributors to factors causing climate change, the Pacific Islands region is one of the most vulnerable to the impacts of climate change including its exacerbation of climate variability, sea level rise and extreme weather events;

Recognising that societies pursuing a path of sustainable development are likely to be more resilient to the impacts of climate change through enhanced adaptive capacity;

Noting that despite Pacific Island countries' low greenhouse gas emissions, they are taking significant steps towards reducing their reliance on fossil fuel, with its detrimental economic and environmental consequences;

Recognising the importance of retaining the Pacific's social and cultural identity, and the desire of Pacific peoples to continue to live in their own countries, where possible;

Recalling Forum Leaders' recognition that climate change is a long-term international challenge requiring a resolute and concerted international effort, and stressing the need for urgent action by the world's major greenhouse gas emitting countries to set targets and make commitments to significantly reduce their emissions, and to support the most vulnerable countries to adapt to and address the impacts of climate change;

Welcoming in this context the progress made by the international community in the "Bali Road Map" towards a comprehensive global climate change agreement under the United Nations Framework Convention on Climate Change and its Kyoto Protocol, along with the accompanying Bali Action Plan covering mitigation, adaptation, financing and technology, and a dramatic enhancement of effort under the Kyoto Protocol;

Recalling the Mauritius Strategy for the Further Implementation of the Programme of Action for the Sustainable Development of Small Island Developing States, which calls for Small Island Developing States, with the necessary support of the international community, to establish or strengthen and facilitate regional climate change coordination mechanisms, and calls on the international community to facilitate and promote the development, transfer and dissemination to Small Island Developing States of appropriate technologies and practices to address climate change;

Recalling also the Pacific Islands Framework for Action on Climate Change (2006–2015), adopted by Forum Leaders in 2005, which establishes an integrated, programmatic approach to addressing the interlinked causes and effects of climate change-related impacts in the region;

Stressing the importance of cooperating towards the establishment of an effective post-2012 framework in which all major economies will participate in a responsible manner, underlining the need to achieve both emissions reductions and economic growth in working toward climate stability;

Welcoming the resources and technical assistance from PFD Partners' including new initiatives such as the Japanese Cool Earth Promotion Programme and the European Union's Global Climate Change Alliance (GCCA) that support practical measures to address the impacts of climate change in Pacific Island Countries, while noting that the priority of Pacific SIDS is securing sustainable financing for immediate and effective implementation of concrete adaptation programmes on the ground;

Hereby

Commit Forum members to continue to develop Pacific-tailored approaches to combating climate change, consistent with their ability to actively defend and protect their own regional environment, with the appropriate support of the international community;

Further commit the members of the Pacific Islands Forum to continue to advocate and support the recognition, in all international fora, of the urgent social, economic and security threats caused by the adverse impacts of climate change and sea level rise to our territorial integrity and continued existence as viable dynamic communities; and of the potential for climate change to impact on intranational and international security;

Request the Secretariat of the Pacific Regional Environment Programme (SPREP) – working in cooperation with other regional and international agencies and bilateral climate change programmes – to continue to meet the individual needs of its member countries through its mandated role of:

a.   strengthening meteorological services,
b.   consolidating and distributing information on climate change,
c.   strengthening adaptation and mitigation measures, and

d.   increasing Pacific Island countries' capacity to manage their engagement in the United Nations Framework Convention on Climate Change; and to secure new and additional financial and technical resources to do this work;

Encourage the Pacific's Development Partners to increase their technical and financial support for climate change action on adaptation, mitigation and, if necessary, relocation, while welcoming the pledged increases in resources to address the climate change challenge; and to ensure their assistance aligns with regional and national priorities and supports existing regional and national delivery mechanisms (in accordance with the Pacific Aid Effectiveness Principles 2007, and the Paris Declaration on Aid Effectiveness 2005);

Further encourage development partners to increase investment in and support for Pacific Island Countries' efforts to move towards alternative and renewable energy sources, which reduce the emissions of our region and improve energy efficiency, as well as help to address the growing unaffordability of fuel;

Call on international partners to assist our development by undertaking immediate and effective measures to reduce emissions, use cleaner fuels, and increase use of renewable energy sources;

Agree that the high sensitivity and vulnerability of Pacific Island countries to climate change, including its exacerbation of climate variability, sea level rise and extreme weather events, means that adaptation is a critical response for Pacific governments, and requires urgent support from regional agencies and development partners alike;

Encourage all Pacific Island countries to act on the ability and information they have now, with the assistance of development partners, to continue to address the impacts of climate change through 'no regrets' or 'low regrets' actions in affected sectors that are already facing development challenges, including food and water security, health, and the capacity to deal with extreme events such as tropical cyclones, flooding and droughts, thereby simultaneously delivering on sustainable development aims;

Agree that the exacerbating effects of climate change in the region will require Pacific Island countries to incorporate adaptive strategies into their national sectoral planning, and that this integration will require a high degree of whole-of-government coordination and leadership; and

Direct the Forum Secretariat to work with relevant CROP agencies and Forum members to support the implementation of the commitments made in this Declaration, consistent with the Pacific Plan, the Pacific Islands Framework for Action on Climate Change, and other existing regional and international initiatives, including examining the potential for regional climate change insurance arrangements, and building regional expertise in the development and deployment of adaptation technologies; and to report on progress to the 2009 Forum Leaders' meeting.

The Pacific Islands Forum, 'The Niue Declaration on Climate Change', Press Statement (92/08), 26 August 2008.

PART 3

Proposed new legal standards

Chapter 24

Roger Zetter

LEGAL AND NORMATIVE FRAMEWORKS

How the legal and normative frameworks are addressed will be critical to the security of people threatened by climate change

A dominant theme of rights-based discourse is that rights should not be violated by displacement. There are, accordingly, well established international, regional and national legal instruments, covenants and norms to protect the rights of people forcibly displaced by conflict, persecution, natural disasters and development projects. It is therefore surprising that a similar framework to protect the rights of people forced to move because of climate-induced environmental change does not exist.

The key questions explored in this article are, first, the case for developing the capacity of domestic and international legal apparatus to support the needs of people vulnerable to displacement induced by climate change. The second is to what extent these legal and normative frameworks could support the capacity of local and regional governance and civil society structures to implement adaptation and resilience strategies in order to avert population displacement.

The aspiration is not to promote a case for developing binding conventions but to initiate a bottom-up process – much as the debate on the Guiding Principles on Internal Displacement did in the early 1990s – which might afford firmer support for the rights of those forcibly displaced by environmental change and of those at risk of displacement but who remain behind.

Conceptual and policy questions

Recognising the role of human agency and the need for states to articulate and address the protection of rights in relation to environmentally induced displacement is a pressing issue. What forms of protection for environmentally displaced people currently exist and, more significantly, should be developed as these migratory processes increase? This same question has recently been posed by, for example, the IASC, IOM, EC and NRC, and at the Hague Debate. A number of issues flow from this question.

It is essential to recognise the multi-causality of environmental displacement in which climate change may be only one of the factors triggering forced migration; this raises the question as to the extent to which it is possible to consider specific forms of protection for a migratory process which does not have a clearly established 'cause'.

A second, and related, challenge is to explore the extent to which people forcibly displaced by environmental factors are subject to violations of basic human rights in the way that refugees and IDPs are. It is necessary to establish the particular nature of threats to human rights caused by ecosystem degradation induced by climate change.

Thirdly, in contrast to one of the fundamental factors on which the 1951 Convention and the Guiding Principles are predicated, those who are forcibly displaced by environmental factors will often not return home. Moreover, whilst it is almost certainly the case that the majority will remain internally displaced and will thus fall within the sphere of national norms and legal instruments to protect their human rights, what has enforced displacement is a global process, not a local crisis. This reflects one of the most fundamental issues related to climate change: accountability – the obligation on the polluting countries of the global north to address the needs of countries that will suffer most in the global south. The interplay between national and international frameworks and issues of state sovereignty in applying protection instruments takes on unique meanings in this context.

Fourthly, much of the current discourse treats environmentally induced migration as a reactive response of last resort where migration is seen as failure. However, migration is sometimes a positive and proactive diversification and development strategy that households, individuals and sometimes whole communities adopt to improve their lives and to reduce risk and vulnerability.

Fifthly, and conversely, the focus of much current academic and political debate is on the interests of those forced to migrate because of environmental factors over the equally important rights of those who remain. For some, remaining may be a positive choice – a strategy of adaptation and resilience. This challenges the notion of vulnerable groups as passive victims, highlighting instead people's skills, strategies and agency. Equally, there may be those who are forced to remain because they lack the opportunities, skills and resources to migrate. In either case we need to consider how a rights-based protection regime and the application of principles of human security might support those who remain.

Lastly, it is in the global south where the incidence of climate-induced environmental displacement is, and will be, most severe. Many of these countries and regions have weak governance and civil society structures and are least able, or willing, to protect human rights and security. How can their protection capacity be enhanced? In this context it is important to recognise that environmental factors do not undermine rights and security in isolation from a broader range of socio-economic rights.

A new framework of guiding principles?

Acknowledging the strong resistance of the international community to developing new international instruments but recognising the need to protect the increasing numbers of environmental migrants, what existing norms and instruments might be embraced in a new framework of guiding principles?

I believe the case is very weak for extending the 1951 Convention and 1967 Protocol to include so-called 'environmental refugees' as has recently been advanced by some researchers and humanitarian agencies. Conversely, the 1998 Guiding Principles, however, are not just a fundamental starting point in their own right but also a model for the process of aggregating and adapting the norms and principles from a wide range of international instruments to protect the rights of the 'environmentally displaced'. The 1948 Universal Declaration of Human Rights protects freedom of movement and other social, cultural and economic rights which can be enjoyed under international human rights law and international humanitarian law but which might be threatened when people are forced to migrate by climate-induced environmental degradation.

There are 'subsidiary' norms and instruments which afford different forms of human rights protection for migrant groups either directly or indirectly, for example: the 1966 Covenant on Economic, Social and Cultural Rights and the 1996 International Convention on Civil and Political Rights, as well as a range of international conventions dealing with specific social groups, such as the 1990 International Convention on the Protection of the Rights of All Migrant Workers, the 1989 Convention on the Rights of the Child 1989, the 1981 Convention on the Elimination of All Forms of Discrimination against Women and the 1991 ILO Convention on the Rights of Indigenous People. Given that statelessness is the likely condition for citizens of small island states which will be submerged by rising sea levels, their protection is a critical challenge under the 1954 Convention Relating to Stateless Persons, the 1991 Convention on the Reduction of Statelessness and the protection mandate of UNHCR for stateless people.

Alongside this framework of international human rights and humanitarian law is a substantial body of sovereign state domestic law and regional instruments providing subsidiary and/or temporary protection. Although implementation is limited in precisely those fragile states where protection is most needed, these laws and instruments offer scope for debate and possible expansion to protect the rights of those displaced by, or affected by, environmental degradation.

A number of international bodies, guidelines and standards buttress the protection and security rights of international law and give practical support to them. Although fraught with the same political challenges which accompany development of the framework of principles, in time we might conceive that the protection mandates of a number of international bodies could be extended, for example that of UNHCR or of the Office of the High Commissioner for Human Rights.

Standards and guidelines that could be extended include the UN Inter-Agency Standing Committee's Guidelines on Human Rights and Natural Disasters, the Code of Conduct for the International Red Cross and Red Crescent Movement and NGOs in Disaster Relief, and the Responsibility to Protect of the International Commission on Intervention and State Sovereignty. Equally, the Sphere Project's Humanitarian Charter and Minimum Standards in Disaster Response and the humanitarian clusters under the Humanitarian Response Review process also provide essential features of protection regimes of relevance to those who are environmentally displaced. Interagency coordination, problematic enough now, will be vital.

Policy relevance

Protection and human security instruments and norms will not have the immediate impact of the physical, spatial and developmental strategies and policies needed to respond to climate-induced displacement – but providing and enhancing protection capacity remain essential components of a comprehensive approach to the challenge of displacement at both national and international levels.

Linking the protection discourse to climate-induced environmental displacement and strengthening protection norms and instruments are essential for supporting the potentially very large numbers of people forced to move as well as those who stay behind. Promoting a rights-based perspective of protection and an analysis based on entitlements can also be used as a tool to indicate the parameters for other 'hard' and 'soft' policy responses to the environmentally displaced – for example, rights of access to land and housing, freedom of movement, and participation and empowerment in decision-making on resettlement. Addressing the impacts of displacement as a rights-based challenge inevitably demands that affected populations are fully involved in developing response strategies, and that advocacy tools and processes are enhanced to promote their rights.

Finally, the policy relevance of developing protection norms, instruments and guidelines is emphasised by the extreme cases where ecosystem degradation and the depletion of environmental resources might lead to conflict and violence – and therefore to refugees in the strictest sense of the 1951 Convention. It is necessary to be cautious about these links, for there is little solid empirical research and it is clear that environmental factors do not work in isolation. Nevertheless, given the inevitability of ecosystem degradation and the resulting increase in the numbers of those who will be forcibly displaced, there is a strong case to be made to ensure that the protection machinery does embrace environmental displacement in these specific contexts.

Roger Zetter, 'Legal and Normative Frameworks', *Forced Migration Review*, Issue 31, October 2008, pp. 62–63.

Chapter 25

David Hodgkinson, Tess Burton, Heather Anderson and Lucy Young

'THE HOUR WHEN THE SHIP COMES IN' A Convention for Persons Displaced by Climate Change

[…]

III A Solution: A Global Agreement to Deal with Climate Change Displacement

A. Introduction

Substantive proposals for a new instrument providing for people displaced by climate change have been advanced by Docherty and Giannini,[56] Biermann and Boas,[57] Williams,[58] Betaille et al[59] and the authors of this paper.[60] All of the proposals cite the scale of the climate change displacement problem as a justification for the development of a new international agreement of some kind.[61] Similarly, all of the proposals identify that CCDPs do not fall within the scope of the existing refugee regime created by the Refugee Convention.[62] However, they differ as to the most appropriate instrument to tackle that problem, and the scope and detail of that instrument. Williams proposes regional agreements under an international framework agreement, Biermann and Boas propose a protocol to the UNFCCC, and Docherty and Giannini, Betaille et al and our earlier study propose global, stand alone agreements.

Part B below examines proposals thus far for an agreement to address climate change displacement. Part C introduces our Convention, outlines its comprehensive scope and particular innovations, and makes a number of claims for the Convention as against other proposals thus far to deal with the climate change displacement problem.

B. Convention Proposals thus far to Address the Problem

Williams proposes the formation of regional agreements dealing with climate change displacement under an international umbrella agreement linked to the UNFCCC and drafted as part of a post-Kyoto agreement. She suggests that concerns for state sovereignty and disagreement as to the definition of climate change refugees (particularly the extension of protection to internally displaced persons) would preclude the formation of an international agreement.[63] Although the scope of each regional agreement would likely differ, Williams proposes a sliding scale of protection depending on the severity of the situation which would entail an assessment of the objective necessity of re-location.[64]

A protocol to the UNFCCC to provide for assistance to climate refugees (the term used by Biermann and Boas) is proposed by Biermann and Boas. They identify four factors that distinguish climate refugees from political or economic refugees: they are unable to return to their homes; they are 'likely to migrate in large numbers and collectively'; they are predictable, because the need for relocation as a result of climate change impacts in particular areas is evident; and they have a moral claim for assistance against industrialised countries historically responsible for emissions.[65] Adopting these principles, Biermann and Boas propose a protocol providing for the recognition, protection and permanent resettlement of climate refugees displaced as a result of sea-level rise, extreme weather events, and drought and water scarcity. The protocol would extend protection to internally displaced persons as well as international migrants, and would be focused on long-term, planned, voluntary resettlement.[66]

Docherty and Giannini propose a global, stand-alone convention to provide for climate change refugees displaced across national borders (but not displaced internally).[67] They identify three priorities in the design of the proposed instrument: providing guarantees of human rights protections and humanitarian assistance for designated climate change refugees; spreading the burden of providing assistance across affected states and the international community; and establishing institutions to administer the regime including a global fund, a coordinating agency and a body of scientific experts.[68]

A convention proposed by Betaille et al aims to guarantee the rights of environmentally displaced persons (their term) and arrange for their reception and return.[69] The principles of the draft convention are those of common but differentiated responsibility; proximity, 'which requires the least separation of persons from their cultural area'; proportionality, which relates to an international system of financial aid; and effectiveness, which provides that in order to make the rights conferred by the convention effective, an agency created by the convention and the state parties shall develop policies to encourage temporarily resettled persons to 'establish normal conditions of life'.[70] Their convention applies to all environmentally displaced persons rather than the subset of CCDPs.

Finally, in addition to our earlier study,[71] a 2008 report, *Climate Change as a Security Risk*, by the German Advisory Council on Global Change summarizes the state-of-the-art of science on climate change and assesses likely impacts of climate change on societies and states.[72] Although a small part of its report, 'environmentally induced migration' is briefly considered. The report proposes 'an interdisciplinary, multilateral convention to regulate the legal status of environmental migrants', which could be linked with the Refugee Convention.[73] At a minimum such a convention would include the following and 'involve the entire international community': acknowledgement of environmental damage as a cause of environmentally induced migration; protection of environmental migrants through the granting of at least temporary asylum; establishment of a formula for the distribution of environmental migrants which ensures that among potential host countries no individual states are overburdened; establishment of an equitable formula for the distribution of the costs of receiving refugees; equalization of the financial burdens of climate-related environmental degradation.[74]

C. Why Another Proposal? Why Our Convention?

As we make clear throughout this article there is some consensus among the authors of convention proposals set out above to address the climate change displacement problem. We acknowledge at various points the contributions made by these authors – in particular, those by Docherty and Giannini,[75] and by Biermann and Boas[76] – and the extent to which we have taken account of and incorporated aspects of their proposals in our own. No proposal, however, has offered a comprehensive, global solution to the displacement problem; our proposal, which builds on our earlier 2008 study, attempts to provide such a solution.

Any convention designed to deal with the climate change displacement problem on a comprehensive basis must include internal displacement – that which occurs within state borders and the most likely form of displacement[77] – and international (or trans-border) displacement. Our Convention, while it necessarily distinguishes between internal and international displacement, provides for both.[78] For those displaced within state borders it institutes a mechanism for the provision of principled, non-discriminatory assistance. It also provides an original definition of 'climate change displaced persons' and a 'climate change event'.

The Convention attempts to prioritise climate change displacement solutions without shaping instruments not designed to deal with the displacement problem.

Our Convention proposal provides specifically for the populations of small island states which may become uninhabitable due to the effects of climate change,[79] and differentiates such states from others which may be affected by large-scale displacement; it treats small island populations as a discrete group. It suggests that certain principles – proximity, for example, and the preservation of intangible culture – be applied to 'bilateral displacement agreements' to be made between small island states and host states.

Finally, our proposal sets out in some detail a sophisticated Convention governance and organisational structure and the roles and obligations of Convention participants and constituent bodies. The mechanics of the Convention's institutional operation and processes are examined and outlined both in narrative and diagrammatic form. The procedures through which offers of and requests for assistance are made are also described.

IV. A Convention for Persons Displaced by Climate Change

A. Scope of Proposed Convention: 'A Global Governance Architecture'

As Docherty and Giannini state, '[b]ecause the nature of climate change is global and humans play a contributory role, the international community should accept responsibility for mitigating climate-induced displacement'.[80] We propose a single, stand alone convention to address the problem of climate change displacement, the scope of which – like the problem, both in terms of causation and consequences[81] – is global; parties to the convention would include both developed and developing states. The convention, to use Biermann and Boas' term, provides for 'a global governance architecture for the protection and resettlement' of CCDPs.[82]

Many proposals for some kind of legal instrument designed to address the problem of climate change displacement seek, in various ways and for various reasons, to link that instrument with the

UNFCCC.[83] In our view, however, which reflects that of Docherty and Giannini, the UNFCCC has limitations as a framework for dealing with climate change displacement. Displacement is not its focus; its concerns lie elsewhere. Its structure and institutions are not designed to address displacement and the issues associated with it. Moreover, 'it does not discuss duties that states have to individuals or communities, such as those laid out in human rights or refugee law'.[84]

Our Convention is a stand alone one (noting, however, that it would draw upon and adapt provisions of other instruments to adequately provide for, assist and protect those displaced by climate change). There has been no coordinated response by governments to address human displacement, whether domestic or international, temporary or permanent, due to climate change. Given the nature and magnitude of the problem which climate change displacement presents, ad hoc measures based on existing domestic regimes are likely to lead to inconsistency, confusion and conflict. The international community has an obvious interest in resolving the problem of human displacement in an orderly and coordinated fashion.

2. Internal and International Displacement

Migration experts state that most persons displaced by climate change will be unlikely to cross an international border;[85] 'most climate refugees are expected to remain within their home countries'.[86] Kniveton et al emphasise

> a broad theoretical consensus that it is generally not the poorest people who migrate overseas because international migration is an expensive endeavour that demands resources for the journey and for the crossing of borders.[87]

The German Advisory Council on Global Change refers to 'environmentally induced migration' and its nature:

> It is thought likely that most such migration currently takes place within national borders and that this will continue to be the case in future. Environmental migrants are therefore more likely to be internally displaced persons rather than migrants who cross national borders.[88]

In our view, if – as migration experts assert – the reality is that most persons displaced by climate change will be relatively unlikely to cross an international border[89] (or, put another way, the people most vulnerable to climate change are not always the ones most likely to undertake trans-border migration[90]), the 'means for facilitating ... assistance' to such persons cannot be 'beyond the scope' of proposals seeking to address climate change displacement.[91] Further, it may be that perpetuating a narrow focus on 'displacement', particularly trans-border displacement, is to act implicitly within the preoccupations of the 'developed' world, with all of the attendant security concerns – and perhaps even the xenophobic reactions – that such a stance entails.

Adopting a multifaceted, cooperative and international approach[92] to providing for, assisting and protecting CCDPs, our Convention encompasses those displaced internally (that is, within a country) and those who cross international borders – thus, both internal and international displacement. While it is necessary to distinguish between internally and internationally displaced persons in the context of climate change displacement,[93] and to make the same distinction in drafting a convention for CCDPs, the provisions of a convention would encompass and reflect careful consideration of both categories of displacement.

Docherty and Giannini note that 'ideally, at some point international law would provide the same assistance for both climate change refugees [those crossing international borders] and IDPs [internally displaced persons]';[94] the convention proposed by these authors provides assistance only for refugees on the basis that

> adopting the Refugee Convention's distinction acknowledges international law's current emphasis on state sovereignty. It recognises that host states, to which refugees flee, are more likely to accept outside assistance than are home states, which may not want interference from the international community.[95]

It seems to us, however, that under a convention in which requests for assistance can come from state parties to the convention, whether such parties be 'home' or 'host' states, and in which assistance can be offered by the climate change displacement organisation under the convention to home or host states (such offered assistance by the organisation to either be accepted or declined by the relevant state), contentious issues of sovereignty should be minimised – although we acknowledge that this may be an overly optimistic view of the convention's operation.

3. Those in Need of Displacement Assistance will be Developing States Parties to the Convention

As the International Council on Human Rights Policy notes, 'the most dramatic impacts of climate change are expected to occur in the world's poorest countries'; indeed, these countries already experience such impacts.[96] For the developing world generally, the magnitude of the impacts – as the World Bank notes – are 'sobering': with regard only to sea level rise, displacement due to climate change will affect hundreds of millions of people.[97]

It is developing state parties to the Convention – with economies dependent on the natural environment, but without resources to mitigate and adapt to the effects of climate change – who will be most in need of displacement assistance. Moreover, as mentioned above, the majority of displacement will be internal rather than across national borders; developing states, which already experience environmental degradation and natural disasters, will bear the additional burden of displacement. For these states, a 'vicious cycle links precarious access to natural resources, poor physical infrastructure … [and] vulnerability to climate change-related harms'.[98] As the German Advisory Council on Global Change notes, '[m]ost cross-border environmentally induced migration will probably take the form of south-south migration; no trend towards large south-north migrations has been identified'.[99]

4. Temporary and Permanent Relocation

Both temporary and permanent relocation would be provided for under the proposed convention. In common with Biermann and Boas[100] we see little value in making a distinction, commonly made in the literature on refugees in the context of environmental-related migration, on whether relocation (or displacement) is temporary or permanent. The need for relocation assistance and protection arises whether or not the relocation is temporary or permanent. It should be noted that the Refugee Convention similarly makes no distinction between temporary and permanent relocation; protection is afforded refugees no matter whether displacement is short-term or permanent.[101]

5. Causation: Climate Change Events and Displacement

A number of issues of causation[102] arise with respect to the provision of protection and assistance for persons displaced by climate change. The first is the extent to which climate change causes the event giving rise to the displacement. At the moment it is not possible, as Docherty and Giannini note, for science to determine whether a particular environmental event was caused by climate change.[103] It is possible, however, as the Intergovernmental Panel on Climate Change (IPCC) shows, to identify certain phenomena and trends as consistent with climate change. So, for example, the IPCC identifies (a) increased incidence of extreme high sea level (excluding tsunamis); (b) intense tropical cyclone activity increases; and (c) areas affected by drought increases, as 'likely', that is, with more than sixty-six percent probability.[104] Further, scientific progress in understanding how climate is changing in space and in time has been gained through improvements and extensions of numerous datasets and data analyses, broader geographical coverage, better understanding of uncertainties, and a wider variety of measurements'.[105]

Climate change science continues to evolve. Any instrument that seeks to address migration induced by climate change events must be based on scientific evidence as to whether those events are consistent with climate change and sufficiently flexible to reflect developments in scientific understanding over time.

A second issue is the extent to which humans contribute to particular climate change events (humans have contributed to climate change generally and, it is argued, should accept and bear some responsibility for dealing with displacement caused by, or resulting from, climate change). Just as science can't determine whether a particular environmental event was caused by climate change, neither can science determine the extent to which humans contributed to specific climate change events. Docherty and Giannini, however, argue that science can determine the likelihood that humans 'contributed to a type of disruption'. Their definition references the IPCC's 'likelihood ranges' and adopts the IPCC's 'more likely than not' standard in order to encompass the range of environmental disruptions most commonly associated with climate change and related displacement.[106] The 'more likely than not' standard means a probability greater than 50%.

While the focus of our Convention is persons displaced by climate change, the convention would recognise problems with (a) determining the extent to which climate change causes an event giving rise to displacement; (b) identifying certain phenomena and trends as consistent with climate change; and (c) the extent to which humans contribute to particular or general climate change events. Our Convention would adopt a 'very likely' standard[107] to identify certain phenomena and trends as consistent with climate change, and human contribution. The reasons for this higher standard are that it provides increased certainty and targeted resource allocation in the context of a convention that could apply to hundreds of millions of people.

By adopting the standard of 'very likely, and in light of the current state of climate change science, we anticipate that requests from state parties attracting the operation of our Convention would overwhelmingly concern slow-onset, gradual displacement, which is more likely to be able to be established as induced by anthropogenic climate change than a sudden disaster. However, as more fully set out below in Part IV, section C, 'Operation of the Convention', the Convention's proposed Climate Change Displacement Environment and Science Organisation (CCDESO), established in part to conduct climate change research, would continually assess the development of climate change science, and would advise the Convention's Climate Change Displacement Organisation (CCDO) of matters which would affect the standard adopted by the Convention, and related matters. With input from the CCDESO, and with the benefit of that organisation's links with the IPCC, the UNFCCC's Subsidiary Body for Scientific and Technological Advice, and climate change science and other research organisations, each biannual meeting of the CCDO Assembly would carefully consider the evolving climate change science, both generally and specifically with regard to sudden onset climate change events. For instance, scientific knowledge may evolve such that increased frequency and severity of sudden disasters can be attributed to anthropogenic climate change, in which case the operation of the Convention would be triggered by the prospect of future displacement as a result of such events. Again, these matters are more fully set out below.

Issues of causality also arise with respect to the question of whether climate change necessarily leads to migration, as previously discussed in Part II of this article. While there seems to be a consensus that sea level rise will result in population movements, there is considerable uncertainty amongst migration researchers as to whether drought and sudden weather events result in displacement.[108] Kniveton et al summarise the state of the research on whether drought induces migration as follows:

> The empirical results…reveal that the nexus between drought and migration is not straightforward…the conceptualisation of drought-affected people as helpless victims who are left with no choice but to flee seems to be false. Depending on their socio-economic position, they might have the choice between a variety of coping strategies, including migration. On the other hand, they might be too poor to migrate at all.[109]

As Kniveton and his coauthors suggest, adaptive capacity is integral to displacement decisions – and both adaptation and displacement are determined as much by available resources as by environmental factors. As discussed below in 'Definition and Designation of CCDPs', our Convention would address the causality issues associated with the multi-factorial nature of population movements by adopting an objective rather than a subjective approach. In addition, our Convention would largely operate prospectively; assistance to CCDPs would be based on an assessment of whether their environment was likely to become uninhabitable due to events consistent with anthropogenic climate change such that resettlement measures and assistance were necessary.[110] In other words, displacement is viewed as a form of adaptation that creates particular vulnerabilities requiring protection as well as assistance through international cooperation.

Because of the necessity of integrating complex issues of causality and evolving science into decision-making in respect of climate change migration, our proposal involves the creation of a sophisticated institutional architecture for designating a particular population as 'climate change displaced people'. While the scope of our Convention includes those persons displaced as a result of sudden climate change events (or impacts), as a practical matter the proposed Convention's machinery or 'architecture' may not be suited to immediately reacting to an unforeseen disaster. Indeed, attempts to apply the Convention in such situations may prevent the operation of existing and more effective disaster relief and management programs. Multiple channels through which aid, assistance and protection are provided to those displaced by sudden environmental impacts already exist in the international arena.[111] Moreover, as discussed above, the science of climate change is currently unable to attribute a particular sudden climatic event to anthropogenic causes with any degree of certainty.

B. Definition and Designation of CCDPs

A key focus of the literature on climate change displacement has been on the question of whether and how persons displaced by climate change should be defined. It is a concern that stems in part from a broader debate on the nature and legitimacy of the concept of 'environmental refugees'.[112] The term 'environmental refugees' came to prominence with a 1985 report by the United Nations Environment Programme (UNEP). The UNEP report defined environmental refugees as those people who have been forced to leave their traditional habitat, temporarily or permanently, because of a marked environmental disruption, natural or triggered by people, that jeopardized their existence and/or seriously affected the quality of their lives.[113]

The concept has been further popularised by Myers, who refers to environmental refugees as those 'who can no longer gain a secure livelihood in their erstwhile homelands' and 'who in their desperation feel that they have no alternative but to seek sanctuary elsewhere'.[114]

Subsequent to the UNEP report, various proposals have been made for a definition of 'environmental refugees'. A series of recurrent themes commonly structure such proposals. Firstly, proposals for definitions of 'environmental refugees' will generally consider the nature of the migration, such as whether it was forced or voluntary; temporary or permanent; and within or external to state boundaries. Secondly, they are also likely to be informed by the character of the environmental change giving rise to the migration, whether the change is sudden or gradual, and, in some cases, whether the change is natural or induced by humans.[115]

A number of proposals have now been made that seek to define persons displaced by climate change. Biermann and Boas define climate refugees as people who have to leave their habitats, immediately or in the near future, because of sudden or gradual alterations in their natural environment related to at least one of three impacts of climate change: sea-level rise, extreme weather events, and drought and water scarcity.[116]

Docherty and Giannini define a climate refugee as an individual who is forced to flee his or her home and to relocate temporarily or permanently across a national boundary as the result of sudden or gradual environmental disruption that is consistent with climate change and to which humans more likely than not contributed.[117]

Williams proposes a definition incorporating gradual recognition and correlating levels of protection on a sliding scale, allowing differing degrees of protection to be accorded depending on the severity of the situation.[118]

Attempts to instigate regimes for the protection of both 'environmental' and 'climate change' displaced persons raise the issue of whether and how to include migration due to gradual environmental change, as opposed to sudden disasters. The core question is when migration as a consequence of slow-onset environmental events can be said to be 'forced' rather than 'voluntary'; in other words, at what point is a person compelled to move, as opposed to exercising their own free choice, when they leave a situation of encroaching desertification or sea-level rise? The underlying concern, which is less often stated, is associated with denial by developed countries of the legitimacy of 'economic migration' and a related fear of 'opening the floodgates' to immigrants from developing nations. However, as a given economy is bound up with its environment, it may be overly ambitious to attempt to distinguish between economic 'pull' and environmental 'push' explanations for population movement, especially where the attempt is based on an individualised definition with reference to subjective criteria.[119]

Proposals for definitions of 'environmental' and 'climate change' displaced persons have advanced a range of strategies for addressing the question of how to distinguish forced from voluntary migration in cases of gradual environmental degradation. For instance, the sliding scale of protection proposed by Williams corresponds to an assessment of the severity of existing environmental damage. CCDPs relocating because an area becomes uninhabitable would be accorded 'acute' status and receive the highest level of protection, whereas CCDPs who relocate but 'could remain within that same environment albeit under increasingly onerous and challenging conditions' would receive a lower level of protection.[120] Williams' sliding scale definition of climate change displacement affords rights based on the degree to which displacement is 'forced' by existing environmental circumstances.

Similarly, Renaud et al distinguish between three categories of environmentally displaced people: (a) environmentally motivated migrants who 'may leave' a steadily deteriorating environment in order to pre-empt the worst (examples include the depopulation of old mining and industrial areas); (b) environmentally forced migrants who 'have to leave' in order to avoid the worst (examples include sea-level rise and desertification); (c) environmental refugees who flee the worst (examples include floods, droughts and hurricanes). The distinction is based on the point of departure and would include an assessment of the 'coping capacity' of the affected community as well as the nature of the environmental event.[121]

In our view, prospective migration based on the likely consequences of climate change is as coerced as migration in response to climate change impacts that immediately render a particular area uninhabitable. In other words, population movements based on the conclusion that due to the effects of climate change, a region *will* no longer be inhabitable constitute 'forced' migration. Further, when developing criteria to determine when a movement is no longer voluntary, the point of departure should not be the subjective motives of individuals or communities for their decision to move but rather the question as to whether in light of the prevailing circumstances and the particular vulnerabilities of the persons concerned it would be appropriate to require them to go back to their homes.[122]

The salient question then becomes institutional rather than definitional and is focused on constructing and administering a set of processes to determine the likely contribution of climate change to both prospective and responsive climate change movements.

There has, historically, been a focus on definitions in literature on climate change and environmental displacement. This focus may be partly explained as an effort to disentangle 'environmental' from 'economic migration'. However, the preoccupation with definitions may also have arisen as a consequence of the centrality of the definition of a 'refugee' to the assignment of protection under the Refugee Convention. In the context of that instrument, the definition of a 'refugee' gives expression to the fundamental principle that international protection has been triggered on the basis that a relationship between an individual and their nation has been damaged or severed in certain prescribed ways. The construction of a convention for CCDPs around definitional criteria or the individual processing of 'asylum' claims is not the optimal solution for the problem of climate change displacement for two reasons. The first is that, as discussed in Part II, it is unlikely to be possible to attribute an individual decision to migrate solely or perhaps even directly to climate change. The second is that climate change impacts and any corresponding displacement are likely to be felt on a mass-scale, as opposed to an individualised basis. Climate change is better understood as affecting entire communities rather than certain individuals, although some persons in those communities will no doubt be more affected than others.

Rather than assigning rights and protections on the basis of the individual satisfaction of definition-based criteria, we propose *en masse* designations of the status of CCDPs through a process of request and determination by states and Convention institutions. Such an approach would nevertheless require a definition of CCDPs because, as Castles observes, 'we cannot get around definitional categories…easily, for definitions are crucial in guiding the policies of governments and international agencies towards mobile people'.[123]

Based on earlier discussion in the 'Causation: Climate Change Events and Displacement' section of this article, and the above discussion, the definition of CCDPs we propose is as follows:

> CCDPs are groups of people whose habitual homes have become – or will, on the balance of probabilities, become – temporarily or permanently uninhabitable as a consequence of a climate change event.

We define a 'climate change event' as 'sudden or gradual environmental disruption that is consistent with climate change and to which humans very likely contributed'.

Under our Convention, protection and assistance would not be triggered solely by fulfilling the requirements of a definition, but rather through an international process of status designation, informed by scientific studies, affected communities, states and international institutions. As a consequence, the definition of a 'climate change displaced person' is thus less pivotal in the context of our proposal than it is to the Refugee Convention.[124] Further, as the designation would be made through prescribed Convention processes, it would obviate the need to construct an elaborate definitional mechanism that would specifically address 'slow-onset' climate change events.

An approach that prioritises *en masse* designations finds support in other publications on climate change induced migration. McAdam argues that the 'traditional western approach of individualised decision-making on technical legal grounds seems highly inappropriate to the situation we are presently facing'.[125] Similarly, Docherty and Giannini state that, in general, 'group status determination of climate change refugees would be preferable, and the climate change refugee instrument should make it the default while still allowing for individual status determination'.[126]

Moreover, the proposals of both Docherty and Giannini and Biermann and Boas move away from a reliance on definitional criteria as a trigger for protection and rather seek to construct an international architecture to determine the status of CCDPs.

C. Operation of the Convention: Institutions, Participants and Obligations

Sections A and B above outlined the scope of our Convention in terms of global coverage, internal and international displacement, temporary and permanent relocation, issues of causation, sudden and slow-onset climate change events, and definition and designation of CCDPs. This section of the Article sets out the operation of the Convention through a focus on its institutional organisation. Through that focus, the obligations of the Convention participants – state parties generally, both developed and developing; home and host state parties; the climate change displacement organisation established under the Convention and its organs; and civil society – become clear.

Docherty and Giannini[127] and Betaille et al[128] provide some detail regarding administrative bodies as part of their proposals for an agreement to address the climate change displacement problem. Biermann and Boas (proposing a network of agencies including the United Nations Development Programme, the United Nations Environment Programme, the United Nations High Commissioner for Refugees and the World Bank)[129] and Williams (proposing a subsidiary body located within the UNFCCC[130]) address administrative bodies only briefly.

For Docherty and Giannini, a convention for displaced persons would create a global fund for the provision of financial assistance,[131] together with a body of scientific experts and a coordinating agency, modelled on the UNHCR.[132] The purpose of the co-ordinating agency would be to work with states to prevent refugee crises, assist host states to provide humanitarian aid and fulfil human rights obligations, and assist with the repatriation or permanent resettlement of refugees.[133]

Betaille et al's draft convention provides for each state party to create a national commission to make status determinations at first instance.[134] The convention provides for the creation of a World Agency for Environmentally Displaced Persons (WAEP) to oversee the application of the convention, including conducting studies on displacement and providing assistance directed to preventing and limiting displacement and promoting the rapid return of environmentally displaced people.[135] A High Authority, a Scientific Council and a Secretariat are created to assist the WAEP.

While these proposals describe discrete coordinating agencies or agencies to administer the convention, no proposal attempts to set out in detail the operation of a climate change displacement organisation and its constituent bodies, and how those bodies work together and interact. Our proposal attempts to remedy this gap by putting forward a climate change displacement organisation structure and, within that structure, setting out the roles and obligations of convention participants both in narrative and diagrammatic form. Although we envisage a single, stand alone convention with a global scope, the convention's governance structure also contemplates a role for regional committees and multi-disciplinary collaborations across developed and developing states, and including government and non-government organisations.

1. Climate Change Displacement Organisation

The Convention would create a Climate Change Displacement Organisation (CCDO), the proposed operation of which is set out in detail below. The CCDO would consist of four core bodies: an Assembly, a Council, a Climate Change Displacement Fund and a Climate Change Displacement Environment and Science Organisation. Climate Change Displacement Implementation Groups would also be formed to facilitate resettlement. The operation of each of these bodies and groups, their roles and functions, and relationship with each other is set out below. [...].

Operational guidelines for the implementation of the Convention would also be drafted, the guidelines to be periodically revised to reflect the decisions of the CCDO Council and changes arising as a result of the operation of the Convention and practices adopted. These operational guidelines could be modelled on the Operational Guidelines for the Implementation of the World Heritage Convention or similar guidelines.[136]

2. State Parties: Offers of and Requests for Assistance, and Designations

The Convention framework contemplates that states providing displacement assistance funding and states receiving such funding would be parties to the Convention. A developing state party (a home or a host state) to the convention would make a request, using documentation prepared by the CCDO, to the CCDO for internal or international resettlement assistance, as the case may be, referencing a relevant home state request, if any (in the case of a host state). That state party would, at the time of making a request for assistance, also request *en masse* designation of the status of CCDPs.

Developing state parties may also make financial and other requests for assistance in fulfilling their Convention obligations. These obligations would include collecting data on climate change displacement; integrating protection of and assistance to CCDPs into future planning programs; and consulting with potential or actual CCDPs and their representative organizations in the collection of data, the development of policy and its implementation.

It would be open to the CCDO (through the CCDO Council) to offer assistance to developing state parties without a request from those parties. It is also open to civil society within state parties to the Convention to request CCDO assistance, but such requests must be authorised and submitted by the relevant state party.

State parties, developed and developing, would form national climate change displacement committees (their title, form and function to be determined by the relevant state) to inform CCDP decision making at the national level as a basis for participation in the Convention. Parties would also provide a delegate (an 'ambassador' or a 'representative') to the CCDO Assembly, an assembly of state parties to the Convention which would meet every 2 years. State parties would also provide members, as applicable, on a rotating basis to the CCDO Council.

3. CCDO Assembly

The governing body of the CCDO would be the CCDO Assembly, comprising representatives of all state parties to the Convention. The Assembly would meet biennially (every two years), and elect a president for a term or terms and on conditions to be determined. It would ratify developed state party's financial contributions to the Climate Change Displacement Fund after advice from (a) the Climate Change Displacement Fund; and (b) the Climate Change Displacement Environment and Science Organisation, such advice received through the Council. Funding from developed state parties would be provided on a biennial basis to coincide with Assembly meetings. Developed state parties would also fund the CCDO itself.

The Assembly would confirm (a) the CCDO chief executive officer and representative state parties, both developing and developed, to the CCDO Council; and (b) state party appointments to the Climate Change Displacement Fund (CCDF).

Finally, the Assembly would, with advice and recommendations from the Climate Change Displacement Environment and Science Organisation (CCDESO), review the state of climate change science at every biennial meeting, with particular focus on implications for sudden disruption and displacement (see Part IV, section A(5) above on sudden and slow-onset events).

APPEAL COMMITTEE

The CCDO Assembly would determine members of its Appeal Committee to hear appeals from developing state parties against funding/assistance decisions of the CCDO Council.

4. CCDO Council

The CCDO Council's main functions would be to (a) assess developing state party (home or host state),[137] requests for internal and international resettlement assistance, including the making of *en masse* designations; (b) confirm the level and terms of that assistance, and assistance to be provided to Climate Change Displacement Implementation Group organisations; and (c) set the direction and guide the operations of the CCDO. Determinations of the Council, including those related to designations, would be informed by information and guidance provided by the CCDF and the CCDESO.

The Council would be chaired by a Chief Executive Officer. Membership of the Council would comprise a small number of representatives from developed and developing state parties to the convention, together with individuals with relevant institutional or subject matter experience, from the private sector, civil society and academic institutions. These individuals would be appointed by the developed and developing state members of the Council. The number of individuals so appointed could equal the total number of developed and developing state members of the Council.

REGIONAL COMMITTEES

Williams[138] outlines the virtues of regional initiatives and cooperation to address the plight of persons displaced by climate change, including the ability to assess 'current vulnerabilities to climate change impacts, along with information exchange on traditional coping practices, diversified livelihoods, and current government and local interventions'.[139] Williams also notes that a regional structure would allow 'for various levels of engagement and development by states, depending on the individual capacity of each country involved and the (perceived) severity of the problem in that area'.[140] While we have not taken up regional initiatives as a formal part of our Convention, there are clear virtues in formalising regional arrangements such that any climate change displacement organisation can be fully informed and take account of regional developments. McAdam notes that responses on a regional basis might 'be more appropriate and culturally sensitive…taking into account the particular features of the threatened population'.[141]

Accordingly, on terms to be established by the Council and the relevant regional committees themselves, we propose the establishment of regional Council committees which would (a) inform Council and CCDO decision-making of regional perspectives, views and developments; (b) work closely with Climate Change Displacement Implementation Groups (see below); and (c) enable the unique situation of small island states to be addressed, in part through involvement with concluding small island states' bilateral displacement agreements.

5. Climate Change Displacement Fund

The CCDF, with members appointed by the CCDO Council, and with oversight from a committee of developed and developing state parties to the Convention, would propose to the Council (a) the level of developed state parties' biennual contributions to the Fund; and (b) the level of assistance funding to developing state parties and Climate Change Displacement Implementation Group (CCDIG) organisations (see 7 below), and provide related advice on funding displacement assistance and requests for funding. After a Council decision to provide assistance, the Fund would (a) disburse funds based on the agreement concluded between the recipient state party and the CCDO; and (b) work with the recipient and the relevant CCDIG on the deployment of the funds.

6. Climate Change Displacement Environment and Science Organisation

As discussed earlier problems exist regarding the extent to which climate change causes an event giving rise to displacement, identification of certain phenomena and trends as consistent with climate change ('more likely than not', 'likely', 'very likely' etc), and the extent to which humans contribute to particular climate change events.[142] Further, '[t]he recognition of human contribution must work within the parameters of existing and evolving science'.[143]

The CCDESO, with members appointed by the Council, would advise (a) the CCDO on climate change science matters as they affected displacement (both sudden, given the discussion above,[144] and slow onset); and (b) the Fund on developed state parties' contributions to the Fund and the level of funding assistance to developing state parties and CCDIGs. The CCDESO would monitor state parties' emissions levels to assist in the determination of such party's financial contributions to the Fund. Given these advisory, assessment and monitoring roles, the work of the CCDESO is crucial to decisions made by the CCDO.

The CCDESO would be established as a significant climate change science and displacement research organisation, itself conducting climate change research and contributing to the evolution of climate change science, with research output available to convention parties. It would establish links with the Intergovernmental Panel on Climate Change, the UNFCCC's Subsidiary Body for Scientific and Technological Advice, and climate change science and other research organisations around the world, thus contributing to what McAdam refers to as 'institutionalized cooperation'. This is discussed again below with regard to the CCDIG.

7. Climate Change Displacement Implementation Groups

We propose dedicated CCDIGs under the Convention to effect resettlement. Members of each CCDIG would consist of the affected state (in the event of internal displacement); the home and host state (in the event of international resettlement); local and state (if applicable) governments; international organisations; relevant UN bodies, again if applicable; and the CCDO, representatives of which would lead each CCDIG. CCDIGs would also work with civil society in home and host states, and work closely with regional committees.[145]

Each CCDIG would represent a form of what Jane McAdam refers to as 'institutionalized cooperation'.[146] McAdam notes that

[b]ecause there are numerous cross-cutting and intersecting issues raised by climate-induced displacement which relate to a variety of institutional different mandates ... the concept risks being dealt with in an ad hoc and fragmented manner.[147]

In our view the formalised CCDO-led CCDIG approach would result in 'a comprehensive and coherent multilateral framework'[148] within which assistance and protection would be provided to CCDPs around the world.

8. Permanent Secretariat

In order to discharge their functions, both the Assembly and the Council (the Council in particular) would be supported by a CCDO secretariat consisting of a permanent staff and supplemented by contract and other staff as required. The secretariat would also support the CCDF, the CCDESO and each CCDIG.

D. Financing

1. Existing Proposals

Existing proposals, no matter the level of detail of those proposals, all set out ways in which provision of displacement assistance can be financed, or funded. Biermann and Boas,[149] Betaille et al,[150] and Docherty and Giannini[151] make proposals for a global fund to manage international financial assistance with regard to climate change displacement. Biermann and Boas propose the creation of an independent Climate Refugee Protection and Resettlement Fund.[152] All funds would be provided on a grant basis and would be additional to grants already promised to existing funds. Eligibility for assistance under the fund would be determined according to the principle of 'full incremental costs', whereby in situations where the causal links with climate change are undisputed, such as displacement caused by sea-level rise, the fund would reimburse the full agreed incremental costs incurred. Where climate change is only one factor causing environmental degradation, the fund would pay for only part of the protection and relocation costs.[153] Biermann and Boas further propose innovative income-raising mechanisms such as an international air travel levy which could be used as a source of funds; this suggests that states would not be required to make mandatory payments under the proposal.

Docherty and Giannini's proposed convention would determine obligatory state contributions, collect payments and distribute funds to states in need.[154] Funds would be provided to home and host states for mitigation as well as assistance following displacement.[155]

The convention drafted by Betaille et al advocate similarly creates a World Fund for the Environmentally Displaced (WFED). The purpose of the WFED is to provide assistance for the settlement and repatriation of environmentally displaced people. Under the draft convention, funds can be granted to home states and host states as well as non-governmental organisations and local governments.[156] The WFED would be supported by 'voluntary contributions from states and private actors', and 'mandatory contributions funded by a tax based principally on the causes of sudden or gradual environmental disaster susceptible of creating environmental displacements'.[157]

2. Our Convention

We propose that developed state parties to the Convention make mandatory, or binding, financial contributions to the Climate Change Displacement Fund (the operation of which was set out earlier), and that such contributions are made on the basis that states and state parties to the Convention have common but differentiated responsibilities. In determining the hard issue of the level of specific state party contributions to the Fund, the CCDESO would advise the Fund with regard to those contributions, with reference to emissions levels (whether historical or current, per capita etc), the capacity of states to pay, and other matters. The Fund would then propose the level of state parties' biannual contributions to the Council for ratification by the Assembly.[158]

The principle of 'common but differentiated responsibilities', the basis upon which developed state parties make contributions to the Fund, recognises historical differences in the contributions of developed and developing states to global environmental problems, and differences in their respective economic and technical capacity to tackle these problems.[159]

As the McGill Centre for International Sustainable Development Law notes, the principle has two basic elements: 'the common responsibility of states for the protection of the environment, or parts of it, at the national, regional and global levels' and 'the need to take into account the different circumstances, particularly each State's contribution to the evolution of a particular problem and its ability to prevent, reduce and control the threat'. This has a number of implications, including the imposition of environmental obligations which differ between states. This principle underpins the obligations of state parties to the Convention.

State practice and legal precedent supports the principle of common but differentiated responsibilities. One hundred and ninety-four state parties to the UNFCCC[160] acknowledge that:

> the global nature of climate change calls for the widest possible cooperation by all countries and their participation in an effective and appropriate international response,

in accordance with their common but differentiated responsibilities and respective capabilities ...[161]

and agree that, in stabilising the level of greenhouse gas concentrations in the atmosphere, they should:

protect the climate system for the benefit of present and future generations of humankind, on the basis of equity and in accordance with their common but differentiated responsibilities and respective capabilities.[162]

The Rio Declaration on Environment and Development similarly incorporates the common but differentiated responsibilities principle.[163]

Our proposal that funding should be apportioned on the basis of the common but differentiated responsibilities principle finds support in other major proposals for climate change displacement instruments. The principle and its application are considered by Docherty and Giannini, Biermann and Boas, and Muller[164] in varying detail. Docherty and Giannini argue that the liability of each state should be determined according to it.[165] Biermann and Boas state that the principle of common but differentiated responsibility should be adopted into a new agreement, but do not discuss the standard as a way of apportioning payments to the fund or in the context of the fund specifically.[166] And Muller notes, with reference to the binding nature of contributions, that

[t]he acknowledged common but differentiated responsibilities for climate change phenomena make the funding of climate-related disaster relief a prime candidate for a transformation from relying on voluntary charitable donations to being based on binding contributions ... The contributions could be proportionate to the parties' differentiated responsibilities and their ability to pay ...[167]

Finally, although not examined here in any detail, the German Advisory Council on Global Change in its report on the security risks of climate change considers 'environmentally induced migration' in its 'conflict constellation' and suggests the application of the 'common but differentiated responsibility' principle as a way of creating 'an equitable formula for distributing the costs of receiving refugees among the whole international community'.[168]

E. Protection and Assistance

1. Introduction

The vulnerability of displaced people has long been recognised by the international community and is reflected in the Refugee Convention, the Guiding Principles on Internal Displacement (the 'Guiding Principles'),[169] international humanitarian, human rights and customary law. According to the International Council on Human Rights Policy, climate change is likely to exacerbate existing human rights vulnerabilities and resource inequities:

The worst effects of climate change are likely to be felt by those individuals and groups whose rights protections are already precarious. This is partly coincidence. As it happens, the most dramatic impacts of climate change are expected to occur (and are already being experienced) in the world's poorest countries, where rights protections too are often weak. But the effect is also causal and mutually reinforcing. Populations whose rights are poorly protected are likely to be less well-equipped to understand or prepare for climate change effects; less able to lobby effectively for government or international actions; and more likely to lack the resources needed to adapt to expected alterations of their environmental and economic situation. A vicious cycle links precarious access to natural resources, poor physical infrastructure, weak rights protections and vulnerability to climate change-related harms.

At another level, the close relation between climate change and human rights vulnerability has a common economic root. Rights protections are inevitably weakest in resource-poor contexts. But resource shortages also limit the capacity (of governments as well as individuals) to respond and adapt to climate change. Worse, where governments are poorly resourced, climate change harms will tend to impact populations unevenly and unequally, in ways that are de facto discriminatory because the private capacity of individuals to resist and adapt differs greatly.[170]

Recognition that CCDPs will require protection and assistance addressed to their particular circumstances and special needs can be found in recent proposals for legal solutions to climate change migration. Docherty and Giannini argue for the establishment of protections for the human rights of those who fall within the ambit of their proposed instrument, which pertains only to trans-border displacement. Their convention would ensure a minimum standard of treatment for CCDPs, such treatment to be provided in a non-discriminatory manner. In addition, Docherty and Giannini assert that an instrument for CCDPs should go beyond the Refugee Convention to require humanitarian aid in order to guarantee that basic survival needs are met.[171]

As opposed to Docherty and Giannini's global convention for transborder displacement with rights protections modelled on the Refugee Convention, Williams emphasises that most displacement will be internal and argues for regionally-based protocols to the UNFCCC; she does not seek to pre-empt the rights which might be contained within those instruments. However, consistent with her emphasis on the predominantly internal nature of climate change displacement, she contends that a regionalised approach could have the benefit of implementing the existing framework for internally displaced persons (IDPs), who have not crossed an internationally recognised state border, into discourse on climate change displacement.[172]

Betaille et al's proposal for an international convention for environmentally displaced people (EDPs) would institute a comprehensive framework of human rights protections consisting of three tiers: global rights for all EDPs; rights for temporary EDPs and rights for permanent EDPs. Under the Betaille et al model, all EDPs would be entitled to rights to information and participation, assistance, water and food aid, housing, health care, juridical personality, to retain the civil and political rights of their State of origin, respect for the family, education and training, work and family unity. In addition, those EDPs that are temporarily displaced would have rights to safe shelter, reintegration, return and prolonged shelter. Further rights for permanently displaced EDPS would include rights to resettlement and nationality.[173] By contrast, while Biermann and Boas refer to the 'protection' of persons displaced by climate change, they conceptualise that protection as involving the provision of 'international assistance and funding for domestic support and resettlement programs of affected countries that have requested such support'.[174] For Biermann and Boas, the emphasis of a global CCDP instrument is less on the protection of persons outside their own State, and more to do with supporting governments, local communities and agencies in protecting people within their own territory.[175]

By extrapolating from existing norms, our Convention would address gaps in the international regime of human rights protections and humanitarian assistance as it currently applies to CCDPs. In a recent submission to the Office of the High Commissioner for Human Rights, the Representative of the Secretary-General on the Human Rights of Internally Displaced Persons concluded that existing human rights norms and the Guiding Principles provide sufficient protection for those forcibly displaced inside their own country where their place of origin has become uninhabitable or been declared too dangerous for habitation. However, there is 'a need to clarify or even develop the normative framework applicable to' situations including persons moving outside their country as a consequence of slow onset disasters or in the wake of designation of their place of origin as a high risk zone too dangerous for human habitation. Further, there is currently no provision at international law for persons leaving 'sinking island states' and moving across internationally recognised state borders.[176] While the proposed Convention provides a general framework for the provision of assistance to CCDPs, regardless of the nature of their displacement, it addresses gaps in existing protections by articulating a normative framework for the protection of those persons displaced across international borders, as well as identifying principles which should apply to the resettlement of persons from 'sinking' small island states (see Part IV, section F below).

Most climate change displacement is likely to occur within state borders.[177] The Guiding Principles provide a coherent statement of the matrix of human rights and humanitarian protections that are applicable to IDPs. However, it is our view that because climate change is a global problem, the international community has an obligation to provide assistance CCDPs regardless of whether their movement has a trans-border dimension. We therefore propose a Convention that incorporates a mechanism for the provision of non-discriminatory aid to internal CCDPs. Such a mechanism would recognise that in order to adequately address the special needs of CCDPs both broad-based assistance and human rights protections are required.[178]

Our Convention would apply to both internal and trans-border displacement. However, unlike the proposals of Biermann and Boas, and Betaille, we take account of the clear distinction that is drawn in international refugee law between refugees and the internally displaced. Refugees must be outside their country of origin and unable or unwilling to take advantage of the protection of that country in order to be eligible for protection and assistance by the international community.[179] By contrast, IDPs who have not crossed an internally recognised state border remain primarily the responsibility of their own nation. The distinction arises primarily from the principles of state sovereignty and non-interference, which are fundamental to the international legal system.[180] Our Convention would recognise the principle of state sovereignty and non-intervention by operating within the existing distinction at international law between trans-border and internal displacement.

Our Convention would distinguish between the explicit provision of rights and protections on the basis of internal and international displacement. Certain ambit provisions, however, should be applicable to persons displaced both internally and across recognised state borders. The principle that non-discriminatory international assistance should be provided is equally applicable to both categories of displacement.

A Convention for CCDPs will be most effective if it recognises the agency of persons who are displaced by climate change or likely to become so displaced and facilitates regionally and locally-based planning and action. Such an approach also accords with recognition in both human rights and international environmental law of rights to information and public participation.[181] The Rio Declaration on Environment and Development includes the principle that:

> Environmental issues are best handled with the participation of all concerned citizens... At the national level, each individual shall have appropriate access to information concerning the environment that is held by public authorities ... States shall facilitate and encourage public awareness and participation by making information widely available.[182]

Our Convention would require state parties to collect reliable and relevant data on climate change displacement; adopt a general policy which integrates the protection of and assistance to CCDPs into future comprehensive planning programs; disseminate information on likely climate change displacement and its general policy to potential or actual CCDPs; and closely consult with and actively involve potential or actual CCDPs and their representative organizations in the collection of data, the development of policy and its implementation. Rights of information and participation with respect to matters relating to climate change displacement should be equally applicable to persons displaced across State borders as a consequence of global warming.

In addition to the provisions outlined above, the Convention would provide a framework for the provision of protection and assistance to persons internally displaced due to climate change, in which obligations are shared between the home State and the international community. In the case of 'international' CCDPs, our Convention would outline rights of the CCDP and obligations of the host State, borrowing from the Refugee Convention. Rather than distinguishing between temporary and permanent displacement, as is the case in Betaille's proposal, the Convention we envisage would allow rights to be gradually accrued based on the duration of the displacement. In this way, State obligations to CCDPs would remain flexible and responsive to environmental changes. Finally, due to their unique circumstances, persons from small island states should be accorded treatment based on a further set of principles which include proximity, self-determination and the preservation of their culture, as discussed in Part IV, section F below.

2. Internal Displacement

(A) INTERNAL DISPLACEMENT AT INTERNATIONAL LAW

According to Hathaway, IDPs are excluded from the refugee definition firstly, because the problem of internal displacement remains the primary responsibility of the state and secondly, because intervention would constitute a violation of national sovereignty as the problems raised by internally displaced persons are invariably part of the internal affairs of state.[183] Rather than a binding convention or United Nations Security Council resolution, protection and assistance to IDPs is provided through the Guiding Principles.[184]

The Guiding Principles were developed by the Special Representative of the Secretary General on the human rights issues related to internally displaced persons and submitted to the Human Rights Commission in 1998.[185] They apply to 'persons or groups of persons who have been forced or obliged to flee or to leave their homes or places of habitual residence' for reasons including 'natural or human made disasters' and 'who have not crossed an internationally recognised State border'.[186] The Guiding Principles recognise that national authorities have the primary duty and responsibility to provide protection and assistance to IDPs. Concerns regarding sovereignty and unwanted action by international agencies were addressed through the requirement that international intervention be requested by the Secretary-General or competent principal organs of the UN and consented to by the State concerned.[187]

The Guiding Principles are not binding,[188] but have been recognised by the UN General Assembly as 'an important international framework for the protection of internally displaced persons'.[189] They are expressed to be reflective of and consistent with international human rights law and international humanitarian law.[190] It is claimed that the Guiding Principles do not extend existing humanitarian or human rights principles, but rather codify them and clarify their application to situations of internal displacement and natural disaster.[191] It is therefore possible to 'cite a multitude of existing legal provisions for almost every principle' in the Guiding Principles.[192] The result, suggests international legal scholar, Catherine Phuong, is that the status of the Guiding Principles is 'confusing' at international law:

> On the one hand, [the Guiding Principles are] clearly a non-legally binding instrument to which state consent to be bound has never been expressed. On the other hand ... the Guiding Principles are a restatement of binding norms contained in existing international treaties and/or customary international law.[193]

It has been argued that the non-binding status of the Guiding Principles is their most significant weakness.[194] States cannot be held accountable if they disregard the Guiding Principles and therefore the Guiding Principles cannot be invoked in legal proceedings at the domestic level.[195] However, Kalin asserts that:

> One should, however, not overestimate this weakness as it is always possible to invoke the hard law that lies behind the Guiding Principles where necessary. The Representative's experience has shown that it is much easier to negotiate with governments if the questions of violations does [sic] not loom in the background but, instead, problems can be approached by looking at what kind of guidance is provided by international standards.[196]

The dilemma represented by the 'soft law' status[197] of the Guiding Principles is that while their flexibility as to the introduction of substantive commitments can result in tokenistic policies at the national level, it is precisely their lack of formal legal status which enables their intervention in the first place, while creating the potential for the development of 'bottom up' national policies and customary law norms.[198]

(B) PROTECTION AND ASSISTANCE PROVISIONS IN THE CONVENTION

Our Convention proposal draws on a range of international law frameworks and precedents, including the underlying logic of the Guiding Principles and the distinction between internal and trans-border displacement in refugee law. It would provide a framework for the provision of protection and assistance to persons internally displaced due to climate change. It adopts a model in which the primary responsibility for CCDPs rests with their own State. However, in recognising that climate change displacement is a global problem and a shared responsibility, our Convention draws on 'the accepted legal principle of international cooperation and assistance' in acknowledging the duty of the international community to assist in the provision of protection and aid in respect of climate change displacement occurring within State borders.[199]

Our Convention would be founded on the recognition in international and human rights law that, while 'different States have different capacities', states are responsible for caring for their own people.[200] Thus, international refugee law incorporates an 'ambulatory principle … obliging States to exercise care in their domestic affairs in light of other States' legal interests, and to cooperate in the solution of refugee problems'.[201] The Convention would therefore recognise that national authorities have the primary duty and responsibility to provide protection and humanitarian assistance to CCDPs within their jurisdiction.[202] Further, as Docherty and Giannini argue, the instrument should require the home State to help prevent a refugee crisis, by either attempting to eliminate the need for migration or by preparing to handle it in an organised way. Such a requirement is consonant with the international law principle that states are responsible for preventing forced migrations.[203]

Climate change-induced displacement is also a matter of global responsibility, regardless of whether such displacement crosses a national border. The international community should therefore be obliged to cooperate in the provision of protection and assistance to internally displaced CCDPs. Upon request by the home State and in accordance with the Convention, other parties to the Convention would be required to provide assistance in cases of internal displacement. The international and customary law right to offer institutional protection would also be applicable.[204]

Of relevance to the proposed Convention is the conclusion underpinning the formulation of the Guiding Principles that IDPs are already entitled as individuals to protection and assistance under a range of human rights and humanitarian law instruments.[205] Thus, as the Inter-Agency Standing Committee Operational Guidelines on Human Rights and Natural Disasters states:

> Persons affected by natural disasters, including those displaced by such events, remain, as residents and most often citizens of the country in which they are living, entitled to the protection of all guarantees of international human rights subscribed to by the State concerned. They are also entitled to, if applicable, the protection of the guarantees of international humanitarian law or customary international law. People do not lose, as a consequence of their being displaced or otherwise affected by the disaster, the rights of the population at large.[206]

The United Nations High Commissioner for Human Rights notes that

> [p]ersons affected by displacement within national borders are entitled to the full range of human rights guarantees by a given State, including protecting against arbitrary or forced displacement and rights related to housing and property restitution for displaced persons.[207]

An area of ambiguity in the application of the obligations codified in the Guiding Principles arises in relation to the question of whether slow-onset disasters due to climate change constitute 'forced displacement'.[208] As discussed at Part IV, section B, our Convention rests on an understanding of forced migration that encompasses gradual environmental degradation due to climate change.

As well as being responsive to existing principles of international law, our Convention would be targeted to the particular needs and characteristics associated with persons internally displaced due to climate change. As Biermann and Boas argue, CCDPs are not displaced as a consequence of political, religious, ethnic or otherwise discriminatory persecution or violence. Rather, in principle, they continue to enjoy State protection.[209] It was similarly argued with respect to the inclusion of the 'man made or natural disaster criterion' in the Guiding Principles that persons displaced by natural disasters are generally likely to face problems associated with economic and social rights, rather than those associated with more conventional refugee situations.[210] Further, government authorities usually routinely appeal for international assistance for the victims of natural disasters, rather than attempting to hinder non-discriminatory provision of aid, as is more likely to occur in cases of civil war and internal strife.[211] As Biermann and Boas also contend,

> [t]he protection of climate refugees is therefore essentially a development issue that requires large-scale, long-term planned resettlement programs for groups of affected people, mostly within their own territory'.[212]

Therefore, the emphasis of a global CCDP instrument should be less on the protection of persons outside their states, and more to do with supporting governments, local communities and agencies in protecting people within their own territory.[213]

Nevertheless, displacement makes people vulnerable, including to infractions and abuses of their human rights.[214] Docherty and Giannini identify the Convention on Cluster Munitions as providing a model framework that requires the provision of 'tangible assistance as well as protection of abstract rights'.[215] Based on the Convention on Cluster Munitions precedent, state parties would be required, with respect to CCDPs in areas under their jurisdiction and control, to provide (to the extent practicable) age and gender-sensitive assistance, including emergency services, evacuation and relocation, medical assistance, housing, food, measures necessary for social and economic inclusion, property restitution where possible, and the facilitation of family reunion. Moreover, in fulfilling their obligations under the Convention, each State Party should be precluded from discriminating against or among CCDPs.

The operation of our Convention would, in many ways, be prospective. As discussed at the introduction to this Part IV, section E, state parties to the Convention would be obliged to collect reliable and relevant data on climate change displacement; adopt a general policy which integrates the protection of and assistance to CCDPs into future comprehensive planning programs; disseminate information on likely climate change displacement and its general policy to potential or actual CCDPs; and closely consult with and actively involve potential or actual CCDPs and their representative organizations in the collection of data, the development of policy and its implementation. The Convention would, therefore, provide a forum for the provision of pre-emptive adaptive resettlement to populations most vulnerable to the impacts of climate change. The principles of information dissemination, public participation and civil society involvement would assist in ensuring that adaptive solutions were 'bottom up', generated by communities in order to maximise their effectiveness and appropriateness.

Assistance under the Convention would, thus, have an adaptive quality, rather than simply facilitating the provision of humanitarian aid. Such an approach correlates with the insight that migration is best understood as a form of adaptation to climate change.[216] Situating displacement within the spectrum of adaptive possibilities enables assistance to address the needs of those most vulnerable to climate change who may otherwise lack the resources to move themselves from climate change threats.

3. International Displacement

The protection of a comprehensive set of human rights and humanitarian entitlements for different categories of migrants is by no means a new phenomenon in international law. Both the Refugee Convention and customary international law recognise that, in certain circumstances, the international community is responsible for the provision of rights and assistance for persons displaced across international borders and unable to rely upon their own nation for protection. However, as previously discussed, there currently exists no provision at international law for most persons migrating across borders as a consequence of climate change.

Docherty and Giannini identify the Refugee Convention as a useful model of what kinds of human rights protections to include in a new instrument for CCDPs because it provides the 'most comprehensive codification of the rights of refugees yet attempted on the international level'.[217] They justify borrowing from the 1951 Convention as it provides an international law precedent 'which is as applicable to climate change refugees as to traditional refugees, because it is well-established and difficult to challenge'.[218]

We endorse Docherty and Giannini's argument that any treaty for CCDPs should be premised on the rights and protections that states have already agreed to accord to traditional refugees. As such, our Convention should guarantee a range of civil, political, economic, social and cultural rights, based on a principle of non-discrimination. Further, CCDPs should be guaranteed a minimum standard of treatment, at least equivalent to aliens in the host country. However, analogously to the 1951 Convention, certain rights afforded to CCDPs should be those of nationals in the host country and in some cases, rights should be afforded based on an absolute standard, rather than being contingent on existing rights in host nations.[219] Rights relating to movement are especially significant to CCDPs, and in particular, CCDPs should enjoy the right to *non-refoulement*, a core principle of refugee law under Article 33 of the 1951 Convention. In the context of the new instrument, *non-refoulement* would prohibit the forcible return of a refugee to a situation if 'climate-induced environmental change would threaten the refugee's life or ability to survive'.[220] To Docherty and Giannini's proposal we would add a principle of proximity that requires the least separation of persons from their cultural area.

Following the Refugee Convention, the rights of CCDPs displaced across international borders should expand on an incremental basis, with rights accruing the longer CCDPs remain in a host nation. Hathaway describes the operation of the 1951 Convention as follows:

Most fundamentally, the refugee rights regime is not simply a list of duties owed by state parties equally to all refugees. An attempt is instead made to grant enhanced rights as the bond strengthens between a particular refugee and the state party in which he or she is present. While all refugees benefit from a number of core rights, additional entitlements accrue as a function of the nature and duration of the attachment to the asylum state.[222]

Adoption of the Refugee Convention model of a gradually deepening set of rights enables the Convention to flexibly adapt to changing environmental conditions and scientific knowledge. As proposed by Docherty and Giannini, international CCDPs would remain eligible for assistance until they acquired a new nationality; voluntarily returned to their home country; or refused to return when it was safe for them to do so.[223]

We concur with Docherty and Giannini that persons fleeing climate change-induced disasters or degradation require humanitarian aid as well as human rights protections. Any instrument for CCDPs should 'go beyond the Refugee Convention to guarantee that basic survival needs are met'.[224] Again, a duty of international cooperation and assistance, based on the principle that climate change is a global problem, is equally applicable to both CCDPs who cross state borders and to those who remain within their own nations. Provision should be made for the international community to render assistance to host states in the protection of the rights afforded under the proposed Convention, where necessary.[225]

F. The Plight of Small Island States

Rising sea levels and the submersion of islands are perhaps the most publicly recognisable consequences of climate change. The populations of small island states may not only be displaced but will likely see the disappearance of their homelands.[226] As a result, although they will amount only to a fraction of the total number of likely CCDPs, the interests and expectations of the populations of these small island states have a high profile. Indeed, just prior to the UNFCCC climate change conference in Copenhagen in December, 2009, the Alliance of Small Island States (AOSIS) issued a communiqué in which it demanded increased financing for the mitigation of climate change. The AOSIS made it clear that failure by developed states to act on these demands would be viewed as 'benign genocide'.[227] With its slow-onset focus, our Convention is well suited to render the international assistance required to accommodate the very particular needs of small island state CCDPs.

Small island states have been identified by the IPCC as particularly vulnerable to the effects of climate change.[228] McAdam suggests that 'their small physical size, exposure to natural disasters and climate extremes, very open economies and low adaptive capacity make them particularly susceptible and less resilient to climate change'.[229] This vulnerability is, however, coupled with their low emissions, creating a situation whereby some of the smallest contributors to climate change face the largest risk of displacement. In this way, the plight of small island state populations symbolises the imperative of ensuring that there is an effective international framework for the protection of, and provision of assistance to, CCDPs.

The very real prospect of 'entire nations disappearing'[230] differentiates the plight of small island states from other regions in which there is likely to be large-scale displacement, and requires specific consideration. The complete loss of a physical territory signals the practical end of those states' national sovereignty and the particular protections and rights of their people. More broadly, it signifies the end of unique ways of life which are intimately connected to precarious physical landscapes. Such a scenario is unprecedented, and existing legal regimes do not adequately articulate the rights that should be accorded to CCDPs in order to recognise this loss. We propose that the principles of proximity, self-determination and the safe-guarding of intangible culture should be applicable to bilateral displacement agreements between small island states and host states in relation to climate change displacement, such agreements negotiated under the aegis of the CCDO, with involvement of CCDO regional committees. These principles are discussed below.

1. Proximity

Betaille et al suggest that a climate change convention should be implemented with recourse to a principle of proximity which requires the least separation of persons from their cultural area.[231] Proximate resettlement of small island state nationals may be particularly appropriate given their strong connections to both land and seascapes. As such, developed nations including Australia and the United States may be preferential host states for small island state nationals. A principle of proximate resettlement should guide bilateral agreements between AOSIS states and developed states in the region.

2. Self Determination

A second guiding principle which should inform small island state resettlement is self-determination. Self-determination is enshrined in the United Nations International Covenant on Economic, Social and Cultural Rights. Article 1 states that

> All peoples have the right of self-determination. By virtue of that right they freely determine their political status and freely pursue their economic, social and cultural development.[232]

In the context of small island state relocation, this articulation of self-determination is particularly useful. We consider self-determination of various forms of development as referring to two distinct elements of relocation: (a) *when* will people abandon their territories; and (b) *where* will people choose to resettle.[233] First, the proposed Convention recognises that small island state nationals may want to remain in their home states for as long as it is practicable. This is certainly suggested by the AOSIS's calls for further financing to mitigate the effects of climate change. Anthropological studies have also indicated that small island state nationals desire to remain, despite the impthreat.[234] The capacity of the proposed Convention's fund to provide prospective resources for persons likely to be displaced by climate change would enable adaptation funding to be provided to small island state CCDPs.

The principle of self-determination should also apply to the second aspect of relocation: the destination to which small island state CCDPs relocate. While proximity is also relevant to this discussion, collective self-determination should inform agreements between small island states and host states. Small island states' preferences may relate to existing migration patterns or proximity and should, to the extent possible, be adhered to. The preferences of small island state CCDPs should inform agreements between State parties as to the location of resettlement.

3. Preservation of Intangible Culture

Our Convention recognises the effects of climate change on populations as well as individuals. This is reflected in our inclusion of *en masse* CCDP designations. Finding, as Docherty and Giannini do, that 'group status determinations … increase opportunities to formulate solutions that would keep the integrity of a group intact, which could help preserve cultures and national identities',[235] our Convention emphasises the importance of the maintenance of the CCDP group's social and cultural cohesion in their host country.

For small island state CCDPs, the protection of cultural autonomy is of particular concern given their territory may cease to exist. Although there exists no explicit protection of refugees' cultural autonomy in current international legal regimes, the protection of social and cultural rights in the Refugee Convention may be invoked as a useful precedent. The Refugee Convention accords refugees treatment 'at least as favourable as that accorded to their nationals with respect to freedom to practice their religion and freedom as regards the religious education of their children'.[236] Significantly, the Refugee Convention here provides the standard of protection to be that of a national (thereby pointing to its importance in the eyes of the international community). It also provides for intergenerational transmission, and the non-interference of this transmission. This extension is further supported by the Refugee Convention's provision of a right of association,[237] freedom of movement[238] and a right to the same protection as is accorded to nationals of that country of their cultural and intellectual property (such as inventions, designs or models, trade marks, trade names, literary, artistic, and scientific works).[239] Betaille et al also argue that any international instrument for environmental refugees should provide for populations to 'constitute themselves collectively and maintain their collective identity'.[240]

As well as the Refugee Convention, there exists another international agreement for the protection of culture which might be usefully deployed in the context of small island states, the Convention for the Safeguarding of the Intangible Cultural Heritage. This Convention was adopted at the General Conference of the United Nations Educational, Scientific and Cultural Organisation (UNESCO) in October 2003 and has 118 State parties. Its purpose is:

a.   to safeguard the intangible cultural heritage;
b.   to ensure respect for the intangible cultural heritage of the communities, groups and individuals concerned;
c.   to raise awareness at the local, national and international levels of the importance of the intangible cultural heritage, and of ensuring mutual appreciation thereof;
d.   to provide for international cooperation and assistance.[241]

The Convention for the Safeguarding of the Intangible Cultural Heritage defines 'intangible cultural heritage' as the practices, representations, expressions, knowledge, skills – as well as the instruments, objects, artefacts and cultural spaces associated therewith – that communities, groups and, in some cases, individuals recognize as part of their cultural heritage.[242]

Intangible cultural heritage is recognised as intergenerational and closely tied to a population's identity. Safeguarding this heritage involves measures aimed at ensuring the viability of the intangible cultural heritage, including the identification, documentation, research, preservation, protection, promotion, enhancement, transmission…as well as the revitalization of the various aspects of such heritage.[243]

Globalisation is sometimes referred to as a threat to intangible cultural heritage because its processes, such as the rapid transfer of information, tend towards the homogenisation of different cultures. Climate change could similarly be conceived as a threat to intangible cultural heritage. Indeed, rising sea levels attributed to climate change may force populations to relocate from disappearing island nations. In these circumstances, CCDPs risk losing a unique way of life (their intangible culture) which is no longer preserved by the laws and institutions of their home state. People, rather than physical artefacts, serve to transmit intangible culture. As such, the legal protection of CCDPs' cultural autonomy may be regarded as a mechanism of preservation under the UNESCO Convention. Safeguarding intangible culture in the circumstances faced by small island states, therefore, can further support the relocation of entire populations and social groups.

G. Civil Society in Negotiation of the Convention

As discussed, like Docherty and Giannini,[244] we envisage civil society (non-governmental organisations, business and other professional associations, advocacy groups, and other non-profit organisations and coalitions) playing a significant role under the Convention, both generally in the context of displacement and specifically in terms of the operation of CCDIGs. We also envisage a significant role for civil society in the negotiation of the Convention and in the design of the framework within which it would operate; recent experience suggests that the role of civil society can be vital not only in the process of drafting and negotiating a treaty but also in securing its ratification.

The 'Ottawa Process', launched in October 1996, the main elements of which included a government–civil society partnership, non-traditional diplomacy with an extensive role for NGOs, a focus on shared humanitarian aims and not narrow national interests, and the intense dedication of numerous individuals inside and outside of government,[245] led to the signing of the Mine Ban Treaty[246] just over a year later; the treaty entered into force on 1 March 1999. It 'became binding international law more quickly than any other multilateral disarmament or humanitarian law instrument'.[247] The work of the Mine Ban Treaty continues to be driven by both government and civil society.

The 'Oslo Process', based on the Ottawa Process, has been similarly successful, and resulted in the Convention on Cluster Munitions in May 2008. The first global conference on cluster munitions had been held in Oslo just over a year earlier, in February 2007. A main feature of the Oslo Process was the active participation in discussions and final negotiations of civil society, represented by UN organisations, the International Committee of the Red Cross and the Cluster Munitions Coalition. Over 100 civil society and NGO representatives participated in the February, 2007 meeting. 'NGOs were given high visibility speaking slots and intervened on the same basis as states',[248] and draft treaty texts were developed in consultation with NGOs.

It is clear from an examination of the Ottawa and Oslo processes that the participation of states and civil society was vital. These processes, it seems to us, also provide useful models for the inclusion of not only civil society but also those states most at risk from the effects of climate change, most vulnerable to the displacement of populations, such as small island states, in the negotiation and formation of the Convention. The Ottawa Process demonstrated that change is most likely to be effected through concerted action. The mine ban movement also demonstrated that it is possible for small and medium size countries, acting in concert with civil society, to provide global leadership and achieve major diplomatic results, even in the face of opposition from major powers. It showed that it is possible to work outside of traditional diplomatic forums, practices, and methods and still achieve success multilaterally.[249]

H. Implementation and Management

In their 2007 paper 'Control, Adapt or Flee: How to Face Environmental Migration', Renaud et al propose the implementation of a number of policy suggestions to address the relationship between environmental change and forced migrations. Their policy suggestions include programmes to achieve a better understanding of the cause and effect between environmental degradation and forced migration, and a framework to recognise environmental refugees either in existing environmental treaties or in a separate convention.[250] Further, individuals who are clearly displaced by environmental degradation processes...should be protected adequately by an international mechanism that would afford them certain rights. Bilateral agreements...should be systematised...for the most pressing environmental degradation issues.[251]

Renaud et al also propose an international mechanism to provide aid to environmental refugees, especially in the context of the displacement of whole communities, and the creation of institutions to assist the flux of forced environmental migrants nationally and internationally.[252]

McAdam notes that, despite calls such as those from Renaud and his co-authors for 'a multifaceted, or cooperative or international approach, the literature does not spell out what this would look like or how it would be achieved'.[253] Detailed proposals that do set out mechanisms to address displacement ignore or don't deal in detail with important facets. Again, that by Docherty and Giannini, for example, ignores what will very likely be the most common form of climate change displacement – that which occurs internally within a country – as 'beyond the scope' of their study.[254]

Our proposal for a convention for CCDPs does deal with internal displacement. It does, as Renaud et al suggest, propose bilateral agreements. It does adopt a multifaceted, international approach, and we do spell out what such a convention would look like.

The International Organization for Migration sets out four principles for effective environmental migration management. They are proactive policy and early action; comprehensive and coherent policies; bilateral and regional cooperation; and multi-stakeholder partnerships involving civil society.[255] Our Convention implements these principles.

V. Summary of Convention

We propose in this article a single, stand alone convention, which is global in scope; parties to it would include both developed and developing states. Our Convention would encompass those displaced internally (and most climate change displacement is likely to occur within state borders) and those who cross international borders, either temporarily or permanently, with a particular focus on those displaced as a result of small island states becoming uninhabitable. It would create an institutional architecture for designating a particular population as CCDPs; we propose *en masse* designations of CCDP status through a process of request and determination by states and Convention processes. The scope of the Convention encompasses displacement as a result of both sudden climate change events, or impacts, and slow-onset, gradual displacement.

Our Convention would create a Climate Change Displacement Organisation (CCDO) with obligations for developed and developing state parties to the Convention and for 'home' and 'host states', together with civil society. It sets out in detail the architecture of the CCDO and the operation of its constituent parts. It also sets out a role for regional cooperation and multidisciplinary collaborations. One of the CCDO bodies contemplated by the Convention is a Climate Change Displacement Fund. Under the Convention developed state parties would make mandatory financial contributions to that Fund. Such contributions would be made on the basis of parties' common but differentiated responsibilities.

A general framework for the provision of CCDP assistance, as well as addressing gaps in the international regime of human rights protections and humanitarian assistance as it currently applies to CCDPs, would be provided under our Convention. The instrument we propose would operate within the existing international law distinction between internal and trans-border displacement. It would institute a mechanism for the provision of principled, non-discriminatory assistance to CCDPs displaced within national borders. Persons migrating across state boundaries due to climate change would be entitled to a set of rights modelled on the Refugee Convention.

Finally, our Convention contemplates the provision of pre-emptive resettlement to those most at risk in terms of the impacts of climate change. Provision of assistance under the Convention would, thus, have an adaptive quality.

It has been suggested that Australia should take the lead in international efforts to develop a framework for responding to climate change displacement. The broader region in which Australia is situated accounts for 60% of the world's population; it's a region that will be significantly affected by the effects of climate change, perhaps most dramatically by sea level rise. As Corlett notes, to take such a lead 'is not to concede that mitigation and adaptation efforts are beyond us'. Rather, planning for a future of mass displacement due to climate change gives us the opportunity – before millions of people are on the move throughout the world because of climate change; before we, and other nations, become tempted to erect walls to keep them at bay; before we start to say as though as a reflex that 'we will decide who comes and the circumstances in which they come' – to develop frameworks and institutions that might not only be politically realistic, but also based on principles that promote human rights and dignity.[256]

VI. Conclusion: 'Will the tiger get me?'

Our Convention is, again, in many ways prospective. It would establish a framework within which adaptive assistance to those vulnerable to climate change impacts could be provided. As Boas and Biermann state, the protection of CCDPs requires large-scale, long-term planning.[257] This in itself presents challenges because, for example, like many others:

Bangladeshis think mainly of tomorrow. Will there be enough rice? Enough clean drinking water? Will the tiger get me? All of us have the same human tendency to plan for the next day, next week, next year. Projecting ... developments 10, 20, 50 years into the future is a chancy business, as imprecise a science in its way as the modelling of climate change. But those are undoubtedly the terms, and the timescales, on which we now have to think.[258]

David Hodgkinson, Tess Burton, Heather Anderson and Lucy Young, '"The Hour When The Ship Comes In"': A Convention for Persons Displaced by Climate Change', *Monash University Law Review*, Volume 36, Issue 1, 2010, pp. 69–120.

Notes

53   Docherty and Giannini, above n 29, 358. They also note that attaching any 'climate change refugee protocol to the UNFCCC has three significant shortcomings: 'the limits of the UNFCCC's mandate, which is not focused on remedies; the historical reluctance to incorporate human rights issues explicitly into environmental treaties; and the UNFCCC's track record of inaction' (at 394).

54   McAdam, above n 21, 5.

55   Docherty and Giannini, above n 29, 397.

56   Docherty and Giannini, above n 29.

57   Biermann and Boas, above n 41.

58   Williams, above n 33.

59   Julien Betaille et al, 'Draft Convention on the International Status of Environmentally-Displaced Persons' (2008) 4 Revue Européenne de Droit de L'Environnement 395.

60   Hodgkinson et al, above n 12.

61   Docherty and Giannini, above n 29, 352–354; Biermann and Boas, above n 41, 21; Williams, above n 33, 506.

62   Docherty and Giannini, above n 29, 359; Biermann and Boas, above n 41, 17–21; Williams, above n 33, 507–510.

63   Williams, above n 33, 517–518.

64   Ibid, 522.

65   Biermann and Boas, above n 41, 15–16.

66   Ibid, 25–30.

67   Docherty and Giannini, above n 29, 361.

68   Ibid, 350.

69   Betaille et al, above n 59, Article 1.

70   Ibid, Article 4.

71   Our study provided for a Convention for the long-term resettlement of CCDPs, both internally within affected countries (as a priority) and internationally, including prior to displacement. Convention parties would provide assistance on the basis of equity and, as a principle, in accordance with their common but differentiated responsibilities; developed state parties to the proposed convention should take the lead in the provision of assistance. More specifically, such parties would assist CCDPs on the basis of the parties' historical greenhouse gas emissions by volume. The convention would provide for displacement flowing from both sudden events and slow-onset events. State parties would contribute to a fund to assist internal resettlement; enable responses to specific climate change events; and assist adaptation and mitigation by affected parties. An international organisation would be created under the Convention with responsibility for climate-induced displacement: Hodgkinson et al, above n 12.

72   German Advisory Council on Global Change, Climate Change as a Security Risk (2007).

73   Ibid, 206.

74 Ibid, 129.
75 Docherty and Giannini, above n 29.
76 Biermann and Boas, above n 41. See also Biermann and Boas, 'Protecting Climate Refugees: The Case for a Global Protocol' (2008) 50 Environment 8.
77 See Part IV, section A(2) below.
78 'The issue of climate change IDPs [internally displaced persons] is beyond the scope' of Docherty and Giannini's article: Docherty and Giannini, above n 29, 360.
79 See Part IV, section F below.
80 Docherty and Giannini, above n 29, 349.
81 Biermann and Boas, above n 41, 26. See also Biermann and Boas, above n 76, 13.
82 Biermann and Boas, above n 41, 31.
83 Biermann and Boas, above n 41 and 76; Williams, above n 33.
84 Docherty and Giannini, above n 29, 358. They also note that attaching any 'climate change refugee protocol to the UNFCCC has three significant shortcomings: 'the limits of the UNFCCC's mandate, which is not focused on remedies; the historical reluctance to incorporate human rights issues explicitly into environmental treaties; and the UNFCCC's track record of inaction' (at 394).
85 Kniveton et al, above n 17, 29; Christian Aid, above n 16, 6; Brown, above n 16, 13; and German Advisory Council on Global Change, above n 72, 118.
86 Biermann and Boas, above n 76, 11.
87 Kniveton et al, above n 17, 28.
88 German Advisory Council on Global Change, above n 72, 118.
89 Kniveton et al, above n 17, 29; Christian Aid, above n 16, 6; and German Advisory Council on Global Change, above n 72, 118.
90 Brown, above n 16, 13. See also Kniveton et al, who emphasise, 'a broad theoretical consensus that it is generally not the poorest people who migrate overseas because international migration is an expensive endeavour that demands resources for the journey and for the crossing of borders': above n 17, 20.
91 Docherty and Giannini, above n 29, 370. Docherty and Giannini state that '[t]he issue of climate change IDPs is beyond the scope of this Article, but it deserves attention as the international community develops ways to deal with climate change migration' (at 360).
92 McAdam states that '[d]espite the common call for a multifaceted, or cooperative or international approach, the literature does not spell out what this would look like or how it would be achieved': McAdam, above n 21, 28.
93 Docherty and Giannini note that '[m]ost authors who define environmental refugee do not distinguish between people who migrate across or within borders': Docherty and Giannini, above n 29, 369 (emphasis added).
94 Ibid, 369–370. Docherty and Giannini also note that 'the means for facilitating such assistance are beyond the scope of this Article' (at 370).
95 Docherty and Giannini, above n 29, 369.
96 International Council on Human Rights Policy, Climate Change and Human Rights: A Rough Guide (2008) 1.
97 Dasgupta et al, above n 15, 44.
98 International Council on Human Rights Policy, above n 96.
99 German Advisory Council on Global Change, above n 72, 118.
100 Biermann and Boas, above n 41, 6.
101 Refugee Convention, above n 43, Article 1(A)(2).
102 As the word is generally understood.
103 Docherty and Giannini, above n 29, 370.
104 Intergovernmental Panel on Climate Change, Fourth Assessment Report: Climate Change 2007: Synthesis Report (2007) 53. 'Virtually certain' means a probability greater than 99%; 'extremely likely' means a probability greater than 95%; and 'very likely means greater than 90% (at 27).
105 Intergovernmental Panel on Climate Change, Climate Change 2007: The Physical Science Basis,

Summary for Policymakers, Contribution of Working Group I to the Fourth Assessment Report of the IPCC (2007) 5.

106   Docherty and Giannini, above n 29, 371.

107   Greater than 90% probability.

108   See Part II, section B above. Some, but not all, of this uncertainty relates to long-term or permanent displacement only.

109   Kniveton et al, above n 17, 34.

110   See Part IV, section E.

111   These channels include the Office for the Coordination of Humanitarian Affairs, a UN Disaster Relief Co-ordinator, an Inter-Agency Standing Committee, a UN Central Revolving Fund and the International Red Cross Movement: Benito Muller, 'An FCCC Impact Response Instrument as Part of a Balanced Global Climate Change Regime', 2002 http://www.oxfordclimatepolicy.org/publications/iri.pdf.C.R.N. at 19 December 2009.

112   The debate about the legitimacy of the term 'refugee' relates to its appropriateness when applied to cases of environmental exodus: Brown, above n 16, 6; Biermann & Boas, above n 41, 6–8; Docherty and Giannini, above n 29, 363; Keane, above n 49, 215; and Kolmanskog, above n 17, 8 –9. This article follows the recommendation of the UNHCR that the term 'refugee' is inappropriate in an environmental context and adopts the terminology of 'climate change displaced person or 'CCDP'.

113   Essam El-Hinawi, Environmental Refugees (1985) 4.

114   Norman Myers, 'Environmental Refugees in a Globally Warmed World' (1993) 43 BioScience 752.

115   Biermann and Boas, above n 41, 3 – 6; Docherty and Giannini, above n 29, 363 – 6; Williams, above n 33, 506–7.

116   Biermann & Boas, above n 41, 8. Biermann and Boas do not make distinctions based on the character of the migration; they do not distinguish between internal and international migrants; they do not view a distinction between permanent and temporary migrants as relevant; and they do not include voluntariness as a criteria. Their definition is instead restricted with reference to the cause of relocation. They exclude climate change impacts that have a marginal link with forced migration, such as heat waves; migration caused by measures related to the mitigation of or adaptation to global warming, such as construction of dams; migration related to other types of environmental degradation (such as pollution) or disasters unrelated to human activities (such as volcano eruptions); and migration caused by secondary or indirect impacts of climate change, such as conflicts over natural resources.

117   Docherty and Giannini, above n 29, 361.

118   Williams, above n 33, 522–3.

119   Brown, above n 16, 5; Kolmannskog, above n 17, 11–2.

120   Williams, above n 33, 522. Williams suggests that each 'regional association' could agree on its own definition of CCDPs to identify the scope of individual regional agreements, but also suggests a 'definitional approach' allowing a 'certain degree of flexibility'.

121   Fabrice Renaud et al, 'Control, Adapt or Flee: How to Face Environmental Migration' (Intersections Publication Series of United Nations University Institute for Environment and Human Security, No.5/2007), 29 –31.

122   Representative of the Secretary General on the Human Rights of Internally Displaced Persons, 'Displacement Caused by the Effects of Climate Change: who will be affected and what are the gaps in the normative frameworks for their protection' (Background Paper to the Office of the High Commissioner for Human Rights Study on the Relationship Between Climate Change and Human Rights, 10 October 2008), 7.

123   Castles, above n 17, 9.

124   A shift away from individual rights towards mass recognition creates new dilemmas, however, particularly the prospect of discrimination against particular populations by host countries and the attenuation of rights that often accompanies acceptance of refugees en masse by recipient states. We attempt to address these dilemmas in the section on 'protection and assistance'. Moreover, a system premised on individual rights determined by definitional criteria arguably may allow the individual

asylum seeker more agency – to determine whether to flee and where to seek asylum. We try to compensate for this by advocating active public participation in all levels of decision-making under our Convention.

125   McAdam, above n 16, 9.
126   Docherty and Giannini, above n 29, 374.
127   Docherty and Giannini, above n 29, 384–391.
128   Betaille et al, above n 59, ch 3.
129   Biermann and Boas, above n 41, 28–29.
130   Williams, above n 33, 519.
131   Docherty and Giannini, above n 29, 383.
132   Ibid, 384.
133   Ibid, 388.
134   Betaille et al, above n 59, Article 12. The national commission consists of 9 members named by the judiciary.
135   Betaille et al, above n 59, Article 11.
136   United Nations Educational, Scientific and Cultural Organisation, Operational Guidelines for the Implementation of the World Heritage Convention (2008).
137   Host states almost always, in our view, developing state parties.
138   Williams, above n 33, 517–523.
139   Ibid, 519.
140   Ibid, 521.
141   McAdam, above n 21, 7–8.
142   See note 104.
143   Docherty and Giannini, above n 29, 371.
144   See Part IV, section 5.
145   For issues associated with coordinating efforts to address environmental displacement see, generally, Tracey King, 'Environmental Displacement: Coordinating Efforts to Find Solutions' (2006) 18 Georgetown International Environmental Law Review 543. King proposes an international coordinating mechanism for environmental displacement which would coordinate the work of organisations that currently focus on such displacement. It is not necessary, in King's view, to create a new organisation to specifically deal with environmental displacement.
146   McAdam, above n 21, 23.
147   Ibid.
148   Ibid.
149   Biermann and Boas, above n 41, 29.
150   Betaille et al, above n 59, Article 11(4).
151   Docherty and Giannini, above n 29, 385. Docherty and Giannini note that, while the proposals vary as to the role and management of the fund, the proposals 'illustrate growing support for establishing such a mechanism'.
152   Biermann, above n 41, 29.
153   Ibid, 30.
154   Docherty and Giannini, above n 29, 385.
155   Ibid, 387.
156   Betaille et al, above n 59, Article 11(4)(a).
157   Ibid, Article 11(4)(b). They do not provide further detail about the proposed tax so it is not clear if it would also be determined based on the principle of common but differentiated responsibilities.
158   The detailed operation of the Climate Change Displacement Organisation, of which both the Fund and the CCDESO form a part, is set out above at Part IV, section C.
159   Centre for International Sustainable Development Law (CISDL), 'The Principle of Common But Differentiated Responsibilities: Origins and Scope', 26 August 2002 < www.cisdl.org/pdf/brief_common.pdf> at 6 December 2009.
160   UNFCCC, above n 52.

161 UNFCCC, above n 52, recitals.

162 Ibid, Article 3(1).

163 Principle 7 of the Rio Declaration on Environment and Development, 'State Cooperation to Protect Ecosystem', provides that 'States shall cooperate in a spirit of global partnership to conserve, protect and restore the health and integrity of the Earth's ecosystem. In view of the different contributions to global environmental degradation, States have common but differentiated responsibilities. The developed countries acknowledge the responsibility that they bear in the international pursuit of sustainable development in view of the pressures their societies place on the global environment and of the technologies and financial resources they command': A/CONF.151/26 (Vol I) - Rio Declaration Principle 22 and Agenda 21, Chapter 26.4 <http://www.un.org/documents/ga/conf151/aconf15126-1annex1.htm> at 21 December 2009.

164 Benito Muller, 'An FCCC Impact Response Instrument as part of a Balanced Global Climate Change Regime' (Paper presented at the Tata Energy Research Institute, New Delhi, May 2002) 3.

165 Docherty and Giannini, above n 29, 386.

166 Biermann and Boas, above n 41, 26 and 28–29.

167 Muller, above n 164, 3.

168 The report states that '[i]n order to achieve a fair and efficient distribution of costs, this distribution formula should be based on the international legal principle of common but differentiated responsibilities. According to this principle, the principal responsibility for bearing the costs lies with the countries that contribute most to causing global greenhouse gas emissions and that also have the greatest financial resources … At the same time, because environmentally induced migration is multi-causal, additional criteria should be defined in order to achieve a burden-sharing arrangement that is as fair as possible': German Advisory Council on Global Change, above note 72, 206.

169 United Nations Refugee Agency (UNHCR), Guiding Principles on International Displacement, E/CN.4/1998/53/Add.2)(1998).

170 International Council on Human Rights Policy, Climate Change and Human Rights: A Rough Guide (2008) 1.

171 Docherty and Giannini, above n 29, 378.

172 Williams, above n 33, 521.

173 Betaille et al, above n 59, ch 2.

174 Biermann and Boas, above n 76, 13.

175 Biermann and Boas, above n 41, 26.

176 Representative of the Secretary General on the Human Rights of Internally Displaced Persons, above n 122, 6.

177 See Part IV, section A(2) above.

178 Docherty and Giannini, above n 29, 379.

179 Guy Goodwin-Gill & Jane McAdam, The Refugee in International Law (3rd ed, 2007) 37.

180 Catherine Phuong, The International Protection of Internally Displaced Persons. (1st ed, 2004) 22.

181 Universal Declaration of Human Rights, Articles 19 and 21, GA Res. 217 (III), UN GAOR, 3d Sess., Supp. No. 13, UN Doc. A/810 (1948) 71, Articles 19, 22 and 25; International Council on Human Rights Policy, above n 170, 49.

182 Rio Declaration on Environment and Development, above n 163, Principle 10. See also the Convention on Access to Information, Public Participation in Decision-Making and Access to Justice in Environmental Matters, opened for signature 28 June 1998, 2161 UNTS 447 (entered into force 30 October 2001); the Universal Declaration of Human Rights, above n 181, Articles 19, 21; and the International Covenant on Civil and Political Rights, opened for signature 19 December 1966, 999 UNTS 171, (entered into force 23 March 1976), Articles 19, 22, 25.

183 James Hathaway, The Law of Refugee Status, (1st ed, 1991) 30 –1 cited in Phuong above n 180, 22.

184 UNHCR, above n 169.

185 Ibid.

186 See also the London Declaration of International Law Principles on Internally Displaced Persons, International Law Association, Declaration of International Law Principles on Internally Displaced

Persons, 29 July 2000, available at: http://www.unhcr.org/refworld/docid/42808e5b4.html at 18 December 2009.

187   Williams, above n 33, 511, referring to United Nations General Assembly Resolutions 47/105, UN Doc A/RES/47/105 (16 December 1992); 48/116, UN Doc A/RES/48/116 (24 March 1994), UN Doc A/RES/49/169 (24 February 1995).

188   The Guiding Principles are neither a convention nor a UN General Assembly declaration.

189   United Nations General Assembly Resolution 60/1, UN Doc A/RES/60/1 (24 October 2005).

190   United Nations Refugee Agency (UNHCR), Guiding Principles on International Displacement, E/CN.4/1998/53/Add.2)(1998).

191   Phuong, above n 180, 66. Phuong argues, however, that in some cases the Guiding Principles progressively develop internal law, such as including a prohibition on non-refoulement for the internally displaced.

192   Walter Kalin, 'How Hard is Soft Law? The Guiding Principles on Internal Displacement and the Need for a Normative Framework', Brookings/CUNY Project on Internal Displacement, 19 December 2001, 6.

193   Phuong, above n 180, 73.

194   Ibid, 66.

195   Kalin, above n 192, 7.

196   Ibid.

197   The Guiding Principles do not constitute typical soft law in the sense of recommendations that rest on the consensus of States: Ibid, 6.

198   Williams, above n 33, 511–512.

199   Docherty and Giannini, above n 29, 383.

200   Ibid, 380.

201   Guy Goodwin-Gill and Jane McAdam, above n 179, 3; Docherty and Giannini, above n 29, 380.

202   This obligation is drawn from United Nations Refugee Agency (UNHCR), Guiding Principles on International Displacement, E/CN.4/1998/53/Add.2) (1998), Principle 3.

203   Docherty and Giannini, above n 29, 381.

204   See Nils Geissler, 'The International Protection of Internally Displaced Persons' (1999) 11 International Journal of Refugee Law 451, 471.

205   'Internally Displaced Persons, Report of the Representative of the Secretary-General, Mr Francis M Deng, Compilation and Analysis of Legal Norms' UN Doc E/CN.41996/52/Add.2 (1995).

206   Inter-Agency Standing Committee, Protecting Persons Affected by Natural Disasters: IASC Operational Guidelines on Human Rights and Natural Disasters, June 2006, 9. Note that international human rights treaties, and most national constitutions, typically allow for the derogation of many human rights in times of emergency: International Council on Human Rights Policy, above n 170, 5. Some human rights, however, are non-derogable under any circumstances, including the right to life, the prohibition of cruel, inhuman and degrading treatment or punishment, the prohibition of slavery and the prohibition of the retroactive application of penal law. Difficulties also arise where states have not ratified the relevant treaties or have invoked limitation clauses. For a detailed description of gaps and areas of ambiguity in international law concerning IDPs see Geissler, above n 204; 'Internally Displaced Persons, Report of the Representative of the Secretary-General, Mr Francis M Deng, Compilation and Analysis of Legal Norms' UN Doc E/CN.41996/52/Add.2 (1995) [410] – [415].

207   Office of the High Commissioner for Human Rights, Report of the Office of the United Nations High Commissioner for Human Rights on the Relationship Between Climate Change and Human Rights, A/HRC/10/61, 15 January 2009, 19.

208   Ibid.

209   Biermann and Boas, above n 76, 11.

210   Geissler, above n 204, 455–6.

211   'Internally Displaced Persons, Report of the Representative of the Secretary-General, Mr Francis M Deng, Compilation and Analysis of Legal Norms' UN Doc E/CN.41996/52/Add.2 (1995) [368]. This generalisation does not always hold true, as was demonstrated in May 2008 when the Burmese

government obstructed the provision to its citizens of international aid after areas of the country had been devastated by Cyclone Nargis. However, given that the purpose of the proposed Convention is to address a gap in international law in relation to a certain category of persons, the proposal presented here does not provide the appropriate forum in which to institute or extend general innovations in international law to facilitate forcible interventions by the international community on humanitarian grounds in order to address the situation exemplified by Burma.

212 Biermann and Boas, above n 76, 11.

213 Biermann and Boas, above n 41, 26.

214 Geissler, above n 204, 452; Phuong, above n 180, 43.

215 Docherty and Giannini, above n 29, 378; Convention on Cluster Munitions, opened for signature 3 December 2008 (see <http://www.clustermunitionsdublin.ie/pdf/ENGLISHfinaltext.pdf> at 19 December 2009).

216 Brown, above n 16, 25.

217 Docherty and Giannini, above n 29, 376.

218 Ibid.

219 Each of the rights and principles enunciated here are explored in greater depth in Docherty and Giannini, above n 29, 376–8. James Hathaway explains that under the Refugee Convention, 'the standard of treatment owed to refugees is defined through a combination of absolute and contingent criteria. A few rights are guaranteed absolutely to refugees and must be respected even if the host government does not extend these rights to anyone else, including its own citizens. More commonly, the standard for compliance varies as a function of the relevant treatment afforded another group under the laws and practices of the receiving country': The Rights of Refugees Under International Law (1st ed, 2005) 155. As is the case with the Refugee Convention, rights would be accorded under a Convention for CCDPs on both an absolute and contingent basis.

220 Docherty and Giannini, above n 29, 377.

221 Ibid, 398.

222 Hathaway, above n 219, 155

223 Docherty and Giannini above n 29, 369.

224 Ibid, 378.

225 The mechanisms by which such assistance would be provided are discussed [reference to DH's section].

226 A number of legal considerations are raised here. The background paper submitted by the Representative of the Secretary General on the Human Rights of Internally Displaced Persons, 'Displacement Caused by the Effects of Climate Change', above n 176, considers that those Islanders who move abroad will be left in 'a legal limbo'. It states that 'it is also unclear as to whether provisions on statelessness would apply as it remains to be seen whether those affected become stateless persons under international law. These persons do not become stateless as long as there is some remaining part of their territory of their State, and even where a whole country disappears it is not certain that they become stateless in the legal sense. Statelessness means to be without nationality, not without state'. In any event, the international law of statelessness does not provide adequate protection (ibid, 5).

227 Linda Mottram, 'Climate change negligence will lead to "benign genocide"', ABC News Radio, 22 September 2009 <http://www.abc.net.au/news/stories/2009/09/22/2693588.htm> at 20 December 2009.

228 McAdam, above n 21, 10.

229 Ibid.

230 Docherty and Giannini, above n 29, 355–356.

231 Ibid, 398.

232 United Nations International Covenant on Economic, Social and Cultural Rights, opened for signature 16 December 1966, 993 UNTS 3 (entered into force 3 January 1976), Article 1.

233 It is probable that the former would operate on an individual basis while the latter would be on a population basis.

234   C Mortreux and J Barnett, 'Climate change, migration and adaptation in Funafuti, Tuvalu' (2009) 19 Global Environmental Change 105–112.

235   Docherty and Giannini, above n 29, 375.

236   Refugee Convention, note 43 above, Article 4.

237   Ibid, Article 15.

238   Ibid, Article 29.

239   Ibid, Article 14.

240   Betaille et al, above n 59, Article 8(2).

241   *Convention for the Safeguarding of the Intangible Cultural Heritage*, Article 1, opened for signature 17 October 2003, MISC/2003/CLT/CH4 (entered into force 20 April 2006).

242   Ibid, Article 2 (1).

243   Ibid, Article 2 (3).

244   See Docherty and Giannini, above n 29, 398–400, for a detailed consideration of the role of civil society.

245   Stephen D Goose, Mary Wareham and Jody Williams, 'Banning Landmines and Beyond' in Jody Williams, Stephen D Goose and Mary Wareham (eds), Banning Landmines: Disarmament, Citizen Diplomacy, and Human Security (2008) 6.

246   The Convention on the Prohibition of the Use, Stockpiling, Production and Transfer of Anti-Personnel Mines and on Their Destruction, opened for signature 18 September 1997, 36 ILM 1507 (entered into force 1 March 1999).

247   Alexander Kmentt, 'A Beacon of Light: The Mine Ban Treaty Since 1997', in Jody Williams, Stephen D Goose and Mary Wareham (eds), above n 245, 18.

248   Stephen D Goose, 'Cluster Munitions in the Crosshairs: In Pursuit of a Prohibition', in Jody Williams, Stephen D Goose and Mary Wareham (eds), above n 245, 227.

249   Jody Williams and Stephen D Goose, 'Citizen Diplomacy and the Ottawa Process: A Lasting Model?' in Jody Williams, Stephen D Goose and Mary Wareham (eds), above n 245, p 182.

250   Renaud et al, above n 121, 34.

251   Ibid.

252   Ibid, 35.

253   McAdam, above n 21, 28.

254   Docherty and Giannini, above n 29, 370. Docherty and Giannini do, however, welcome in-depth analysis, research and examination regarding the issue of climate change internally displaced persons: ibid, 360, note 65.

255   International Organization for Migration, *Discussion Note: Migration and the Environment*, MC/INF/288, Ninety-Fourth Session, 1 November 2007, 7.

PART 4

Affected countries

The Pacific

Chapter 26

Pacific Islands Forum

THE PACIFIC ISLANDS FRAMEWORK FOR ACTION ON CLIMATE CHANGE 2006–2015

In this framework, Pacific Island Countries and Territories (PICTs) refers to American Samoa, Cook Islands, Fiji Islands, French Polynesia, Guam, Kiribati, Commonwealth of the Northern Marianas, Marshall Islands, Federated States of Micronesia, Nauru, New Caledonia, Niue, Palau, Papua New Guinea, Samoa, Solomon Islands, Tokelau, Tonga, Tuvalu, Vanuatu, Wallis and Futuna.

The timeframe for this Framework is 2006–2015. This Framework builds on The Pacific Islands Framework for Action on Climate Change, Climate Variability and Sea Level Rise 2000–2004.

In this Pacific regional framework, climate change refers to any change in climate over time both as a result of human activity and natural variability.[1]

The adverse effects of climate change and sea-level rise present significant risks to the sustainable development of Pacific Island Countries and Territories (PICTs) and the long-term effects of climate change may threaten the very existence of some of them. This was agreed to generally by Small Island Developing States together with the international community most recently in the Mauritius Strategy for the Further Implementation of the Barbados Programme of Action for Sustainable Development of Small Island Developing States.

PICTs' priorities and needs in the area of climate change are reflected in international documents such as the Mauritius Strategy. These are also reflected in national communications, the outcomes of the UNFCCC Conferences of the Parties and the outcomes of related international meetings.

(i) Scope

I. Preamble

At the regional level, PICTs priorities and needs have been reiterated for over a decade in relevant documents such as Forum Leaders Communiqués, regional policy frameworks and related action plans together with the strategic plans of the regional intergovernmental and non-governmental organizations.

At the national level, PICTs are also taking action to address climate change through their national sustainable development strategies or their equivalent which are linked to national budgetary and planning processes.

PICTs recognize their commitment to sustainable development is a national responsibility but realise that this cannot be achieved without development partner support. Within this context the Framework identifies broad priorities for PICTs. It provides a strategic platform not only for use by policy and decision makers at all levels, but also for the development and strengthening of partnerships for implementation of national and regional initiatives.

The Framework runs from 2006–2015 and is consistent with the timeframes of the Millennium Declaration, the Johannesburg Plan of Implementation and the subsequent work of the UN Commission on Sustainable Development. It does not create legal rights or impose obligations under international law.

The Framework is intended to promote links with, but in no way supercede, more specific regional and national instruments and plans across specific sectors that link to weather and climate including: water; agriculture; energy; forestry and land use; health; coastal zone management; marine ecosystems; ocean management; tourism and transport.

Addressing the issues of climate change requires an integrated, multi-stakeholder approach. Furthermore, a strategic programmatic approach is required rather than an increase in stand-alone project initiatives.

II. Pacific Context

PICTs experience a high level of risk from the effects of extreme weather and climate variability. Climate models suggest the tropical Pacific region will continue to warm. This warming has the potential to alter and indeed increase such risks, through changing the frequency and/or intensity of extreme weather or climate variability phenomena or through accelerated sea-level rise. The impacts of these climate events will exacerbate already stressed marine, freshwater and terrestrial environments.

Reducing the risks associated with the impacts of extreme weather and climate variability is a fundamental developmental challenge faced by PICTs. This must be urgently addressed in order to contribute to improving livelihoods, economic wellbeing and health as well as maintaining biodiversity and culture.

An integrated and multi-stakeholder approach that considers the complete cycle of interlinked causes and effects, within the context of risk management across all sectors, is vital. A high priority is the need to develop and strengthen community-centered initiatives.

III. Vision

Pacific island people, their livelihoods and the environment resilient to the risks and impacts of climate change.

IV. Goal

Ensure Pacific island people build their capacity to be resilient to the risks and impacts of climate change with the key objective to deliver on the expected outcomes under the following principles:

- implementing adaptation measures;
- governance and decision making;
- improving understanding of climate change;
- education, training and awareness;
- contributing to global greenhouse gas reduction; and,
- partnerships and cooperation.

V. Principles

Principle 1. Implementing adaptation measures

Building resilience through adaptation to climate change, climate variability and extreme weather events has been identified as the key priority for PICTs. All PICTs agree that they are already witnessing the adverse effects of climate change. Atoll states in particular believe that their very survival is threatened.

The ecological fragility, economic and social vulnerability, and the remoteness of many Pacific Island countries makes recovery from extreme weather events very difficult.

Adaptation now will greatly increase our capacity to better adapt to future climate change impacts. Appropriate adaptation measures using a multi-stakeholder approach need to be integrated into national/sectoral sustainable development strategies or their equivalent.

PICTs will encourage adaptation measures based on the principles of risk management and, where this is not possible, the "no regrets" or precautionary approach with a focus on improving the livelihoods of its people including safety and security.

EXPECTED OUTCOMES BY 2015:

1.1   Adaptation measures to the adverse effects of climate change developed and implemented at all levels.

1.2   Identification of vulnerable priority areas/sectors and appropriate adaptation measures using available and appropriate information recognizing that such information may be incomplete.

1.3   Adaptation measures in vulnerable priority areas supported by existing data sets and traditional knowledge, or new data developed in some instances as necessary.

1.4   Appropriate adaptation measures integrated into national/sectoral sustainable development strategies or their equivalent and linked to the budgeting process.

Principle 2. Governance and decision making

PICTs recognize that they have a national responsibility for addressing the risks and effects of climate change in the context of their national sustainable development strategies and reflecting principles of sustainable development and good governance.

All stakeholders have a role to play in developing individual and collective resilience through adapting, preventing and/or mitigating the adverse effects of climate change. Climate change and its effects are a shared responsibility, which also requires effective partnership with all relevant stakeholders in decision-making and implementation of strategies and actions at all levels.

Recognizing the presence of limited technical and financial resources and institutional capacity at the national and regional levels, collaboration and partnerships between CROP agencies in support of national efforts, consistent with the Pacific Leaders' vision, is critical, harnessing key disciplinary skills and expertise across the region.

Good governance ensures the adoption of core principles of accountability and transparency by all stakeholders and at all levels, which is critical for cost effective adaptation and greenhouse gas reduction activities against the risks of climate change.

EXPECTED OUTCOMES BY 2015:

2.1 Climate change considerations mainstreamed into national policies, planning processes, plans and decision-making at all levels and across all sectors.

2.2 Partnerships and organizational arrangements between government agencies, private sector, civil society, community and other stakeholders strengthened.

2.3 CROP agency partnerships coordinated, harmonized and strengthened to ensure country- and outcome-focused delivery of services.

2.4 Good governance by all stakeholders in climate change activities management at regional, national and local levels strengthened.

Principle 3. Improving our understanding of climate change

Better understanding of climate change, variability and extreme weather events is needed to inform local, national and regional responses. This will mean enhancing human resource capacity for generating, analyzing and managing climate related data sets; sustaining and upgrading existing observation and application systems; developing and strengthening technical data sets and tools for climate observations; establishing baseline data in different sectors and maintaining the collection of the latest information on sea level rise.

As a basis for improving our understanding of climate change, is the ongoing need to engage research into improving understanding in the variations, circulations and climatic patterns in the Pacific region.

Translating climate change science into applicable information products through user-friendly materials and tools is necessary to inform the decision making process at all levels.

EXPECTED OUTCOMES BY 2015:

3.1 Existing meteorological, hydrological, oceanographic and terrestrial institutional capacity including data collection systems sustained and upgraded.

3.2   Technical data sets integrated with relevant climatic, environmental, social and economic information and data sets, and traditional knowledge for risk management.

3.3   Analytical frameworks, models and tools for projections of regional climate change and variability, risk assessment and management strengthened.

3.4   Development, strengthen where necessary datasets and information required to underpin, strengthen and monitor vulnerable priority areas, sectors and adaptation measures.

Principle 4. Education, training and awareness

PICTs' capacity to use economic, scientific and traditional knowledge to monitor, assess and predict environmental, social and economic risks and effects of climate change needs strengthening. This is critical for developing and implementing viable and sustainable national programmes on cost effective adaptation and greenhouse gas reduction measures.

Concerted efforts need to be undertaken to enhance human capacity in the assessment of the risks and impacts of climate change, climate variability and extreme weather events. A pool of informed resource persons conversant with development and application of practical steps in adaptation tools and methods is critical. Increased awareness and understanding of risks and effects of climate change is particularly important at the community level to increase their resilience.

EXPECTED OUTCOMES BY 2015:

4.1   Strengthened human capacity to monitor and assess environmental, social and economic risks and effects of climate change.

4.2   Strengthened human capacity to identify, analyse and implement cost effective adaptation measures as well as greenhouse gas reduction measures and creation of a pool of informed resource persons conversant with development of practical steps in adaptation tools and methods.

4.3   Strengthened human capacity to identify and integrate economic, scientific and traditional knowledge into adaptation and greenhouse gas reduction practices.

4.4   Better informed public on climate change issues.

Principle 5: Contributing to global greenhouse gas reduction

PICTs' contributions to the total global emission of greenhouse gases are insignificant compared to the rest of the international community. Nonetheless, PICTs wish to contribute to the global effort to reduce emissions. As part of their national policies, PICTs will promote cost effective measures to reduce greenhouse gas emissions, including increased energy efficiency and increased use of appropriate low carbon and renewable energy technologies.

There may be the opportunity to work with developed countries on the Kyoto Protocol Clean Development Mechanism projects to support these efforts. Complementing the effort will be national plans and policies to ban the use of ozone depleting substances.

EXPECTED OUTCOMES BY 2015:

5.1   Energy efficiency actions and cost effective technologies promoted and implemented.

5.2   Cost effective renewable energy technologies and local sources promoted, shared and implemented.

5.3   Commitments met on ozone depleting substances.

5.4   Clean Development Mechanism initiatives developed and implemented, where appropriate.

Principle 6. Partnerships and cooperation

Partnerships and cooperation provide an enabling environment and are an essential part of PICTs' efforts to build resilience to the adverse effects of climate change.

PICTs will continue to advocate for the reduction of greenhouse gas emissions and to advance adaptation internationally. Networks and partnerships to inform policy development for harmonized regional, national and local responses to climate change is necessary.

Additional resources will need to be accessed through multilateral and bilateral funding. One of the roles of regional organizations is to support national efforts to access this assistance and to coordinate existing and new innovative projects and programmes, including the Pacific Partnership Initiative for Adaptation to Climate Change launched by Pacific leaders at the World Summit on Sustainable Development. Efforts will be taken to ensure climate change partnerships are strategic and well coordinated.

EXPECTED OUTCOMES BY 2015:

6.1   Existing and emerging international partnerships for the Pacific islands region on climate change and related issues strengthened and established.

6.2   Enhanced coordination of regional action on climate change issues.

6.3   Climate change related assistance from development partners coordinated and harmonized to maximize benefits to PICTs.

6.4   Access by PICTs to secure increased resources from funding mechanisms related to climate change instruments optimized.

6.5   Promote significant international support through advocacy for further reduction in greenhouse gases and securing resources for adaptation.

PICTs recognise that the implementation of this Framework, the Mauritius Strategy, Agenda 21 and the Johannesburg Plan of Implementation, as well as the achievement of the internationally agreed development goals, including those contained in the Millennium Declaration, are mutually reinforcing.

The implementation of this Framework will be further elaborated in the Pacific Islands Action Plan on Climate Change 2006–2015. It will require more focused and substantially increased effort by PICTs and appropriate support from their regional organisations and the international community. PICTs recognize that each country has primary responsibility for its own development and that the role of national policies, development strategies and the allocation of dedicated financial resources cannot be overemphasized.

[…]

VII. *Monitoring Progress and Updating this Framework*

Targets and indicators will be established within the Action Plan linked to the Framework and set at the appropriate levels. The framework will be subjected to a mid-term review in 2010 to determine overall progress.

Evaluating progress towards achieving the outcomes of this Framework will be measured every two years against the agreed national and regional indicators with the support of regional organizations and the international community. This will require PICTs to identify progress towards achieving the principles contained in this Framework, and to identify emerging gaps requiring priority action and adjustment of priorities in future. The regional organizations will where necessary provide support and a coordinating role, for regional and international reporting.

From Pacific Islands Forum, 'The Pacific Islands Framework for Action on Climate Change 2006–2015', 12 June 2005.

Note

1   Refer to Intergovernmental Panel on Climate Change (IPCC) and the United Nations Framework Convention definition of climate change.

Chapter 27

Oxfam

THE FUTURE IS HERE
Climate Change in the Pacific

[...]

Executive summary

> For a highly vulnerable country like Tuvalu, we cannot just sit back and watch our homeland slowly disappear. If necessary, we will use whatever legal means available to us to seek the necessary restitution for all damages created by climate change. Hopefully, the international community will respond before such action is necessary. But time is running out fast. Climate change could well be the greatest challenge that humanity has ever known. I make a very strong plea to all to act quickly and responsibly, to ensure that countries like Tuvalu do not disappear.[1]
>
> (Tuvalu Prime Minister Apisai Ielemia)

Developing nations in the Pacific are at the frontline of global climate change. Livelihoods and food and water sources that have sustained communities over generations are being threatened. People are losing land and being forced from their homes. Unless wealthy, developed countries like Australia and New Zealand take urgent action to curb emissions, some island nations in the Pacific face the very real threat of becoming uninhabitable.

Impacts of climate change in the Pacific

Climate change has the potential to affect almost every issue linked to poverty and development in the Pacific. In a region where half the population lives within 1.5 kilometres of the sea, few people will be untouched by the consequences of climate change. Natural weather variability in the region means developing island countries in the Pacific already face severe threats to human security and economic losses arising from extreme weather events like storm surges, cyclones and king tides. Projected sea-level rise and increases in the intensity of natural disasters like cyclones will exacerbate these problems. Scientists have also projected an increase in diseases such as malaria and dengue fever, together with significant soil and coastal erosion as a result of climate change.

Combating climate change in the Pacific

The fairest and most cost-effective way of dealing with climate change in the Pacific is to ensure that the most extreme climate impacts are avoided altogether.

Preventing catastrophic climate change in the Pacific means keeping warming as far below 2°C as possible compared to pre-industrial temperatures. To achieve this, wealthy, polluting countries such as Australia and New Zealand must reduce their emissions by at least 40% by 2020, and at least 95% by 2050.

The current emissions reductions targets set by Australia and New Zealand fall short of their international obligations and do not go far enough to contribute to a safe, fair global climate agreement.

Pacific island communities must also be supported to implement low carbon development plans. Australia and New Zealand can play a critical role in supporting local efforts to explore and access a range of renewable energy sources and protect forests across the region.

[...]

Climate change in the Pacific

Adapting to the impacts of climate change in the Pacific

Pacific communities urgently need support to adapt to the impacts of climate change they are already experiencing. Adaptation efforts range from planting mangroves in order to reduce coastal erosion to building rainwater tanks to maximise fresh water supplies.

Governments, civil society and local communities have a critical role to play in planning and implementing adaptation strategies in the Pacific. This will help to ensure the best use of adaptation funds and the effective use of traditional knowledge.

While initial support from the Australian and New Zealand governments has been welcome, the scale of the problem means that much more money is needed. Moreover, financial support for adaptation in the Pacific must be in addition to existing aid commitments so that crucial efforts to alleviate poverty and promote development across the region are not compromised.

At least double the current level of adaptation funding is required simply to address the most urgent adaptation needs. Meeting these needs will require between AUD $365 million/NZD $455 million and AUD $668 million/NZD $834 million.

Of course, it is critical to ensure these funds are spent effectively. To this end, a greater proportion of adaptation support for the Pacific must be allocated to basic resilience programs at a community level. There is also a need to safeguard access to food and water for Pacific communities.

Climate displacement and migration

By the year 2050, about 75 million people could be forced to leave their homes in the Asia-Pacific region due to climate change. Pacific island governments are already tackling climate change-related relocation and resettlement.

Given the significant implications of these population movements for our region, it is vital that Australia and New Zealand governments hold discussions with Pacific island governments about this issue now. Planning for climate displacement will require looking at the most effective ways to support Pacific islanders who are forced to move from their homes, including through appropriate immigration policies.

Time to act

Climate change is affecting our region now. Its impacts are profound. The lives, homes, livelihoods, food and water of many Pacific communities are under threat. Wealthy, developed countries like Australia and New Zealand must act urgently to:

> reduce their own emissions; support Pacific communities to follow low carbon pathways to development; help those communities adapt to the impacts of climate change they are already experiencing, and work with Pacific island governments to plan for displacement that is likely to be caused by climate change.

Australia and New Zealand not only have a responsibility to take such action, it is also in our best interests. Acting now will save money and save lives. Most importantly, it will help to create a safer, more sustainable and more peaceful future for the Pacific.

Recommendations

Recommendation 1:

The Australian and New Zealand governments must set higher medium and long-term emissions reduction targets to reduce greenhouse gas emissions.

Preventing catastrophic climate change in the Pacific means keeping warming as far below 2°C as possible compared to pre-industrial temperatures. To achieve this, wealthy, polluting countries such as Australia and New Zealand must reduce their emissions by at least 40% by 2020, and at least 95% by 2050.

The current emissions reductions targets for both countries do not go far enough to meet their international obligations and contribute to a safe, fair global climate agreement.

The impacts of climate change are already undermining the lives of millions of the world's poorest people, including people in the Pacific. Developed countries like Australia and New Zealand have a critical role to play in tackling climate change and must act urgently to safeguard the rights of poor communities across our region.

Recommendation 2:

The Australian and New Zealand governments must support developing countries to follow low-carbon pathways to development.

Catastrophic climate change can only be avoided through cooperative efforts in which rich countries like New Zealand and Australia take responsibility for both reducing their own emissions and providing vital support to developing countries to pursue low-carbon development pathways. In the Pacific, there is potential to develop new areas of work, such as village level renewable energy initiatives, through networks such as Pacific Energy and Gender (PEG), which is working with government and community organisations to promote new solar technologies such as solar cookers.

Recommendation 3:

The Australian and New Zealand governments must provide new and additional money for adaptation focused on the Pacific.

As high per-capita emitters of greenhouse gases, Australia and New Zealand are among the developed nations collectively responsible for the damage that climate change has done to the Pacific. Consistent with the "polluter pays" principle,[2] Australia and New Zealand have a responsibility to fix the problem they have helped to create. But adaptation funding has fallen well short of requirements in the Pacific.

Pacific island countries are calling for new and additional money for adaptation in the region, rather than donors continuing a pattern of reallocating existing official development assistance. Adaptation funds should be provided as grants, not loans.

Globally, Oxfam has estimated that at least AUD $187 billion/NZD $233 billion is needed each year, to finance emissions reduction and adaptation efforts in developing countries. As developing countries cannot afford this on their own, wealthy counties, which have contributed three-quarters of the carbon in the atmosphere now, and who have grown wealthy by burning fossil fuels, must provide this finance, as they promised to do at the United Nations climate negotiations in Bali in 2007.

Australia's fair share of this is AUD $4.3 billion a year; New Zealand's fair share is NZD $792 million. Oxfam also calls on Australia and New Zealand to support the establishment of a single Global Climate Finance Mechanism. This mechanism would minimise effort wasted by Pacific island nations on red tape by replacing funding currently provided through numerous multilateral and bilateral mechanisms.

Recommendation 4:

More adaptation resources should be directed towards local communities and draw on local knowledge in developing responses to climate change.

In assisting Pacific communities to adapt to the impacts of climate change, a greater proportion of funding needs to be allocated to basic resilience programs at a community level, rather than on more consultants and scientific testing. Efforts should also be made to promote the use of local knowledge and local history to help adapt to the impacts of global warming. Older Pacific islanders have knowledge of changes over many decades to coastlines, forests and access to water, as well as practical adaptation information like how to find food in times of disaster.

Recommendation 5:

Adaptation efforts should be focused on livelihoods, food and water security.

Ensuring access to sustainable sources of food and water must be a central priority in efforts to assist Pacific communities adapt to climate change. It will also be important to ensure

Pacific islanders are able to pursue sustainable livelihoods at a community level. Focusing on these key areas will help to develop community resilience in the face of climate impacts and natural disasters.

Recommendation 6:

The Australian and New Zealand governments should prepare for climate displacement.

By 2050, approximately 150 million people may be displaced because of climate change.[3] Seventy-five million of these are likely to be in the Asia-Pacific region, with that number growing to around 150 million[4] by 2100. Many people will resettle within their own country, and Pacific island governments are already tackling climate change-related relocation and resettlement. But not all people forced to leave their homes will have the option of moving within their country. Both those people who relocate internally and those who are forced to relocate to another country will require assistance. The Australian and New Zealand governments have been slow to address this issue.

The potential for forced displacement among the Pacific islands' population of about 8 million people demands urgent debate on what future resettlement and relocation might involve. It is vital that local communities have the opportunity to participate in this debate.

Australia and New Zealand need to engage in dialogue with Pacific island governments, plan to address issues of climate displacement, and develop immigration policies which support Pacific island communities that are forced to move from their homes.

Recommendation 7:

Pacific governments must ensure that women and men participate equitably in all decision-making about climate change and that their differentiated needs are reflected in adaptation efforts.

Women are disproportionately affected by climate change, because they tend to depend more on the natural environment for their livelihoods than men and they also bear the brunt of the impact of climate-related disasters and diseases like malaria. Yet women are often left out of the conversation about climate change. An effective climate change strategy requires governments to recognise that women have specific needs in climate change policies and to insist on greater participation by women in decision-making at all levels.

[…]

Case study: Climate-proofing communities in Fiji

Sitting round the kava bowl, the Fijian men of Korotarase village tell of the chaos on the day their village was flooded. Ratu Emosi Rokotuibua, the turaga-ni-koro (village chief) at Korotarase, says many houses and food gardens were damaged and the community's playing field was 1.5 metres under water.

Korotarase is located on low-lying swampy land alongside a river and beach on Fiji's northern island of Vanua Levu. In March 2007, the coincidence of heavy upstream rainfall and a king tide from the ocean led to the village being flooded.

Working together

Now, the people of Korotarase have joined with five other Fijian villages, in an innovative program of community climate adaptation. The villagers are working to climate proof their homes and communities, in preparation for future impacts caused by tidal surges, coastal erosion or flooding caused by heavy rainfall after cyclones. They are trialling salt-resistant varieties of staple foods such as taro, planting mangroves, native grasses and other trees to halt coastal erosion, protecting fresh water wells from salt-water intrusion, and relocating homes and community buildings away from vulnerable coastlines.

The program, supported with funding from Australia, is coordinated by Lavinia Tawake and Patrina Dumaru at the Institute for Applied Science (IAS) through the University of the South Pacific (USP) in Suva.

The 2007 flood at Korotarase village greatly increased erosion along the riverbank, and today some houses and the community hall are at risk of collapsing into the river. The problems are increased because sedimentation from upstream logging operations is changing the river's path.

"There has been erosion of the river bank and in the longer term we need to stabilise the land and prepare for future floods," Ratu Emosi says. "The erosion and siltation are changing the course of the river and affecting the fishing at the river mouth. The village was fearful of further flooding, so we decided to seek help to develop a long-term plan."

Human and natural causes

At Buretu village, located on the Rewa Delta alongside the Navolau River, villagers also face serious problems of coastal erosion. The natural changes to the river flow are impacted by changing rainfall patterns and also human-induced effects (including foot traffic along the riverbank and waves generated by outboard motors). With falling coconut trees, risk to buildings like the community's church, and the long-term danger of flooding due to tidal surges or cyclone-induced rainstorms, the community is acting to address the erosion.

Faced with the options of retreating to higher ground or expensive dredging works in the river, the community has chosen to plant appropriate trees and grasses along the bank. By planting rows of vetiver hedgerows along the eroded bank and mangroves and creepers along the water's edge, the community hopes to slow the erosion of the riverbank.

Community planning

Increasingly, the villagers are drawing on the expertise of outside researchers or government staff to complement local initiatives. At Korotarase village, project staff and the local provincial liaison officer organised a forestry workshop with Eliki Senivasi, a forester from the Ministry of Primary Industry. The villagers discussed options such as developing a community watershed plan, establishing a community nursery and working with local logging firms to limit adverse effects from forestry operations upstream of the village.

The project is based on a belief that local knowledge can help document the changes to the environment that are occurring over longer periods. Villagers have noticed, for example, that the seasonality of certain nuts and fruits is changing, with the ivi nut appearing in September rather than December.

For researchers Lavinia Tawake and Patrina Dumaru, current efforts at climate adaptation build on previous community initiatives, like creating marine reserves to control overfishing.

"Rural communities have lots of ideas on how to protect their local environment," Patrina says. "Our role is to support them to analyse the vulnerabilities and develop environmental management plans that can help with future climate change.

"Overseas donors need to direct more resources to community level to ensure that climate adaptation efforts are really helping the people that need the greatest assistance."

[…]

Australian support for adaptation in the Pacific

In the 2008–2009 budget, the Australian Government followed through on a pre-election commitment to spend AUD $150 million on adaptation over three years as part of an already committed 9% increase in the overall aid budget to AUD $3.7 billion (0.32% of Gross National Income).

Whilst this initial investment is welcome, the scale of the problem means that much more money is needed. Australia's efforts to support adaptation in the Pacific also need to be in addition to the aid program so that they do not diminish broader efforts to alleviate poverty and promote development in the region.

The focus of the International Climate Change Adaptation Initiative (ICCAI) funding has also been heavily on science and research, which while important, does not address the urgent on-the-ground needs for adaptation in the Pacific.

So far, investments of around AUD $35 million have been announced under the ICCAI including:

- AUD $3 million for a Pacific Future Climate Leaders program to train future Pacific climate change leaders through scholarships, exchange programs and community education;
- AUD $6 million over three years to the Global Environment Facility's Small Grants Program, to support community-based adaptation programs in the Asia-Pacific region;
- AUD $5 million to help deliver and coordinate scientific and technical assistance to tackle climate change;
- AUD $800,000 towards strengthening Pacific meteorological services in partnership with New Zealand and Pacific island countries, and to support efforts under the UNFCC to identify new and innovative tools to limit the financial risks of climate change in developing countries;
- AUD $20 million for scientific research and building a regional science network.

New Zealand support for adaptation in the Pacific

The New Zealand Government supports a range of activities on climate change in the region but, like Australia, it has largely failed to commit sufficient new and separate funds to adaptation. In 2008–2009, NZAID allocated NZD $6.5 million for its Pacific Regional Environment and

Vulnerability program.[38] NZAID provides separate assistance of approximately NZD $10 million to Pacific regional organisations that work on sustainable natural resource management, disaster risk reduction, renewable energy and climate change.

NZAID is providing NZD $1.5m towards the Kiribati Adaptation Program.[39] Like Australia, New Zealand contributes bilaterally with coordinated disaster relief immediately after natural disasters. NZAID also has multi-year funding arrangements with SOPAC; the Ministry of Civil Defence and Emergency Management (MCDEM); the New Zealand Meteorological Service (weather forecasting, cyclone tracking and some risk mitigation work in PICTs); and Radio New Zealand International. NZAID has also supported a regional environmental education initiative.

The need for greater adaption support in the Pacific

As high per-capita emitters of greenhouse gases, Australia and New Zealand are among the developed nations collectively responsible for the damage done to the Pacific by climate change now and in the future. Consistent with the "polluter pays" principle,[40] Australia and New Zealand have a responsibility to fix the problem they have helped to create. Equitable adaptation funding cannot be achieved by redirecting existing financial efforts to alleviate poverty and promote development in the region.

A recent study by the Overseas Development Institute (ODI) stresses that this will be a central issue in the 2009 climate negotiations in Copenhagen:

> Northern countries should be expected to assist where countries in the South are unable to meet present financing needs, but not through a donor-recipient relationship, but rather in terms of proportionate payments for damage already inflicted on global public goods.[41]

Pacific island countries have argued for greater funding to be allocated for adaptation programs by major donors like Australia and New Zealand, as they have insufficient resources to address the extensive adverse effects of global warming. A central message coming from Pacific governments is that this funding should be new and additional money rather than the reallocation of existing ODA funds. The Pacific Islands Forum leaders' communiqué in 2008 stressed:

> The priority of Pacific SIDS [Small Island Developing States] is securing sustainable financing for immediate and effective implementation of concrete adaptation programmes on the ground.[42]

Oxfam estimates that AUD $187 billion/NZD $233 billion is needed every year to fund adaptation and emissions reductions in developing countries on top of existing aid commitments. Australia's fair share of this is AUD $4.3 billion a year; New Zealand's fair share is NZD $792 million.

Oxfam argues this should be delivered through a single Global Climate Change Finance Mechanism to avoid wasting money and resources in a myriad of bilateral and multilateral schemes.

Over the next three years Pacific Island countries will need at least double Australia's current commitment to adaptation funding, just to complete the most urgent adaptation tasks, based on the National Adaptation Programmes of Action [NAPAs] submitted by Pacific island countries to date.[43] In the Australian Journal of International Law, McGoldrick estimated that between AUD $365 million/NZD $454 million and AUD $668 million/NZD $830 million would be required. The AUD $35 million that Australia is due to spend in the Pacific in 2008–2009 would not even cover the most pressing adaptation needs of three of the region's most vulnerable and least developed countries — Kiribati, Tuvalu and the Solomon Islands. [...]

Conclusion

Pacific island countries are clearly feeling the early effects of global climate change, with environmental migration and loss of land already occurring in some parts of the region.

Without a significant effort by developed countries now, some island nations in the Pacific face the very real threat of becoming uninhabitable in the decades ahead.

Developed countries such as Australia and New Zealand should take urgent action to tackle climate change and safeguard the rights of Pacific island communities to life, livelihoods, food, water, health and security.

It is in Australia and New Zealand's best interests to take this action now. The more frequent disasters caused by climate change will require Australia and New Zealand to respond, and the displacement of people in the Pacific due to rising sea levels will force them to look for new homelands.

Moreover, it makes financial sense to act now, given that for every $1 spent on disaster preparedness and risk reduction, two to ten dollars is saved in disaster response.

The key first step is to prevent further climate damage to the Pacific by urgently adopting tougher targets — reducing emissions by at least 40% on 1990 levels by 2020, and at least 95% by 2050.

Equally, there is an urgent need to increase support for adaptation in the Pacific and to plan for forced migration and climate displacement. People across the Pacific are already adapting with the resources they have, but they cannot do it alone. More adaptation resources need to be directed towards local communities and draw on local knowledge in developing responses to climate change. Adaptation efforts should be focused on livelihoods, food and water security and developing countries need to be supported to follow low-carbon development pathways.

These steps simply represent Australia and New Zealand's fair share of what is needed to address climate change in the region and support our neighbours to become resilient climate change survivors.

Oxfam, 'The Future is Here: Climate Change in the Pacific' Briefing Paper', 2009.

Notes

1   "A Threat to Our Human Rights — Tuvalu's Perspective on Climate Change", *UN Chronicle*, Issue 2, 2007.

2   In environment law the polluter-pays principle provides that the party responsible for producing

pollution is responsible for paying for the damage done to the natural environment. In international environmental law it is mentioned in Principle 16 of the Rio Declaration on Environment and Development. See Phillippe Sands, *Principles of International Environmental Law*, 2nd edn, Cambridge University Press, 2003.

[...]

40   In environment law the polluter pays principle provides that the party responsible for producing pollution is responsible for paying for the damage done to the natural environment. In international environmental law it is mentioned in Principle 16 of the Rio Declaration on Environment and Development. See Phillippe Sands, *Principles of International Environmental Law*, 2nd edn, Cambridge University Press, 2003.

41   Bird Neil and Leo Peskett: "Recent bilateral initiatives for climate financing: are they moving in the right direction?, ODI Opinion No.112, Overseas Development Institute, September 2008.

42   Forum communiqué, Pacific Islands Forum, Niue, August 2008.

43   National Adaptation Programmes of Action (NAPAs) submitted to-date, identify the minimum requirements for adaptation in Samoa, Vanuatu, Kiribati, Tuvalu and Solomon Islands. These nations averaged urgent funding requirements of more than USD $11 million each.

AUSTRALIA

Chapter 28

Peter Kinrade and Benjamin Preston

PEOPLE, PROPERTY AND PLACES
Impacts of Climate Change on Human
Settlements in the Western Port Region

Executive Summary

Introduction

Since the mid-19th century, the average temperature at the Earth's surface has increased by approximately 0.7°C (IPCC, 2007a). This warming is "very likely due to the combined influences of greenhouse gas increases and stratospheric ozone depletion (IPCC, 2007a)."

While much focus has traditionally been placed on building scientific understanding of such global changes in climate and their attribution, increasing attention is now focused on how communities, enterprises and governments should respond. This necessitates building understanding about the local-scale implications of climate change that can be used to guide adaptive decision-making.

This report examines the nature and extent of potential impacts of climate change to the Western Port region of Victoria. The report's focus is on the impacts of climate change on the built environment, the social and economic implications of the impacts and the vulnerability of different localities and groups. It is one part of a wider 'integrated assessment' of climate change in the region that covers:

- regional climate changes and biophysical impacts;
- socio-economic and infrastructure impacts (this report);
- risk assessment; and
- adaptation response.

The Western Port region is well suited to a study of the impacts of climate change on human settlements – it has a large, diverse and growing population and a number of key implications of climate change are pertinent to the region including coastal and inland flooding, wildfires and drought.

In using this report it is important to be mindful of a range of uncertainties and limitations associated with the various analyses of exposure and impacts.

Profile of the Western Port region

The Western Port region, which encompasses the local government areas of Bass Coast, Cardinia, Casey, Frankston and Mornington Peninsula Shire, has a coastal climate with relatively mild temperatures and high rainfall compared to other parts of Victoria.

The region is comprised of a diverse economy and demography consisting of key residential hubs for metropolitan Melbourne, as well as thriving business and industrial sectors. The region's population is projected to grow by approximately 45% by 2031, presenting many opportunities for economic expansion and diversification, but also increasing the exposure of people, buildings and infrastructure to climate variability and change.

Temperatures in the Western Port region have risen by approximately 0.05°C to 0.15°C per decade since 1970, while annual rainfall has declined by approximately 50 mm per decade.

Climate models suggest temperatures in the region will increase by 0.5 to 1.1°C by 2030 and 0.9 to 3.5°C by 2070, while average annual rainfall will change by −4 to 0% by 2030 and −23 to 0% by 2070. Meanwhile, sea-level rise projections are uncertain, but the IPCC has estimated a range of 6 to 17 cm by 2030 and 15 to 49 cm by 2070, with the potential for additional sea level contributions from acceleration of glacial ice melt.

While these represent some of the projected changes in average climate conditions, climate extremes are projected to change as well, and it should be recognised that it is the changes in such extremes that are likely to pose the most significant threat to the region's environments, infrastructure and communities in the years ahead.

The Western Port region is significantly exposed to climate extremes and natural hazards such as storm surge and coastal inundation, floods, bushfires and extreme temperatures. These hazards are projected to increase in frequency and/or severity.

Increased frequency or severity of climate hazards are projected to have a broad range of direct impacts on urban settlements (both market and non market). These include impacts on land use and management, damages and maintenance costs to public and private property and infrastructure, human health and water availability. Increases in natural hazards will also have indirect and intangible consequences including disruptions to economic activity, the costs of emergency service provision, public amenity and quality of life.

Some of these impacts will be pervasive throughout the region; for others the extent of impact will be situation specific depending upon the spatial distribution of hazards and human assets, populations and infrastructure across the region and in local government areas.

Specific climate consequences for the region's settlements are discussed in more detail below.

Impacts associated with coastal inundation

Sea-level rise in future decades will undoubtedly affect the coastlines of the Western Port region and drive progressive erosion in many locations. The effects of sea-level rise will however be most pronounced during storm events. For example, storm surge inundation simulations for the region, undertaken by CSIRO for this assessment, suggest that a current 1 in 100 year storm surge could become a 1 in 1 to 1 in 4 year storm surge by 2070.

Furthermore, the land area subject to inundation during a 1 in a 100 year storm surge event may increase by 4 to 15% by 2030 and 16 to 63% by 2070.

The fact that only a narrow strip of land is exposed to coastal processes such as storm surge means that the exposure of associated property, populations and infrastructure is inherently constrained. Nevertheless, such inundation would impinge upon over 2,000 individuals, over 1,000 dwellings, and approximately $780 million in improved property value.

Public infrastructure is also at risk, including major thoroughfares such as the Nepean and South Gippsland Highways, and boating facilities. Beaches, foreshore reserves and coastal wetland areas throughout the region, as well as the amenities they provide, are likely to be affected as well. […]

In the absence of adaptation measures, the economic and social consequences of impacts to the region's beaches and foreshore areas could be substantial. These include disruptions to the region's tourism industry and a major loss of social, cultural and environmental amenity values.

Areas most at risk include townships on Phillip Island in Bass Coast Shire, coastal townships in the City of Casey including Tooradin and Warneet, and the township of Hastings in Mornington Peninsula Shire.

Impacts associated with intense rainfall and inland flooding

Flood events typically represent the most costly type of disaster in Australia, contributing to property damage and disruption of services and businesses as well as injury and death. At least 619 km^2 (18%) of the Western Port region lie in land areas subject to inundation or overland flow paths. This highlights the present risk the region faces in regard to flood hazard, typified by the Koo-Wee-Rup 'Super Flood' of 1934.

While significant advancements in flood protection have been made over the past century, future climate change poses an additional challenge.

Simulations of extreme rainfall in 2030 suggest increases of up to 25% in extreme rainfall from events of 1 to 24 hours in duration in at-risk areas of Western Port region. By 2070, extreme rainfall is projected to increase by up to 70%, depending on location. Such increases in extreme rainfall could drive increases in the frequency or magnitude of flood events or flood heights.

Flood mapping in the region is an ongoing process, with some flood prone areas still to be mapped. Based on current information though, an estimated 18,000 properties with a total capital improved value of almost $2 billion are vulnerable to flood events. Approximately 13,000 of the properties are residential, about 40% of which contain dwellings that are vulnerable to above-floor flooding.

Despite the total value of assets in harm's way, the potential implications of climate change for above-floor flooding suggests that increases in the magnitude of damages to residential and commercial properties from changing rainfall extremes may be relatively modest. However, there could well be a significant increase in the frequency of any given flood event and associated damage costs.

A range of Melbourne's and the State's major transport corridors also occur in at-risk areas including over 1,500 km of roads and 125 km of rail, and dozens of bridges. A range of businesses, industries, public services and utilities are also vulnerable to disruptions from flooding. Indirect economic costs associated with flooding of this infrastructure could be significant relative to direct damages.

Areas most at risk include much of southern Cardinia Shire (the Koo-Wee-Rup Swamp), southeast and central Casey, northwest Frankston and the Frankston Central Activity District (CAD).

Impacts associated with changes to fire weather conditions

Bushfires are a major economic, social and environmental hazard in Southeast Australia. The 1983 Ash Wednesday bushfires represents one of the worst bushfire events in Victoria's history, with communities in the present Cardinia Shire being amongst the worst affected.

Recent studies have confirmed a trend toward worsening fire weather conditions in southeast Australia in recent decades, and projections of a warmer and drier future climate translate into even greater risk in the future. Modelling of forest fire risk at locations in proximity to the Western Port region indicate the number of days of 'very high' or 'extreme' forest fire risk will increase by 1 to 2 days by 2030 and by 2 to 7 days by 2050, a potential increase of 60% or more.

At present, an estimated 710 km^2 (21%) of the Western Port region are in bushfire prone areas. Given the current distribution of property, people and infrastructure, over 73,000 individuals, approximately 35,000 properties (including 28,000 dwellings), with a capital improved value of $7.6 billion lie in at-risk areas.

A range of major transport corridors also occur in at-risk areas including over 1,600 km of roads and 75 km of rail, including the Nepean, South Gippsland, and Bass Highways as well as heavily travelled rail lines. Electricity transmission lines from the Latrobe Valley that traverse the Western Port region and provide electricity to Melbourne and Gippsland are also exposed to bushfire in multiple locations.

Overall, those areas most at risk include the Gurdies in Bass Coast Shire, much of northern Cardinia including the townships of Emerald, Cockatoo and Gembrook, central and southeast Frankston and bushland areas in Mornington Peninsula Shire, particularly around the townships of Dromana and Mornington as well as around HMAS Cerberus.

Impacts associated with changes to average and extreme temperatures

Extreme heat events represent one of the leading causes of climate-related mortality in the developed world, even in coastal locations that often benefit from milder climates than inland areas. With temperatures projected to rise from climate change, the increased incidence of extreme heat days and heat waves, in conjunction with a growing and aging population, is projected to contribute to significant mortality in future decades. These are likely to outweigh reductions in winter mortality in response to increases in temperatures.

At least one study suggests the heat-related annual death rate among people 65 and older in the Melbourne region would increase from 289 at present to 484 to 636 by 2100. On a proportional basis (but not allowing for local differences in climate and/or physiological differences between the population groups), the number of additional deaths in the region, among this group due to heat stress, could be approximately 30 to 53 annually by 2100.

While the occurrence of extreme temperature events may be similar throughout the Western Port region, different areas and populations may be more or less vulnerable based upon the age distribution of the population, type of housing and access to climate control.

Elderly people (>65 years of age) are likely to be the most vulnerable group in the community to extreme temperatures. Significant concentrations of elderly occur in Bass Coast and on the Port Phillip Bay coastline of Mornington Peninsula Shire.

Higher temperatures in both summer and winter will also affect energy use. Greater demands for air conditioning to maintain building thermal comfort, for example, is likely to drive increased electricity consumption during summer months. Meanwhile, energy demand during the winter may decline.

The economic implications of changes to energy use patterns may vary depending upon how electricity is priced – a characteristic that is likely to change in the future based upon the balance of supply and demand. While new buildings built to '5 Star' energy efficiency ratings may be able to manage temperature increases with little increased cost, these represent a small fraction of the region's housing stock at present.

Impacts associated with changes to average rainfall

Average rainfall in the Western Port region is projected to decline by up to 8% in 2030 and 23% by 2070, with reductions potentially coming in all seasons but especially in winter and spring. Drought frequency and intensity are projected to increase.

Streamflows in the Bunyip River and South Gippsland basins and in Melbourne Water's catchments, from where most of the region's water is sourced, are projected to decline, perhaps substantially. This has major implications for the region's water supply/demand balance.

The nature and level of regional economic and social impacts to changes in the water supply/demand balance are expected to depend on government policy response. It is likely though, that all water users in the region will face significantly higher water prices in the future.

Reduced average rainfall and reduced streamflow will also adversely impact on ecological, amenity and recreational values in the region. Values impacted are likely to include: waterway, wetland and coastal estuarine health; the viability and/or cost of maintaining domestic gardens and municipal gardens, parks and sporting fields; and streetscapes.

Some of the region's infrastructure exposed to drying soils will be vulnerable to increased degradation or structural failure.

Cross sectoral issues

Groups vulnerable to the impacts of climate change

Low income earners and the elderly are especially vulnerable to the impacts of climate change in the region.

Assessment of exposure of these groups to coastal inundation, flooding and bushfires suggests that low income earners are over represented in many of the localities exposed to coastal inundation but not so much in the localities exposed to floods and bushfires.

Elderly residents, on the other hand, are over represented in most localities exposed to coastal inundation and also in many of the localities exposed to floods and bushfires.

As previously noted, elderly residents throughout the region are particularly vulnerable to an increase in extreme temperatures.

Land use planning

Climate change in the Western Port region presents particular challenges for future land use planning and decision-making. In particular, climate change adds to the imperative for efficient use of this land and careful monitoring of growth in Casey-Cardinia corridor, since much of the land designated for growth is exposed to either flooding or bushfires.

Climate change also adds to the complexity of land use planning decisions associated with population growth and development along the region's coastline and in rural and semi-rural (Green Wedge) areas, since much of it will occur in areas exposed to coastal inundation, flooding or bushfires.

These challenges need to be considered in the context of the current planning and policy environment in Victoria including the proposed review of the Planning and Environment Act 1987.

Emergency management and volunteer support

Climate change poses significant challenges for emergency service planners and providers in the Western Port region including the SES, CFA and the Victorian Ambulance Service. Although those bodies already have comprehensive emergency management planning arrangements in place at the regional and sub-regional levels, plans may need to be reviewed in anticipation of more frequent or extreme weather events.

Local councils in the Western Port region also have an important role in emergency management and to that end have in place comprehensive municipal emergency management plans. However, the increased frequency and/or intensity of natural disasters due to climate change could add additional cost burdens on councils' emergency management role.

The cumulative effects of more frequent and/or more severe extreme weather events also poses significant challenges for resourcing of volunteer and community groups in the region, highlighting the importance of strengthening the capacity of community groups and volunteer organizations to respond to climate change.

Local government financial impacts

Generally speaking, climate change could have substantial implications for council finances and, particularly, the allocation of financial resources to various operational activities and services.

A preliminary review of capital and operating expenditure by local councils in the Western Port region indicates that approximately 19% of council's annual expenditure can be classified as being directly 'climate exposed', with this expenditure consisting of capital and maintenance expenditure on roads, drains, open space and buildings. In addition, there are a range of council expenditures that are indirectly exposed to climate events and climate change, including emergency management, health and aged care.

Opportunities from climate change

Climate change in the Western Port region is likely to create many opportunities, some region wide and some specific to particular industries or groups. Some of these opportunities will stem from responses to the impacts of climate change discussed in earlier chapters (e.g. improved housing design), while others will flow from favourable climate changes (e.g. tourism and recreation opportunities due to higher average temperatures and/or reduced rainfall).

Research needs

This report represents one preliminary step in a long-term process of adaptation among the communities of the Western Port region. Building upon this assessment to enhance specificity and clarify uncertainties in regional impacts will require the acquisition of more robust data sets, more detailed assessments of smaller subregions or a more constrained suite of climate changes and hazards.

In particular, research could be profitably targeted at specific at-risk locations in the region, for example, to conduct more rigorous hydrological modelling of extreme rainfall and flood risk as well as the effects of sea-level rise, storm surges and waves on both coastal erosion and inundation. Similarly, more robust spatial modelling of fire risk or vulnerability to extreme heat events could also be conducted.

To maximise the utility of such work for adaptive decision-making regarding climate change, such modelling should also seek to incorporate plausible scenarios of changes in population, land use, building stocks, economic activity and environmental management.

[...]

Peter Kinrade and Benjamin Preston, '*People, Property and Places: Impacts of Climate Change on Human Settlements in the Western Port Region,*' CSIRO, June 2008, pp. i-ix.

Chapter 29

Commonwealth of Australia

ENGAGING OUR PACIFIC NEIGHBOURS ON CLIMATE CHANGE
Australia's Approach

[...]

Projected climate change in the Pacific

The Fourth Assessment Report of the 2007 Intergovernmental Panel on Climate Change (IPCC AR4) Working Group II details future trends in climate and weather for small island states, including those in the Pacific. While this is the most comprehensive and widely accepted synthesis of climate change science for the region, climate change science is constantly evolving.

Under a business as usual scenario ('A1FI'), the IPCC AR4 projects that by 2100 global sea levels will rise by between 0.26 and 0.59m, with an additional 0.1 to 0.2m for ice sheet dynamics on a 1990 baseline. In March 2009, scientists at the International Scientific Congress on Climate Change in Copenhagen reported that projections based on current trends indicated that a sea level rise of 1m or more by 2100 could not be discounted. Limited scientific understanding of the dynamics of ice sheets in Antarctica and Greenland creates uncertainty in sea level rise projections.

Studies in the southwest Pacific suggest that the proportion of intense tropical cyclones may increase. Uncertainty remains about the trends in the frequency of cyclones in the region, closely related to uncertainty in El Niño Southern Oscillation (ENSO). The IPCC AR4 reports that precipitation from tropical cyclones is likely to increase in a warmer climate. The potential increases in peak wind speeds and the intensity of precipitation in tropical cyclones, coupled with sea level rise, could worsen the impacts of storm surges and flooding in the Pacific.

Key potential impacts from climate change

The geographical, social, institutional and economic characteristics of Pacific island countries make them susceptible to a range of climate change impacts, some of which are already being felt. National Adaptation Programmes of Action and initial National Communications prepared by Pacific island countries for the UNFCCC outline these impacts from individual countries' perspectives. Many potential impacts are common throughout the region, although their extent can differ.

- Storm surges, flooding and coastal erosion threaten coastal settlements and the transportation, water and sanitation infrastructure that supports them.
- Climate change may have severe economic impacts, including reduced income from agricultural exports, tourism and fisheries.
- Climate change could exacerbate existing water security challenges faced by countries with a dependency on rainfall, underdeveloped water infrastructure (including in urban areas), saltwater intrusion and rising water demands.
- Food security could be further threatened as traditional subsistence agriculture is undermined by extended droughts or loss of fertility due to increased rainfall and extreme events. Projected declines in the viability of fisheries—likely to arise from warmer oceans, ocean acidification and coral bleaching—may have significant implications for people's protein intake.
- Increasing temperatures could change the distribution of disease bearing vectors like mosquitoes, potentially exacerbating existing threats to human health, such as limited access to clean water and the restricted availability of public health services.
- The diverse and resource rich coastal systems of the Pacific are already under pressure and are expected to be further threatened by more severe weather events, and, in the longer term, sea level rise. It is very likely that projected future increases in sea surface temperature of about 1 to 3°C will result in more frequent coral bleaching events and widespread coral mortality, if corals cannot acclimatise or adapt. Disintegration of degraded reefs following bleaching or reduced calcification may exacerbate the impacts of storm surge in coastal regions.
- Regional biodiversity and vital ecosystem services — such as pollination and soil enrichment — could be degraded.

Temperature, rainfall and extreme event changes may affect the timing of reproduction in animals and plants; the migration of animals; the length of the growing season, species distributions and population sizes; the availability of food species; and the incidence of pest and disease outbreaks. Marine biodiversity may be damaged by flooding in wetlands, increasing acidification and declining coral ecosystem health.

The value of Pacific fisheries

Fisheries are a major source of food and income for Pacific island countries. Therefore, the health of freshwater, coastal and oceanic ecosystems and the long term sustainability of fisheries are key issues in the region. Coastal and oceanic fisheries are vital to all Pacific island countries. Freshwater resources are also particularly important in communities engaged in subsistence fishing and small scale pond culture in inland Papua New Guinea, Solomon Islands and other large islands of Melanesia, and they are locally important elsewhere in the region.

(Source: Commonwealth of Australia (2007) *Valuing Pacific Fish: A framework for fisheries related development*.)

Resilience

The people of the Pacific have a long history of resilience in the face of often hostile climatic conditions. Pacific island countries and communities are actively building their resilience to climate change through a combination of traditional and modern practices that reduce vulnerability to climatic extremes and variability. These include growing staple crops that are more resilient to climate variability (for example, yam and taro rather than cassava), diversifying crops, using forest food resources, and practising traditional food storage and preservation methods. In the past, social tenets governing marine resource use and indigenous local knowledge have proved effective tools for protecting marine ecosystems.

Resilience is enhanced by non-government organisations and church organisations that spread information and awareness about climate change, and implement concrete adaptation projects. Kinship networks and strong communities also underpin resilience by creating social safety nets providing and international remittances flowing back to Pacific communities, which can be used to support local adaptation actions. Regional knowledge-sharing can provide further gains, particularly in exchanging information about successful adaptation practices.

Mitigation

Australia and the Pacific share a strong interest in effective global action to reduce greenhouse gas emissions and stabilise atmospheric concentrations at levels sufficient to avoid dangerous climate change.

Australia is committed to playing a full, fair and constructive part in building global solutions to climate change, and, as a developed country, is committed to taking the lead in reducing emissions. Domestically, Australia is taking strong action to put emissions on track to meet our emission reduction goals. Australia does not expect the poorest and most vulnerable developing countries to take on stringent, unsupported mitigation actions. Australia is making a significant contribution to mitigation solutions globally, through our active participation in international negotiations and through a suite of bilateral, regional and multilateral partnerships.

Although Pacific island countries collectively account for less than one per cent of global emissions, they have expressed a desire to contribute to global mitigation efforts. Climate change responses should aim to facilitate opportunities for growth and innovation. Australia and the Pacific have the resources to exploit developments in clean and renewable energies and energy efficient technologies, which will play an important role in reducing emissions and enhancing energy security.

Climate change policy in the Pacific

A lasting and equitable response to the complex problem of climate change requires policy which is adapted to local and regional circumstances. A clear understanding of the issue at the grassroots level can provide the impetus for action.

The ability of Pacific island governments to address climate change issues—particularly adaptation—is often constrained by limited capacity and a focus on more immediate development challenges. This is recognised in the Pacific Islands Framework for Action on Climate Change

2006–2015. The Framework also recognises that there are opportunities to deal with climate change more effectively by incorporating it into broader decision making and national planning strategies. Experience shows that a 'silo' approach to climate change policy is ineffective. The problem must be tackled across all sectors and government portfolios, including infrastructure, fisheries, agriculture, energy, health, foreign affairs and trade. Recognising this, Australia is working to integrate climate change considerations across the development assistance program.

Better awareness and more reliable and accessible information are required at every level of society to achieve necessary policy outcomes. Communities, decision makers, the public sector and the private sector will all benefit from relevant, comprehensible and culturally appropriate information.

An ambitious global outcome on climate change and the potential for increased funding for adaptation in particularly vulnerable developing countries should focus higher level political attention on climate change. However, elevating climate change as a policy issue places additional strain on human resources. In the Pacific, climate change policy-making is resource intensive and climate change officers are obliged to 'wear many hats'. Decision makers across all portfolios and sectors may benefit from skills that enable them to identify barriers to mitigation and adaptation efforts and set achievable priorities for action in their own policy areas.

Adapting Pacific agriculture

A range of viable adaptation options are available to reduce the vulnerability of Pacific agriculture to climate change.

Practical farm level actions include the choice of crops and varieties, diversification, greater use of food preservation techniques, changes in planting dates and local irrigation. Potential national and regional adjustments include developing new cultivars and expanding irrigation systems on a large scale.

Many of these changes are simple and low cost but could reduce yield losses by at least 30–60% compared with no adaptation.

[…]

Climate change and displacement

The potential for climate change to displace people in the Pacific is increasingly gaining international attention. Australians are aware of and concerned about this issue. Sea level rise leading to inundation of low lying atolls is often viewed as the main driver of potential migration. But, particularly in the shorter term, the relationship between climate change and migration is more complex.

While climate change is a new and significant threat, it is the potential for climate change to exacerbate existing problems which poses the greatest immediate challenge. Water supply systems, already under pressure due to high population growth, will be further challenged by possible changes in rainfall patterns. Food security, which may be threatened by over fishing and stress on coral reef ecosystems, could be further affected as climate change impacts on fisheries and agricultural productivity. If such challenges become greater due to climate change, more people may consider leaving their homes.

The Niue Declaration on Climate Change, agreed by leaders of all Pacific Islands Forum countries in August 2008, recognises the desire of Pacific peoples to continue to live in their own countries, where possible. Australia is committed to assisting our neighbours in the Pacific realise their stated desire to stay in their countries by responding to the challenge of climate change.

Australia believes that the most effective way to reduce the likelihood of climate change-induced displacement is to reach a strong and effective global agreement to reduce greenhouse gas emissions. For this reason, Australia is committed to playing its full and fair role in strong and decisive action to avert dangerous climate change.

Given that some impacts of climate change are unavoidable and are already being felt by Pacific island countries, building communities' resilience to climate change impacts is vital. Sustainable development activities and measures directly aimed at climate change adaptation are vital to securing livelihoods and helping people have the choice to remain in their homes wherever possible. The Australian Government will continue these substantial efforts to simultaneously support sustainable development while building resilience to climate change.

While climate change may become an increasingly significant factor in decisions to move, the IPCC has noted that "the reasons for migration are often multiple and complex, and do not relate straightforwardly to climate variability and change". For the lifetime of this document, incremental climate change impacts are unlikely to cause widespread migration. However, there is always potential for natural disasters to cause sudden dislocating impacts to populations, requiring efforts to support rebuilding or relocation.

The Niue Declaration on Climate Change encourages the Pacific's Development Partners (including Australia) to increase their technical and financial support for climate change action, including on relocation if it becomes necessary. Building resilience to climate change is a key priority for Pacific island countries. In the longer term, the possibility remains that permanent migration could become an option for some Pacific islanders. In these circumstances, Australia will work in close consultation with the region to ensure that Pacific islanders' vital interests— economic, social and cultural—are paramount.

Australia has long helped the Pacific region deal with its challenges, and will continue to play its role in helping to find solutions.

[...]

Commonwealth of Australia, 'Engaging our Pacific Neighbours on Climate Change: Australia's Approach', 2009, pp. 9–12.

Chapter 30

Bob Sercombe and Anthony Albanese

OUR DROWNING NEIGHBOURS
Labor's Policy Discussion Paper on Climate Change in the Pacific

Executive Summary

This policy discussion paper proposes a Pacific Climate Change Strategy. The proposals it makes are informed by two key themes:

1. Australia must help our Pacific neighbours to meet the challenge of climate change – from assisting our neighbours to adapt to the effects of climate change now, to taking climate change refugees when countries are finally overcome by rising sea levels.
2. Australia must also do its part locally and globally to combat climate change – Australia should be a leader on climate change issues, both domestically in promoting clean energy, and globally through the Kyoto Protocol process. Unfortunately, the Howard Government's approach has badly sullied Australia's reputation in the Pacific and more broadly.

This discussion paper builds on Labor's first discussion paper on the Pacific, Towards a Pacific Community. Australia needs to establish a more genuine partnership with our Pacific neighbours if we are to establish a Pacific Community – and assisting our vulnerable Pacific neighbours to deal with the impact of climate change should be part of that effort.

A Pacific Climate Change Strategy should involve seven key elements:

1. A Pacific Climate Centre to ensure proper measurement and monitoring of the effects of climate change.
2. Assistance for mitigation, adaptation and emergency response efforts, such as protecting fresh water sources from salt water contamination, and dealing with infrastructure decay caused by coastal erosion.
3. Assistance with intra-country evacuations when citizens have to be moved from low-lying areas to higher ground.
4. Training to help the citizens of countries that have to be fully evacuated.
5. Establishing an international coalition to accept climate change refugees when a country becomes uninhabitable because of rising seas levels.
6. Assistance to preserve the cultural heritage of those who are evacuated.
7. Establishing a Pacific Climate Change Alliance to add greater momentum to global efforts to deal with climate change. But for Australia to credibly be part of such an alliance, it must ratify the Kyoto Protocol and commit to cutting greenhouse gas emissions.

Labor believes Australia can and should be a leader on climate change issues – locally, regionally and globally. This requires a long-term vision of what Australia and the region need to do to prepare for the impact of climate change. This discussion paper aims to set out such a vision.

Chapter One: Challenge

There is scientific consensus that human activity has increased the levels of greenhouse gases in the atmosphere and that this is making our climate change. As a result, the atmosphere is warming at a rate not seen for 10,000 years. 2005 was the hottest year ever recorded in Australia and the second hottest ever recorded globally.

Climate change is real and it is hurting the planet now. Small island countries in the Pacific are particularly vulnerable to climate change, through rising sea levels and extreme weather events.

Even if we were to stop all greenhouse gas emissions right now, Pacific island countries would still be vulnerable to climate change because it is already built into our atmospheric systems.

Our Pacific neighbours are facing a monumental challenge in climate change. This section examines the four key threats to our Pacific neighbours:

- rising sea levels;
- extreme weather events;
- collapsing ecosystems; and
- the contamination of fresh water with salt water.

Rising sea levels

The Pacific has some of the smallest, and lowest-lying, countries in the world. It has been predicted that climate change will lead to a rise in sea levels of 14–32cm by 2050.[1] However, a smaller rise would still have a devastating effect on many Pacific countries. Tuvalu faces the prospect of total inundation by rising sea levels, as do islands in Vanuatu, Kiribati, the Marshall Islands, the Federated States of Micronesia and Papua New Guinea. King tides are already flooding islands across the region.[2]

The SEAFRAME Project attempts to secure data on sea level change that is absolute (this means ignoring the role of land movement as part of sea level shifts). The early data sourced from this study shows a rise in sea levels of 5.9mm a year at the Tuvalu measuring station, 8.1mm a year at the Manus island station (in Papua New Guinea) and 15.5mm a year at the Tonga station, compared to the global average of a rise of 1–2mm a year.[3] This is an imminent catastrophe for atoll states such as Tuvalu. Some Pacific leaders are declaring that it is already 'too late' for their countries to be saved.[4]

Extreme weather events

The increasingly volatile weather patterns associated with climate change[5] are an immediate and rising threat to our Pacific neighbours. Deaths from weather related disasters have already increased in Oceania by 21% since the mid 1970s.[6]

Cyclone wind speeds are predicted to increase by 10–20% over the next few years because of their complex relationship with sea temperature.[7] The projected increase in the power of tropical storms is compounded by the increased volume of tropical storms that has occurred over the last 30 years.[8] We have recently seen the devastation that a Category 5 hurricane, such as Hurricane Katrina, wreaked on the United States. Many of our neighbours have to plan for the economic, environmental and social devastation of a natural disaster because of the increasingly violent cyclones and storm surges throughout the Pacific.

As the UN Intergovernmental Panel on Climate Change has stated, these extreme weather events pose a greater threat to atoll states, as storm surges will cause greater structural damage and have ongoing adverse effects.[9] Aside from the deaths, injuries and financial cost associated with storm surges, they cause widespread coastal erosion. Land space is already at a premium on the smaller Pacific islands, so they cannot sustain an ongoing loss of land.

Collapsing eco-systems

Pacific societies are highly dependent on their eco-systems, as their economies are often a mixture of formal exchange and subsistence practices. Unfortunately, Pacific island countries' eco-systems are particularly vulnerable to climate change.

One effect of climate change is the bleaching of reefs. Reefs are the foundation of atoll states, as they break the impact of violent weather on land. They protect against coastal erosion from severe weather and also generate the material that replenishes coastal areas. Reefs are a secure location for fish stocks to breed and feed. Tuvalu provides an example of this interdependence – its inner reef areas (along with the lagoon) provide most of Tuvalu's food.

Finally, reefs are a sink, or absorber, of carbon dioxide. However, increases in sea surface temperature have resulted in significant coral bleaching in the Pacific.[10] The increase in sea level temperature causes reef algae to be expelled or reduced, resulting in the death of the coral building animals.[11]

Reefs are vital to many Pacific island eco-systems – so their bleaching is a threat to the sustainability of their entire social-eco-systems.

Contamination of fresh water with salt water

Rising sea levels have the potential to make a number of our Pacific neighbours uninhabitable long before they are completely flooded.[12]

Rising sea levels, coupled with storm surges, contaminate fresh water with salt water. *The Sydney Morning Herald* recently profiled Tegua island, which is part of Vanuatu, and will shortly be evacuated because of the impact of climate change:

The biggest problem for Tegua islanders has been the lack of fresh water. There are no rivers or creeks, so they relied on two small freshwater springs. One has dried up and the other is covered by the sea. For the past decade or more, islanders have had to rely on rainwater they get for six months of the year...For the other six months there is only what water they have saved in small plastic containers.[13]

Citizens on the Carteret islands in Papua New Guinea are also currently being moved because 'their health has been steadily deteriorating because they are losing access to fresh water, and gardens are being destroyed by advancing salt water.[14]

The UN Intergovernmental Panel on Climate Change believes salt water contamination is one of the most potentially devastating symptoms of climate change. The contamination of groundwater is a significant threat to our Pacific neighbours, reducing agricultural production and the availability of drinking water.

It has been estimated that the fresh water supplies of some Pacific island countries could drop by up to 50%.[15] While this reduction obviously threatens the viability of low lying states, elevated states will also have to deal with the contamination of coastal agricultural land. According to the UN Intergovernmental Panel, many tropical root plants, such as taro, have a low tolerance for salt.[16]

The challenge – and why it matters for Australia

These are the four key threats – rising sea levels, extreme weather events, collapsing eco-systems and the contamination of fresh water with salt water – that our Pacific neighbours are facing. But there are others – climate change also poses significant public health risks. Pacific island countries will need to adapt to the increase in vector and water borne diseases (for example, malaria) caused by warmer temperatures. In Australia, for instance, the Australian Medical Association and the Australian Conservation Foundation have estimated temperature related deaths could double in Australia to 2,500 deaths per year by 2020.[17]

These threats represent a considerable challenge to individual countries, but they also represent a challenge to regional stability and security. Climate change has the potential to destroy development gains in these countries. According to a recent Pentagon report, climate change has the potential to destroy food systems and living conditions, and lead to considerable instability, disruption and conflict.[18]

This is why climate change in the Pacific is an issue for Australia for security reasons, not just for environmental or altruistic reasons. If Australia is committed to the stability and security of the Pacific as a precursor for its own security – this is, after all, why the Howard Government undertook the $1 billion Regional Assistance Mission to Solomon Islands (RAMSI) – it is essential that Australia adopt a proactive, strategic approach to climate change in the Pacific.

Australia also has international obligations under the UN Framework Convention on Climate Change (UNFCCC) to assist Pacific island countries and other developing nations adapt to the impact of climate change. Under the UNFCCC, all Parties are required to formulate and implement national and, where appropriate, regional mitigation and adaptation programs. Developed countries like Australia are required to assist developing countries 'that are particularly vulnerable to the adverse effects of climate change in meeting the costs of adaptation to those adverse effects.'

Chapter Two: A Pacific Climate Change Strategy

A Pacific Climate Change Strategy needs to assist our Pacific neighbours to deal with the current impacts of climate change, while also preparing for the long-term effects, which include the flooding of entire nations. Labor believes an effective Pacific Climate Change Strategy must include the following elements.

1. A Pacific Climate Centre

The Fijian Director of Meteorology has said that the 'current Pacific observing and monitoring systems are among the worst in the world.'[19] In Towards a Pacific Community, Labor proposed that a Pacific Environment and Resource Agency should be created, building on the current lead regional agency for environmental issues, the South Pacific Regional Environment Programme (SPREP). Given the threat climate change poses to Pacific island countries, and given the current dire state of Pacific monitoring systems, Labor believes a dedicated Pacific Climate Centre should be created as part of the Pacific Environment and Resource Agency.

The Centre should ensure proper measurement and monitoring of the effects of climate change, and model the impacts of climate change on Pacific island countries. The Centre would coordinate and build on the work of the South Pacific Sea Level and Climate Monitoring Project (SPSLCMP) which Australia created in 1991, the Pacific Island Global Climate Observing System (PI-GCOS), the South Pacific Applied Geoscience Commission (SOPAC), SPREP, the CSIRO and other international meteorology institutions.

The Centre should also assess the vulnerability of island countries to extreme weather events, and institute an early warning system for such events. The 2005 Indian Ocean tsunami demonstrates the devastation that can occur when early warning systems are not in place. Unfortunately the Pacific currently lacks such a comprehensive early warning system, and is reliant on the few national meteorological services in the region. For the sake of all Pacific Community members, Australia included, a warning system is needed that can coordinate information and provide a complete picture of weather patterns.

Extreme weather events pose a rising threat – the cost in human lives and infrastructure from lack of preparation is too great to ignore.

2. Assistance for mitigation, adaptation and emergency efforts

Australia should assist Pacific island countries with their efforts to mitigate, or adapt to, the effects of climate change, such as protecting fresh water sources from salt water contamination, adapting to land losses and dealing with infrastructure decay caused by coastal erosion.

Labor believes Australia should work with other donors to assist each Pacific country to develop a national adaptation strategy, by providing funding and expertise and by assisting with specific adaptation measures. Canada for instance has worked with some Pacific island countries on appropriate national responses. Facilitating local community consultation will also be vital – to ensure that communities fully understand the nature of the problem, and to collaborate with them in adaptation measures.

Consideration should also be given to developing a regional plan for climate change emergency responses. This could cover issues such as capacity building for Pacific island countries to respond to climate related events such as storm surges, as well as developing a regional capacity to assist Pacific island countries prepare for, and respond to, climate related natural disasters.

Australia's development assistance must also take account of climate change considerations. For example, it would not make sense to provide funds for an infrastructure project in an area which is vulnerable to climate change related events.

3. Assistance with intra-country evacuations

Some Pacific island countries will not have to be evacuated entirely, but their citizens will have to evacuate from one part of the country to another. Labor believes Australia should assist in these efforts as needed.

4. Training

Labor believes Australia should assist the citizens of Pacific island countries that have to be fully evacuated with training initiatives. Tuvalu, for example, is expected to become uninhabitable in the next decade. Australia should invest in vocational training in Tuvalu, so that:

- Tuvalu citizens can meet the skilled migration requirements in a variety of countries, should some choose to emigrate prior to the full evacuation of Tuvalu; and
- the remaining Tuvalu citizens can successfully adapt and find work in receiving countries when the country is fully evacuated.

The Tuvalu government has previously said that it wants to establish a Tuvalu Technical Education Centre, which would provide training in particular trades.[20] Labor believes Australia should work with other aid donors to support this initiative.

5. An inter3national coalition to accept environmental refugees

In time, it is likely that one or more Pacific island countries will have to be completely evacuated – that is, their citizens will not have the option of moving to a different part of the country because the whole country is being flooded, or has been made uninhabitable because of salt water contamination. On current projections, Tuvalu is likely to be the first country that is fully evacuated due to climate change, but ultimately Kiribati, the Marshall Islands and others may also have to confront this scenario.

Tuvalu has twice approached the Howard Government for assistance with climate change refugees. It has been refused both times.

Labor believes that Australia should, as part of an international coalition, do its fair share to accept climate change refugees as part of our humanitarian immigration program. Yet Australia needs to work with our Pacific neighbours to prepare for such contingencies now. Firstly, Australia should help to develop a coalition of Pacific Rim countries willing to accept climate change refugees. Secondly, Australia should be working at the UN to ensure appropriate recognition of climate change refugees in existing conventions, or through the establishment of a new convention on climate change refugees.[21]

6. Assistance to preserve cultural heritage

When islands, and indeed entire countries, have to be evacuated, it is important that their cultural heritage is not lost. The Pacific should engage in a collation and recording exercise that would assist in securing the cultural heritage of those peoples that lose their homelands.

In Towards a Pacific Community, Labor proposed the creation of a Pacific Cultural Centre to protect and promote the cultures of the Pacific. This Centre could assist with the documentation of the cultural practices, traditional technologies and history of Pacific island countries, and areas within countries, most under threat.

7. A Pacific Climate Change Alliance

Pacific island countries have led the world in calling for strong action on climate change. At international conference after conference, they provide the clarion call for urgent action. Obviously, they have the most to lose in the short term from climate change, but Australia also has a lot to lose from worsening climate change – as our recent drought has demonstrated. Australia is also vulnerable to the extreme weather events that batter our Pacific neighbours.

Unfortunately, Australia has not developed a shared agenda for tackling climate change with our Pacific neighbours. Australia is currently not invited to preparatory meetings with our Pacific neighbours prior to major international environmental conferences, such as the recently held UN Climate Change Conference in Montreal. Australia does not lobby collectively with Pacific island countries. There is no Pacific regional negotiating strategy which includes Australia.

This is a failure of leadership by the Howard Government. The failure goes back to 1997, when the Howard Government arrogantly insisted at the Pacific Islands Forum that the Forum Communiqué should not contain any references to climate change, despite the devastating effects on our Pacific neighbours.

The Howard Government's heavy-handed approach undermines the great potential of a genuine partnership between Australia and our Pacific neighbours, if we worked together to combat climate change. The development of cooperative relationships between all Pacific island countries and others in the broader Asia Pacific region would provide greater momentum to international climate change negotiations.

Labor believes Australia should work with our Pacific neighbours to establish a Pacific Climate Change Alliance. A Labor government has previously demonstrated the effect a committed coalition of countries can have on the international system. During the Uruguay Round of trade negotiations, the Australian-led Cairns Group succeeded in opening up agricultural markets for the first time (of course, there is still work to be done). The Pacific Climate Change Alliance should likewise lead the world in calling for effective global action on climate change. It could use the voices of Pacific Islanders to tell the story of how climate change is hurting now, and the dramatic impact it will have in the future. Such an alliance would provide an important voice for those nations that have the most to lose from climate change.

For Australia to credibly be part of such alliance, though, the Howard Government must ratify the Kyoto Protocol, commit to cutting greenhouse gas emissions and do more to provide economic incentives for the use of clean technology.

The potential impact of a Pacific Climate Change Alliance can be made clearer by considering two existing global and regional mechanisms for tackling climate change.

Firstly, an Adaptation Fund was established when the Kyoto Protocol came into force, to support the implementation of adaptation projects and programs (see the Appendix for further details). Unfortunately, because Australia has not ratified the Kyoto Protocol, it is not able to shape the rules and structure of the Adaptation Fund.

However, if Australia did ratify the Kyoto Protocol and also established a Pacific Climate Change Alliance, the Alliance could ensure the Fund played an important role in assisting Pacific island countries.

Secondly, the Alliance could endeavour to ensure the Asia Pacific Partnership on Clean Development and Climate also benefits the Pacific. The Partnership was recently established to exchange clean technology – a positive, but limited, approach to tackling climate change. Unfortunately, unlike the UNFCCC and the Kyoto Protocol, the Partnership does not extend to our Pacific neighbours, who are crying out for such technology. The Pacific Climate Change Alliance could work to ensure this particular weakness is corrected. A third possible area of action for the Alliance could be in lobbying for appropriate insurance arrangements to cover Pacific island countries affected by climate change. The Howard Government has dealt Australia out of the global debate over climate change. It has ignored the pleas of our Pacific neighbours for developed countries to act responsibly, and pursue environmental sustainability. Labor believes it's time to re-establish Australia as an agenda-setter on environmental issues through a Pacific Climate Change Alliance.

Chapter Three: Helping Our Neighbours

This policy discussion paper builds on Labor's first discussion paper on the Pacific, Towards a Pacific Community. It has proposed a Pacific Climate Change Strategy, to help our Pacific neighbours deal with the potentially devastating effects of climate change, now and into the future. Such a strategy should also be part of Australia's efforts to establish a more genuine partnership with our Pacific neighbours.

The proposals made in the paper have been guided by two themes:

• Australia must help our Pacific neighbours meet the challenge of climate change; and
• Australia must also do its part locally and globally to combat climate change.

Next steps

We will continue to consult with our Pacific neighbours about these proposals in coming months. A possible road-map for realising this Pacific Climate Change Strategy would involve the following steps:

• The Pacific Islands Forum could call a special Leaders meeting to discuss climate change, and the climate change issues facing each country
• Leaders could commit to the establishment of a Pacific Climate Centre
• Leaders could commit to the establishment of a Pacific Climate Change Alliance, to help set the agenda on climate change issues globally

- Australia could start working with international partners to develop an international coalition to accept climate change refugees from the Pacific
- Australia could work at the UN to ensure that climate change refugees are appropriately recognised in international conventions
- Australia must ratify the Kyoto Protocol, and commit to cutting its greenhouse gas emissions
- Australia could provide more support for clean energy projects, both domestically and through the aid program – in 2004–05, only $238,000 was devoted to renewable energy projects out of an aid budget of $2.25 billion.[22]

Labor believes Australia can and should be a leader on climate change issues, locally, regionally and globally. This requires a long-term vision of what Australia and the region need to do to prepare for the impact of climate change.

It's time to begin establishing a genuine partnership with our drowning Pacific neighbours to combat climate change.

It's time for Australia to be a leader on environmental issues once again.

[…]

Bob Sercombe MP Shadow Minister for Overseas Aid and Pacific Island Affairs and Anthony Albanese MP Shadow Minister for Environment and Heritage and Water, 'Our Drowning Neighbours: Labor's Policy Discussion Paper on Climate Change in the Pacific', 2006.

Notes

1   See J Barnett and N Adger, Climate Dangers and Atoll Countries, Tyndall Centre for Climate Change Research, Norwich (2001).
2   Ben Bohane, 'A village flees for safer ground', The Sydney Morning Herald, 23-25 December 2005, 9.
3   See R Jones, A Pittock and P Whetton, 'The Potential Impacts of Climate Change' in A Gillispie and W Burns (eds) Climate Change in the South Pacific: Impacts and Responses in Australia, New Zealand and Small Island States, Kluwer Academic Publishing, Dordrecht (2000) 7-32.
4   L Falcam et al, Statements by World Leaders at the World Summit on Sustainable Development 2002 Available at http://www.un.org/events/wssd/statements/.
5   See W Burns, 'The Impact of Climate Change on Pacific Island Developing Countries in the 21st Century' in A Gillispie and W Burns (eds) Climate Change in the South Pacific: Impacts and Responses in Australia, New Zealand and Small Island States, Kluwer Academic Publishing, Dordrecht (2000) 233–250.
6   See M Canisbee and A Sims, Environmental Refugees: The Case for Recognition, NEF Pocketbooks, London (2003).
7   See W Burns, 'The Impact of Climate Change on Pacific Island Developing Countries in the 21st Century' in A Gillispie and W Burns (eds) Climate Change in the South Pacific: Impacts and Responses in Australia, New Zealand and Small Island States, Kluwer Academic Publishing, Dordrecht (2000) 233-250, 6; M Canisbee and A Sims, Environmental Refugees: The Case for Recognition, NEF Pocketbooks, London (2003).
8   E Spouga, Statement by Ambassador Enele S. Sopoaga, Permanent Representative to the United Nations at the 60th Session of the General Assembly (Second Committee: Agenda Item 52: Sustainable Development) New York (2005).
9   See L Nurse and G. Sem, 'Small Island States', in IPCC (ed), Climate Change 2001: Impacts, Adaptation and Vulnerability (2001).

10   See J Barnett and N Adger, Climate Dangers and Atoll Countries, Tyndall Centre for Climate Change Research, Norwich (2001).

11   See W Burns, 'The Impact of Climate Change on Pacific Island Developing Countries in the 21st Century' in A Gillispie and W Burns (eds) Climate Change in the South Pacific: Impacts and Responses in Australia, New Zealand and Small Island States, Kluwer Academic Publishing, Dordrecht (2000) 233–250.

12   See L Nurse and G. Sem 'Small Island States' in IPCC (ed) Climate Change 2001: Impacts, Adaptation and Vulnerability (2001).

13   Ben Bohane, 'A village flees for safer ground', The Sydney Morning Herald, 23–25 December 2005, 9.

14   Ben Bohane, 'A village flees for safer ground', The Sydney Morning Herald, 23–25 December 2005, 9.

15   See W Burns, 'The Impact of Climate Change on Pacific Island Developing Countries in the 21st Century' in A Gillispie and W Burns (eds) Climate Change in the South Pacific: Impacts and Responses in Australia, New Zealand and Small Island States, Kluwer Academic Publishing, Dordrecht (2000) 233–250.

16   See L Nurse and G. Sem 'Small Island States' in IPCC (ed) Climate Change 2001: Impacts, Adaptation and Vulnerability (2001).

17   Australian Medical Association and Australian Conservation Foundation, Addressing the public health impacts of climate change (September 2005).

18   See Peter Schwarz and Doug Randall, An Abrupt Climate Change Scenario and its Implications for United States' National Security, Pentagon, Washington (October 2003).

19   'Fiji Weatherman raises climatic woes for the region', Fiji Times, 8 November 2005, 4.

20   See S Paape, Education For All: 2000 Assessment Tuvalu Report, UNESCO (2000). Available at http://www2.unesco.org/wef/countryreports/tuvalu/contents.html.

21   See T Collins, As Ranks of 'Environmental Refugees' Swell Worldwide, Calls Grow for Better Definition, Recognition and Support, in UN Day for Disaster Reduction, United Nations University: Institute for Environment and Human Security (2005); D Falstrom, 'Stemming the Flow of Environmental Displacement: Creating a Convention to Protect Persons and Preserve the Environment' (2001) 1 Colorado Journal of International Environmental Law and Policy 2–32.

22   AusAID, Answers to Questions on Notice (June 2005), 121.

Chapter 31

Katharine Murphy

WELCOMING PACIFIC MIGRANTS 'IN OUR SECURITY INTERESTS'

Australia needs to throw open its emissions trading scheme to neighbours in the Pacific — particularly Papua New Guinea and the Solomons — and welcome climate change refugees, a think tank says.

The Lowy Institute has urged the Rudd Government to be more active on a regional solution to climate change given the issue has now been acknowledged as a security threat by Australian and Pacific leaders.

In a paper prepared for a Senate inquiry into economic and security challenges in the Pacific, the institute says the Federal Government must, for national security reasons, acknowledge the vulnerable position of Kiribati and Tuvalu.

To ignore the threat would result in security and aid agencies having to deal with mass migration forced by rising sea levels, it says.

"Australia has to acknowledge that the only viable future for the people of low-lying atoll states, like Kiribati and Tuvalu, lies in migration," policy analysts Jenny Hayward-Jones and Fergus Hanson say.

"Given that Australia will be at the centre of plans to address the forced relocation of the populations of the atoll states, it would be in Australia's interests to develop a plan now to manage their migration."

The institute proposes a two-stage migration plan under which the Government would increase the number of scholarships available to students in Kiribati and Tuvalu for study in Australia. Scholarship students should be allowed to find full-time employment in Australia if they complete their education and then be eligible for permanent residency or fast-tracked family reunion visas.

"This approach would have the advantage of providing incentives for young people to study in Australia, encourage an ordered and voluntary rather than forced migration process from Kiribati and Tuvalu and ultimately lessen pressures on aid and on the welfare system in Australia," the institute says.

The Government's climate change adviser, Ross Garnaut, and environmentalist Tim Flannery are strong advocates of involving PNG in climate change efforts.

Professor Flannery supports a scheme in which Australians would be able to buy online credits to preserve PNG's endangered forests as part of an effort to sequester carbon.

Earlier this year, Prime Minister Kevin Rudd and his PNG counterpart, Sir Michael Somare, signed the PNG-Australia Forest Carbon Partnership as a first step towards a joint effort on global warming.

The institute endorses this but says the Government should also allow PNG and the Solomon Islands to take part in the emissions trading scheme.

Katharine Murphy, 'Welcoming Pacific Migrants "In our Security Interests"', *The Age,* 30 December 2008.

Kiribati

Chapter 32

Ministry of Environment, Land, and Agricultural Development, Kiribati

REPUBLIC OF KIRIBATI NATIONAL ADAPTATION PROGRAM OF ACTION (NAPA)

[...]

Preface

Kiribati is one of the most vulnerable countries to the adverse impacts of climate change. The atolls of Kiribati rise 3–4 metres above mean sea level and are an average of a few hundred metres wide. These atolls are the home of nearly 90,000 Kiribati people with their distinct culture. Inundation and erosion destroy key areas of land, and storm surges contaminate the fresh groundwater lens which is vital for survival. An economic evaluation of the costs of climate change related risks has been estimated to be 35% of Kiribati GDP. The estimate takes into account only the potential impacts of climate change on coastal zone (US$7–$13 million a year) and water resources (US$1–$3 million a year). In 1998 the GDP was US$47 million (WB, 2000).

The United Nations Framework Convention on Climate Change entered into force as an international agreement on March 1994 and entered into force specifically for Kiribati in May 1995. The Convention sets out blueprints for the common but differentiated responsibilities of parties to address the global concern of climate change. The cause of climate change is largely blamed on past emissions of greenhouse gases by industrialized countries, and these countries have acknowledged their leadership role in responding to climate change. In line with this leadership role, industrialized countries are assisting adaptation measures in countries that are considered most vulnerable to adverse impacts of climate change.

Least developed countries (LDCs) and Small Island Developing States are among the countries that are considered most vulnerable to climate change. They are so, because in the case of the former their special circumstances make them unable to meet the costs of adaptation, and the latter because of their physical susceptibility to the effects of climate change. Kiribati is in both of these groups.

National Adaptation Programmes of Action (NAPA) is an approach to enable LDCs to communicate their immediate and urgent needs for adaptation to the Conference of the Parties.

The process involved in the development of the NAPA is designed to ensure the principles of stakeholder participation, country driven-ness, multidisciplinary input, complementarity to other projects, and cost effectiveness.

Concurrent with the NAPA, which has been implemented through the United Nations Development Programme, is the Kiribati Adaptation Project (KAP), initiated by the World Bank. The Kiribati government is equally committed to both projects. Separate management and accountability for each project is maintained whilst coordination, collaboration, and harmonizing of activities of the projects has been achieved. This has been done by adopting a single structure for overseeing adaptation activities in the form of a multidisciplinary technical team reporting to a steering committee consisting of senior level government and NGO officials.

The experience of implementing the two projects on adaptation gave rise to the need to have a policy statement and a strategy on adaptation. This policy statement stresses that Kiribati needs to be prepared for adaptation, piloting small scale adaptation projects, and collecting data useful for designing adaptation measures that achieve climate proofing aims. NAPA is consistent with this policy and intends to make visible adaptation efforts through undertaking work on upgrading and protection of essential physical assets that are increasingly being exposed to risks of climate change impacts, particularly from droughts, storm surges, storm variability and sea level rise.

The process of the NAPA preparation included national consultations, CCST's work, and some external review on technical information. Kiribati is particularly grateful for very constructive comments from the LEG following their review of the NAPA Draft at our request.

These comments guided further work on the NAPA Draft which on completion enabled the CCST and the NASC to present to Cabinet in September 2006 a final version. This version, as it is, was endorsed by Cabinet at its meeting on the 10th January 2007. This document seeks financial assistance to follow through with the firm foundation for action established by the NAPA process.

[…]

Executive Summary

This document sets out a 3 year plan for urgent and immediate actions in the Republic of Kiribati to begin work in adapting to climate change. These actions forming the project profiles are developed through the NAPA process.

Kiribati is situated in the Central Pacific Ocean and consists of 33 atolls with a total land area of about 800 sq km. The atolls have a maximum height of 3 to 4 m above mean sea level and support an estimated population in 2005 of about 95,000 people. Most people live a subsistence lifestyle.

In 1979 Kiribati gained independence and is now a democratic republic state under its own constitution. Kiribati is politically stable.

In 2000 only 9200 people of 84,494 were employed – about 66% in finance services and public administration sectors. Per capita GDP is quite low by international standard, and there is a concern about financial sustainability given that the gap between import and export continues to widen.

Kiribati has a National Development Strategy for 2004–2007, ministry operational plans for each ministry, and a single year budget. Adaptation activities will be managed within the framework created by these documents. The Kiribati government has also approved a Climate Change Adaptation Policy and Strategy. All of these documents have been considered in preparing this National Adaptation Programmes of Action (NAPA).

This project, the NAPA, has been implemented concurrently with the Kiribati Adaptation Project (KAP I). Both projects share the same co-coordinating bodies. The Kiribati Adaptation Project focuses on long term planning for adaptation while the NAPA focuses on urgent and immediate needs.

Environmental issues and problems that have been identified as part of this project include:

- emerging "unacceptable level of inequity" (2004–2007 NDS p.28);
- increasing population;
- deteriorating states of coastal zones, coral reefs, fisheries, fresh ground water, health and biodiversity;
- inadequate urban services such as water supply and sanitation;
- overexploitation of natural resources in urban Tarawa; and
- difficultly in enforcing land use management strategies and controls.

With warmer temperatures, sea level rise, increased storm surges, climate variability and the increase of associated adverse effects such as erosion, past adaptation practices in Kiribati are no longer found to be effective.

[…]

2.2 Settlement, land and coastal area

Land in the traditionally inhabited atolls of Kiribati is privately owned, in many cases by multiple owners of clear lineal connection, but registered under the most senior among them.

Traditional settlement areas or "kaainga" that were originally scattered anywhere along the island were relocated at existing village sites that now spread out along one edge of the island, usually at a sheltered side from the prevailing easterly winds. These village sites were formally instituted in the early 1900s and each site consisted of a few smaller settlement areas. Households living in any of these smaller settlement areas are generally descended from a common ancestor. The villages remain significant units of the whole island community.

Most likely climate change affects the processes of coastal erosion and accretion and these in turn threaten the village institution. A few cases of relocation of part of the village have occurred, with implications on the uses and sometimes conflicting claims over resettled land.

Both processes – erosion and accretion – are not new observations to the people. What is new is the observation that the traditional methods of checking erosion appear to no longer be effective as coastal land erosion becomes more extensive, intensive, and persistent. Erosion is an expected impact of sea level rise but difficult at this time with short data series to prove that sea level has in fact risen and caused the observed extensive erosion. The process of accretion is observed in the deposition of sediments to parts of the beach, or in sand bars formed on the lagoon platform. The processes of erosion and accretion have more serious impacts in urban

South Tarawa where seawall protection, land reclamation, accreted land, uprooted coconut trees by shoreline erosion, dilapidated buildings that are undermined through erosion, and sand mining form mixed features of the shoreline.

Erosion threatens existing roads and buildings, which requires them to be protected. For public assets this is done by the Government, and for private assets it is done by their owners. The protection is in the form of sea defence of various designs and constructions. None of the sea defences have been totally effective and remnants of seawalls can be seen around South Tarawa and outer islands. Existing causeways and seawalls need upgrading and strengthening if they are to last any longer. New causeways and seawalls, however, continue to be constructed. It was estimated in 1999 that coastal structures covered 45% of Betio shoreline in South Tarawa (GOK,1999). For the whole of South Tarawa, the total number of constructed seawalls was 60 in 2005.

Storm surges or extra high spring tide have caused flooding of residential areas. Traditional houses have raised floors, and this design has proved appropriate in times of flooding. Where flooding leads to erosion, or when persistent, people have to relocate themselves or retreat. Few incidents of this form of adaptation have occurred.

More specific to South Tarawa, increasing urbanization are key drivers to changes to the shoreline. Other problems are squatters, unclear rights between landowners and land leased by the Government, enforcement of multilayered and sector requirements of the planning system, and relaxing of tenants conditions for Kiribati Housing Corporation houses. These are the challenges of development, but they act too as exacerbating factors to the vulnerability and barriers to the adaptation of South Tarawa to the impacts of climate change. Conversely adaptation projects will benefit the problems that arise from urbanization and other socio-development driving forces.

As people are responsible for much of the causes of the deteriorating urban environment, they are also the very resources to effect the change to sustainable urban environment. This change can come from increased public awareness of climate change impacts and rationales of any guidelines on coastal zone uses. This awareness can hopefully influence individuals in how they pursue their developmental lifestyles whilst at the same time avoid causing irreversible adverse effects on the environment-ecosystems and their functions.

[...]

Chapter 4. Identification of key adaptation needs

4.1 Past adaptation practices

Adaptation to climate change adverse impacts is a new experience for Kiribati. In the past extreme weather conditions such as droughts, storms and storm surges, rainfall flooding, and sea water flooding as well as spread of fatal diseases have occasionally occurred but people have lived and in dignity through them. Past practices to cope with these extreme weather conditions included the preservation of traditionally prepared foods from crops that were regarded of great social values. This conventional lifestyle may have dissipated partly because of increasing integration of Kiribati into the world economy with corresponding erosion of traditional values.

These socio-economic facts exacerbate the vulnerability of the people to extreme weather events because the change in their social values does not encourage preparation based on natural resources for these events, but these resources were what they had depended on in the past. At the same time, people are more exposed to economic vulnerability.

There had been no traditional way to ameliorate the problem of water getting brackish during drought. When the closest well within the village became brackish, people would turn to the known next closest well of lesser salinity, usually at further distance away. Whilst climate change scenarios project higher precipitation on multi-year basis, seasonal variability is unknown and droughts will still be expected in the climate change related weather pattern. Inter annual variability of rainfall is unknown but La Niña as part of the ENSO will continue to bring dry conditions to Kiribati. Furthermore, sea level will increase and erosion is anticipated as a consequence, thus the risk of reduced groundwater lens exists. As climate change impacts progress, there might not be ground water lenses remaining as they are now. All could turn in time to be very brackish. The now traditional coping strategy for drought that involves getting water from the next nearest source of fresh ground water will increasingly become inadequate under climate change.

Storms and storm surges can occur for several weeks, but are particularly strong for a few days. During such times, people might camp out at sheltered side of the island, and houses in the village had to be propped up with extra timber supports. A cooking shed usually had a supply of firewood that was preserved for this occasion. Preserved fish such as sun dried salted fish, or dried shark meat and skin were likely to be the only protein food. These traditional ways including the type of housing reduce the costs of vulnerability to storm and storm surges impacts. They could remain adequate but in the case of housing the demands on local building materials can increase; and with climate change that may lead to greater stress on trees and plants, local housing could deteriorate in quality. Firewood abundance could decrease as well.

More residential houses and other buildings are of more permanent imported materials. For these buildings, there is no traditional adaptation for them and they just have to meet the costs of damage from storms and storm surges should damage be occasioned. These forms of economic development bring out the need for new forms of adaptation such as in the form of coastal zone management and land use management. Hard structural options protecting them from the shoreline erosion has been going on for while but none has been totally effective and all require upkeep.

Timber seawalls and seawalls of stones built perpendicular to the shoreline, particularly the first, were traditional ways of protecting houses and house plots within the village from storm surges and from coastal erosion. These are no longer effective for that purpose. There are now more complaints about sea water over-wash getting into houses, wells, and bwabwai pits, and coastal erosion occurring unchecked.

The traditional purpose for which preserved fish were valued may have lost its significance for the same reasons that preserved agricultural products have lost their significance. This fact could only increase the vulnerability to climate change. Increasing storminess, sea temperature can result in decreased fisheries productivity. This means impoverishment of the life of the people even if people can afford to buy tinned fish.

Strengthening traditional values and practices in particular cases can therefore be considered as adaptation measures. Economic development contribute to national adaptive capacity if at the same time environmental impacts of related activities to the economic development do not result in a worse state of the environment than what it had been. For example, from income that flows from economic development people will be able to live on imported food during storms, build permanent housing structures and protect them from storm surges or coastal erosion. If however this economic development is based on excessive removal of sand and aggregates from the shoreline then it might leave the atoll environment worse off for providing particular services and functions that support the livelihood of the people. Environmental legislation can provide guidelines to screen and ensure that development activities and traditional practices are not barriers to adaptation.

Certain parts of any atolls are clearly more vulnerable than others. A village area is considered to be more vulnerable due to the risk of inundation of human settlement than the lower land outside the village. However, the latter will be more vulnerable for agriculture purposes. Limited resources only allow a few of the competing vulnerable sites to get attention for adaptation. In the small area of the atoll, this situation adversely affects the livelihood of the people.

Rainfall pattern and trend are expected to change, whilst temperature and sea level are expected to increase. People claim that these changes are already being experienced. In addition, they claim that tree crops productivities have changed. There are limits to the extent to which traditional adaptation practices can cope with these impacts.

Through the National Consultations, technical reports provided under KAPI, and the Initial National Communication, a broad range of coping strategies were developed. These strategies formed the basis from which the Climate Change Study Team has identified adaptation projects that are proposed in this NAPA document. More detailed information on the Consultations and prioritization processes are given in section 5.

Climate variability, climate change related hazards and risks, and impacts on the livelihood of the people, are readily recognized as requiring immediate responses, even without taking into consideration their long term impacts.

Resources such as water resources, coastal structures, roads and other physical structures, land space, agricultural resources, and terrestrial and marine resources are at risk. Communities who are living with these risks are not equipped to solve the problems they experience as they do not have the financial and technical capabilities to do so. A UNDP project executed by the Ministry of Internal and Social Affairs may assist in building up the technical capabilities of the communities for project planning and management, including good governance at the local government level.

Governance and services will include institutional arrangements for managing and mainstreaming adaptation strategy, and improvement and upgrading of the Kiribati weather observing system. A National Strategic Risk Management Unit is being established and through the NASC and CCST will coordinate adaptation activities. The NEPO will assume extra responsibility for the integration of adaptation activities into the national planning and budget management processes. This is to be achieved through NAP A activities proposed for the NEPO.

The ECD has acquired technical information and knowledge of local and global environmental issues and linkages through implementation of MEAs. NAPA proposals will support and strengthen ECD capability to implement the MEAs. All levels of governing and decision making play important roles in the undertaking of adaptation.

Population size and growth rates, as pointed out during the National Consultations, have significant impacts on the state of the environment, aggravating vulnerability and adaptation needs. In this respect, population policy is an important consideration of adaptation strategies.

External financial and technical assistance make international cooperation for adaptation possible, and facilitate technical capacity building of ministries in addressing climate change.

Ministries will implement NAPA project activities that fall within their respective responsibilities. They will follow the established budget management system, the MOPs mechanism, and the monitoring and reporting mechanisms. In addition, MELAD in close collaboration with the OB will provide oversight through the NASC and the CCST of all the NAPA project profiles. The NEPO will additionally facilitate mainstreaming and monitoring of NAPA project profiles.

[…]

Environment and Conservation Division, Ministry of Environment, Land, and Agricultural Development, Government of Kiribati, 'Republic of Kiribati National Adaptation Program of Action (NAPA)', *Tarawa*, January 2007, pp. iii-iv, 1, 10–12 and 27–30.

Chapter 33

Jane McAdam and Maryanne Loughry

WE AREN'T REFUGEES

Over the past decade a new term has entered the lexicon of policy makers and the media: climate change refugees. Human movement caused by environmental factors – drought, land degradation or significant climatic events (like cyclones) – is not new; what is new is the number of people now thought to be susceptible to such pressures. In a recent report, *The Anatomy of a Silent Crisis*, Kofi Annan describes millions of people suffering – and ultimately being uprooted or permanently on the move – because of climate change. According to the UN High Commissioner for Refugees, it is becoming difficult to categorise displaced people because of the combined impacts of conflict, the environment and economic pressures.

The spectre of climate change has also focused attention on the small Pacific nations of Kiribati and Tuvalu. Straddling the equator, these former British colonies – once known as the Gilbert and Ellice Islands – gained their independence only three decades ago. In recent years they have featured in media reports as "sinking islands" that will be uninhabitable by the middle of this century, with their people becoming the world's first "climate refugees." Kiribati has a population of around 100,000, while Tuvalu, with only 10,000 people, is the world's smallest state apart from the Vatican.

As researchers with an academic interest in forced migration, we recently visited Kiribati and Tuvalu. As we drove along the main road on the central Kiribati atoll of Tarawa, with the lagoon on one side and the ocean on the other, the sense of vulnerability was palpable – and that feeling is magnified when there is an extreme event like a cyclone or king tide. But are the apocalyptic projections about these two nations, and their people, accurate?

While they face very similar challenges, Tuvalu and Kiribati are not identical in their approach to the pressures generated by climate change. But in both countries we found a wholesale rejection of the "refugee" label, at both the political and community levels. For the people of these small Pacific nations, the term refugee evokes a sense of helplessness and a lack of dignity that contradicts their very strong sense of pride.

It's important to remember that people at risk of environmental or climate displacement are not "refugees" as a matter of law. A refugee is someone who has a well-founded fear of persecution on account of their race, religion, nationality, political opinion or membership of a particular social group, who is outside their country of origin and whose government is unable or unwilling to protect them.

Even as a merely descriptive term, the refugee label is at best pre-emptive, and at worst offensive, for those to whom it is applied. In both Kiribati and Tuvalu people resoundingly reject it. At first, we could not grasp why, despite its legal inaccuracy. As scholars of forced migration, we view refugees as very resilient, capable people who have overcome adverse situations to secure protection for themselves and their families. Many refugees make a significant contribution to their new societies, including some who have achieved fame in their fields of endeavour: politician Madeleine Albright, Czech writer Milan Kundera, scholar Edward Said and artist Judy Cassab, for example.

But in these small Pacific countries, which are not themselves signatories to the Refugee Convention and whose languages do not even have a word for refugee, the term has many negative associations. We wondered whether this was partly a response to the draconian policies implemented under the Howard government, particularly the Pacific Solution. (Tuvalu, for example, was approached as a site for an offshore detention centre, but declined.) But the reaction seems to run far deeper than that. Interestingly, it highlights some of the central failures of the international system for providing protection to refugees and other displaced people, most notably the failure of many countries to provide assistance and accommodation, leaving millions of refugees in camps or leading precarious lives in bordering countries.

In part, the discomfort with the refugee label stems from the fact that, by definition, refugees flee their own government, whereas the people of Kiribati and Tuvalu have no desire to escape from their countries. (The exceptionally low rate of visa overstayers from Kiribati in Australia suggests that these are people who like being at home.) They say it is the actions of other countries that will ultimately force their movement, not the actions of their own leaders. (These countries have some of the lowest emissions rates in the world.) There are also fears that the refugee label connotes victimhood, passivity, and a lack of agency. In Tuvalu and Kiribati refugees are viewed as people waiting helplessly in camps, relying on handouts, with no prospects for the future. To be a refugee, in their eyes, is to lack dignity. Some men told us that for them, being in such a situation would signal a failure on their part to provide for and protect their family. Tuvaluans and i-Kiribati people don't want to be seen like that – they want to be viewed as active, valued members of a community.

For this reason, the president of Kiribati is trying to secure options for labour migration to Australia and New Zealand so that those who want to move have an early opportunity to do so, and can gradually build up i-Kiribati communities abroad. This long-term strategy would see gradual, transitional resettlement so that, if and when the whole population has to relocate, communities and extended family networks would already be functioning in their new home countries. The president hopes that in this way, i-Kiribati culture and traditions will be kept alive, and that his people will also be able to adapt to new cultures and ways of life.

Nevertheless, the government of Kiribati also recognises that migration schemes will eventually need to be accompanied by humanitarian options. It is keen to secure international agreements in which other governments recognise that climate change has contributed to Kiribati's predicament and acknowledge that they are obliged to assist with relocation. The government of Tuvalu, on the other hand, does not want relocation to feature in international agreements. It fears that industrialised countries may simply think that they can solve problems like rising sea levels by relocating affected populations rather than reducing carbon emissions, which would not bode well for the world as a whole.

Interestingly, despite the rejection of the refugee label, governments in both Tuvalu and Kiribati recognise that certain principles of refugee law might have a place in addressing questions about legal obligation, and that some features of the refugee protection regime are pertinent to the present problem. One government official in Kiribati conceded that a framework akin to that of the Refugee Convention, minus the refugee tag, would be welcomed.

Both governments have at various times sought relocation options in other countries, most notably Australia and New Zealand. Just this month it was reported that Indonesia had offered to rent out islands to countries affected by climate change. And the Maldives, another island state, has talked about buying up land in India or Sri Lanka. But there is much more to relocation than simply securing safe territory; those who move need to know that they can leave and re-enter the new country, enjoy work rights and health rights there, have access to social security if necessary, and be able to maintain their culture and traditions. They also want to know the citizenship status of children born there. At present, there is little in international law to prevent a host country from expelling people should it wish to do so. If en masse relocation to another country is being considered as a permanent solution, then the situation will be far more complex than a standard immigration plan. In the absence of territory formally being ceded to the affected country, there needs to be security of title and immigration and citizenship rights for those who move.

There is also a real question about what role climate change plays in this scenario. Neither of us is a climate change skeptic, but we recognise that movement from these countries involves multiple pressures that may affect individuals and families in different ways at different times. As one government official in Kiribati observed, climate change overlays pre-existing pressures – overcrowding, unemployment, environmental factors and economic development – which means that it may trigger a tipping point that would not have been reached in its absence.

The view is somewhat different in Tuvalu. There, some government officials worry that highlighting the complex and multifaceted dimensions of movement will shift the focus away from climate change, and that the magnitude of that problem warrants maintaining attention solely on its impacts. Yet we were intrigued to hear that whereas ten years ago any community meeting related to climate change would draw a large crowd, interest has died down in recent years. Perhaps this was because the doomsday scenarios advanced a decade ago have not eventuated. While people could describe recent changes to the environment, weather patterns and local resources – changes they attributed to climate change – they also felt that they could adapt to them over time. We were also confronted with the belief that God had promised Noah that there would be no more floods and God could be trusted to keep this promise today.

Certainly the impact of climate change on low-lying atolls must not be underestimated. But there is a risk that solely focusing on climate change obscures other social changes that together provide a more realistic explanation of why people act in particular ways, including whether they stay on their atolls or seek to migrate elsewhere. Such an approach acknowledges the role of climate change in exacerbating existing social, economic and environmental pressures.

This is not to suggest that complacency is an appropriate response. Rather, it is important to work with the people of Kiribati and Tuvalu to develop sustainable solutions that enable those who wish to remain to do so for as long as possible, and that facilitate the gradual movement of families and communities to help ease the transition for movement over time. While raising the profile of these countries and their future is crucial, so too is garnering international support, creating sustainable and harmonised solutions, and listening to the voices of those concerned so that migration can become part of a broader adaptation response.

Talk of sinking islands and climate change refugees may stimulate media interest and the popular imagination, but it is an oversimplification of the reality and suggests a situation of movement (or preparation for movement) which simply does not exist. The process itself is likely to be far less dramatic than the Atlantis-style predictions. People are not fleeing, but they are reluctantly recognising that at some point in the future their home may no longer be able to sustain them. Many have family or friends living abroad, and we quickly learnt that mobility is, as in many Pacific countries, a historical pattern that continues to this day.

Jane McAdam and Maryanne Loughry, 'We Aren't Refugees ', *Inside Story,* 30 June 2009.

New Zealand

Chapter 34

Ministry of the Environment, New Zealand

ADAPTING TO SEA-LEVEL RISE

Observations of sea-level rise

Global average temperatures have warmed by over 0.7°C over the last 100 years, as reported in the Intergovernmental Panel on Climate Change's Fourth Assessment Report on the science of climate change. The ocean is absorbing 80 per cent of the heat added to the climate system and average ocean temperatures have increased down to a depth of 3000m. This warming is causing an expansion of ocean water which, in combination with water from the melting of land-based ice, is causing sea-levels to rise. Consequently, sea-levels in New Zealand rose by 17 centimetres last century and they have risen on average 1.8 mm/year over the last 40 years.

During 1961–1993 global average sea level has risen at an average rate of 1.8 mm per year and during 1993–2003 average sea level rise was approximately 3.1 mm per year. It is still not clear if this is a temporary increase in the rate of sea-level rise or if it reflects a change to significantly higher trends.

The impacts of climate change on our coast

Much of New Zealand's urban development and infrastructure is located in coastal areas, making it vulnerable to coastal hazards such as coastal erosion, inundation and sea-level rise.

The following changes are likely to occur as a result of climate change:

- coastal defences are overtopped by waves or high tides more often
- severe storms increase in intensity and storm surge levels rise
- sandy beaches, like in the Manawa, continue to accrete, but more slowly
- gravel beaches, like Haumoana in Hawke's Bay, are more likely to erode
- waves in Wellington could be 15 per cent higher by 2050 and 30 per cent higher by 2100
- areas with smaller tidal ranges, like Wellington, the Cook Strait area and the East Coast, may have bigger problems with the high tide mark exceeded more often.

Guidance on planning for future sea-level rise in New Zealand

It is important that we start planning for future sea-level rise now. Planning decisions on infrastructure made by councils and engineers often have consequences that extend over 50–100 years or more. Sea-level may rise significantly in this time.

The Ministry for the Environment provides a range of guidance materials and publications to help identify climate change impacts and adaptation responses. The publication "Preparing for Coastal Change" provides an overview of the impacts that climate change is expected to have on coastal hazards, not only through sea-level rise but also through storm surge, wind and waves.

The Ministry recommends planning for the following projection of future sea-level rise:

- For planning and decision timeframes out to 2090–2099, a base value sea-level rise of 0.5m relative to the 1980–1999 average be used along with an assessment of potential consequences from a range of possible higher sea-level rise values. At the very least, all assessments should consider the consequences of a mean sea-level rise of at least 0.8m relative to the 1980–1999 average.
- For planning and decision timeframes beyond the end of this century an additional allowance of 10mm per year be used.

Central government response to planning for sea-level rise and increasing coastal risk

The New Zealand Coastal Policy Statement 1994 provides further direction on planning for development in the coastal zone. The Department of Conservation is currently reviewing the New Zealand Coastal Policy Statement as required under the Resource Management Act.

A proposed national environmental standard on future sea-level rise is also being investigated. Such a standard would:

- reduce the time and cost to local authorities for selecting a projection to plan for
- reduce the risk of litigation.

Local government response to planning for sea-level rise and increasing coastal risk

Most of local government has already started to plan for sea-level rise. Many have completed coastal hazard assessments, with maps showing areas expected to be affected over the next 50–100 years. Other activities being undertaken include:

- restricting development in coastal erosion areas
- planning for managed retreat
- rejecting consents for alterations or extensions to existing buildings in the coastal zone
- discouraging the construction of defences such as sea walls.

Ministry of the Environment, New Zealand, 'Adapting to Sea-Level Rise', 14 December 2009.

Papua New Guinea (Carteret Islands)

Chapter 35

Ursula Rakova

HOW-TO GUIDE FOR ENVIRONMENTAL REFUGEES

For some time now, Carteret Islanders have made eye-catching headlines: "Going, going … Papua New Guinea atoll sinking fast". Academics have dubbed us amongst the world's first "environmental refugees" and journalists put us on the "frontline of climate change."

So perhaps you have heard how we build sea walls and plant mangroves, only to see our land and homes washed away by storm surges and high tides. Maybe you can even recognise the tragic irony in the fact that the Carterets people have lived simply (without cars or electricity) — subsisting mainly on fish, bananas and vegetables — and have therefore not had much of a "carbon footprint".

You might know that encroaching salt water has contaminated our fresh water wells and turned our vegetable plots into swampy breeding grounds for malaria-carrying mosquitos. Taro, the staple food crop, no longer grows on the atoll. Carteret Islanders now face severe food shortages, with government aid coming by boat two or three times a year.

However, the story you have not likely read is the one of government failure and the strategy we developed in response, so as to engineer our own exile from a drowning traditional homeland.

Carterets people are facing, and will continue to face, many challenges as we relocate from our ancestral grounds. However, our plan is one in which we remain as independent and self-sufficient as possible. We wish to maintain our cultural identity and live sustainably wherever we are.

While we call on the Papua New Guinea government to develop policy, we are not sitting by. Instead, we now want to see the media headlines translate into practical assistance for our relocation program. And we hope our carefully designed and community-led action plan can serve as a model for communities elsewhere that will be affected by climate change in the future.

Six islands become seven

Situated 86 km northeast of Bougainville, the main island in the autonomous region of which the Carterets form part, our atoll is only 1.2 meters above sea level. They say evacuation of the islands was inevitable as for many, many years erosion has been doing its work.

"King tides", or particularly high tides, are now doing worse. Originally the Carterets were six islands, but Huene was split in half by the sea and so now there are seven. In 1995 a wave ate away most of the shorelines of Piul and Huene islands. Han island, as shown in the video-brief, has suffered from complete inundation.

What climate change's exact role is, even experts are hard put to answer. Debate has raged over whether the islands are sinking, if tectonic plates play a role, and whether sea levels are in fact rising.

We do not know much about science, but we watch helplessly as the tides wash away our shores year in and year out. We also know that we are losing our cultural heritage just as the sea relentlessly wipes out our food gardens.

We do not need labels but action.

Association and resettlement

To relieve the land shortage caused by eroding shorelines, in 1984 the government resettled 10 families from the Carterets to Bougainville, but they returned to the atoll in 1989 in flight from what began as a protest by landowners against a mining company and escalated into civil war. Since that time, and despite many promises, very little has been done by the Bougainville or PNG government to assist Islanders' relocation efforts.

Tired of empty promises, the Carterets Council of Elders formed a non-profit association in late 2006 to organise the voluntary relocation of most of the Carterets population of 3,300.

The association was named Tulele Peisa, which means "sailing the waves on our own". This name choice reflects the elders' desire to see Carteret Islanders remain strong and self-reliant, not becoming dependent on food handouts for their survival.

After much hard work, the first five fathers moved to Tinputz in April, onto land donated by the Catholic Church. These fathers are already building gardens so that their wives and children can join them later when there is food.

"I have volunteered to relocate as I would like my family to be able to plant food crops like taro, banana, casava, yams and other vegetables that we cannot grow on the island," said Charles Tsibi. "I also want my family to grow some cash crops like cocoa to sustain our future life here in Marau, Tinputz."

According to a recent Tulele Peisa survey, 80 other families would like to move immediately and 50 wish to move later on. Twenty families have already relocated on their own. Thirty families remain unsure about relocating.

Our plan

Tulele Peisa's plan is for Carteret Islanders to be voluntarily relocated to three locations on Bougainville (Tinputz, Tearouki and Mabiri) over the next 10 years. Our immediate need is for funding so that we can accomplish the initial 3-year phase of our Carterets Integrated Relocation Programme (download the paper for details).

The list of objectives is long and challenging but our plan is holistic so we have faith it will succeed.

Firstly, the three host towns have a population of 10,000 and we are cognisant of the many complexities involved in integrating the Carterets people into existing communities that are geographically, culturally, politically and socially different. Therefore exchange programs involving chiefs, women and youth from host communities and the Carterets are in progress for establishing relationships and understanding.

While this is going well, the next urgent steps include securing more land and surveying and pegging site boundaries. Next comes constructing housing and infrastructure for 120 families. With the help of the Catholic Church in Bougainville, the relocation programme aims to provide design and carpentry services and local materials for basic housing for these families.

We also need to get on with implementing agricultural and income generation projects (like the rehabilitation of cocoa and coconut blocks), as well as education, health and community development training programmes.

"The plan is slow to achieve but covers all areas dealing with human relations and has adaptation alternatives, such as small cash income activities for relocated families," said elder Tony Tologina, chief of the Naboin clan.

On the long term, we want to build the capacity of Tulele Peisa to be certain it can carry out its objectives and also develop it as a resource agency for the Carterets and host communities on Bougainville.

"Tulele Peisa is our own initiative and will continue to co-ordinate and facilitate the relocation of our island people. After the relocation, TP will continue to provide monitoring and evaluation skills and further focus on development options available to our people," said Rufina Moi, woman chief.

Forever linked to our reefs

An important part of the programme is that it will also set up a Conservation and Marine Management Area that will let Carteret Islanders make sustainable use of our ancestral marine resources.

To keep the links between the relocated Carterets people and their home island, sea resources and any remaining clan members (who are not yet relocated), the plan includes developing an equitable sea transport service for freight and passengers.

"In the future, we will keep coming to these reefs and manage them as our fishing ground," explained community youth leader Nicholas Hakata. "When our children come back, they will have a connection to their heritage."

Documenting a model

"We have fully documented our process since beginning our plan and will continue to, for the sake of developing a model relocation programme," said Thomas Bikta, a chief from Piul Island.

"At the same time, we are developing and formulating a Carterets relocation policy that we will advocate to the Autonomous Bougainville Government and the rest of the world," Bikta added.

We also intend to build an alliance of vulnerable Pacific communities impacted by climate change who can lobby and advocate for justice and policies that recognise and support those affected.

We think the Papua New Guinea government must set an example of such policies by re-developing the Atolls Integrated Development policy and beginning a recognized financing mechanism similar to REDD (Reducing Emissions from Deforestation and Forest Degradation in Developing Countries). The committee or board must include all relevant stakeholders, including community representation and other expertise, not just government officials.

Ursula Rakova, 'How-to Guide for Environmental Refugees', *Our World*, 16 June 2009.

Tuvalu

Chapter 36

Ministry of Natural Resources, Environment, Agriculture and Lands, Tuvalu

TUVALU'S NATIONAL ADAPTATION PROGRAMME OF ACTION (NAPA)

[...]

Executive Summary

Tuvalu is one of the most vocal countries in the world at the international arena for a solution to the global issue of climate change and how it will affect low-lying countries like Tuvalu. Dependence on natural resources, inadequate infrastructure and human resources, low economic base and social development, and lack of institutional capacity make Tuvalu more vulnerable to adverse impacts of climate change, variability and extreme events.

The National Adaptation Programme of Action (NAPA) for Tuvalu, prepared initially under the Office of the Prime Minister (OPM), and completed under the Ministry of Natural Resources and Environment, is the Government of Tuvalu's response to COP 7 decisions. It is also an opportunity for Tuvalu to communicate its priority activities to address urgent and immediate needs of Falekaupule and stakeholders in Tuvalu for adaptation to adverse effects of climate change. The basic approach for the Tuvalu NAPA preparation is to be in line with the development aspirations of the government of Tuvalu as stipulated in "Te Kakeega II National Strategy for Sustainable Development 2005–2015". This is framed around the Millennium Development Goals (MDGs), the national sustainable development goals embodied in the Malefatuga Declaration, sector plans, other multilateral environmental agreements, the challenges Tuvalu is facing at present, and those that the nation will face in the coming future. Since climate change will directly impact Tuvaluan communities, families and individuals, it is important that different stakeholders at every level of society are engaged as part of the NAPA preparation process in the selection of adaptation measures and ranking of project activities.

The enumerated population of Tuvalu is 9,359 (Census 2002). The average population growth for the resident population is 0.6% from 1991–2002. About 42% of the enumerated population resides on Funafuti – the capital and only urban center of Tuvalu. It also has the highest population density of 1,610 as compared to the outer islands population density of 222 people per km^2. Internal migration is high due to increasing changes in lifestyle and dependence on imported foods.

Stakeholders pointed out during the NAPA stakeholders' consultation that coastal erosion is a major problem; and for some families, lands have been lost as a consequence. Flooding, inundation and salinity intrusion especially into pulaka pits, shortage of potable water, destruction to primary sources of food and increasing frequency of natural disasters are other problems attributed to climate change, variability and extreme events. However, the problems are not limited to this list. Flooding and inundation provide suitable medium for vector breeding, and salinity intrusion enhanced by the porosity of soil in Tuvalu destroys pulaka crops and decreases fruit trees' yields of coconut, banana and breadfruit – a major concern to food security. Increasing numbers of low rainfall days, prolonged droughts, high extreme temperature and evaporation are major problems for the agriculture and water sector, especially for the densely populated areas (Funafuti) and islands closer to the equator (northern islands). The frequency of extreme events like cyclones, storms and surges are increasing and exacerbating climate risks.

It is envisaged that the above adverse effects of climate change, variability and extreme events noted by stakeholders will be disparaging to the development of Tuvalu, unless they are effectively addressed. The most damaging effects of climate change are tropical cyclones, coastal erosion, salinity intrusion and drought. These have been noted to affect crops, fruit trees and human livelihood. The current challenges (as listed below) that stakeholders are facing at present are exacerbated by climate change:

1. Coastal erosion, saltwater intrusion and increasing vector and water borne diseases due to sea level rise;
2. Inadequate potable water due to less rainfall and prolonged droughts;
3. Pulaka pit salinisation due to saltwater intrusion; and
4. Decreasing fisheries population.

The severity of the impacts of climate change on communities is identical on all islands since sources of staple food and village locations are similar. Furthermore, sources of subsistence food production (agriculture and fisheries) are both severely impacted by climate change, variability and extreme events. Therefore, food security will be at risk in the future.

Adaptation measures are required to enhance community livelihood and promote sustainable development by reducing adverse effects of climate change, variability and extreme events. These adaptation measures are selected from stakeholders' and sector expert suggestions, and must be based on the ability of stakeholders and sectors concerned to easily implement adaptation measures to limit adverse circumstances of climate change. Some adaptation measures have already been undertaken in Tuvalu at community level on some islands such as coastal protection and increasing household water storage facilities. These have shown some successes on some islands and failures on others. Future climate change challenges are complex. Therefore, suggested approaches and technologies acceptable to Falekaupule and communities concerned are mostly required.

The NAPA understands that adaptation measures will relatively reduce severity of adverse impacts of climate change, but it will not absolutely solve existing problems.

[…]

The NAPA conducted a countrywide public consultation with stakeholders on all the nine Island Communities of Tuvalu to identify stakeholders' urgent and immediate adaptation needs through:

Identification of existing problems observed by stakeholders due to climate change, variability, sea level rise and extreme events; Identification of local coping strategies to existing problems; and Articulation of key adaptation needs based on ideas from stakeholders and sectoral experts.

Information from the synthesis of the Tuvalu vulnerability and adaptation report, initial national communication, including other relevant national reports was pooled to identify critical sectors in Tuvalu impacted by climate change. These critical sectors are also the sectors that the communities and stakeholders suggested are already impacted by climate change. The impacts on these sectors are likely to increase in the future. NAPA observation and stakeholders' suggested damages in these critical sectors, on respective islands, further strengthen the need for adaptation intervention. Some of the most common vulnerabilities stated by stakeholders are listed below:

1. Coastal erosion and loss of family land is evident on all islands of Tuvalu. Since the majority (more than 90%) of the communities live close to the coast, including important religious infrastructures, coastal erosion is therefore a priority stakeholders' urgent need. Sea level rise, overland flooding, storm surges, tropical cyclones and major hurricanes are the main causes of coastal erosion, including also that destruction on coastal coconut tree plantations. However, anthropogenic causes such as building aggregates excavation and coastal development activities are not exempted. Overland erosion due to heavy rainfall also resulted in sedimentation in central and coastal areas, affecting coastal and lagoon fisheries.

2. Flooding and inundation in February 2006 became the worst ever on Funafuti. It resulted in the evacuation of some families including the call from the Funafuti community for compensation for damages to family pulaka pits due to saltwater intrusion. Other islands are also experiencing flooding and inundation in new areas of their lands. Furthermore, this problem will be exacerbated by climate hazards such as tropical cyclones and storm surges.

3. Growth in population increases public demands for potable water. Since the main source of potable water for Tuvalu comes from rainwater, the quantity, quality and accessibility to water resources are very important. Vulnerability for the water sector is caused by the lack of household water storage facilities and changes to rainfall patterns due to climate change and variability. Water shortage enhances skin diseases and other health problems.

4. Destruction to primary sources for Tuvalu subsistence such as terrestrial crops and coastal fisheries is of major concern to the islanders. With respect to crops, the current increasing occurrence of new diseases and pests, including fruit-fly infestation, is attributed to climate change and variability. Furthermore, tropical cyclones, storm surges and coastal flooding destroys coastal coconut plantations. On the other hand, coastal fisheries are affected by the sea surface temperature changes, human intrusion and increasing frequency of extreme events.

5. Climate Hazards such as tropical cyclones, storm surges, droughts, and fires results in damages to individual and community assets. Coastal infrastructures such as harbours, church buildings, cooperative shopping centers, clinics and dispensaries, tar sealed roads (Funafuti only) and household properties are all exposed to the destructive forces of extreme events. Past experiences of a storm surge in the early 1990s totally dismantled the Vaitupu multi-million dollar harbour. A fire in 2000 completely obliterated a girls' dormitory with the loss of sixteen lives at Motufoua secondary school. El Niño-driven drought of 1997 and declara-

tion of the state of emergencies for two of the Northern Islands (Nanumaga and Niutao), and later declared for Funafuti resulted in the importation of desalination plants.

The participation of stakeholders and communities is an integral part of the Tuvalu NAPA development process in identifying adverse impacts of climate change on vulnerable sectors; the selection of potential adaptation measures and ranking of the projects to be included in the NAPA. Gender balance was also encouraged throughout the NAPA development process.
[…]

2.2 Key Environmental Stresses
[…]
2.2.1 Soils and Coastal Areas

Coastal erosion is severe and predominantly active on western coastlines of the islands of Tuvalu. On some islands, several important infrastructures on the edge of severely eroded areas, urgently need attention. Erosion on one side results in accretion on other parts of the island coastline. However, in some areas such as Tepukasavilivili on Funafuti, a total loss without accretion of eroded sediments: loss of land resources and agricultural lands has been witnessed (INC 1999). Limited land resources make many terrestrial and near-shore resources vulnerable to overexploitation, and discrete dumping of wastes on land. Saltwater intrusion and water logging due to climate change cause the deterioration of chemical and biological properties of soils, and that has rapidly decreased productivity of agricultural lands and pulaka pits. Incidences of saltwater intrusion and water logging have increased over time, and in combination with aridity of the soil, make soil parameters attractive to non-food, salt-tolerant shrubs and trees.

2.2.2. Water Resources

Tuvalu is poorly gifted, with no surface water. Therefore, it is currently dependent on rainwater as its main source collected from iron roofs of houses, and stored in concrete cisterns or tanks. In the past, the people also tapped the groundwater resources for household use. But, groundwater resources have been polluted by saltwater intrusion and waste leachate. Therefore, no longer suitable for human consumption. Water resources availability is a challenge that is exacerbated by climate change, resulting in frequent water shortages. Consequently, desalination plants were introduced into the country in 1999, to relieve public water demands during the water crisis, caused by the 1999/2000 El Niño.

In general, water resources are not centralized because rainwater is collected and controlled by private households, thus restricting its use. Funafuti, with the highest population density, water scarcity is a common problem: not only during the dry season (Jun–Sept), but also occasionally during the wet season (Oct–Mar). A water resource survey conducted on Funafuti in 2006 concludes that household water storage facilities are insufficient in meeting household water demand and needs. The government of Tuvalu has ensured that major water reserves are constructed in the basement of major buildings such as the New Princess Margaret Hospital and the Government Central Office. Although these are designed as water reserves including desalinated water, the water collected is being daily trucked to meet public water demand.

[...]

3.3.7.5 Sea Level Rise

Tuvalu's land area is becoming smaller. In recent years, the country has lost a lot of land around the circumference of the largest atoll due to erosion. At its widest point, Tuvalu only spans about 200 meters, so any rise in sea level is of concern to the people. Tuvalu installed a tidal gauge in 1993 to monitor sea level, pressure, temperature, wind speed and direction.[4] From 1993–1999, the average sea level rise was 22 mm per year, rather confounded by a drop of 36.6 mm in 1997–1998 due to El Niño and La Niña effects.

3.3.7.6 Coastal Erosion

The islands of Tuvalu are geologically young, having poorly developed, infertile, sandy and coralline soils. The atolls are dynamic and are subject to continued erosion and deposition, some of this occurring over long periods, and some occurring rapidly as a result of major storms. For example, Cyclone Bebe, which struck Funafuti in 1972, produced a rubble and rock rampart nearly 18 km long on the eastern ocean side of the island of Funafuti. Typically, the lagoon side is eroded while the ocean side builds up, meaning that care needs to be taken in locating buildings and infrastructure on the more popular and attractive lagoon side.

Increased extreme storm events, rising sea levels, and more intensive land use along the coastal zone combine to make Tuvalu more vulnerable to coastal erosion. Previously constructed sea walls have not been adequately maintained and may also contribute to coastal erosion in areas not protected by sea walls. Sea walls cut off the landward supply of sand during storm events, resulting in waves attacking unprotected areas to a greater extent than they would have done prior to the sea wall construction. Soft structures, which absorb wave energy and tree planting are more suitable for erosion control. For example, the Tuvalu Council of Women implemented a NZAID-funded tree planting project and the NGO's plant-a-tree project on Niutao with local species like Calophyllum inophyllum and Pandanus species, which could be used as a green-belt model.

In 1984, a beach profile and bathymetric survey was carried out by SOPAC to assist in the identification of problem areas along the shoreline and provide estimates of seasonal sediment transport. In addition, other studies were also carried out on Vaitupu and Nukulaelae (1993), Fongafale (1995), Amatuku (1996) and Nukufetau (1996) to address coastal erosion, sand transport and sedimentation problems in order to advance coastal management on the islands.

3.3.8 Overview of Stakeholders' Observations of Climate Change Impacts

The national stakeholder consultative consultation was undertaken in three phases especially to raise public awareness on climate change and to record stakeholders' perceptions of the adverse impacts of climate change, and most importantly, to record the communities' and stakeholders' urgent and immediate needs on their respective islands. The consultation convened members from different organizations on the island and people from different forms of livelihoods. The consultation also requested island Falekaupule's where women's voices are not heard in decision making to allow women's views during discussions for gender balance purposes. Stakeholders pointed out a lot of sectoral changes that they had observed and these changes could be attributed to climate change, variability and sea level rise. Despite the fact that stakeholders' understanding of the climate change issue is limited, stakeholders could identify direct impacts of climate change on their livelihood which need urgent action: such as changes in rainfall patterns resulting in frequent water shortages, saltwater intrusion resulting in salinization of groundwater and pulaka pits, decrease in easy access to coral reef fisheries resources due to lower coral fisheries population, and low agricultural yields due to increasing fruit destruction by fruit flies. The stakeholders also did not forget traditional knowledge of resource conservation that has been practiced in the past but rarely practiced at present. Prevalence and frequency of tropical cyclones, storms and drought was reported by stakeholders to be increasing on all the islands. However, they could not measure the enormity of the adverse effects of future climate change on their livelihood. The notion from stakeholders on all the islands is that climate change is already occurring in Tuvalu. Climate change adverse effects are increasing for the worse, and our collective ability to give our children a livable Tuvalu is decreasing over the coming years.

3.3.8.1 Historical Information

Historical information is generally passed down from father to son. However, some strategic information related to traditional knowledge is strictly passed down in the family line. The NAPA questioned the elders on some of the impacts of climate change and most importantly inundation and flooding of low-lying lands by up-welled saline waters. The Falekaupule reported that now there is an increasing percentage of land flooded due to inundation, something that has never occurred in the past.

Furthermore, the local church foundation of the first Mission school in Tuvalu that was build on Amatuku, Funafuti in 1904, and currently the duty officer's hut for the Tuvalu Maritime Training Institute is more than 30 cm below saline water level. It is unbelievable to think that the ancient builders of Tuvalu will build a house foundation that is submerged under the water level. Similarly, more and more pulaka pits are being permanently destroyed by saltwater intrusion. Despite the fact that overtopping waves has in the past destroyed pulaka pits; they usually recover after a few months. Currently, saltwater intrusion has destroyed some pulaka pits permanently on the islands.

Fisherman also reported that coral reef fish are usually caught with ease due to the high abundance of reef-fish population. But now, more time and effort is needed to catch a fish. Some islands have already reported that bivalves were not replicating and may lead to extirpation. In addition, ocean fish like tuna rarely come closer to shore recently, and therefore; fishermen go farther out to catch tuna, thus, causing fish price inflation.

3.3.8.2 Impacts of Climate change on Livelihood

The disparity on the degree of climate change impacts that the people will experience on the islands will depend on the island's location, economic condition and population size. It is also important to know the degree of climate change impacts on different sectors as these affect the lives and livelihoods of the people. Stakeholders pointed out during the consultations the observed adverse impacts of climate change on sectors such as agriculture, fisheries, health and water, which are the basic necessities of their livelihood.

3.4 Relationships of the NAPA Framework and the Kakeega II

In 2005, Tuvalu finalized it national strategy for sustainable development 2005–2015 known as "Te Kakeega II". It reflects the views of all stakeholders from all walks of life expressed during a National Summit. The vision for the Kakeega II recognizes the importance of sustainable development, and not compromising the ability of future generations of Tuvalu to meet their needs. Each of the sectoral strategic development priorities of the Kakeega II will contribute to the achievement of its vision, which is: "To achieve a healthier, more educated, peaceful and prosperous Tuvalu". The NAPA activities will ensure compatibility with the Kakeega II development priorities and other plans such as the National Action Plan on Desertification and Land Degradation for the United Nations Convention to Combat Desertification (UNCCD), and should also be mainstreamed into other on-going programmes to enhance synergy and cost effectiveness of programmes at sectoral level. It is also important as a way forward, to improve the way climate change is considered in national decision making processes.

The NAPA consultation pulled together a lot of valuable information for communities to develop informed decisions on adverse impacts of climate change. However, communities should also consider how they could better adapt to climate change impacts. Adopting a "risk management" policy development approach could be an ideal solution.

[…]

3.5 Adaptation Measures, Urgency and Immediacy for Tuvalu

Stakeholders' adaptation needs were identified and assessed based on the degree of adverse effects of climate change on Tuvalu including stakeholders' expertise in determining urgency and immediacy of adaptation needs. Adverse effects of climate change on ecological and human systems including the frequency and intensity of natural disasters were also considered. It is envisioned that without adaptation, most terrible adverse impacts are expected. Furthermore, there is solid evidence that prevention pays.

3.6 Complementaritys of NAPA with the Kakeega II and other MEAs

The NAPA has developed adaptation measures as activities to address the immediate and urgent needs of stakeholders, complementary to the national vision of the Kakeega II as well as other multilateral environmental agreements. The Kakeega II has identified some of the major environmental risk issues such as sea level rise, rising population on Funafuti, decline in traditional resources management and unsustainable use of natural resources to name a few. All these risks will be enhanced by a warmer climate. Minimizing climate change impacts through implementation of adaptation measures identified under the NAPA will effectively address these risks. Furthermore, the integration of climate change adaptation into sectoral policies, programmes and development projects is vital. Enhancing this process is through increasing the awareness of stakeholders at all levels of society.

3.6.1 Potential Barriers and Constraints to Implementation

The barriers to implementation of adaptation measures to climate change will be discussed in three levels, national awareness, integration of climate change impacts into national development plans and sectoral policy objectives, and implementation of adaptation measures. One of Tuvalu's great strengths is the ability of individuals, families, communities, islands and the nation to work together cooperatively.

3.6.1.1 Lack of National Awareness

The climate change issue is new to Tuvalu. The awareness is also low, at all levels, from national level policy makers down to the Falekaupule and civil society. It is important that all key groups and stakeholders understand the climate change issue, and how it will adversely affect their livelihood. The NAPA national consultation also included a brief stakeholders' awareness on the agenda. An awareness booklet in the local vernacular, to raise awareness of the people of Tuvalu at all levels of the civil society including primary schools, has also been completed. However, awareness on new issues of climate change should also be strengthened.

3.6.1.2 Lack of Climate Change Impacts Integration into Policies

Tuvalu has no environmental law in place that will guide the appropriate treatment and protection of the environment, or to control the degradation of the environment. It is important that climate change impacts are incorporated into national development plans, especially plans and programmes for the most climate-sensitive sectors such as: water, coastal zone, agriculture, disaster, etc. Although, the current Kakeega II has included some environmental priorities and strategies on climate change and impacts, yet, the actual implementation of these policies and strategies is still in the distance, therefore, this process needs to be enhanced.

3.6.1.3 Implementation of Adaptation Measures

Implementation of adaptation measures requires appropriate tools, knowledge and methodologies to guide the stakeholders to make informed decisions. Therefore, due to the lack of appropriate tools, knowledge and methodologies both at the technical and grassroots levels little adaptation actions will be initiated at the local level. In addition, religious misconception of the sea level rise issue is a major issue that the churches need to look into and clarify. Building the capacity and knowledge to assess climate change impacts, including appropriate tools and methodologies to identify adaptation activities needed, may lead to important adaptation activities in the near future.

4.0 Identification of Key Adaptation Needs

4.1 Past and Current Practices for Adaptation to Climate Change and Variability

The cooperative assistance of individuals and families helped to lower the number of deaths during the 1972 destructive Hurricane Bebe on Funafuti. Over the past years the Government and Non-Governmental Organizations have implemented activities and projects to minimize adverse effects of climate change such as: development of communities disaster plan, plant-a-tree programme, transplanting local fruit trees between the islands, and community water tanks to name a few.

The Government assisted the construction of the seawall on Nanumaga, with community participation through the provision of labour, to minimized intrusion of saline water from the sea into the pulaka pit. However, the seawall has deteriorated and the problem has returned. The government recently assisted victims whose homes were severely inundated with saline water during the king tides of February 2006. This is also the first time that the new Disaster plan was put into action for the evacuation of king tide victims. A community disaster response plan will be useful in local communities.

Intrusion of saline water has also increased groundwater salinity thus exacerbating water problems in highly populated areas such as Funafuti. Additionally, insufficient family water storage facilities almost resulted in a water crisis and third declaration of the state of emergency on Funafuti in May 2006. Even though the government and NGOs are active on providing tanks for household and community, water conservation technologies and awareness raising are limited, though vitally important for conserving water.

Shellfish is a delicacy and a high economic earner for families. The NAPA consultation found that Tridacnidae species populations are not reproducing but decreasing at an alarming rate. The government recently undertook a coral survey and NGOs replants the corals and disseminate the technology to the islands. It was also found that people understood the destructive impacts on climate sensitive sectors were linked to climate change.

4.2 Future Potential Adaptation Strategies

Potential Adaptation Strategies (PAS) have been identified from existing practices highlighted by stakeholders during the NAPA consultation and expertise of personnel from sectors concerned. The primary goal and objectives of potential adaptation strategies are to reduce adverse impacts of climate change, variability and extreme events, thus, enabling ecological and human systems to adapt accordingly.

Some initial activities on adaptation strategies for coastal erosion such as seawalls have been successful on some islands but not on others. Stakeholders suggested that for coastal erosion, it is best to establish coastal current breakers (constructed from concrete) to decrease coastal current speed and to be installed between islets of atoll islands. On the agricultural sector, transplanting local crops between islands to increase biodiversity and food security on all islands, the NAPA is mindful not to introduce the coconut pest (currently present on two islands) into the remaining seven pest-free islands. Therefore, it is vital to assess limitation of particular adaptation strategy to achieve good results rather than re-creating a new problem.

The rate and frequency of climate change, variability and extreme event is likely to intensify in the coming future, hence, severity of adverse impacts will also intensify as well. Therefore, new and improved PAS will be needed including capacity building in the future. Although, NAPA noted that no single PAS could absolutely remove adverse impacts of climate change, variability, sea level rise and extreme events in Tuvalu, the adverse impacts of climate change, variability, sea level rise and extreme events could be minimized to a manageable level. Therefore, adverse impacts of climate change not covered by PAS will be the cost borne by the community and stakeholders concerned.

The selected potential adaptation strategies and measures were selected from stakeholders' suggestions during the NAPA consultation and selected NAPA sector task team ideas.

Several adaptation strategies and measures have been undertaken on several islands of Tuvalu, either independently, or with assistance from the government of Tuvalu and donor countries. Since coastal erosion is a major problem, several measures have been initiated as adaptation measures such as seawalls and coastal tree planting programmes. With respect to seawalls, in some cases, they have been very successful but failed in others. As observed, current designs of seawalls are more effective on coastal areas where the wave force and coastal currents are weak, as with central lagoons on atoll islands. On the other hand, planting of local trees on the coastline is still being monitored for successes.

The Tuvalu NAPA is mindful that addressing future adverse effects of sea level rise is a complex issue for Tuvalu. A system approach acceptable to the community concerned must be suggested. In addition, NAPA is also aware that no PAS could absolutely eliminate adverse impacts of climate change, variability, sea level rise and extreme events. Therefore, stakeholders'still have to bear some of the cost of climate change.

[...]

Ministry of Natural Resources, Environment, Agriculture and Lands, Department of Environment, *Tuvalu's National Adaptation Programme of Action (NAPA)',* under the auspices of the United Nations Framework Convention on Climate Change, May 2007, pp. 6–7, 12, 19, 32–36.

Note

4.   Tuvalu Initial National Communication under UNFCC, October 1999.

Chapter 37

Stephanie Long and Janice Wormworth

TUVALU: ISLANDERS LOSE GROUND TO RISING SEAS[3]

On the string of coral reefs and atolls that form Tuvalu, the highest ground is just 4.5 metres above mean tide level, and most is well below that. Each year Islanders nervously await king tides, the year's highest tides. That's when salty ocean water overcomes shorelines, and bubbles up through the islands' porous limestone. Crops, homes and roads are flooded. Many scientists believe that king tides, naturally driven by a combination of short- and long-term tidal cycles, are now becoming more extreme due to sea level rise from global warming. Tuvaluans' ability to grow food has already been affected, and mid-range UN projections warn that the very continuity of their island life is at risk.

Impacts

> Even recently, one of the islands by (our) main island capital just disappeared.[1]
>
> (Former Tuvalu Prime Minister Maatia Toafa, in Japan, May 2006)

A remote Pacific nation, Tuvalu is made up of nine islands lying halfway between Hawaii and Australia, scattered across 560 kilometres of the Pacific Ocean. Just 400 metres across at its widest point, Tuvalu's homes, infrastructure, and commercial activities are never far from the seafront. Half the population of about 11,500 lives just three metres above sea level.

This fact demonstrates why the lives and health of those on Tuvalu and other small island developing states (along with the North African region) have already been declared by the UN IPCC (Intergovernmental Panel on Climate Change) as being the most vulnerable to climate change.[2] A full metre's width of land has been lost to the sea from around Tuvalu's largest atoll, and record levels of flooding and spring tide peaks have occurred in recent years.

A level of controversy

Just how much the sea level has risen, and will rise in future, is still debated. Some scientists assert the rate of rise is 2 mm per year at Tuvalu, and global averages indicate sea levels rose 10–20 cm over the 20th century.[3]

The latest UN mid-range figures predict a global sea level rise of 20–43 cm over the next century. However, these predictions don't include possible meltdown of ice sheets such as

Greenland's, which would lead to increases measured in metres instead of centimetres, experts say. Furthermore, Pacific atolls may be naturally subsiding, which makes Tuvaluans even more vulnerable to climate change.

Fragile foundation

The coral that forms Tuvalu's reefs and atolls (ring-shaped coral islands enclosing lagoons) provides natural breakwaters that shield shorelines from waves and storm surges. Coral reefs also provide habitat for fish and wetlands. Yet coral's vulnerability to climate change is another concern in Tuvalu, where the coral grows relatively slowly and likely won't keep up with sea level rise. Nor will coral tolerate changes in water surface temperature or rising ocean CO_2 levels.[4]

Cyclone danger

Tuvalu is also located near the cyclone belt.[5] "We are already experiencing increased frequency of cyclones, tornados, flooding, and tide surges many of which unexpectedly hit us outside the usual climatic seasons of the islands,"[6] Tuvalu's former Prime Minister told the UN in 1997. Sea level rise coupled with increased cyclone activity threatens extreme flooding events. Islanders have not forgotten 1972 Cyclone Bebe, which left 800 homeless.

> "Moving away from Tuvalu is not good for our culture and values. We want to live in our own land, our home and where our forefathers have lived. Tuvaluan people don't like to be called refugees."
> (Annie Homasi, Coordinator, Tuvalu Association of Non-Governmental Organisations)

Freshwater is another growing concern

For many Pacific Island nations, especially atoll nations such as Tuvalu, rainwater is the major water source. An important backup is the thin layer of underground freshwater (called a freshwater "lens") which sits atop the heavier, deeper saltwater. In Tuvalu both these freshwater resources are at risk. The El Niño Southern Oscillation (ENSO), which is expected to become more frequent and persistent, has caused droughts for Tuvalu,[7] while sea level rise threatens the freshwater lens.[8]

Poisoned paradise

Tuvaluans have always relied on locally-grown food and fishing to meet their needs. Yet on six Tuvalu islands, rising sea level is already making some soils too salty, poisoning gardens. For example, puluka, a giant swamp plant and Tuvalu's main source of taro, is grown in deep pits to tap the fresh water lens. Puluka cannot tolerate salt and is very vulnerable to saltwater intrusion. These problems, along with land shortages, and increased purchasing power due to employment, have led to a shift from local to imported foods.

Siuila Toloa, teacher, former Tuvalu Red Cross secretary, and board member of Island Care, a Tuvalu environmental group.

On the global politics of climate change

Countries are denying the fact that it is our business to address climate change. I say it is OUR business. Tuvalu is a small country. We are looking at facing a big problem there … We can barely save our people's lives if the story comes true and Tuvalu is sinking. … The small island states contribute insignificantly to global emissions, but suffer most.

On the prospect of Tuvaluans becoming climate refugees

Climate change is an environmental issue that leads to the complete obliteration of Tuvalu. … Tuvaluans become climate change refugees when the land of Tuvalu becomes uninhabitable. With this last resort adaptation to climate change we Tuvaluans lose our sovereignty, our traditional customs. I think you all know how important these are to us as native landholders.

On local impacts

Tuvalu lives off a subsistence income and therefore is heavily dependent on its immediate surroundings: the marine and terrestrial resources. The people are noticing a marked decline in their traditional crops and marine resource harvest. In other words, there is a decline in local food security.

Once in 2003, the most amazing thing, we really don't know why — it was just a really calm day and high waves came, affecting the coastal areas and really damaged the people's gardens. And then frequent drought — it's so frequent … three months of drought, it is really bad for us.

On water

The decrease in locally grown food is the product of an increase in areas that are degraded by salt water intrusion. This reduces the land's productive capabilities. … It has affected traditional crops of six of Tuvalu's eight islands and it will increase. Some family residences have been affected. It has also increased ground water salinity. … This [ground water] is the main source of potable water for Tuvalu and we have lost a valuable resource.

Unhealthy trends

Tuvaluans' dietary shift from local to imported food is already associated with lifestyle diseases such as high hypertension and diabetes. Another important health risk relates to water. Malaria and dengue fever can become more prevalent with warming and flooding. Failed sewage and water systems due to flooding can also increase the prevalence of other diseases.[9]

Adaptation

> We live in constant fear of the adverse impacts of climate change. For a coral atoll nation, sea level rise and more severe weather events loom as a growing threat to our entire population. The threat is real and serious, and is of no difference to a slow and insidious form of terrorism against us.
>
> (Former Tuvalu Prime Minister Saufatu Sopoanga at the UN General Assembly, New York, 2003)

Pacific islanders contribute little to the problem of global warming, producing only 0.03 percent of the global emissions (from burning fossil fuels), though they are home to 0.12 percent of the world's population.[10] Yet Tuvaluans' losses will be great indeed if international action is not taken to arrest global warming; they will ultimately be forced to abandon their homeland.

Small but vocal

Thus one major Tuvalu survival strategy has been a strong presence at international climate change negotiations, starting in 1992. Though a major expense for the word's second-smallest independent country (by population), Tuvalu joined the United Nations in 2000, to further spotlight climate change. UN membership allows Tuvalu to position itself as a conscience and to be the most vocal voice in this crucial work, with and on behalf of the Association of Small Island States.

Grassroots approaches

On the home front, the strongly-Christian nation can draw on the church, which can play a major role in civil society and grassroots approaches to climate change. In terms of practical measures, so far Tuvaluans have been adapting by planting crops in buckets, rather than in the ground, as a response to saltwater intrusion. Introducing salt-tolerant crops is another logical step, one of many that scientists and government officials urge as "no-regrets" policies — those that make sense regardless of whether the seas rise. Tree-planting programs to protect beaches from erosion have been led by NGOs, and sea-walls have been constructed to protect from storm surges.

Another no-regrets policy would be addressing local pollution caused by population growth and poor environmental management. Garbage is dumped on beach areas and in "borrow pits" dug by the US Army during WW II. Garbage and liquid waste threatens to pollute underground drinking water, and sea water, and could thus harm corals.[11] Yet another step would be to curtail beach mining which speeds up coastal erosion. Though illegal, this is done to provide material to build homes; yet construction material could be obtained in less destructive ways. Awareness projects on climate change are also needed, especially for inhabitants on outer islands who lack access to information and are less likely to speak English.

Annie Homasi, Coordinator, Tuvalu Association of Non-Governmental Organisations.

On local impacts

The weather changes and heat affects people, but also sea-level rise. ... My own experience is that during spring tides in March, my house concrete foundation is now half in the water. This is what I have seen and based on my own markings of the water level at my house.

On global politics of climate change

The Australian government has not been willing to consider environmental refugees, and is not very friendly. New Zealand has been more flexible and a work scheme has been negotiated between New Zealand and Tuvalu. People in Tuvalu are thinking that they will need to make a move because of global warming. People living in Melbourne, Australia, who have moved there 30 or 40 years ago are very concerned about where people of Tuvalu will be able to go.

Moving away from Tuvalu is not good for our culture and values. Where we live now, we know how to behave and live within our means. It will not be comfortable to live in another place. We want to live in our own land, our home and where our forefathers have lived. Tuvaluan people don't like to be called refugees.

> We don't want to leave this place. We don't want to leave, it's our land, our God given land, it is our culture, we can't leave. People won't leave until the very last minute.
>
> (Former Assistant Secretary, Tuvalu Ministry of Natural Resources, Energy and Environment, Paani Laupepa)

The last resort

Climate refugees are likely to be the largest and fastest-growing category of ecologically displaced people.[12] Tuvalu is the first country forced to evacuate residents because of rising sea levels; many Tuvaluans have also migrated internally, to the larger atoll of Funafuti from outer islands.

The Tuvalu government has actively pursued migration options. One result is New Zealand's Pacific Access Category programme, which accepts 75 Tuvaluans each year. Yet applicants must be of "good character and health, have basic English skills, have a job offer in New Zealand, and be under 45 years of age".[13] Tuvalu government representatives have so far met with no success in attempts to discuss immigration with Australia.

Conclusion

What's more, the people of Tuvalu want to see a positive response from you people on the issue of climate change. We need to work together as friends to address the climate change issue. To do this all developed countries must ratify the Kyoto Protocol ... If you love us, please, sign the Kyoto Protocol for Tuvalu's sake.

(Tuvaluan teacher and environmentalist Siuila Toloa)

Climate change is a risk to environments worldwide, but on Tuvalu and other small island states, whole nations and cultures are in jeopardy. Small size and limited access to capital, technology, and human resources compound the difficulties atoll countries such as Tuvalu face in adapting to climate change.

Ultimately, however, Tuvalu will be unable to adapt in the face of the relentless sea level rise and extreme weather events that would follow a global failure to curtail emissions. This fact explains the urgency with which Tuvalu's leaders plead their case in international climate change fora. At risk is a nation's unique culture — traditional skills, knowledge, social networks and agricultural practices that have allowed Tuvaluans to survive on their island paradise for 3,000 years. The loss of Tuvalu, possible within this century, would make our human community so much the poorer.

Stephanie Long, Friends of the Earth Australia and Janice Wormworth, 'Climate Change: Voices from Communities Affected by Climate Change', Friends of the Earth International, November 2007, pp. 32–34.

Notes

1   www.planetark.com/dailynewsstory.cfm/newsid/36546/story.htm.
2   IPCC (2001): Climate Change 2001: Impacts, Adaptation, and Vulnerability. Cambridge. www.ipcc.ch/pub/reports.htm, p. 847.
3   The 7.1 million citizens in 22 Pacific island countries, including Tuvalu, are responsible for emissions of approximately 6.816 million tons of CO_2 per year. www.germanwatch.org/download/klak/fb-tuv-e.pdf.
4   UNFCCC (Eds., 1999): Tuvalu Initial National Communication Under the United Nations Framework Convention of Climate Change; www.unfccc.int/resource/docs/natc/tuvnc1.pdf p.28.
5   UNFCC 1999, p. 13.
6   www.tuvaluislands.com/kyoto-panieu.htm.
7   IPCC (2001), p. 861.
8   www.foe.org.au/resources/publications/climate-justice/CitizensGuide.pdf/view.
9   IPCC (2001), p. 864.
10   IPCC (2001), p. 867.
11   UNFCCC (1999), p. 29.
12   www.foei.org/publications/pdfs/island.pdf.
13   www.germanwatch.org/download/klak/fb-tuv-e.pdf.

[Sources omitted]

Chapter 38

Ilan Kelman

ISLAND EVACUATION

Global environmental change is expected to have particular impact on islands around the world

Islanders from Vanuatu and the Bay of Bengal have already been forced to move as a result of sea-level rise while many island communities in Alaska – in the face of fierce storms and rapid coastal erosion – are contemplating a move inland. Five main climate change-related factors, some interlinked, threaten the viability of living on some islands, especially low-lying atolls: sea-level rise, increased severity and frequency of storms, changes to marine resources, increasing acidity of oceans and changes to freshwater resources.

Examples of entire island countries which are threatened by sea-level rise are Kiribati, the Maldives and Tuvalu. Additionally, if ice sheets collapse, much of their land could end up under water. With the expectation of tropical cyclone tracks changing while storms might become more frequent and more severe, islands which had previously experienced few extreme events might have to deal with them more regularly.

Chemical, rather than geomorphological changes, could also reduce low-lying islands' habitability. Oceanic absorption of atmospheric carbon dioxide is leading to ocean acidification,[1] damaging coral reefs which in turn exposes islands to increased wave energy while changing the nature of near-shore fisheries.

For marine resources, the possible impacts are uncertain. In some places, numbers might decline and species might become extinct but many others could migrate. Some islands might gain more plentiful fish or other marine resources, while some could lose the food supply on which they have relied. Similarly, for freshwater resources – often already in limited supply on islands – many places will experience drier conditions; even if more tropical storms bring more water, the damage wrought by them could offset the benefits of increased water supply.

Depending on the exact impacts on specific islands, permanent displacement may be the only viable long-term option. Severe environmental change has led to islander displacement in the past. Approximately 700 years ago, sea-level fall and regional changes in the Pacific climate forced many Pacific islanders to abandon their settlements.

Decision making

If an island community decides that displacement or evacuation of an entire island is an appropriate option, the first decision is the timing of that migration. Should the evacuation happen as soon as possible, before severe impacts of environmental changes are felt? This would enable migration to be properly planned. Or would it be easier to convince people to go only after a major disaster? Then they risk loss of life and loss of possessions (including cultural/ community artefacts). The main disadvantage with long-term planning is that an extreme event could strike at any time. A combination of both solutions could be sought, perhaps planning to leave quickly as soon as an extreme event threatens or strikes.

After the timing of migration has been determined – or left to extreme events to decide – the second decision is where people should go in order to create a new community. Two options exist. They could abandon their identity and their community and integrate elsewhere. The 12,000 Tuvaluans still on Tuvalu, for example, could easily disperse among the millions of Sydney, Tokyo, Los Angeles or other large cities.

Rather than losing a culture, language and identity, however, island communities could instead be re-created. Resettlement on land (especially islands) similar to, but more secure than, their current location would be preferable but might not be feasible because most low-lying areas would suffer similar fates as the islands being evacuated. As well, many potential island candidates for re-creating island communities are protected as environmental, tourist and/or scientific havens – or are uninhabitable due to their size or resource constraints.

Such resettlement could also require another state to cede territory. For the Pacific region, Australia and New Zealand are usually suggested as the most likely candidates to provide land. Other possibilities could be Indonesia, the Philippines, the Solomon Islands, Vanuatu, the US or Japan. Another option would be to create new land (perhaps through land reclamation) but this would involve legal ramifications – such as delineating the new state's territorial waters.

Sovereignty

Whether existing land or new land is used for resettlement, more decisions have to be made concerning levels of sovereignty or autonomy. Should sovereign states and non-sovereign territories be entirely re-created or should these governance regimes be adjusted? There are different options, including:

- Joint access to an island's resources, as is the case of Svalbard in the Arctic
- a level of autonomy involving parallel and complementary justice systems, such as those for indigenous people in Canada and New Zealand
- a level of autonomy involving, for example, parallel currency systems.

Once a governance model is approved by all concerned parties, many practical and ethical questions remain. Who pays for the move and the construction of new communities or new land? How will any territorial or jurisdictional disputes be resolved? How will those to be displaced retain significant control over these aspects? If an island country is entirely evacuated but the islands are submerged only at the highest tides, who owns the fishing rights in the surrounding seas? Could those rights be sold, with oil and other mineral resources potentially being more valuable than fish? If a state is disbanded because of displacement rather than re-created, how do the answers to these questions change?

Security questions also emerge regarding locations where islanders are resettled. Could a country claim a security threat from potential future sovereignty demands if an entire island country population is settled there? Could resettlement be used to reduce enmity and to galvanise international cooperation in solving environmental issues? (Studies in disaster diplomacy that have investigated this last point conclude that such opportunities are usually squandered.)[2]

These issues are not unique to islands. Many coastal settlements could suffer similar displacement for similar reasons. Although non-island coastal settlements have an 'inland' to which they can move, some islands also have that option, especially larger hilly islands such as Puerto Rico and Fiji's largest island, Viti Levu. Yet that would still result in significant changes, both for the people who must move and for the people already living 'inland'.

Learning from experience

Island evacuation due to global environmental change may be unique in living memory but island evacuation due to environmental change is not new. There have been many island evacuations, for example, related to volcanic activity. However, there are differences between evacuation because of volcanic activity and evacuation as a result of global environmental change. Most islanders evacuated after a volcano starts erupting expect the evacuation to be temporary; in many cases, they return home even earlier than recommended. For global environmental change, many islands are expected to experience such severe and irreversible changes that resettlement would not be possible for centuries to come. Temporary displacement is very different from leaving one's land, home and identity for ever. We need to learn from mistakes made in the past, especially regarding who makes decisions and who pays.

There is time now to draw on the experience of previous environmentally-induced displacement, both island and non-island. Precautionary planning now would be prudent, rather than reacting after it is too late.

Ilan Kelman, 'Island Evacuation', *Forced Migration Review*, Issue 31, October 2008, pp. 20–21.

Notes

1   www.royalsoc.ac.uk/document.asp?id=3249.
2   www.disasterdiplomacy.org.

Vanuatu

CHAPTER 39

Peter Boehm

GLOBAL WARNING
Devastation of an atoll

Villagers on the South Pacific island of Tegua are packing up and leaving their homes for good – the first real victims of increasing sea levels caused by climate change.

As a Pacific island destination, Lateu struggles to sell itself. A typhoon wiped away its only beach a few years ago and today a handful of squalid thatched huts stand forlornly on its coastline. It will soon be a deserted village, its population the first real victims of rising sea levels brought about by global warming. Even the village's palm trees are dying, their roots washed away by inexorably rising seas. The roofs of its thatched huts are leaking, there are gaping holes in the palm frond walls. All that remains of several are a few pathetic looking poles, braced against the prevailing wind from the vast expanse of the Pacific.

Lateu's people don't bother to patch their huts anymore, and an unpleasant mould covers the ground of every dwelling as a result of frequent flooding. For a while the islanders tried to rise above the surging seas by putting their huts up on makeshift foundations of coral, but they soon gave up. Now they are preparing to move to higher ground, some 300 metres inland, where they have already built six communal structures with financial aid from Canada.

Lateu is the only village on Tegua island, a half-moon-shaped speck of land less than four miles long and 10 miles wide in the South Pacific. It is one of five coral atolls in the Torres Group, 650 miles north of Vanuatu's main island, Efate. Getting here involves a flight in a small plane that goes once a week to the Torres' main island, Loh, and a prohibitively expensive, wet and frightening 40-minute ride over the open sea in a fisherman's boat.

The impact climate change has had on Tegua atoll is hard to ignore. Everywhere on the windward side of the island, palm trees are immersed by the sea. Some have survived the ordeal, others have fallen, littering the shallow waters of the coast line. "At the end of the Eighties, our village was flooded for the first time," says Reuben Seluin, 63, Lateu's village head. "Nowadays it happens every other month."

When the islanders resettled Tegua in the Sixties, they built Lateu directly next to the bay. During low tide, a tiny pond of sweet fresh water appears between the sea's coral ground. The pool gave Lateu its name. Aside from a rain water tank, the only thing Vanuatu's administration ever built on this island, it used to be the only source of drinking water for the atoll. Now the women use it to do their laundry.

During high tide, waves crash on a low pile of coral that separates the village from the sea, but Mr Seluin says that Lateu used to have a white sand beach. Part of it, he says, was washed away by a tsunami, triggered by an earthquake on Torres in 1997. "With every big tide that followed, the sea came a bit closer. I can't tell you whether this is due to climate change. I just know that we used to have a beach, and that it's gone."

The effects of climate change and rising sea levels can be seen on many islands in Vanuatu. In 1993, Australian scientists set up the Sea Level and Climate Monitoring Project. They recorded the sea level at 12 points in the South Pacific and detected a rise of, on average, 6mm per year, or 7.8cm (3.1 inches) in total. Vanuatu's Meteorological Department monitors the number of storms that have hit the nation since 1941. In the Forties, says the department's head, Jotham Napat, the number in their records was five, but in the past few years, the average has been 15.

Some climatologists, such as Stephen Koletti from the University of Southern California, say it is still too early to jump to conclusions. He warns that washed out beaches alone are not sufficient evidence for a rise in the sea level. "The ocean and the beach are part of a dynamic system," he says. "During the storm season, the ocean washes out the beaches, while in the summer they are replenished." But the fact that the coastlines of Vanuatu don't seem to recover has worried him.

Mr Seluin is not interested in the intricacies of scientific climate discussions, and when he recounts Tegua's fate, there is no bitterness in his voice. No, he is not angry, he says while thoroughly examining a shell on the ground in front of him, when it is put to him that the pollution in the developed world is responsible for the rise in sea level but the impact is felt on their island. He mumbles eventually: "I would never say something like that."

Life is not all bad on this tiny atoll. It is after all remarkably fertile. The islanders grow fruit and vegetables in their plots of land in the rainforest, a half hour walk straight up the hill. If they want meat with their meals, they go for a walk to pick up some crimson-coloured homade crabs, or, a little deeper into the forest, coconut crabs, which are a sought-after delicacy in restaurants of far-away Port Vila, on Efate. When the sea is calm, they go fishing in crudely carved dugouts, or dive at night with a waterproof flashlight to rouse some lobsters.

From time to time, a group of four to six paddle two hours in their kayaks to the main part of the island to sell a few coconut crabs and buy merchandise such as matches, batteries, and other material. Little else gets through to Tegua. One islander owns a shortwave radio. A one-man news agency, if he hears something interesting, he spreads the news. Neither the World Cup nor the war in Lebanon made the cut.

Mr Seluin's feet look wide and flat, from walking barefoot his whole life. He sports the same washed-out shorts every day and a T-shirt with the slogan: "Life's a bitch... And then you dive."

The crew of a passing yacht donated it to him a few years back. Mr Seluin is not only the head of administration, the sole policeman, and only judge on the island, he is the clan chief as well, because all of the 60 inhabitants on Tegua belong to one extended family. Aden Seluin, the village head's eldest son, mans the tiny health station. With its shiny blue wooden walls, it comes closest to what a regular house looks like. Anyone suffering from something more severe than malaria has to be taken to the main island.

Godwin, Seluin's middle son, acts as a catechist for the small Anglican church in Lateu. He says that when a government delegation came to Tegua in 1998 and told them that the flooding was due to global warming, the pastor from the main island joined them. "He wanted to give us some comfort. We were really shocked. He told us that, yes, we were experiencing climate change, but that God would help us and not everybody was going to die."

When the islanders go to their gardens in the morning, they leave their children with Bettina Seluin, the village head's niece, in Lateu's tiny kindergarten.

As the children play with cardboard toys, she says: "In school we were never told about climate change. Because of the tsunami, a government delegation came to Tegua. That's when we heard about it for the very first time."

The government delegation that visited Tegua in early 1998 actually came because Vanuatu had become eligible for relief funds, having just signed the United Nations Framework Convention on Climate Change. Delegations visited all of Vanuatu's 83 islands to assess what impact global warming had had and would be likely to have in the future. That is how they learned about Lateu's plight.

After the tsunami, the villagers had decided to move their village inland, and the administration agreed to use part of a Canadian relief fund to support them. They settled for a clearing 300 meters inland, but up a steep hill from their old village, Lirak. With $50,000 (£27,000), they built six shacks. They will drink rain water, collected from the roofs of the communal buildings, and they will get a satellite-based radio system, through which the Meteorological Department will send weather updates.

Only two families have built their private huts in Lirak so far. Titus Woilami, the village head's brother-in-law, with his wife and four daughters, is one of them. He says: "During full moon and new moon Lateu was always in danger. Then we have king tides. That really put a strain on me. I could barely sleep, because it constantly swirled around my head: Will the water run into our hut again?"

The village head says moving the whole village with the church, kindergarten and health station will take a few more months. He is, of course, grateful for the new communal buildings. "Everything we ever owned, we put into Lateu," he says. But his clan will not leave Tegua. "We love our island. We have our gardens, we have fish in the sea, we have crabs to feed us. It doesn't matter that we don't have a radio, or a boat. We will stay here."

Peter Boehm, 'Global Warning: Devastation of an Atoll', *The Independent*, 30 August 2006.

Asia

Bangladesh

Chapter 40

Ministry of Environment and Forest, Bangladesh

BANGLADESH NATIONAL ADAPTATION PROGRAMME OF ACTION (NAPA)

[...]

Executive Summary

It is well recognized both in the scientific and negotiating community that Bangladesh would be one of the most adversely affected countries to climate change. Low economic strength, inadequate infrastructure, low level of social development, lack of institutional capacity, and a higher dependency on the natural resource base make the country more vulnerable to climate stimuli (including both variability as well as extreme events).

The National Adaptation Programme of Action (NAPA) is prepared by the Ministry of Environment and Forest (MOEF), Government of the People's Republic of Bangladesh as a response to the decision of the Seventh Session of the Conference of the Parties (COP7) of the United Nations Framework Convention on Climate Change (UNFCCC). The preparation process has followed the generic guiding principles outlined in the annotated guideline prepared by LDC Expert Group (LEG). The basic approach to NAPA preparation was along with the sustainable development goals and objectives of the country where it has recognized necessity of addressing environmental issue and natural resource management with the participation of stakeholders in bargaining over resource use, allocation and distribution. Therefore, involvement of different stakeholders was an integral part of the preparation process for assessing impacts, vulnerabilities, adaptation measures keeping urgency and immediacy principle of the NAPA. Policy makers of Government, local representatives of the Government (Union Parishad Chairman and Members), scientific community members of the various research institutes, researchers, academicians, teachers (ranging from primary to tertiary levels), lawyers, doctors, ethnic groups, media, NGO and CBO representatives and indigenous women contributed to the development of the NAPA for Bangladesh.

The stakeholder consultation workshops pointed that the erratic behavior of weather some times first time in their memory such as fogs in places where these were never heard of during summer time, drought, salinity intrusion far from the sea, floods including flash flood, and cyclone and storm surges as major problems they are facing in different parts of the country.

Problems related to floods include water logging and drainage congestion, early and untimely floods, localized to reduction of freshwater flow from upstream, salinization of groundwater and fluctuation of soil salinity are major concerns. Continuous and prolonged droughts, extreme temperature and delayed rainfall are major problems that the agriculture sector is facing. Storms, cyclones and tidal surges appear to have increased in the coastal areas.

It is revealed that adverse effects of climate stimuli including variability and extreme events in the overall development of Bangladesh would be significant and highly related to changes in the water sector. Most damaging effects of climate change are floods, salinity intrusion, and droughts that are found to drastically affect crop productivity almost every year. Climate change-induced challenges are: (a) scarcity of fresh water due to less rain and higher evapo-transpiration in the dry season, (b) drainage congestion due to higher water levels in the confluence with the rise of sea level, (c) river bank erosion, (d) frequent floods and prolonged and widespread drought, (e) wider salinity in the surface, ground and soil in the coastal zone. It was found that the population living in the coastal area are more vulnerable than the population in other areas. The agricultural sector will face significant yield reduction. Thus food-grain self sufficiency will be at risk in future.

The strategic goals and objectives of future coping mechanisms are to reduce the adverse effects of climate change, including variability and extreme events, and promote sustainable development. Future coping strategies and mechanisms are suggested based on existing processes and practices keeping main essence of adaptation science which is a process to adjust with adverse situation of climate change.

Sharing knowledge and experiences of existing coping strategies and practices to other area that would come under similar problems related to climate change. Development of techniques for transferring knowledge and experiences from one area/ecosystem is also necessary.

Some initial activities have already been pioneered in Bangladesh in adaptation to climate change at the community level. Several workshops at national and international level have been organized to develop better understanding on approaches on adaptation and increase negotiation skills in the Conference of the Parties to the UNFCCC. Notable among them are the International Workshop on Community Level Adaptation to Climate Change organized by the Bangladesh Centre for Advanced Studies in association with IIED and IUCN, Dialogue on Water and Climate Change organized by IUCN Bangladesh. Pilot level activities are being carried out to increase resilience of individual and community. For example the Reducing Vulnerability to Climate Change (RVCC) project in a number of flood-prone villages in coastal Bangladesh implemented by the CARE international in partnership with local NGOs has generated much useful knowledge in how to communicate climate change (and adaptation) messages at the community level.

Addressing future problems related to sea level rise appears to be a complex issue for Bangladesh and therefore an integrated approach both in terms of sectors and technologies needs to be analyzed along with acceptability by the communities for whom the technologies would be suggested.

It is evident from the science of climate change and impacts studies that severity of impacts and frequency will increase in future and therefore limitation of existing coping strategies needs to be assessed. Moreover, preparation for this on a regular basis will reduce impacts but will not solve the problem. Insurance as a mechanism may be considered for which further analysis is necessary.

Considering the above background, the following adaptation measures have been suggested for Bangladesh to address adverse effects of climate change including variability and extreme events based on existing coping mechanisms and practices. The suggested future adaptation strategies are:

1. Reduction of climate change hazards through coastal afforestation with community participation.
2. Providing drinking water to coastal communities to combat enhanced salinity due to sea level rise.
3. Capacity building for integrating climate change in planning, designing of infrastructure, conflict management and land-water zoning for water management institutions.
4. Climate change and adaptation information dissemination to vulnerable communities for emergency preparedness measures and awareness-raising of enhanced climatic disasters.
5. Construction of flood shelter, and information and assistance centre to cope with enhanced recurrent floods in major floodplains.
6. Mainstreaming adaptation to climate change into policies and programmes in different sectors (focusing on disaster management, water, agriculture, health and industry).
7. Inclusion of climate change issues in curriculum at secondary and tertiary educational institution.
8. Enhancing resilience of urban infrastructure and industries to impacts of climate change.
9. Development of eco-specific adaptive knowledge (including indigenous knowledge) on adaptation to climate variability to enhance adaptive capacity for future climate change.
10. Promotion of research on drought, flood and saline-tolerant varieties of crops to facilitate adaptation in future.
11. Promoting adaptation to coastal crop agriculture to combat increased salinity.
12. Adaptation to agriculture systems in areas prone to enhanced flash flooding in North East and Central Region.
13. Adaptation to fisheries in areas prone to enhanced flooding in North East and Central Region through adaptive and diversified fish culture practices.
14. Promoting adaptation to coastal fisheries through culture of salt-tolerant fish special in coastal areas of Bangladesh.
15. Exploring options for insurance and other emergency preparedness measures to cope with enhanced climatic disasters.

1 Introduction

The National Adaptation Programme of Action (NAPA) of Bangladesh draws upon the understanding gathered through discussion with relevant stakeholders in 4 sub-national workshops and one national workshop, prior research, background papers prepared for this report as well as research carried out for these background reports, and expert judgments. In the course of the preparation of the report it has been clear that climate change will exacerbate many of the existing problems and natural hazards that the country faces. But there are various coping mechanisms, formal and informal, already in place. What is new is the urgency of the matter to be integrated within the development process so that when the climate change impacts become more clearly discernible, the nation shall be ready to handle that as almost a routine affair in its development process.

2 Context

Bangladesh, except for the hilly regions in the northeast and southeast and terrace land in northwest and central zones, is one of the largest deltas in the world, formed by the dense network of the distributaries of the mighty rivers, namely the Ganges, the Brahmaputra and the Meghna. The country is located between 20°34' to 26°38' north latitude and 88°01' to 92°42' east longitude. The total land area is 147,570 sq. km. and consists mostly of low and flat land. A network of more than 230 major rivers with their tributaries and distributaries crisscross the country.

2.1 Characteristics

This section provides a brief description of the biophysical, social, economic, technological and political context of the country. These characteristics depict exposure, sensitivity, adaptive capacity and vulnerability of these systems to climate variability and change.

2.1.1 PHYSIOGRAPHIC CONDITION

The land area of the country may be divided broadly into three categories, i.e. floodplain (80%), Pleistocene terrace (8%), and tertiary hills (12%) based on its geological formation. The floodplain comprises a succession of ridges (abandoned levees) and depressions (back swamps or old channels). Differences in the elevation between adjoining ridge tops and depressions range from less than 1 meter on tidal floodplains, 1 meter to 3 meters on the main rivers and estuarine floodplains, and up to 5 to 6 meters in the Sylhet Basin in the north-east. Only in the extreme north-west do land elevations exceed 30 meters above mean sea level. The tertiary hill soil occupies the Chittagong hills in the south-east, and the low hills and hillocks of Sylhet in the north-east. The two major uplifted blocks (Pleistocene terrace) are known as Madhupur (in central Bangladesh) and Barind tracts in the north-west.

The land type of the country has been classified according to depth of inundation with seasonality. All land types except highlands are exposed to monsoon flooding for part or whole of the year.

Floodplains located in the north-western, central, south-central and north-eastern regions are subject to regular flooding at different frequency and intensity while the coastal plain is subject to cyclones and storm surges, salinity intrusion and coastal inundation. Pleistocene terrace land is characterized by moisture stress while flash flood is common in the hilly areas and the piedmont plains in the north-east and north-western parts of the country.

[...]

3.2 Actual and Potential Adverse Effects of Climate Change

3.2.1 PRESENT IMPACT OF CLIMATE VARIABILITY AND EXTREME

Most damaging effects of erratic behavior of present climate and extreme events are flood, drought, and heat stress that are found to drastically adversely affect crop productivity in almost every year. About 1.32 m ha of cropland is highly flood-prone and about 5.05 m ha moderately flood-prone. Besides crops, perennial trees and livestock are damaged by flood every year. In two severe flood years of 1974 and 1987, the shortfalls in production from trend were about 0.8 and 1.0 Mmt of rice, respectively. During 1984, flood affected both Aus and Aman rice crop and the shortfall was about 0.4 Mmt.

Drought of different intensities in Kharif, Rabi and pre-Kharif seasons caused damage to 2.32 m ha of T. aman and 1.20 m ha of rabi crops annually. Yield reductions due to drought vary from 45–60% in T. aman and 50-70% in rabi crops in very severe drought situations. In the severe drought of 1979 the shortfall was about 0.7 million tons. During 1981 and 1982 drought affected the production of the monsoon crop (Aman) and the shortfalls from the trend were 0.5 and 0.3 Mmt, respectively.

3.2.2 POTENTIAL FUTURE VULNERABILITY

Over the last decade a number of studies have been carried out on impacts, vulnerability and adaptation assessment for Bangladesh to climate change and sea level rise. Regional stakeholder consultation workshops have identified vulnerability of different sectors in the context of climate variability and change. Summary of vulnerability of different sectors is given below. Much of the future vulnerability due to climate change will not necessarily add any new climate related hazards to the already well known ones of floods, droughts and cyclones, but will enhance both the frequency as well as intensity of such climatic events in future. Particularly, the areas prone to the floods, cyclones and salinity intrusion all may increase in future. The climate related hazards will in turn be compounded by other factors including land use patterns, water management and control of river flows upstream. Some of the specific vulnerabilities due to climate change impacts are described below.

3.2.2.1 WATER RESOURCES

Water related impacts due to climate change and sea level rise are likely to be some of the most critical issues for Bangladesh, especially in relation to coastal and riverine flooding, but also in relation to the enhanced possibility of winter (dry season) drought in certain areas. The effects of increased flooding resulting from climate change will be the greatest problem faced by Bangladesh as both coastal (from sea and river water), and inland flooding (river/rain water) are expected to increase. In addition, changes of the riverbed level due to sedimentation and changes in morphological processes due to seasonal variation of water level and flow are also critical for Bangladesh.

Sedimentation and River Bed Rise. The process of sedimentation may rise as water level gradients due to higher downstream water levels at sea result in lower flow velocities.. The morphologically highly dynamic rivers in Bangladesh are expected to adapt to such changes in water levels.

Box 1. Knowledge Base on Scenario Building in Bangladesh

General Circulation Model (GCM) used by the US Climate Change Study team for Bangladesh reported that the average increase in temperature would be 1.3°C and 2.6°C for the years 2030 and 2070, respectively. It was found that there would be a seasonal variation in changed temperature: 1.4°C change in the winter and 0.7°C in the monsoon months in 2030. For 2070 the variation would be 2.1°C and 1.7°C for winter and monsoon, respectively. For precipitation it was found that the winter precipitation would decrease to a negligible rate in 2030, while in 2075 there would not be any appreciable rainfall in winter at all. On the other hand, monsoon precipitation would increase at a rate of 12 % and 27 % for the two projection years, respectively (Ahmed et al., 1999).

It was found that there would be excessive rainfall in the monsoon causing flooding and very little to no rainfall in the winter forcing drought. It was also found that there would be drastic changes in evaporation in both winter and monsoon seasons in the projection for year 2075. It was inferred from the GCM output that moderate changes regarding climate parameters would take place by 2030, while severe changes would occur by 2075.

The results also reveal a trend of a general increasing temperature. In 2030, the increase is much more pronounced in winter months, although the maximum change is observed for post-winter months, i.e., April, May and June. However, in 2075, the increase in temperature during April and May is much higher; about 4.0°C (Ahmed et al., 1999).

OECD has recently carried out 17 General Circulation Models for Bangladesh in order to assess changes in average temperature and precipitation using a new version of MAGICC/SCENGEN. It has selected 11 out of the 17 models which best simulate current climate over Bangladesh. The models were run with the Intergovernmental Panel on Climate Change (IPCC) B2 scenario of Special Report on Emissions Scenarios (SRES) (Agarwala et al., 2003).

The climate models all estimate a steady increase in temperatures for Bangladesh, with little inter-model variance. Somewhat more warming is estimated for winter than for summer. With regard to precipitation, whether there is an increase or decrease under climate change is a critical factor in estimating how climate change will affect Bangladesh, given the country's extreme vulnerability to water related disasters. The key is what happens during the monsoon. Most of the climate models estimate that precipitation will increase during the summer monsoon because air over land will warm more than air over oceans in the summer. This will deepen the low pressure system over land that happens anyway in the summer and will enhance the monsoon. It is notable that the estimated increase in summer precipitation appears to be significant; it is larger than the standard deviation across models. This does not mean that increased monsoon is certain, but increases confidence that it is likely to happen. The climate models also tend to show small decreases in the winter months of December through February. The increase is not statistically significant, and winter precipitation is just over 1% of annual precipitation. However, with higher temperatures increasing evapo-transpiration combined with a small decrease in precipitation, dry winter conditions, even drought, are likely to be made worse (Agarwala et al., 2003).

Forecasts show that at the bifurcations of the Jamuna river with its distributaries Dhaleswari river and Old Brahmaputra river, the bed level will rise 0.08, 0.12 and 0.41 m at the mouth of the Dhaleswari river, and 0.05, 0.08 and 0.27 m at the mouth of the Old Brahmaputra river for the years 2015, 2025 and 2095 respectively (BCAS/RA/ Approtech, 1994). This will probably result in a considerable increase in the discharges in the distributaries and a small decrease of the discharges in the Jamuna and Padma rivers. The discharge distribution at the tributaries of the Ganges and the Padma rivers (Gorai and Arial Khan rivers) will change also due to the considered sedimentation. These changes might be of important consequences for the course of the main river channels in Bangladesh.

Change of Land Type. Bangladesh Climate Change Country Study (1997) assessed vulnerability of water resources, considering changes in flooding conditions due to a combination of increased discharge of river water during the monsoon period and sea level rise for the two projection years, 2030 and 2075.

From the analysis it is found that much of the impact would be for F0 land followed by F1 land in the year 2075 where embankment played an important role in restricting the extent of flood affected areas. Again, it is the F0 land followed by F1 land in 2030 which would experience much of the changes in the north-central region in 2030. A combination of development and climate change scenarios revealed that the Lower Ganges and the Surma floodplain would become more vulnerable compared to the rest of the study area. On the other hand, the north-central region would become flood free due to embanking of the major rivers (Alam et al., 1999).

3.2.2.2 COASTAL ZONE

Several studies indicate that the coastal zone vulnerability would be acute due to the combined effects of climate change, sea level rise, subsidence, and changes of upstream river discharge, cyclone and coastal embankments (BCAS/RA/Approtech, 1994, WB, 2000). Four key types of primary physical effects, i.e. saline water intrusion; drainage congestion; extreme events; and changes in coastal morphology have been identified as key vulnerabilities in the coastal area of Bangladesh (WB, 2000). A relationship between agents of change and primary physical effects in the coastal zone of Bangladesh is given in Table 7.

- The effect of saline water intrusion in the estuaries and into the groundwater would be enhanced by low river flow, sea level rise and subsidence. Pressure of the growing population and rising demand due to economic development will further reduce relative availability of fresh water supply in future. The adverse effects of saline water intrusion will be significant on coastal agriculture and the availability of fresh water for public and industrial water supply will fall.
- The combined effect of higher sea water levels, subsidence, siltation of estuary branches, higher riverbed levels and reduced sedimentation in flood-protected areas will impede drainage and gradually increase water logging problems. This effect will be particularly strong in the coastal zone. The problem will be aggravated by the continuous development of infrastructure (e.g. roads) reducing further the limited natural drainage capacity in the delta.

Increased periods of inundation may hamper agricultural productivity, and will also threaten human health by increasing the potential for water-borne disease.

- Disturbance of coastal morphological processes would become a significant problem under warmer climate change regime. Bangladesh's coastal morphological processes are extremely dynamic, partly because of the tidal and seasonal variations in river flows and run off. Climate change is expected to increase these variations, with two main (related) processes involved:

- Increased bank erosion and bed level changes of coastal rivers and estuaries. There will be a substantial increase of morphological activity with increased river flow, implying that river-bank erosion might substantially increase in the future.

- Disturbance of the balance between river sediment transport and deposition in rivers, flood plains and coastal areas. Disturbance of the sedimentation balance will result in higher bed levels of rivers and coastal areas, which in turn will lead to higher water levels.

- Increased intensity of extreme events. The coastal area of Bangladesh and the Bay of Bengal are located at the tip of northern Indian Ocean, which has the shape of an inverted funnel. The area is frequently hit by severe cyclonic storms, generating long wave tidal surges which are aggravated because the Bay itself is quite shallow. Cyclones and storm surges are expected to become more intense with climate change. Though the country is relatively well equipped, particularly in managing the aftermath of cyclones, the increased intensity of such disasters implies major constraints to the country's social and economic development. Unless proper adaptive measures are undertaken, private sector investment in the coastal zone is likely to be discouraged by the increased risks of cyclones and flooding.

The salinity intrusion, for different sea level rise scenarios, has been estimated using mathematical models. From the analysis it is found that the area under salinity level of 5 ppt under the Business as Usual scenario is increasing. The 5 ppt line move from the lower tip of Sundarbans to the point of lower Meghna river at Chandpur by year 2100 under an assumed SLR of 88 cm. The salinity front will move about 60 km to the north in about 100 years. The SLR will increase the salinity level in the Tentulia River, at present the only fresh water pocket in the estuary.

Due to backwater effect the water levels around the polders are also likely to be affected. A hydrodynamic model shows that high water levels at the surrounding rivers of polders may increase in the range of 30 to 80 cm for sea level rise in the range of 32 to 88 cm. This rise will eventually hamper the proper functioning of a number of polders.

[...]

Ministry of Environment and Forest, Government of the People's Republic of Bangladesh, 'Bangladesh National Adaptation Programme of Action (NAPA)', November 2005, pp. xv–xvi, 1–2 and 10–13.

Chapter 41

Care, CIESIN, UNHCR, UNU-EHS and the World Bank

THE GANGES DELTA
Temporary migration as a survival strategy

[...]

Including the Ganges, Bangladesh contains seven major and over two hundred minor rivers, all of which define the delta geography of Bangladesh and the way of life of its people. Bangladesh is one of the most densely populated countries in the world, and a large part of its people depends on natural resources for their livelihoods. Although flooding is a part of the livelihood structure and culture, climate change will accelerate change in this already dynamic environment and leave millions of Bangladeshis exposed to increased flooding, severe cyclones, and sea level rise impacts.[75]

More than 5 million Bangladeshis live in areas highly vulnerable to cyclones and storm surges, and over half the population lives within 100 km of the coast, most of which is less than 12 meters above sea level.[76] Flooding currently displaces about 500,000 people every year. In 2007, two extreme weather events devastated the country: flooding caused 3,363 deaths and affected 10 million people as well as reducing crop yields by 13 percent. Just months later, Cyclone Sidr destroyed 1.5 million houses, large areas of cropland and mangrove forests, and affected 30 out of 64 districts in the country. Millions experienced food insecurity (monga) and required evacuation, shelter and relief assistance.[77] As devastating as these cyclones were, early warning systems were successful in preventing the deaths of many thousands more. In 1970, a cyclone caused the deaths of an estimated 300,000, and in 1991 another 140,000 died.[78]

The Bangladesh EACH-FOR case study found that flooding and bank erosion are a complex mix of natural and socioeconomic processes contributing to population displacement.[79] Combined with sea level rise, storm surges linked to cyclones could temporarily inundate large areas of Bangladesh—one study suggested that up to 25 percent of the country could experience such a scenario.[80]

Temporary migration linked to flooding and other disasters, frequently to Dhaka and other urban centers, is viewed as both a coping and survival strategy to escape riverbank erosion, the devastation of cyclones, and food insecurity. Almost all areas in Bangladesh are densely populated and under cultivation, and many locations are vulnerable to similar environmental risks. There are no guarantees of finding employment or housing in the place of destination.

For coastal fishing villages, cyclones, storm surges, and sea level rise pose a formidable adaptation challenge. One fisherman interviewed by a journalist during the 2008 cyclone season noted, "The sea has been coming closer and closer," then added in Bengali, "Allah jane ke hobe. Sahbi shesh ho jabe." [God only knows what will happen. Everything will come to an end.] In spite of accelerated erosion related to stronger and higher tides, villagers are determined to stay and pursue their livelihoods as long as possible. The same journalist interviewed another fisherman who said, "We can't do anything else, which is why we think twice about migrating from here. We know the end is coming, but what work will we find to feed our families elsewhere?"[81]

Even if the causes of migration are similar from one person to the next, people opt for different strategies in terms of destination and timing of migration. But there might be a moment when they will not be able to adapt any more. In 20 or 30 years Bangladesh may see mass movement of people from flood-prone areas, possibly to urban centers. The current structures and organizations to help the victims of disasters will not be enough to cope with the increase of migration flows in the future. Given the political instability of the region, population movements associated with climate change could become an issue for regional security.

However, adaptation strategies could reduce the environmental vulnerability and increase the resilience of local populations. EACH-FOR research suggests that the population is already working to adapt to the new situation, mainly by leaving agriculture for other livelihoods such as shrimp farming.[82]

The worsening of the environmental situation in the Ganges delta, however, could render migration as one of the most realistic options available for some Bangladeshi people.

Care, CIESIN, UNHCR, UNU-EHS and World Bank Social Dimensions of Climate Change, 'In Search of Shelter: Mapping the Effects of Climate Change on Human Migration and Displacement', May 2009, p. 13.

Notes

[...]
75 Vorosmarty et al. 2009. See note 26 above; and Usapdin, T. 2008. South Asia: Building safer communities. IFRC.org, 27 August. www.ifrc.org/docs/news/08/08082701; and Poncelet, A. 2009. The land of mad rivers. Bangladesh Case Study Report for the Environmental Change and Forced Migration Scenarios Project. http://www.each-for.eu.
76 McGranahan et al. 2007. See note 7 above.
77 Women's Environment and Development Organization (WEDO), ABANTU for Development in Ghana, ActionAid Bangladesh, and ENDA in Senegal. 2008. Gender, climate change and human security: Lessons from Bangladesh, Ghana, and Senegal. http://www.wedo.org/files/HSN%20Study%20Final%20May%2020,%202008.pdf.
78 British Broadcasting Corporation (BBC). 1988. Bangladesh cyclone 'worst for 20 years'. BBC.co.uk, 2 December.http://news.bbc.co.uk/onthisday/hi/dates/stories/december/2/newsid_2518000/2518233.stm.
79 Poncelet 2009. See note 75 above.
80 UNEP-GRID. See note 23 above.
81 IRIN. 2008. Bangladesh: When climate change gives you a sinking feeling. IRIN Print Report Humanitarian News and Analysis, IRINnews.org, 22 October. http://www.irinnews.org/PrintReport.aspx?ReportId=81079.
82 However, alternative livelihoods must be feasible for those most vulnerable to climate change and

other environmental stressors. The EACH-FOR field researcher for the Bangladesh case study noted that some activities like shrimp farming may be too expensive for vulnerable farmers to take up as a livelihood alternative. Poncelet, A. 2009. Alternative livelihoods for vulnerable farmers in Bangladesh. Personal communication 11 May 2009.

India

Chapter 42

Nicola Macnaughton

CLIMATE CHANGE IN INDIA
A Humanitarian Perspective

Last month, in the states of Karnataka and Andrha Pradesh in Southern India, flooding induced by heavy rains left approximately 2.5 million people homeless and more than 250 dead.[1] This devastation occurred just two months after India's finance minister, Pranab Mukherjee, announced that the government was to begin importing food to make up for shortages caused by a drought estimated to be affecting 700 million people throughout the country.[2]

There is little doubt that India is already experiencing some of the impacts of climate change, which are predicted to worsen and intensify over the coming decades. In addition to sporadic monsoon conditions, experts have also expressed concerns about rising sea levels which threaten to submerge low-lying regions, and diminishing water accessibility which will deplete crop production.[3] From a humanitarian perspective, this is worrying: "India is entering a period of severe…vulnerability," says Jayakumar Christian, the National Director of World Vision.

A report released last year by CARE International described India as a "humanitarian hotspot" for climate change risk.[4] The United Nations Intergovernmental Panel on Climate Change (IPCC) has further specified that, because they are most dependent upon natural resources for their food, shelter and livelihoods, the poorest and most marginalized groups in Indian society are likely to be the most affected by environmental changes.[5]

India is home to approximately 800 million such poor people, the vast majority of whom "live on ecologically fragile land… and lack the institutional and financial capacity to protect themselves against climate change."[6] Many of these individuals belong to "adivasi" (tribal) communities, who already struggle with day-to-day burdens such as water shortage, food insecurity and disease.

In response to the floods last month, Karnataka's state government is providing food packages and temporary rehabilitation for those affected. More must be done, however. As the impacts of climate change intensify and increase over time, it is important that the government seeks to take a more proactive role to protect the lives and livelihoods of its citizens. In his speech at the Global Summit on Development and Climate Change in New Delhi in September, Vice President Shri M Hamid Ansari proclaimed "any action on climate change must enlarge, not constrict, the possibilities for development and empowerment of the world's poor." In reality, however, this is far from the case in India – the government appears to be doing very little to protect the poorest groups from the impacts of climate change.

In June 2008, Prime Minister Manmohan Singh proposed a National Action Plan on Climate Change (NAPCC). The Plan focuses on eight "National Missions" that concentrate on adaptation (preparing for foreseeable adverse effects) and mitigation (slowing and reducing harmful effects) strategies for addressing climate change, including energy conservation, energy efficiency, and sustainability through an array of new and existing programs and alternative energy projects.[7] But because building alternative power projects such as dams and hydropower plants physically displace rural communities and destroy livelihoods, human rights scholars have voiced concerns about the implementation of the NAPCC.[8] Similarly, Victoria Tauli-Corpuz, Chair of the UN Permanent Forum on Indigenous Issues (UNPFII) notes people living in rural communities often suffer "from the effects of climate change mitigation measures which are mainly market-based."[9] Large irrigation projects and other alternative energy "solutions" are estimated to have displaced between 80 and 150 million tribal and rural poor in India to date.[10] A representative from an affected village community in the rural state of Orissa told me "every three or four months, the Forest Department tries to clear us out... we were born and brought up here... If we can't stay here, where else can we go?"

In addition to social displacement, the design and implementation of these programs has been criticized for focusing on top-down measures which disempower local communities. Amnesty International has said that "in several [Indian] states... authorities failed to comply with new legislation guaranteeing access to information by denying affected communities information on planned development projects. In most cases communities were excluded from decision-making processes."[11] Market-based mitigation measures also create conditions for land speculation, land grabbing, land conflicts, corruption, and embezzlement of international funds by national elites.

If the Indian government is to tackle the effects of climate change while addressing its citizens' needs, it must recognise the importance of including local communities and empowering them to take part in the decision-making processes. A number of civil society and non-governmental organizations throughout India are working on local level initiatives, seeking to provide citizens with the skills and knowledge they need to cope with the impacts of climate change. Although some collaboration exists between these groups and the Indian government, it is the latter's responsibility to enhance such initiatives through further funding and support. This can be done in a number of ways.

First and foremost, the government should implement a structured public awareness program about climate change, its causes, and its potential impacts. It is vital that the groups likely to be most affected are aware of climate change and have strategies in place for coping with its effects. In order to help communities learn about and cope with climate change, the Centre for Environment Education is currently working in collaboration with the Ministry of Environment and Forests (MoEF) in implementing these types of environmental education programs, including workshops and capacity building programs.

In Bangladesh,[12] Oxfam International is facilitating the establishment of community preparedness committees – local groups which are given the responsibility to inform and help people before, during, and after floods. Hawa Parvin, the President of one of the village committees says "Previously we just reacted. We'd work together, but now we plan before the flood happens. It's meant that, for example, we didn't have to leave here [after the floods in

2004].”[13] It is vital that funding is channelled into similar initiatives in India, providing knowledge and local capacity building which will reduce the vulnerability of rural groups across the country.

Secondly, the government should introduce advocacy and civil society inclusion programs to empower local stakeholders to participate in the development of government plans and strategies. This will enable the identification of potential problems or issues before they arise, combining technical expertise with local knowledge in order to help the government achieve sustainable solutions. Gathering and spreading local community knowledge is likely to have an array of positive outcomes, including devising local agricultural and forestry solutions. A number of communities throughout Asia are already using local knowledge to create climate change adaptation strategies; in Vietnam for example, some communities have planted dense mangroves along the coast to diffuse tropical-storm waves.[14]

In an attempt to take this notion forward, UNPFII is collecting documentation from adivasi groups about their “experiences in terms of their resilience, the adaptation and mitigation measures they themselves are taking and their traditional knowledge.”[15] These reports are then compiled and submitted to the UN Framework Convention on Climate Change, illustrating how “indigenous knowledge can become part of a shared learning to address climate-change.”[16] Dr. Jacob Thundyil, Convenor of the National Advocacy Council for Development of Indigenous People (NAC-DIP), believes that the Indian government should include the opinions of these groups in its policy making: “there should be legal provisions enabling the… communities to patent for their intellectual property and traditional knowledge… [and the] benefits of the mega-projects… should be shared with the community.”

Lastly, in order to support the suggestions above, the Indian government should have in place monitoring and research activities to measure local changes in climate and community-level impacts of climate change. This will develop the country's resilience to climate change by providing valuable information to help local communities create strategic climate change adaptation and mitigation measures. Development Alternatives (DA), for example, is an NGO engaged in a range of environmental monitoring and education activities. They conduct research to develop a southern perspective on climate change – one of the ways in which they do this is by disseminating kits to schools, NGOs, and companies to encourage people to monitor local environmental quality. Proactive steps like these – those which actively seek out local information and engage communities – will ensure the future livelihoods of the people of India.

In Sum

While the Indian Government's National Action Plan on Climate Change represents an important first step in addressing the country's future approach to global warming, it fails to address this increasingly pressing issue from a humanitarian perspective. The implementation of alternative power plants has forcibly displaced millions from their homes, and lack of consultation with local communities is unlikely to create sustainable solutions for the country.

Climate change is happening, and its poor and marginalized groups with the least knowledge about it are likely to be affected the most. The future lives and livelihoods of these citizens at present rests in the hands of social action and humanitarian groups that seek to implement local solutions to this global problem. In order to enhance these programs and to achieve reach across the country, it is vital that the Indian government comes on board to help implement a national people-centred plan, improving access to climate change education, promoting social inclusion, and researching the impacts that climate change is already having across the country. At present, the Indian government's lack of involvement raises serious concerns about equity, justice, and human rights in a country which is widely acknowledged as the world's "largest democracy."

In a country with a population of 1.2 billion, full commitment from the government is the only way India will be able to cope with the impacts of climate change in the imminent future. It must recognize that its own citizens are the key to tackling climate change and implementing sustainable development solutions.

Nicola Macnaughton, 'Climate Change in India: A Humanitarian Perspective', *The Mantle*, 2 November 2009.

Notes

1   Hindustan Times. "Death toll in Karnataka and Andhra Pradesh floods crosses 250" (October 6, 2009).
2   The Economic Times. "Food Import Likely to Tide over Economic Crisis" (August 22, 2009).
3   United Nations. "The United Nations Framework Convention on Climate Change: Facing and Surveying the Problem," (http://unfccc.int/essential_background/feeling_the_heat/items/2914.php).
4   CARE International. "Humanitarian Implications of Climate Change: Mapping Emerging Trends and Risk Hotspots" (2008).
5   R.V. Cruz et al. "Asia: Climate Change 2007; Impacts, Adaptation and Vulnerability," in Fourth Assessment Report of the Intergovernmental Panel on Climate Change, M.L. Parry, et al., eds. (Cambridge: Cambridge Univ. Press, 2007): http://www.ipcc.ch/pdf/assessment-report/ar4/wg2/ar4-wg2-chapter10.pdf.
6   CARE International, 2008.
7   Government of India, Prime Minister's Council on Climate Change. "National Action Plan on Climate Change (NAPCC)" (June 2008).
8   Rahul Goswami. "Blind spots in India's new National Action Plan on Climate Change," Infochange. (September 2008): http://infochangeindia.org/200809237384/Environment/Analysis/Blind-spots-in-India%E2%80%99s-new-National-Action-Plan-on-Climate-Change.html.
9   Victoria Tauli-Corpuz. "UN Permanent Forum on Indigenous Issues Statement on Biodiversity and Climate Change (Agenda Item 4.5)," Conference of Parties, Bonn Germany (May 23, 2008).
10 Anna Pinto. "India in Climate Change," Centre for Organisation Research and Education for the Gender and Climate Change Network (January 22, 2009).
11 Amnesty International. "Report 2009," http://thereport.amnesty.org/en/regions/asia-pacific/india.
12 A neighbouring country similarly affected by climate change.
13 Oxfam International. "Bangladesh: preparing for flood disaster," Retrieved November 1, 2009: http://www.oxfam.org/en/campaigns/climatechange/bangladesh-preparing-flood-disaster.
14 Jan Salick and Anja Byg. Indigenous Peoples and Climate Change (Oxford, Tyndall Centre for Climate Change Research, 2007): 17.
15 Tauli-Corpuz, 2008.
16 Ibid.

Indonesia

Chapter 43

Britta Heine and Lorenz Petersen

ADAPTATION AND COOPERATION

Adaptation to climate change has become an important issue, both at international climate policy level and at the level of practical implementation on the ground.

Adaptation focuses on reducing (poor people's) vulnerability and thereby preventing both displacement and conflicts over scarce resources. Developing countries are especially vulnerable to the consequences of climate change, particularly where their livelihoods are directly dependent on climate and weather conditions. Poverty itself is a major cause of vulnerability to the consequences of climate change. A lack of capacity (technical, human and financial) makes it harder to adapt to changing conditions and to mitigate the risks. In other words, climate change will first and foremost intensify pre-existing problems in developing countries, which will generally have difficulties in coping with and adapting to these additional challenges.

Climate change adaptation means re-examining and, if necessary, modifying our policies, programmes, investments and, ultimately, behaviours in the light of our knowledge about climate change and its impacts. This may mean coping with changing risks but it may equally mean capitalising on positive impacts of climate change.

It is important to distinguish between non-climate drivers, such as government policies or population growth, and actual climate drivers. Climate drivers can be processes with slow-onset changes such as sea-level rise, build-up of salt in agricultural land, desertification and growing water scarcity. Climate events are sudden dramatic hazards such as monsoon flooding, storms and outburst floods from glacial lakes. All contribute to increasing the number of vulnerable people living on marginal land exposed to climate change. While climate processes – being long-term by nature – need to be addressed by long-term adaptation strategies, climate events require measures of disaster risk management. In combination, the application of adaptation strategies and the implementation of successful disaster risk management will lead to less vulnerability.

Adaptation strategies

Adaptation calls for the collective efforts of various actors, working on different levels and across sectoral boundaries. Every adaptation strategy involves three main steps. The first step is to gain a clear picture of the anticipated climate impacts in order to gauge the vulnerability of societies and ecosystems. In contrast to disaster risk management, this extends beyond an appraisal of the immediate hazards and vulnerability; it also encompasses an assessment of future trends or the possible range of anticipated climate changes. The second step is to compare climate impacts with vulnerabilities in order to derive possible adaptation measures. Establishing financial and economic costs by carrying out cost-benefit analyses helps to identify priority measures. The third and final step in this sequence is to determine the governance aspects. Who should most usefully tackle which area, with which risk management intervention? By following this sequence it is possible to develop local, national or regional adaptation strategies.

These steps have been applied within existing projects and programmes of German development cooperation focusing on climate change adaptation and disaster risk management. In view of the dimensions of the problem, however, international development cooperation can only contribute a part of the necessary resources for adaptation measures. Hence, supporting local governments in formulating adaptation strategies and priority setting is an important task for development cooperation.

Disaster risk management – coping with climate risks in Mozambique

The core aim of disaster risk management is to reduce the risk of disaster for societies living in regions threatened by natural hazards (risk management) and to prepare them to cope if disaster strikes (preparedness). In Mozambique, German development cooperation has successfully implemented a community-based programme which exemplifies the important role of disaster risk management for successful adaptation to climate events.

Mozambique is one of the poorest countries in the world. The watershed of the Búzi River in Sofala province is affected by very severe floods and neighbouring communities are also regularly hit by flooding and cyclones. Flooding in 2000 affected some 4.5 million people and claimed the lives of 800. According to reports from the IPCC, a further increase in heavy rainfall events and more intensive and widespread droughts in Mozambique are very likely.

[…]

In 2001 – one year after the flood – GTZ initiated a rural development programme with a disaster risk management component. A participatory risk analysis identified one-third of the district's inhabitants as particularly vulnerable to extreme natural events. The most risk-prone areas have since been mapped in detail, along with areas of higher ground which could be used for emergency evacuation purposes. On the basis of the risk analysis, local Disaster Management Committees were established in nine communities and trained with the help of experts from Costa Rica and Honduras. Villagers and volunteers also established a local early warning system based on relatively simple resources. Readings of daily rainfall and river water levels are taken at seven measuring stations; these trigger early warnings when necessary. The local Disaster Management Committees receive these warnings, mainly through Radio Comunitarió do Búzi (in Portuguese), and then inform the neighbourhood – in the local dialect, Ndau – and organise transport and evacuation. Translation into the local dialect has been an important factor for the success of the project.

Furthermore, a syllabus and lesson plans were developed at four project schools to raise awareness among children and young people about the impacts of climate change and to familiarise them with the theme of disaster risk management at an early age. Both themes have been integrated into local curricula.

As a result of these activities, villagers and local government representatives, teachers and schoolchildren have been sensitised to the issues, and disaster risk management measures for climate change adaptation have been integrated into the district's development plan. The robust disaster risk management system now makes it possible to deliver an early warning of flood events and the system has already proven its efficiency. During the rainy seasons of 2005 and 2007, extreme flooding struck the area once again but most of the inhabitants escaped and survived. In the meantime, the system has been further calibrated and refined.

The success of the project is ascribed to its high level of participation of and ownership by the people of the Búzi River. Moisés Vicente Benessene describes it as a "people-centred early warning system", based on local knowledge, customs and cultural values. As local leaders, doctors and teachers have all volunteered to serve in local Disaster Management Committees, taking on responsibility has become highly respected in the communities. Some challenges remain, however, such as keeping the system running and constantly maintaining people's awareness of climate risks and the use of disaster preparedness.

The experiences of Búzi have been shared with other districts, where plans are now also in hand to establish local disaster risk management mechanisms. The people of Búzi have shown that climate-driven disasters and threats can be effectively met by concerted, decentralised community action and self-organisation at low cost.

Mainstreaming adaptation in Indonesia

Adaptation is a cross-cutting task, requiring the coordinated efforts of different actors within and also beyond the state. Adaptation needs to build on and be supported by activities by relevant ministries (e.g. environment, finance and planning) and those responsible for identifying budgetary priorities as well as specialised agencies such as geological and meteorological services and institutions for disaster prevention. National strategies can help provide a framework for coordinating adaptation activities, enabling informed decision-making, mobilising national and international support and developing appropriate institutional structures for adaptation.

Indonesia faces increased vulnerability to the effects of climate change, especially rising sea levels, changes in precipitation and extreme weather events. Climate projections indicate that the mean wet-season rainfall will increase across most of Indonesia while the length of the dry season is expected to increase, bringing increased risk of floods during the rainy season and drought in the dry season. This will have a particular impact on water resources, agriculture and forestry, health and infrastructure.

Dr Sutardi of Indonesia's Ministry of Public Works and Executive Secretary of Indonesia Water Partnership explains that "most people have not yet integrated the issue of climate change into their everyday life. They still feel there was just a bit 'too much rain' during the rainy season or 'too little' in the dry months." However, adaptation to climate change is now a major concern of the Indonesian government. The Ministry of the Environment has initiated the development of a national strategic approach to adaptation planning. Challenges faced include the availability and dissemination of relevant information and planning tools, awareness of the issue among decision-makers, and the involvement of line ministries and local-level administrations that are key to implementing adaptation.

GTZ is working on a project to help enhance the capacity of policymakers to mainstream climate change issues into development planning. It provides assistance in assessing vulnerability to, and the economic impacts of, climate risks as well as prioritising adaptation options in the water sector. The second focus is on cooperation between different institutions concerning adaptation. The assessment of impacts, vulnerabilities and adaptation options should lead to increased awareness and informed decision-making in water resource management and beyond. The integration of measures into development and financial planning should contribute to the sustainable management of public resources. And, finally, systematic priority setting should improve the efficiency of measures that reduce vulnerability to climate change.

Throughout the project, the importance of providing relevant information at different levels and to different stakeholders has become evident. In the water sector, detailed technical information is necessary to inform decision-making while awareness-raising requires there to be more aggregated knowledge. Improving institutional capacities for coordination is of particular importance at all levels.

Migration as an adaptive response

At some point a region may become no longer capable of sustaining livelihoods. People will be forced to migrate to areas that present better opportunities. Generally, the international adaptation community tends to regard migration as an 'adaptation failure'. However, migration is (and has been for a long time) an adaptive response to climate stress in many areas.

Temporary migration, for example, in times of climate stress can help top up a family's income (from paid work elsewhere) and reduce the drain on local resources. In Botswana, for instance, many of the urban poor rely on livestock and farmland in rural home areas for food and income reserves. Yet, as non-residents in their home area, they are not entitled to drought relief and risk heavy losses without compensation in the event of failure of the rains.

Policies attempting to limit migration while disregarding causes and circumstances are likely to increase poor people's vulnerability. National (and regional) adaptation strategies should therefore incorporate migration as an adaptation option – recognising, for instance, that people often live and keep their assets in more than one place. In this respect, development cooperation can help improve local government's capacity to address migration as an adaptation option and accommodate migration rather than attempting to limit it.

[Footnotes omitted]

Britta Heine and Lorenz Petersen, 'Adaptation and Cooperation', *Forced Migration Review*, Issue 31, October 2008, pp. 48–50.

Maldives

Chapter 44

Ministry of Environment, Energy and Water, Maldives

MALDIVES NATIONAL ADAPTATION PROGRAMME OF ACTION (NAPA)

[…]

Chapter I: Introduction

> …there must be a way out. Neither the Maldives nor any small island nation wants to drown. That's for sure. Neither do we want our lands eroded nor our economies destroyed. Nor do we want to become environmental refugees either. We want to stand up and fight.
>
> (President Maumoon Abdul Gayyoom, Small States Conference on Sea Level Rise, Male', 1989)

The Maldives is among the most vulnerable to predicted climate change and non-action is not an option for the country. The number of scientific and technical assessments undertaken in the country since 1987 has reiterated the need for long-term adaptation to climate change. Since the commencement of sea wall construction around the capital Male' in September 1988 the government has implemented several projects aimed at adaptation to environmental threats.

The Maldives played an important role in the negotiations that led to the United Nations Framework Convention on Climate Change (UNFCCC) and was the first to sign the Kyoto Protocol to the UNFCCC. The Maldives submitted the First National Communication (FNC) to the UNFCCC in 2001 following the implementation of the Maldives GHG.

Inventory and Vulnerability Assessment: A Climate Change Enabling Activity. The FNC contained mitigation and adaptation measures and the project profiles for continuing climate change adaptation and mitigation process.

This is the first National Adaptation Programme of Action (NAPA) developed to communicate the most urgent and immediate adaptation needs of the Maldives as stipulated under UNFCCC Decision 28/CP.7. NAPA was prepared with support from the Global Environment Facility (GEF) and United Nations Development Programme (UNDP). Preparation of NAPA began in October 2004 and the process was halted because of the South Asian tsunami of December 2004. NAPA work recommenced in February 2006. NAPA process was guided by the principles of broad stakeholder engagement, partnership building among focal agencies and ownership by the people of Maldives especially the atoll population. A multidisciplinary National Climate Change Technical Team (CCTT) was established as a first step to foster stakeholder engagement. Community consultations and awareness raising activities were held for representatives from seven atolls of the Maldives and the capital Male'. Targeted awareness raising and activity-based learning was conducted for school children from five secondary schools. Existing climate data for the Maldives was analysed with international expertise culminating in the first Climate Risk Profile for the Maldives. National experts produced vulnerability and adaptation (V&A) related technical papers for priority sectors identified by the NAPA Working Group. Extensive consultations at regional and national level were undertaken based on a prior agreed methodology to identify vulnerabilities and adaptation activities and to prioritize these activities.

The NAPA is intended to be concise as well as brief and contains seven chapters. Following this introduction, Chapter Two presents the NAPA goal and describes the National Adaptation Policy Framework. Chapter Three describes the country characteristics and national development goals. Chapter Four depicts the climate hazards and risks. Chapter Five analyses vulnerabilities and the biophysical impacts of climate change. Chapter Six lists the adaptation needs and priority activities. NAPA concludes with Chapter Seven that contains the project profiles for adaptation to climate change in the Maldives.

[...]

5.1 Land, Beach and Human Settlements

The small size, extremely low elevation and unconsolidated nature of the coral islands place the people and their livelihoods at very high risk from climate change, particularly sea level rise.

Maldives is the sixth smallest sovereign state in terms of land area. The total land area of the Maldives is estimated to be approximately 235km^2, based on the latest satellite and aerial imagery. This land is divided over 1192 coral islands and 96% of the islands are less than 1km^2 in area. Only 10 islands are more than 2.5km^2. The largest island Gan (Laamu Atoll), has an area of 6.1km^2. Land is highly scarce and the 358 islands that are currently in use account for 176km^2. The 834 unutilised islands make up only 59km^2.

[...]

Over 80% of the total land area of the Maldives is less than 1m above MSL. The highest point recorded in the country is a beach ridge at Fuvahmulah (Gnaviyani Atoll), with an elevation of 4m above MSL (MHAHE 2001). As future sea level is projected to rise within the range of 9 to 88cm between 1990 and 2100, the islands of Maldives would be submerged in the projected worst case scenario.

The coral islands that make up the Maldives are morphologically unstable and change in their size, shape, elevation and position on reef platforms over time. The beaches of these islands are particularly dynamic with substantial seasonal changes. At present, the total beach area is estimated at 13km^2 or 5% of the total land area and the coastline of the Maldives is estimated to be 2,300km long (Shaig 2006).

The small size of the islands forces people to live next to the sea. At present, 44% of the settlement footprints of all islands are within 100m of coastline. This translates to 42% of the population and 47% of all housing structures being within 100m of coastline. More than 50% of the housing structures in 121 islands are within 100m of coastline (Shaig 2006). Only Nolhivaramu (Haa Dhaalu), has all its housing structures 100m away from the coastline. Given the close proximity of the settlements to the sea and low elevation of the islands, homes of people are at severe risk of inundation with higher sea levels.

The small size of the islands and their low elevation also makes human settlements defenseless against severe weather events and storm surges. Over the last 6 years more than 90 inhabited islands have been flooded at least once and 37 islands have been flooded regularly or at least once a year (Shaig 2006). The series of swells, between 10–15 feet, which hit the Maldives on 15–17 May 2007, impacted an estimated 68 islands in 16 atolls and more than 500 housing units were damaged. An estimated population of more than 1600 people was evacuated from their homes (OCHA 2007). Since housing designs, structures and materials are not adapted to flooding, it exacerbates the vulnerability (UNEP 2005). The flooring of houses does not have adequate elevation from the ground and because of the poor construction material used for housing structures, higher frequency and increase in intensity of flooding could make these islands uninhabitable.

The beaches that represent 5% of the total land area of the Maldives are of unconsolidated nature and naturally dynamic and unstable. More than 97% of inhabited islands reported beach erosion in 2004, of which 64% reported severe beach erosion. Erosion patterns of inhabited islands have been further complicated due to human intervention in coastal areas. The problem of erosion is not specific to inhabited islands. More than 45% of the 87 tourist resorts have reported severe erosion (Shaig, 2006). Although beach erosion can be attributed to a number of factors, changes in climatic conditions is known to exacerbate erosion (Nurse and Sem 2001). In the Maldives the intensity and duration of northeast and southwest monsoons affect beach erosion patterns. Further aggravation of erosion through elevated sea level and storm surges would cause significant loss and damage to people's property, tourist resorts, valuable land and critical infrastructure. About 85% of the 68 impacted islands during the series of wave surges that hit the Maldives in May 2007 reported having significant erosion of the island coastline (OCHA 2007.)

[…]

Human pressures also increase the vulnerability of the land and beaches to predicted climate change. The population of the country has increased by four-fold since 1911 and overcrowding is a significant problem. Already 34 of the inhabited islands do not have additional land for new housing and another 17 islands will reach their carrying capacity by 2015 (Shaig 2006). Land reclamation work has been carried out to alleviate population pressure on land. Table 5.2 lists some of the major land reclamation projects in the last 30 years. Hulhumale' (Kaafu Atoll) is the largest land reclamation project where approximately 2km² was reclaimed to reduce population pressure on Male'.

Apart from land reclamation, several other human activities have increased vulnerability of land, beaches and human settlements. They include construction of poorly designed coastal infrastructure, poorly engineered coastal protection measures, removal of coastal vegetation, and sand mining. Coral reefs have a critical coastal protection function, yet there have been a number of human stresses on the reef system such as coral mining, reef entrance blasting, dredging, solid waste disposal and sewage disposal that has affected the health, integrity and productivity of reefs.

The scarcity of land in the Maldives, the smallness of the islands and extreme low elevation makes retreat inland or to higher grounds impossible. Building setback has limited utility and beach replenishment may only be a temporary remedy for beach loss. Unless expensive coastal protection measures are undertaken the human settlements face the threat of inundation.

[...]

6.1 Adaptation Needs

The adaptation needs were identified through wide stakeholder consultations. The adaptation needs are classified here under the priority sectors presented in Chapter Five. These sectors are similar to those given in the FNC (2001).

6.1.1 Land, Beach and Human Settlements

1. Consolidate population and development.
2. Acquire support for the speedy and efficient implementation of Safer Island Strategy.
3. Strengthen land-use planning as a tool for protection of human settlements.
4. Build capacity for coastal protection, coastal zone management and flood control.
5. Protect beaches through soft and hard-engineering solutions.
6. Protect house reef to maintain natural defense of islands.
7. Improve building designs and regulations to increase resilience.
8. Integrate climate change adaptation into national disaster management framework.
   [...]

Ministry of Environment, Energy and Water, ' Maldives National Adaptation Program of Action (NAPA)', Republic of Maldives, 2007, pp. 1–2, 19–22 and 40.

Vietnam

Chapter 45

Care, CIESIN, UNHCR, UNU-EHS and the World Bank

MEKONG DELTA
Living with Floods and Resettlement

[…]

Environmental degradation, particularly impacts caused by flooding, is a contributing factor to rural out-migration and displacement in the Mekong Delta of Vietnam. The Vietnamese portion of the Mekong Delta is home to 18 million people, or 22 percent of Vietnam's population. It provides 40 percent of Vietnam's cultivated land surface and produces more than a quarter of the country's GDP. Half of Vietnam's rice is produced in the Mekong Delta, 60 percent of its fish-shrimp harvest, and 80 percent of Vietnam's fruit crop. Ninety percent of Vietnam's total national rice export comes from the Mekong.

Flooding plays an important role in the economy and culture of the area. People live with and depend on flood cycles, but within certain bounds. For example, flood depths of between half a meter to three meters are considered part of the normal flood regime upon which livelihoods depend. These are so-called "nice floods" [ngập nông] by Vietnamese living in the delta, such as upstream in the An Giang Province. Flood depths beyond this such as between three and four meters [ngập vừa], however, challenge resilience capacities of affected people and often have harrowing effects on livelihoods.

Floods exceeding the four meter mark, called "ngập sâu" for severe flooding, have increased in magnitude and frequency in Vietnam in recent decades.[83] In Phnom Penh (Cambodia) one migrant from the Delta noted, "Flooding occurs every year at my former living place. I could not grow and harvest crops. Life therefore was very miserable. Besides, my family did not know what else we could do other than growing rice and fishing. Flooding sometimes threatened our lives. So we came here to find another livelihood."[84]

Another migrant said, "My family had crop fields but in recent years, floods occurred very often so the crop was not stable. In addition, the price of fertilizer increased very fast, and the diseases of the rice plant are too much, so the crop yield was nothing. Even sometimes the yield was not enough to cover the amount required for living".[85]

"Natural hazards, in combination with the stress placed on the environment due to rapid socioeconomic development within Vietnam and upstream South-east Asian countries, overlaid with the threats posed to Vietnam by climate change, places Vietnam's natural resources and those who depend upon them for their livelihoods in a precarious position. In the face of environmental stressors, people in the Mekong Delta adapt in various ways. One type of adaptation mechanism may be migration, particularly in light of the rapid socio-economic changes that Vietnam is currently experiencing, which create stronger pull factors towards urban environments".[86]

Fieldwork from the EACH-FOR project indicated that lack of alternative livelihoods, deteriorating ability to make a living in the face of flooding, together with mounting debt, can contribute to the migration "decisions" in the Mekong Delta. People directly dependent on agriculture for their livelihood (such as rice farmers) are especially vulnerable when successive flooding events destroy crops. This can trigger a decision to migrate elsewhere in search of an alternative livelihood. During the flooding season, people undertake seasonal labor migration and movement towards urban centers to bolster livelihoods. As an extreme coping mechanism, anecdotal information from fieldwork pointed to human trafficking as one strategy adopted by some families who have suffered from water-related stressors.

A migrant interviewee referred to the financial vulnerability of her family related to flooding, "Disasters occurred so often – my family lost the crop, my family had to borrow money to spend. Now, my family is not able to pay off the loan so I have to come here to work to help my family to pay the loan."[87] The government in Vietnam has a program known as "living with floods."[88] This program may become more important as the impacts of climate change become more pronounced. The government, as part of this flood management strategy, is currently resettling people living in vulnerable zones along river banks in the An Giang province.[89] Almost 20,000 landless and poor households in this province are targeted for relocation by 2020. Households are selected for resettlement based on a number of factors related to the environment, such as living in an area at risk of natural calamities (flooding, landslides) or river bank erosion. These resettlement programs allow families to take up a five-year interest-free loan to enable them to purchase a housing plot and basic house frame. Households then often need a further loan to complete building the house.[90] The clusters provide few infrastructure services like access to schools, health, or water and sewage treatment facilities.[91] People planned for relocation are usually the landless who have nowhere else to move if their houses collapse and are often too poor to move to urban areas. For these people, social networks provide the link to livelihoods—most rely on day-to-day employment as laborers. Although the "residential clusters" are usually located only 1–2 kilometers away from the former residence, moving people out of established social networks threatens their livelihoods and contributes to a sense of isolation. The resettlement clusters are not yet planned in a way that allows participation of potential residents.

The Vietnamese strategy of "living with floods" will combine resettlement, shifting livelihoods (i.e. from rice to fishery-based jobs), and some migration. In the future one out of every ten Vietnamese may face displacement by sea level rise in the Mekong Delta.[92]

Care, CIESIN, UNHCR, UNU-EHS and World Bank, Social Dimensions of Climate Change, 'In Search of Shelter: Mapping the Effects of Climate Change on Human Migration and Displacement', May 2009, p. 15.

Notes

83   Lettenmaier 2000 cited in White, I., ed. 2002. Water management in the Mekong Delta: Changes, conflicts and opportunities. IHP-VI Technical Papers in Hydrology No.51, UNESCO, Paris, 11; and Nguyen Thanh Binh, Nguyen Thanh. 2009. Flood levels in Vietnam. Personal communication. UNU-EHS WISDOM Project.

84   Pancelet 2009: 17.

85   Ibid.

86   Zhang, H., P. Kelly, C. Locke, A. Winkels, and W. N. Adger. 2006. Migration in a transitional economy: Beyond the planned and spontaneous dichotomy in Vietnam. Geoforum 37: 1066–1081.

87   Dun, O. 2009. Linkages between flooding, migration and resettlement. Case study report on Vietnam for the Environmental Change and Forced Migration Scenarios Project, 17. http://www.each-for.eu/documents/CSR_Vietnam_090212.pdf Pp.17.

88   Ibid.

89   Le, T., H. Nguyen, H. Nhan, E. Wolanski, T. Tran, and H. Shigeko. 2007. The combined impact on the flooding in Vietnam's Mekong River delta of local man-made structures, sea level rise and dams upstream in the river catchment. Estuarine, Coastal and Shelf Sciences 71:110-116.

90   People's Committee of An Giang Province. 2006. Project: Removal of Canal Houses to Secure Environmental Sanitation of An Giang Province from now to 2020 (English translation). An Giang: People's Committee of An Giang Province.

91   Dun 2009. See note 87 above.

92   Dasgupta, S., B. Laplante, C. Meisner, D. Wheeler, and J. Yan. 2007. The impact of sea level rise on developing countries: A comparative analysis. World Bank Policy Research Working Paper 4136 (WPS4136), World Bank, Washington.

PART 5

Community and **NGO** responses and proposed solutions

Chapter 46

Displacement Solutions

CLIMATE CHANGE DISPLACED PERSONS AND HOUSING, LAND AND PROPERTY RIGHTS
Preliminary Strategies for Rights-based Planning and Programming to Resolve Climate-induced Displacement

Executive Summary

Few global issues have received the levels of persistent global attention afforded to climate change in recent years. The vast majority of this consideration has focused on the mitigation (prevention) dimensions of climate change, with coverage of adaptation (remedial) issues receiving substantially less attention. As a result, the mitigation dimensions of climate change are now found on top of most domestic political agendas, while adaptation issues lag far behind.

Despite this considerable coverage of both of these key responses to climate change, however, only limited attention has thus far been given to the particular issue of forced displacement caused by climate change, and even less, to the particular housing, land and property (HLP) dimensions of this form of displacement. Given the fact that estimates of those to be displaced due to climate change range between 150 million and one billion persons, this dearth of attention is staggering.

This paper, which forms a part of Displacement Solutions' Climate Change and Displacement Initiative, attempts to begin a structural process of closing this gap by focusing squarely on these issues and proposing strategies to States and the international community on how to best address the housing, land and property dimensions of climate change. It outlines the housing, land and property rights of climate change displaced persons, examines the consequences of climate displacement in four selected countries and proposes a range of rights-based housing, land and property legal and policy measures that can be initiated now in support of climate change displaced persons and their HLP rights.

This report urges States to immediately re-double legal and policy efforts within the housing, land and property sectors focused specifically on the rights of climate change displaced persons, and concludes with a series of recommendations to both Governments and the international community designed to improve the HLP rights prospects of climate change displaced persons.

1. Introduction

A plethora of studies, papers, reports, books and films have addressed the question of climate change, its causes, its effects and ways of ameliorating its consequences to people and the planet as a whole. The vast majority of these publications, as well as media coverage given to climate change, have focused on the scientific evidence of climate change, as well as the mitigation (or prevention) measures required to halt future climate change or keep these processes at manageable levels. These issues continue to dominate policy discussions within the climate change arena, with adaptation (or remedial) concerns relegated to a place rather far down the list of political priorities by the world's most powerful States and institutions. The International Panel on Climate Change (IPCC) WGII Fourth Assessment Report in 2007 is fairly typical in this respect, with perhaps the most profound statement on adaptation stating simply that "A wide array of adaptation options is available, but more extensive adaptation than is currently occurring is required to reduce vulnerability to future climate change. There are barriers, limits and costs, but these are not fully understood". Even further down the global political agenda than adaptation are the human rights implications of climate change, although these are beginning to receive much needed and important forms of attention. Still further down the climate change agenda comes the question of the human displacement sourced to the consequences of climate change. Right at the lowest echelon of global attention to these issues comes the topic of finding durable solutions to climate-induced forced displacement.

This is despite the stupefying fact that many millions – almost assuredly tens or even hundreds of millions – of people will face forced displacement and subsequent housing, land and property losses due to climate change in the coming decades. Indeed, human displacement issues remain a sorely neglected aspect of the current climate change debate, even though these raise very real legal, political, economic, human security, human rights, public health and even security and conflict prevention concerns. Understanding the related human rights dimensions of climate change, particularly the very practical issues associated with possible housing, land and property rights infringements affecting climate change displaced persons and their eventual restoration, will assist in providing the conceptual foundations required to develop both practical and durable rights-based solutions for those facing climate-induced displacement.

This report addresses two key questions. Firstly, what specific HLP rights do people forced to flee their homes and lands due to climate change – climate change displaced persons – have under existing international human rights law? And secondly, which general steps should be taken by Governments and the international community as a whole to adequately address these challenges and subsequently fulfil the rights of those concerned? In particular, which precise housing, land and property rights options are (or should be) available to climate change displaced persons in general terms, and in particular within four of the most heavily-affected countries, namely Papua New Guinea, Kiribati, Tuvalu and Bangladesh?

This paper is designed to assist current and prospective victims of climate change, in particular climate change displaced persons and organisations supporting the rights of climate change displaced persons, to better formulate their housing, land and property rights demands in terms of human rights. It is hoped that this analysis will also be of particular use to policy-makers in the run-up to the crucial round of climate negotiations in Copenhagen set for the end of 2009, where it is hoped that vital international agreement on climate change policy will take place and build upon the Kyoto Protocol to UN Framework Convention on Climate Change which expires in 2012.

At the same time, this report aims to assist local and national governments responsible for securing the housing, land and property rights of climate change displaced persons to better understand their current legal obligations under international human rights law, and where to turn in the event that they are no longer capable of respecting and protecting the housing, land and property rights of their citizens. Finally, this analysis equally seeks to assist countries throughout the world, both those generating climate change displaced persons and host countries, in determining the best policy responses to the displacement caused by climate change and the international community in supporting adequate national and local strategies to cope with the climate-induced displacement challenges ahead in a human rights consistent manner through international cooperation.

2. The Housing, Land and Property Rights of Climate Change Displaced Persons

Human rights play a direct role within the context of climate change in a myriad of ways. Rights such as the right to life, the right to water, the right to freedom of expression, the right to health, the right to food, the right to an adequate standard of living, the right to political participation, the right to information, the right to be free from discrimination, the right to equal treatment, the right to security of the person and a host of other rights should have a direct bearing on a wide cross section of climate change decisions made by Governments, and thus influence, how the consequences of these decisions and the impact of climate change will be experienced by individual rights-holders. The normative framework enshrining these rights is very clear, considerable, constantly evolving and ever expanding.

Taking a human rights approach to climate change-induced displacement can provide a clear and globally applicable means of developing viable rights-based solutions to this growing crisis. Grounded as the international human rights regime is, in the principle of the inherent dignity of the human person, implies that each and every person, family and community that is forced from their homes and lands, against their will, must have access to some form of remedy – both substantive and procedural – which respects their rights, protects their rights and, if necessary, fulfils their rights as recognised under international human rights law. In effect, everyone whose HLP rights are affected by climate change needs to have a means of remedying these denials through the provision of appropriate and durable HLP solutions to their status as climate change displaced persons.

Before identifying some of the HLP rights of climate change displaced persons, it is important to note that the 1992 UN Framework Convention on Climate Change (UNFCCC) mentions neither the displacement to be caused by climate change nor the human rights dimensions of global warming. Likewise, the 1951 Refugee Convention does not recognise the particular rights of climate change displaced persons, nor do any of the major human rights treaties. This lack of inclusion of this specific group of climate change displaced persons has led some to propose the amendment of the 1951 Refugee Convention and 1967 Protocol to expand the protection of these instruments to include climate change displaced persons. This initiative does not appear likely at the moment to have the political support required to amend the Refugee Convention and Protocol. Another proposal for a Protocol on the Recognition, Protection and Resettlement of Climate Refugees to the UNFCCC has also been developed, as has a specific convention detailing the rights of those displaced due to climate change. None of these pertinent and worthy initiatives, however, are likely to be approved any time soon.

The human rights dimensions of climate change were rather late arrivals into the global climate change debate. From 2007 onwards a series of efforts have been undertaken in this regard, including the release of publications making the explicit link between human rights and climate change. For instance, after extensive lobbying efforts, following an initiative led by the Maldives, in 2008 the UN Human Rights Council decided to produce a study on the human rights dimensions of climate change. The report was released in 2009, and while not containing many particularly new ideas, was nonetheless important in signifying UN attention to these issues. The OHCHR report notes that the following rights may be affected by climate change: the right to life, the right to adequate food, the right to water, the right to health, the right to adequate housing, and the right to self-determination, and rightfully emphasises further that "persons affected by displacement within national borders are entitled to the full range of human rights guarantees by a given state, including protection against arbitrary or forced displacement and rights related to housing and property restitution for displaced persons". The report further recommends that:

> Within countries, existing vulnerabilities are exacerbated by the effects of climate change. Groups such as children, women, the elderly and persons with disabilities are often particularly vulnerable to the adverse effects of climate change on the enjoyment of their human rights. The application of a human rights approach in preventing and responding to the effects of climate change serves to empower individuals and groups, who should be perceived as active agents of change and not as passive victims. (Para 94)
>
> Further study is also needed of protection mechanisms for persons who may be considered to have been displaced within or across national borders due to climate change-related events for those populations which may be permanently displaced as a consequence of inundation of low-lying areas and island states. (Para 98)

Also at the UN, the Special Rapporteur on Housing Rights, Raquel Rolnick, has recently drawn attention to the human rights issues arising from climate change, with an initial focus in the Maldives, as has the Special Representative to the UN Secretary-General on the Rights on Internally Displaced Persons, Walter Kaelin. These and a range of other initiatives increasingly recognise that housing, land and property rights lie at the core of a series of human rights remedies that are required to find durable solutions for climate change displaced persons in the context of forced climate displacement. Giving substance to these understandings, however, has been the exception to the rule.

What Are HLP Rights?

In terms of HLP rights protections, climate change displaced persons – for the moment at least – will need to rely on the rather vast body of international human rights law (and international law more generally) and the domestic human rights provisions found within the national legal frameworks of all nations as a basis for claiming and asserting their HLP rights. While the implementation of human rights law remains weak in many countries, viewed in its entirety the normative framework offered by human rights law may prove to be more inclusive in terms of protecting the general human rights and housing, land and property rights of climate change displaced persons than many will have imagined.

Combining the sentiments of the Universal Declaration on Human Rights (1948), the International Covenant on Civil and Political Rights (1966), the International Covenant on Economic, Social and Cultural Rights (1966) and a range of other treaties, together with a vast array of equally important instruments and interpretive standards such as the UN Committee on Economic, Social and Cultural Rights General Comment No. 4 on the Right to Adequate Housing (1991), General Comment No. 7 on Forced Evictions (1997), the UN Guiding Principles on the Rights of Internally Displaced Persons (1998), the UN 'Pinheiro' Principles on Housing and Property Restitution for Refugees and Displaced Persons (2005) and many others, reveals a very considerable body of international human rights laws and standards which can and should be used by Governments to build the legal, policy and institutional frameworks required to provide rights-based durable solutions to the displacement caused by climate change.

Human rights law provides that everyone, everywhere possesses a body of HLP rights and every Government in every State is obliged to ensure the protection and enforcement of these rights. When combining together all of the entitlements and obligations inherent within this bundle of rights or HLP (housing, land and property) rights as they are now commonly referred to, people everywhere are meant to be able to live safely and securely on a piece of land, to reside within an adequate and affordable home with access to all basic services and to feel safe in the knowledge that these attributes of a full life will be fully respected, protected and fulfilled. As citizens and rights holders, these rights need to be enjoyed universally by all climate change displaced persons. In this regard, if it can be shown that the HLP losses incurred by climate change displaced persons constitute either direct violations of their HLP rights, or at the very least, result in the inability to access the protections afforded under law by HLP rights, then it is clear that according to human rights law, appropriate forms of reparation and restitution must be accorded to those who have lost access to, use of or ownership over housing, land or property lost due to climate change.

As with all other nations, people in countries most heavily affected by climate change, in particular those who are subject to forced displacement, are entitled under international human rights law to enjoy a series of rights which together constitute housing, land and property rights. In many respects, it will be housing, land and property rights that form the basis of the durable solutions required to resolve displacement sourced to climate change. Problematically, however, domestic HLP solutions – particularly in the long-term – may prove incapable of securing the rights of climate change displaced persons, and thus these rights will need to be fully secured in host countries where these persons eventually resettle.

HLP rights are widely recognised throughout the ever evolving corpus of human rights law, and when we examine those rights which have a direct bearing upon the conditions in which people live in a residential context, e.g. their housing, land or property rights, it is clear that these are far more extensive than commonly assumed. In essence, HLP law constitutes a composite of the following existing rights found within international human rights law:

- The right to adequate housing and rights in housing
- The right to security of tenure
- The right not to be arbitrarily evicted
- The right to land and rights in land
- The right to property and the peaceful enjoyment of possessions
- The right to privacy and respect for the home
- The right to HLP restitution/compensation following forced displacement
- The right to freedom of movement and to choose one's residence
- The right to political participation
- The right to information
- The right to be free from discrimination
- The right to equality of treatment and access
- The right to water
- The right to energy.

According to human rights law, in their totality, these HLP rights should fully inform a wide cross-section of legislative, policy and practical decisions made by Governments. The degree to which these rights and underlying human rights principles such as participation, accountability, non-discrimination and transparency are woven into the contours of State law and policy will greatly affect how these decisions will be supported by individual rights-holders and to what extent they will actually be able to meet the underlying needs of the people affected. The long struggle to define the components of HLP rights and to delineate the corresponding obligations of States to secure these rights has resulted in a clear normative framework of what HLP rights mean in terms of human rights law. It is, for instance, widely agreed that States with HLP rights obligations are not necessarily required to be the primary housing providers within a society, nor are Governments necessarily required to substantively provide a house to all citizens who claim to need a dwelling in which to live. Rather, human rights laws suggest that States are expected to comply with various levels of obligations that emerge from a recognition of HLP rights that lead to the creation of conditions within a given society that are as facilitative as possible for everyone, within the shortest possible time-frame, to secure by various means the full attributes of HLP rights. This takes many forms and involves a series of corresponding obligations.

States are bound by obligations to respect, protect, promote and fulfil these rights. Human rights laws indicate that once such obligations have been formally accepted through the ratification of an international or regional treaty or promulgation of related domestic legislation, the State must endeavour by all appropriate means to ensure everyone has access to HLP resources adequate for health, well-being and security, consistent with other human rights, including those affected by climate change. Governments must, therefore, adopt the policies, laws and programmes required – to the maximum of their available resources – to continually and progressively expand the enjoyment of these rights and simultaneously ensure in policy, legal or other terms, that no deliberately retrogressive measures are taken that lead to the decline in the enjoyment of these basic rights. Of all the HLP rights, the right to adequate housing has advanced the farthest. In 1991, the UN Committee on Economic, Social and Cultural Rights adopted 'General Comment No. 4 on the Right to Adequate Housing' which indicates that the following seven components form the core contents of the human right to adequate housing: (a) legal security of tenure; (b) availability of services, materials, facilities and infrastructure; (c) location; (d) habitability; (e) affordability; (f) accessibility; and (g) cultural adequacy. General Comment No. 4 also reiterates that the right to adequate housing should not be interpreted in a narrow or restrictive sense which equates it with the shelter provided by merely having a roof over one's head or views shelter exclusively as a commodity, but that housing rights should be seen as rights to live somewhere in security, peace and dignity. To achieve these rights, States need to respect these rights by ensuring that no measures are taken which intentionally erode the legal and practical status of this right. Governments need to comprehensively review relevant legislation, refrain from actively violating these rights by strictly regulating forced evictions and ensure that the housing, land and property sectors are free from all forms of discrimination at any time. States must also assess national HLP conditions, and accurately calculate, using statistical and other data and indicators, the true scale of non-enjoyment of these rights, and the precise measures required for their remedy. All of these are directly relevant to climate change displaced persons. Governments need to protect the rights of people by effectively preventing the denial of their rights by third parties such as landlords, property developers, social service providers and others capable of restricting these rights. To promote HLP rights, Governments should adopt targeted measures such as national HLP strategies that explicitly define the objectives for the development of the HLP sector, identify the resources available to meet these goals, the most cost-effective way of using them and how the responsibilities and time-frame for their implementation will be applied. Such strategies should reflect extensive genuine consultation with, and participation by, all those affected, including groups traditionally excluded from the enjoyment of HLP rights. Finally, the obligation to fulfil these rights involves issues of public expenditure, the regulation of national economies and land markets, housing subsidy programmes, monitoring rent levels and other housing costs, the construction and financing of public housing, the provision of basic social services, taxation, redistributive economic measures and any other positive initiatives that are likely to result in the continually expanding enjoyment of HLP rights.

In terms of human rights law, thus, citizens and residents of every country should have direct and sustained access to the full realization of the entire spectrum of housing, land and property rights, while Governments should take seriously their numerous obligations to respect, protect, promote and fulfil HLP rights. The people of all of the countries most heavily affected by climate change, therefore – just as people everywhere – are entitled to live in societies where HLP rights are treated with the seriousness accorded them under human rights law, and in accordance with the maximum of resources available to the State to respect and protect them.

To be effective in this regard, HLP law needs to be internally consistent, conform with all relevant international standards and norms and, above all, reflect the popular will of the population. It is difficult to point to any nation in the world today which has a flawless body of HLP legislation in place, as this area of law remains both immense, complicated and often subject to effective policy as a means to promote its implementation. However, a solid legislative framework supporting HLP rights is indispensable if these rights are to be subject to full respect and protection, in particular when they are affected by climate change. Good HLP law is a vital adjunct to improving the HLP rights circumstances faced by climate change displaced persons, and when Governments awaken to this fact and work together with civil society to build legislation and policy that reflects the popular will and basic human needs of the population and addresses the full range of HLP concerns, then and only then, can societies everywhere begin to experience the reality of what the promise of HLP rights means in everyday life.

3. Forced Displacement Caused by Climate Change

While identifying the precise degree to which climate change may be the cause of particular instances of displacement remains challenging depending on the country and the context, it is clear that displacement due to the effects of climate change will – and indeed already has – affect scores of countries and tens, perhaps hundreds, of millions of people. The very real and extensive displacement and other human rights implications of climate change and the rights-based solutions that will be increasingly required in coming years as adaptation, internal relocation and international resettlement become ever more necessary clearly require greater attention by the international community and those Governments which are most heavily affected.

It is often argued that the spectre of permanent, non-reversible displacement caused by climate change and rising sea levels is a phenomenon that has yet to be clearly defined enough for States and their people to enable them to take the measures required to secure the long-term HLP rights of everyone affected by climate-induced displacement. And yet, climate change-induced displacement will almost assuredly become the most dramatic displacement disaster – ever. Already, Papua New Guinean atolls such as the Carteret, Morelock, Tasman and Nugeria Islands, States such as Tuvalu, Kiribati, Vanuatu, the Maldives and Bangladesh and others have begun to permanently resettle people because of land lost to rising seas, subsidence and salinisation of fresh water supplies. According to a recent 2009 report. The Anatomy of a Silent Crisis: Human Impact Report Climate Change, "an estimated 26 million of the 350 million displaced worldwide are considered climate displaced people. Of these, 1 million each year are estimated to be displaced by weather-related disasters brought on by climate change". Clearly, these and other cases are only the beginning of what will inevitably become an ever larger problem. And yet, very little has been done to date to address these displacement dimensions of climate change, including adequately addressing the HLP issues concerned. The International Panel on Climate Change's (IPCC) highly influential Fourth Assessment Report, for instance, simply notes that "...adaptation is occurring now, but on a limited basis", and otherwise the report almost entirely ignored this vital issue. Indeed, one commentator was sadly not wrong in claiming that "[T]here has been a collective, and rather successful, attempt to ignore the scale of the problem".

Building on the links between human rights and climate change, if we focus solely on the question of forced displacement due to climate change for the moment, estimates range from a global total of 50 million to worst-case scenarios predicting that up to one billion people could face the loss of their homes and lands during the coming century. The first Assessment Report of the IPCC in 1990 estimated that by 2050 some 150 million people could be displaced due to climate change, a figure also echoed in the well-known Stern Report and by Norman Myers of Oxford University who also predicted that, as a conservative estimate, the number of people set to be displaced by climate change would be 150 million, a figure he later expanded to 200 million. Christian Aid estimates a higher figure of some 200–250 million people will face forced displacement. The International Organisation for Migration (IOM) estimates that eventually some one billion people could be environmentally displaced from their original homes and lands. In terms of likely displacement within particular countries, several stand out as particularly dire, including China (est. 30m displaced), India (est. 30m displaced), Bangladesh (est. 20m displaced) and Egypt (est. 14m displaced). Al Gore's book, *An Inconvenient Truth*, speaks of 20 million people being displaced in the Beijing area alone, with an additional 60 million who may be forced to move in Kolkatta and in neighbouring Bangladesh.

But what form will this displacement take? Forced climate displacement does not occur solely due to king tides or land loss due to rising sea levels. Rather, there are four primary types of climate-induced displacement: weather-related disasters, such as hurricanes and flooding; gradual environmental deterioration and slow-onset disasters such as desertification, sinking of coastal zones and possible total submersion of low-lying island States; increased disaster risks resulting in relocation of people from high-risk zones; and social upheaval and violence attributable to climate change-related factors. Each of these may cause people to involuntarily flee their homes and lands and thus be responsible for climate-induced displacement.

In terms of when displacement due to climate change actually takes place, this is likely to manifest in essentially five primary ways. These are:

1. *Temporary Displacement* – People who for generally short periods of time are temporarily displaced due to a climate event such as a hurricane, flood, storm surge or tsunami but who are able to return to their homes once the event has ceased, such as during a larger than usual king tide in Vanuatu;

2. *Permanent Local Displacement* – People who are displaced locally, but on a permanent basis due to irreversible changes to their living environment, in particular sea-level rise, coastal inundation and the lack of clean water and increasingly frequent storm surges. This form of displacement implies that localised displacement solutions will be available to this group of forced migrants, such as higher ground in the same locality. This would include dwellers along Bangladesh's coastline who flee to higher ground in the immediate vicinity.

3. *Permanent Internal Displacement* – People who are displaced inside the border of their country, but far enough away from their places of original residence that return is unlikely or impossible. This would concern a family displaced from one region of a country to another region in a country, for instance, from a coastline to an inland town or city, such as the ongoing resettlement from the Carteret Islands to the larger island of Bougainville in Papua New Guinea.

4. *Permanent Regional Displacement* – People for whom displacement solutions within their own countries are non-existent or inaccessible and who migrate to nearby countries willing to offer permanent protection. This would involve, for instance, a citizen of Tokelau or Tuvalu migrating on a permanent basis to New Zealand.

5. *Permanent Inter-Continental Displacement* – People for whom no national or regional displacement solutions are available, and who are able to receive the protection of another State in another continent, such as a Maldivian who migrates to London.

Each of these five categories will have different policy and legal implications for Governments, the people concerned and whatever international agencies that may be assigned to assist climate change displaced persons to find durable solutions to their plight. Such responses, which can initially be understood in terms of short- and long-term options, have important ramifications for those affected and for those involved in ameliorating the emerging displacement crisis caused by climate change.

Problematically, the record of treatment thus far faced by those who have arguably already been displaced due to climate change does not bode particularly well for the millions yet to be displaced. Of the most well-known cases of what are seen as climate change-related displacements, none have thus far very successfully resettled those displaced, and in virtually all instances it is clear that the HLP rights of those affected have clearly not figured centrally within the remedial policies pursued to date. Community-driven initiatives and emerging alliances of climate change displaced persons are making inroads in some countries, particularly Papua New Guinea and Bangladesh, but the approaches of most Governments to date to climate-induced displacement have been inadequate.

Indeed, the following four brief case studies strengthen the view that unless Governments in countries that generate climate-induced displacement fundamentally improve law, policy and practice in this regard, it is clear that initiatives driven by climate displaced people themselves will be required for rights-based solutions to climate displacement to become part of official national and international strategies designed to protect the HLP rights of climate displaced people. They also show that thus far no explicit use has been made of HLP rights as the normative basis for any of the initiatives presented.

Papua New Guinea – The Integrated Carterets Relocation Programme of Tulele Peisa

The resettlement process that is currently underway from the Carteret Islands in Papua New Guinea to the much larger island of Bougainville (also in PNG) is one of the first organised resettlement movements of climate change displaced persons. When the national PNG Government and the Autonomous Provincial Government of Bougainville decided several years ago to resettle those from the Carterets and other atolls to Bougainville, many expected the relevant governmental bodies to effectively manage this process, by identifying and allocating sufficient land on Bougainville to resettle those fleeing their atolls. After a frustrating period of inaction by the Provincial Government in Bougainville to find durable HLP solutions for the 3,000-strong population of the Carterets – which includes the still unexplained non-expenditure of 2m Kina (+/- US$ 670,000) allocated for these purposes under the national PNG budget – the community-driven initiative Tulele Peisa was founded with a view to actually finding HLP and related solutions for those to be displaced.

The Integrated Carterets Relocation Programme of Tulele Peisa offers unique policy and planning lessons for other resettlement plans in other countries. Tulele Peisa ("Riding the Waves on Our Own") is led by the dynamic Ursula Rakova from the Carteret Islands, and has set out to find permanent housing, land and property solutions for the population of the Carterets on nearby Bougainville Island. The work of Tulele Peisa is truly path breaking and worthy of close inspection by anyone concerned with finding long-term HLP solutions for climate change displaced persons. Working against the odds and with very limited financial resources, Tulele Peisa thus far has been able to amass some 300 acres of land on Bougainville, most of which has been donated by the Catholic Church for the purposes of resettling a portion of the Carteret Islanders. More land is obviously needed, but an important start has been made in developing the methods required to provide sustainable HLP solutions to the atoll dwellers.

With more than 96% of Bougainville's land mass governed by customary land ownership, finding available land for the purposes of resettlement has proven extremely challenging. There are many problems associated with the identification of land for resettlement on Bougainville, however. Most of the land is subject to claims by customary owners or privately owned, with less than 3% held by the government in PNG. The Carteret Islanders do not have the financial resources to buy land themselves for the resettlement process, while the Government also lacks the political will to either purchase or expropriate land. Based on the experience of an earlier resettlement process that had failed in the area, the Islanders feel it is vital that sufficient land will be allocated to each family to enable them to earn a livelihood to ensure that the resettlement will be sustainable. This has led to their conclusion that 5ha per family would be required, in order to provide sufficient land for farming cash crops.

There are also a number of obstacles regarding obtaining clear legal title to land in Bougainville. Most land was subject to competing claims by customary landowners, and establishing clear title was a complex (and often unclear) process. To date, the Carteret communities have only been granted land owned by the Church, but they had also entered into direct negotiations with traditional landowners as they wanted to maintain good relations and integrate with local communities upon resettlement. They still do not, however, have clear title to the land on which resettlement would take place.

Demand for resettlement from Carteret Islanders has increased since the process began due to the fact that the impact of climate change on the atoll is becoming ever more apparent. In 2006, only 3 families wanted to resettle, whereas in 2008 some 38 families wanted to move. In all, some 300 families or more will ultimately need to be relocated as the Carterets disappear into the sea. The identification of land for the resettlement process remains one of the key challenges that needs to be addressed. In terms of the resettlement criteria developed by the Carterets community, some 1,500 hectares of land will be required to accommodate all 300 families (5ha per family). So far, the Catholic Church has provided some 80 hectares for resettlement, with 5 families already relocated to date. The process is in its very initial stages, with some 1400 or so hectares still required to accommodate everyone in the long-term.

The logistics of the relocation process developed by Tulele Peisa first involves a number of steps on the atoll itself. Initially the Council of Elders was mobilised and the relocation plans discussed and approved. The plan was then put before the ABG and endorsed. Once the plan was approved, the group then set out to raise awareness of the issues throughout the islands comprising the atoll and developed a Task Force Committee which became the lead body responsible for elements of the resettlement process. Ceremonial preparations were then carried out, followed by the mobilisation of public and private resources. In terms of activities on Bougainville, the Carterets Integrated Relocation Plan involves a well thought out 14-step process, which when completed, leads ultimately to successful resettlement by those moving to Bougainville. The fourteen steps are: 1) Scoping out available land; 2) Identifying traditional land owners; 3) Negotiating with land title holders; 4) Engaging with landowners; 5) Exchange programmes; 6) Entering into land negotiations; 7) Carrying out social and resource mapping; 8) Planting gardens; 9) Identify families using objective selection criteria; 10) Prepare families for relocation; 11) Prepare host families for relocatee arrivals; 12) Building homes; 13) Moving families to the new resettlement sites; and 14) Exchanging traditionally valuable items such as shell money.

The resettlement experience thus far concerning the Carteret Islanders, even at this very early stage, presents a number of lessons for resettlement exercises in other areas where climate-induced displacement will manifest, including:

1. *The importance of land identification* – The identification and allocation of sufficient land for relocation purposes is central to resolving climate displacement. Much of the available land in Bougainville is privately owned or subject to claims by traditional landowners, leaving very little public land available to the government to earmark for resettlement. While the Catholic Church has provided 81ha of land for resettlement, this will provide only a small portion of the land reserves needed to resettle the entire population from the Carterets. This and all other climate change-related resettlement exercises will need to take a pro-active approach towards land allocation as a prerequisite for successful resettlement.

2. *The central role of the affected communities* – Government efforts in support of relocation have been stalled by the unclear political situation in Bougainville, as well as limited political will to devote the resources required to ensure successful resettlement. This, in turn, necessitated the emergence of Tulele Peisa, which again shows the vital role to be played by affected communities themselves in orchestrating their own resettlement plan.

3. *The need for sustainable and comprehensive resettlement* – Earlier resettlement programmes to Bougainville (and elsewhere) justified by reasons not related to climate change, failed due to the lack of livelihood opportunities for those relocated. Because the mere provision of a new house and garden is never sufficient to restore the lives and livelihoods lost as a result of involuntary resettlement, the comprehensive needs of those to be resettled have been structurally built into the plans of Tulele Peisa.

Much work remains to be done to find HLP solutions for the entire population of the Carterets wishing to resettle to Bougainville, but a start has been made driven by the population to be displaced. It is hoped that other communities throughout the world which will require resettlement will learn from this experience, and also that movements similar to Tulele Peisa emerge on the other three PNG atolls which are equally threatened by rising sea levels.

Tuvalu

As is well known, few of the countries worst affected by climate change are under as dire a threat as Tuvalu. The loss of land in Tuvalu is so severe and potentially catastrophic, in fact, that Prime Minister Apisai Ielemia issued a formal request to the Government in Australia in 2008 to cede to Tuvalu a small piece of territory for the purposes of re-establishing Tuvalu on a minute portion of what is now Australian territory and resettling the entire population of the country there. Australia did not support this request, but in response to the Federal Government's reluctance, and in an act of remarkable islander solidarity, representatives from the Torres Strait Islands in the north of Australia unofficially offered Tuvalu use of one of its islands to re-establish itself there. Could this be an option for Tuvalu or other islanders as things proceed to move from bad to worse? Aside from the fact that some 3000 Tuvaluans are already said to have departed the country on a permanent basis largely for economic and educational reasons, the population density in Tuvalu is over four times higher than Kiribati, which itself has a very high population density.

Unlike the atoll dwellers such as those in PNG which at least can be resettled to Bougainville (which, of course, is within the same country as the atolls), and a similar but less promising situation in Kiribati which, according to the government official responsible for climate change adaptation, sees its largest atoll of Kiritimati as "our version of Bougainville as far as resettlement is concerned", Tuvalu's 10,000 inhabitants have very limited domestic land options available to them. It is becoming increasingly clear that third country resettlement in all likelihood will become the only viable option available to the population should large-scale adaptation measures fail.

At the moment, however, neither Australia nor New Zealand has expressed a willingness to integrate the entire population of Tuvalu into their own territories, although both countries have immigration programmes in place for a small number of Tuvaluans each year. New Zealand's Pacific Access Category enables 75 residents from Tuvalu to immigrate to the country each year. Principal applicants have to meet a series of strict criteria, however, which will exclude large numbers of eventual climate change displaced persons from immigrating to New Zealand. The seasonal labour scheme to New Zealand, in addition to the PAC, will eventually allow 5000 workers from five Pacific nations to work within the agricultural sector. In neighbouring Australia, well over 200,000 immigrants settle in the country each year proving clearly the capacity of the regional superpower to incorporate large numbers of new arrivals every year, including those displaced due to climate change. The recently developed Pacific labour programme in Australia which entitles a small number of Pacific islanders to work in Australia within the agricultural sector is seen by many as a precursor to a larger plan down the road involving ever larger numbers of Tuvaluans to its shores.

Many options are under discussion now, but nothing is as yet clear about the future of Tuvalu as a nation and the collective future of its citizens and their HLP rights. Many pleas have been made by concerned Australians and New Zealanders urging the Governments of those two countries to assist Tuvaluans. One study entitled Our Drowning Neighbours: Labor's Policy Discussion Paper on Climate Change in the Pacific drafted prior to the current Labor Government taking power, proposes that Australia establish an international coalition to accept climate refugees and work at the United Nations to ensure international recognition of climate refugees, however, this has yet to become policy. As a result, the citizens of Tuvalu will in all likelihood migrate in increasingly large numbers from their country to whichever countries are willing to accept them. How the HLP rights they possess under Tuvaluan law and the degree to which their housing, land and property rights are accorded them within the territories to which they emigrate remains to be seen.

Kiribati

Kiribati is severely threatened by rising seas and in all likelihood, the country as a whole will be uninhabitable within a century from now. The Government has tackled this reality head on and in many respects Kiribati is well ahead of the curve in beginning to structurally address many of the displacement dimensions of climate change, many of which stem from severe overcrowding in the capital, Tawara. In 2005, the Government finalised an Integrated Land and Population Development Programme as part of the broader National Republic of Kiribati Climate Change Adaptation (CCA) Strategy which set in motion a series of actions designed to resolve climate-related challenges. In the plan, the Government plans to stablise the national population at 125,000 by 2025 through family planning programmes and large-scale inter-island relocation. The proposed population distribution will be 50,000 in South Tawara, 45,000 spread over the other islands in the Gilberts, and the resettlement of roughly 30,000 to the larger island of Kiritimati. Severe overcrowding in Tawara (with the area of Betio islet having a population density of nearly 8000 persons per square kilometre, giving it a population density similar to Hong Kong) were as much behind the plan as broader climate change concerns. Under the plan, squatters and other vulnerable groups in Tawara will be given incentives to move voluntarily to Kiritimati, an atoll which contains 70% of Kiribati's land mass, but which is located a full 2000km away from the capital.

While the Government of Kiribati has been particularly pro-active in addressing the severe climate-induced displacement crisis to come by focusing on resolving current overcrowding by adopting creative policy measures designed to promote domestic relocation to islands such as Kiritimati, it remains unclear to what extent these measures will be adequate over the longer-term. Problematically, Kiritimati is only four metres above sea level which means that people will eventually be forced to abandon even this island over the longer-term. The current measures will surely go some way towards alleviating overcrowding and growing slums, inadequate housing and landlessness (and do implicitly recognise the HLP rights of the populace), but these may turn out to be only interim measures depending on the speed with which the seas eventually rise.

Like Tuvalu, Kiribati benefits from the Pacific Access Category programme, with New Zealand accepting an annual quota of immigrants from the country. Although the PACs do not expressly mention climate change as a reason or basis for according migrant status to these groups, nor do they apportion responsibility for the displacement of these populations, the PAC is largely seen to be a precursor to a larger climate migration programme should this become inevitable. It may be in the form of regional and inter-continental resettlement that the ultimate durable solutions will be found to Kiribati's displaced population, but far more work needs to be done to develop the concrete plans required to secure the short- and long-term HLP rights of everyone in Kiribati requiring eventual relocation or resettlement.

Bangladesh

Although the Pacific and Indian Ocean island nations receive the bulk of attention in discussions linking climate change and displacement, few countries will actually be more affected than Bangladesh. Already severely affected by land scarcity, overcrowding and slums that grow by two million dwellers each year, and with half of Bangladesh's population living in areas less than five metres above sea level, the country has begun to witness climate-induced displacement across much of its coastline. According to one analysis, "In the severe climate change scenario, sea level rise poses an existential threat that would inundate 18 percent of Bangladesh's total land, directly impacting 11 percent of the country's population. Salt water intrusion from sea level rise in low-lying agricultural plains, along with other hazards, could lead to 40 percent decrease in food grain production and will increase forced migration to the urban slum areas". Estimates show that just a 1 to 2 degree increase in temperature would force physical dislocation of more than 35 million people in Bangladesh. The results of modelling longer-term changes in coastlines as a result of rising sea levels suggest that the Government may be required to support mass movements of coastal population, with perhaps one in every seven Bangladeshis displaced by climate change by the year 2050. In order to address this dramatic crisis, some analysts have argued that for any climate displaced person legal and policy regime to be effective, it must be tailored not to the needs of individually persecuted people but of entire groups of people, such as populations of a village, city, province, or even entire nation, as in the case of small island states. In this regard, some Bangladeshi NGOs have proposed that a new status of 'Universal Natural Person' should be accorded to climate change displaced persons, who in turn should be treated as permanent residents to the regions or countries that accept them.

The recent emergence of the Climate Refugee Alliance, a grouping of affected communities assisted by the Coastal Resource Centre, and the Equity and Justice Working Group are hopeful signs that more concrete moves are underway to find viable HLP options for those threatened with forced climate displacement. Among other things, the Climate Refugee Alliance has pressured the Government to set aside State land for the exclusive purpose of resettling what they are labeling as 'Climate Refugees'. The Alliance has begun to address questions of land purchase and acquisition and the development of community land trusts which may hold promise for the millions to be displaced in the country due to the multiple effects of climate change. Whether these measures will succeed in finding HLP results for all of Bangladesh's climate change displaced persons, however, remains to be seen.

4. Recommendations – The Quest for Housing, Land and Property Solutions for Climate Change Displaced Persons

The challenge in addressing the specific HLP rights requirements of climate change displaced persons lies in not only ensuring that a principled approach to these issues is accepted at legal and policy levels, but that this is given content and clarity in terms of guidance and support for particular communities. Extensive direction can be taken from existing human rights instruments, recent policy work, disaster-response and reconstruction efforts and the general accumulated international legal framework including on socio-economic rights.

For the world's growing population of climate change displaced persons to secure both durable solutions to their displacement and the full enjoyment of all elements of their housing, land and property rights – both prior to and following their displacement – a series of steps need to be taken now by both migrant-generating and migrant-receiving states to identify and implement the legal, policy and programmatic measures required to respect and protect the pre-existing human rights of climate change displaced persons.

It is important to reiterate that these measures must be grounded deeply in laws and values that are already in place within virtually all states; indeed, securing the HLP rights of climate displaced persons will not inevitably require the establishment of new legal regimes or institutional frameworks, although creating specific institutional frameworks for respecting and protecting the HLP rights of climate change displaced persons should be a feature common to all States facing climate-induced displacement. In many contexts, the legal and political tools required to support the implementation of these rights are already in place. In these cases, therefore, what is required will be an augmentation of existing law, policy and practice in recognition of the fact that the displacement consequences of climate change will continue to grow and that they will affect the enjoyment of basic human rights. In many settings all that remains to be achieved in this respect is simply generating the political (and economic) will to bring them to fruition, and to develop specific measures that are designed directly to assist climate displaced persons to enjoy the full spectrum of HLP rights.

Much can be done – now – by States and the international community to strengthen the prospects of climate change displaced persons, twelve of which are outlined below.

Promote Major, Rights-Based Improvements in HLP Law and Policy

Many of the HLP challenges posed by climate change can be effectively addressed within domestic settings by substantially improving existing HLP law and policy. This is particularly true in countries which may face mass displacement, but which will retain the vast majority of existing land mass. For instance, housing laws can be re-written to acknowledge that new land will be required for relocating climate change displaced persons. Insurance arrangements can be expanded to cover losses incurred due to climate change. New national (and regional) bodies can be developed to ensure that adequate protection and housing measures are available to all climate change displaced persons. These and related measures are inherently feasible and can be employed by all States to one degree or another. From a governance perspective, citizens need to be able to identify which specific domestic institutions are responsible for ensuring the rights of climate change displaced persons, and which international institution(s) should be responsible for assisting these victims of climate change. In determining this, climate change displaced persons as beneficiaries of existing HLP rights can reasonably be expected to have provided to them clear answers to several very straightforward questions, including:

- Where do I turn for social, financial and resettlement assistance?
- Which public institution is entrusted with enforcing, respecting and protecting my HLP rights?
- If I am displaced due to climate change, what rights do I have to a new home or new land?
- What laws and rules are in place recognising my HLP rights and how can I best seek to enforce them?
- If I lose my home or land due to climate change, am I entitled to compensation or reparations? If so, where do I find out how to access these remedies?

All Governments should have precise, rights-based answers to these and other questions that may be posed by climate displaced persons. Beyond the improved application of existing HLP law and State obligations to respect and protect them, the effective protection of the HLP rights of climate change displaced persons will require the existence of clearly defined institutions that can provide clear and sustainable answers to all of these questions wherever they may be posed.

Ensure Full and Genuine Consultation with and Participation by Affected Communities, Their organisations and Eventual Host Communities

In order to effectively respect and protect the HLP rights of the world's growing population of climate change displaced persons, mechanisms need to be developed that ensure full consultation and participation with affected communities at all levels, as well as between donors and affected local Governments to ensure the success of planning and response measures to address displacement aspects of climate change. Climate change displaced persons need to be treated as equals in the development of practical solutions to their displacement and require a central place at the negotiating tables where such solutions are identified. Affected communities should be assisted to develop organisations representing their interests and be provided with technical assistance and support, as required, in formulating their demands in a manner consistent with human rights laws. New bodies such as the recently formed Climate Change, Environment and Migration Alliance (CCEMA), which was established to help mainstream environmental and climate change considerations into the migration management policies and practice, should ensure that forced climate migrant communities and their leaders play a central role in the policy development process.

Carry Out High Quality, Long-Term and Rights-Based HLP Planning

Good planning, good institutional frameworks, good laws and good policies will all be required if the HLP rights of climate change displaced persons are to be taken seriously. Human rights law and the growing number of judicial decisions on HLP rights, in particular, (see the Grootboom Case in South Africa in 2000, for instance) indicate that planning is one of the most important roles any local or national Government can play in respecting and protecting HLP rights. Human rights laws require States not only to plan, but to carefully diagnose domestic human rights challenges, develop laws and policies adequate to address these and to ensure that remedies of various sorts are available to individuals and communities unable to or prevented from enjoying the full array of human rights protections. A number of Governments have already started. In early 2008, for instance, local councils in Australia were instructed to carry out comprehensive climate change planning exercises in all communities threatened by flooding and inundation. This and other such examples could act as good models for other nations wishing to successfully mitigate and adapt to the climate changes ahead. To ensure that such plans are adequate, these plans should build in the likely displacement dimensions from the start, and to identify – on a person by person and community basis – how the HLP rights of those affected will be addressed.

There is an urgent need for all States, both those affected by climate change and eventual resettlement and in-migration, to draw-up high-quality, long-term and rights-based plans to address displacement-related aspects of climate change. Human rights law provides a useful framework for such planning. Planning, in this respect, should focus on worst-case scenarios to ensure that adequate response mechanisms are in place, including, in particular the identification of adequate land resources and budgets for resettling those displaced due to climate change. Planning needs to bear in mind that additional problems such as the "lost home syndrome" that commonly affects those facing involuntary resettlement in other contexts, can be reduced if people are afforded appropriate resettlement programmes that fully recognise and integrate their legitimate human rights.

Encouragingly, some 32 national adaptation plans have been completed thus far. However, there is little evidence that land and displacement issues feature prominently in any more than a few of these adaptation plans. Further work is required to review these plans to determine the extent to which and how displacement and HLP issues and rights are addressed. National adaptation plans are vital ingredients in an overall national planning process that respects and protects the HLP rights of climate change displaced persons. Because the vast majority of climate-induced displacement will occur within the national borders of various nation states and not necessarily require international flight to third countries, it will be increasingly important to ensure that the HLP dimensions of forced climate migration find a central place within these national planning processes. As financial costs increase and the likelihood of adaptive success decreases as the effects of climate change worsen, vigorous attention and resources are needed now to appropriately address the HLP challenges facing the world's newly displaced population.

Encourage Land Purchase, Acquisition and Set-Aside Programmes

Land purchase, land acquisition and land set-aside programmes should be undertaken by all countries affected by climate-induced displacement. Such programmes can identify and isolate land parcels for future use by families and communities forced to flee their places of habitual residence because of climate change. States should begin now to review public land holdings and to select possible long-term resettlement sites that will be removed from the land market through land set-aside programmes. These are complex issues with innumerable dimensions, however, few Governments are structurally unable to at least begin the land identification process as a part of the planning process. Politically, it will be significant for small island States to be able to demonstrate that an attempt to resolve displacement issues locally has been made prior to lodging appeals for regional and other resettlement-based solutions. In this regard, States should take immediate measures to identify available land and other appropriate resources for the purposes of relocating and resettling climate change displaced persons, both those displaced internally, as well as those likely to seek resettlement in third countries.

Governments should be encouraged to review domestic legislation as it relates to questions of the compulsory acquisition of land in the public interest for the exclusive purpose of expanding land reserves for the eventual use of permanently resettling climate change displaced persons. Climate change-induced displacement will put immense pressure on urban areas and the slums that surround them, and without appropriate adaptation measures, including the isolation of targeted land reserves for resettlement, the world's slums will grow at a far faster rate, and in turn create health, social, economic and other crises far worse than many would now predict.

To assist in reducing these pressures, Governments everywhere should begin identifying unused land for possible acquisition and then allocation to climate change displaced persons and their communities in a non-discriminatory way should this become necessary. Where appropriate, additional measures to release land for the relocation or resettlement of climate change displaced persons could be carried out through targeted land purchase programmes and the development of community land trusts on such land. Governments should be encouraged to establish land funds and/or to provide support to civil society land funds. Where possible, methodologies should be developed to ensure that climate change displaced persons are involved in such purchases to ensure direct buy-in by these stakeholders. The proposed Papua New Guinea and South Pacific Evacuation, Migration, Protection, Integration and Reconstruction Fund is one model that should be studied by those engaging in land purchase initiatives.

Promote Community Land Trusts

It is clear that identifying new land (and housing) resources lies at the core of the bundle of durable solutions required to simultaneously resolve climate-induced displacement and protect the HLP rights of those affected. To maximise the utilisation of land for climate change displaced persons, efforts should be made to ensure that such land is placed within community land trusts to ensure that such land remains held in common over time. This will ensure that land allocated in this manner will be available for use by new climate change displaced persons as migrants from earlier periods eventually move to new locations elsewhere. Approaches to resettlement on new land that are based exclusively on individual property title approaches are unlikely to provide the basis for community development and infrastructure development required to secure the full spectrum of housing, land and property rights of the communities concerned.

Prevent the Dangers of Poor Resettlement and Strive for Best Practice Outcomes

Given the poor global track record in protecting the HLP rights of affected populations during processes of involuntary resettlement, every effort must be made to prevent the many dangers of poor resettlement. There now exists considerable experience with resettlement around the world which provides valuable lessons in what works and what does not. The need for sustainable resettlement and the prerogative of reconstituting societies in a rights-based way, and not simply building new houses, should be drawn upon to ensure that experiences with resettlement caused by climate change do not lead to the same type of poor results. Some of the key issues necessary for successful resettlement include appropriate site selection, settlement design which was socially and culturally appropriate rather than being driven merely by economic factors, culturally appropriate housing, and community participation in the planning process.

To a limited degree, guidance for grappling with displacement due to climate change can be gleaned from the lessons learned in resolving displacement caused by natural disasters, however, these are premised on the notion that returning home will invariably be the remedy sought by the overwhelming majority of those displaced by disaster. Natural, manmade and environmental disasters including earthquakes, tsunamis, storms and floods and others always result in the destruction of housing that, in turn, invariably results in the large-scale displacement of people from their homes, lands and properties. In many settings, those displaced choose to return home once conditions so permit, and quickly begin the long and difficult task of rebuilding their former lives. A number of important lessons appear to be increasingly recognised by those working in post-disaster contexts, which may be of assistance in guiding thinking on how best to grapple with the displacement dimensions of climate change. Many of these are found within the useful Protecting Persons Affected by Natural Disasters – IASC Operational Guidelines on Human Rights and Natural Disasters, which were released in 2006. For instance, best practice indicates that all displaced persons should have the right to voluntary return (housing, land and property restitution), without discrimination, to the homes from which they were displaced whenever this is not materially impossible. The fact that in situ rehousing efforts have proven the most efficient and effective means of providing relief to victims in other post-disaster settings is clear. International standards now support the rights of disaster-affected populations to return to and recover their former homes and lands should they so wish. Many now appreciate that measures should be taken to remove any discriminatory inheritance and property ownership laws that may prevent the equitable transfer of property to survivors, particularly women and girls, and should ensure that women and girls do not suffer direct or indirect discrimination as a result of the relief and reconstruction efforts and that all reconstruction efforts take fully into account the needs of especially vulnerable or marginalised groups including ethnic minorities, children, the elderly, the disabled, the chronically ill and households headed by single parents or children.

When resettlement is the only option available and all other avenues have been considered, there is growing acceptance of the principle that permanent relocation should never result in homelessness, and that alternative accommodation which complies with international human rights standards on adequate housing should be provided to everyone as a matter of rights. In many respects this is the crux of the matter as far as the housing, land and property rights of climate change displaced persons is concerned. For in contrast to most other cases of displacement where the primary objective is generally return and the restoration of HLP rights, in the case of forced displacement due to climate change, internal relocation or international resettlement may often be the only remedies available to a family or community whose present homes are no longer viable in residential terms. Forced climate displacement, therefore, forces policy makers and human rights advocates into the comparatively novel position of supporting the relocation and resettlement of people as a durable remedy, in lieu of restitution or local integration options.

The Representative of the Secretary-General on Human Rights of Internally Displaced Persons, Walter Kaelin, has backed this point of view, indicating that a person who cannot be reasonably expected to return to his/her place of habitual residence should be considered a victim of forced displacement and be granted at the very least a temporary stay within safe third countries. This is, of course, a minimalist approach, and others have urged receiving states to provide climate change displaced persons with the full spectrum of rights enjoyed by refugees and, to the maximum possible extent, citizens of the country concerned. Under human rights law, climate change displaced persons that are forced by circumstances beyond their control to move across an international border are to be ensured general human rights guarantees in the receiving State, but do not generally possess rights to enter that State.

Mobilize National Financial Resources for HLP Rights Now

Although all Governments will be hard-pressed to allocate the financial resources required to address the needs of climate change displaced persons, efforts should be made now to mobilise national resources for these purposes. While there remains a dearth of international resources that have thus far been devoted to the impact of climate change on affected communities, most national Governments have done just as poorly in earmarking funds to protect the rights of climate change displaced persons. Given the extensive costs associated with addressing displacement-related aspects of climate change, ranging from the allocation of land and housing for resettlement, to compensation and skills training for those relocated, there was a need to work now to acquire and allocate funds for these purposes.

Strengthen the Capacity of the UN Adaptation Fund

Beyond new institutional frameworks and new standards, there can be no doubt that much more needs to be done to augment the Global Adaptation Fund. Because climate change will test the very concept of Government control over and management of territory and will very likely lead to the prospect of multiple State failures, with the burden of displacement falling disproportionately on the poorest of the poor who can never afford proper forms of mitigation or adaptation, now is not the time for the wealthy world to be stingy or cheap. We need to recall that the sentiments of international law are closer to the view that requires the protection of victims of climate change as may be thought. One standard, for instance, clearly indicates that "Victims of natural disasters, people living in disaster-prone areas and other groups should be ensured some degree of priority consideration in the housing sphere". It is clearly time to make this principle a reality through augmenting the UN Adaptation Fund and ensuring that it is directly available to climate change displaced persons and their organisations.

Expand Regional Migration and Labour Programmes

Pre-existing regional migration and labour programmes in New Zealand and Australia need to be expanded, and similar initiatives undertaken in other regions likely to face large-scale forced climate displacement. The Pacific Access Category in New Zealand is a laudable first-step in creating immigration options for a number of Pacific Islanders, but it is doubtful that this measure alone will be sufficient. Australia's more recent labour programme which provides employment visas to nationals from five Pacific nations to work within the agricultural sector is equally an important first step in recognising the eventual need to open resettlement to ever growing numbers of Pacific Islanders, however, it is clearly only a very partial effort towards finding larger rights-based solutions for all who are displaced in the region. These regional measures require re-appraisal and expansion both in terms of the number of annual beneficiaries, as well as in terms of the particular housing, land and property dimensions of these programmes. Other regions, most notably Europe, Asia and North America should also seriously consider similar regional migration and labour programmes.

Establish a Pacific Region Forced Displacement Solutions Initiative

The relative proportional scale of forced climate displacement in the Pacific region demands the presence of a regional institutional framework that can provide support to all of the Pacific Islands nations in the region – both climate displaced person generating and climate displaced person receiving – to ensure that all climate change displaced persons in the region are provided with the best possible HLP rights outcomes to their displacement. A body of this nature could, in principle, be coordinated by a range of institutions with reasonable track records in the region, most notably the UN High Commissioner for Refugees (UNHCR), the UN Economic and Social Council for Asia and the Pacific (ESCAP) or perhaps the UN Development Programme (UNDP). The Coordinating Body of the Pacific Region Forced Displacement Solutions Initiative would be entrusted with assisting all Governments in the region to develop 5, 10 and 25 year plans designed to best grapple with the looming displacement crisis in the region and to ensure Pacific citizens that their human rights will be met in full. Such an institution could be recognised as the primary regional body entrusted with implementing the broader responsibility to protect in the event that national governments are no longer capable of doing so.

The development of a country-by-country database on climate change-induced displacement can eventually be developed with a fair degree of precision, and if undertaken would play a vital role in the Pacific Region Displacement Solutions Initiative. At the same time, to date not enough is known specifically about each country in terms of precise numbers of people affected the size and scope of land thus far lost and likely to be lost; the variations in property law regimes and title issues; the resettlement, compensation and other policies that have been adopted to provide at least some measure of protection to these groups. Developing a country-by-country database outlining the precise displacement effects on each country and the corresponding needs to find durable solutions for all would assist greatly in the formulation of appropriate policy and legal decisions. Such processes would facilitate greater understanding of the attributes of ideal domestic, legal, policy and institutional frameworks designed to meet the needs and concerns of displaced and receiving communities, in an environmentally sustainable, economically viable manner consistent with international human rights standards.

Litigate Strategically and Carefully

Finally, another option open to both heavily-affected States as well as individual and community victims of forced climate displacement would be pursuing litigation at the national, regional and international levels in the pursuit of remedial action in response to their displacement. While it is by all means uncertain that such avenues will provide residential justice to those presenting such claims, cases of this sort may generate public attention leading to eventual policy and legal changes that benefit (or do not, as the case may be) climate change displaced persons. The Inuit Case (2005) at the Inter-American Commission on Human Rights sought to place blame on the United States for causing global warming, which in turn led to violations of the rights of the Inuit people. Some of the rights alleged to have been violated included central HLP rights such as the right to use and enjoy the lands they have traditionally occupied, to use and enjoy their personal property and rights to residence, movement and inviolability of the home. Although this case was never actually adjudicated, it did receive considerable media attention that led to ever vigourous debates as to the origins of many of the consequences of climate change. This vanguard case shows that litigation strategies may be of possible interest to climate change displaced persons to make certain points to the broader public, but it equally reveals the remedial shortcomings of cases of this nature. At the same time, new cases filed with the specific intent of strengthening the HLP rights position of climate change displaced persons may be of considerably more utility, and the idea of developing broader HLP climate litigation strategies towards this end has definite merit.

5. Conclusions – The Need for a New HLP Rights Global Social Contract

The outcome of the 2009 Copenhagen conference will have a massive bearing on the scale of future climate-induced displacement. It has been suggested, for instance, that should the meeting agree to cut CO_2 emission levels to 450 parts per million, instead of the much lower figures suggested by the global environmental movement, this will ensure that far larger numbers of people will be displaced than would have otherwise been the case. This would seem to raise human rights concerns as such a decision would knowingly and consciously create future circumstances that will intentionally violate the housing, land and property rights, as well as other rights, of affected communities. The world needs to be prepared for such outcomes that place HLP rights at the centre of policy and legal development in the coming years to ensure these processes are used in a way that protect rather than ignore the rights of climate change displaced persons everywhere.

All who are affected by climate change – particularly those who will be displaced – must be afforded rights and remedies that protect them, provide them with housing, land and property options consistent with their rights, and ensure them the lives and livelihoods that the essence of human rights and the laws and principles that comprise them have bestowed upon them by virtue of their very humanity, their very essence of being human beings sharing our wounded world.

[Selected resources removed]

Displacement Solutions, '*Climate Change Displaced Persons and Housing, Land and Property Rights: Preliminary Strategies for Rights-based Planning and Programming to Resolve Climate-induced Displacement'*, 2009.

Chapter 47

Displacement Solutions

THE BOUGAINVILLE RESETTLEMENT INITIATIVE
Meeting Report

> Forced climate migrants everywhere – and in particular those being displaced now in PNG – need to feel a sense of justice in the universe; without this we will be assured of living in precisely the type of world that we are seeking to prevent.

1. The above quote from one of the participants at the Bougainville Resettlement Initiative meeting held in Canberra, Australia on 11 December 2008 captures the sentiments of a meeting that was widely felt to be a unique, first of its kind, gathering on finding specific solutions for climate-induced forced migration. The meeting was convened to assist in facilitating resettlement solutions for forced climate change migrants from four atolls which were in the process of being inundated to the island of Bougainville in Papua New Guinea. In 2007 the national government of PNG and the Autonomous Bougainville Government agreed to resettle the 6000 inhabitants of the Carterets and three other atolls to the much larger island of Bougainville. Some 3500 Carteret Islanders and another 2500 island dwellers from three other nearby atolls (the Mortlock, Tasman and Nuguria Islands) will need to resettle on Bougainville due to increasing land loss, salt water inundation and growing food insecurity. The meeting sought to identify the components of a model resettlement process for forced climate change migrants from the atolls, building on the remarkable achievements of Tulele Peisa, an organisation formed to assist the displaced from the Carteret Islands. Tulele Peisa, meaning "Sailing the waves on our own", aims to "maintain our cultural identity and live sustainably wherever we are".

2. The resettlement process from the Carteret Islands to Bougainville is one of the first organised resettlement movements of forced climate change migrants anywhere in the world. As such, it offered unique policy and planning opportunities to learn the lessons of previous resettlement exercises unrelated to climate change, such as involuntary resettlement due to development and infrastructure projects, which have virtually always been unsuccessful. In convening the meeting, Displacement Solutions felt that the fact that displacement was already occurring in the Pacific and that requisite resettlement measures were already underway, would provide a solid basis for analysing the modalities of resettlement thus far and what lessons there might be for use elsewhere in a region where forced climate change displacement will dramatically increase in future. In bringing together for the first time the key stakeholders in the Carterets resettlement process,

namely representatives of the displaced population, officials from the Autonomous Bougainville Government entrusted with the resettlement process, landowners on Bougainville, representatives from AusAid and others involved in various human rights efforts within civil society, the meeting aimed to open political space on Bougainville and more broadly within PNG and Australia to work out the dynamics of a more sustainable and adequately financed resettlement plan.

3. Carteret Islander Ursula Rakova, the Director of Tulele Peisa, gave an in-depth and impressive presentation on the resettlement efforts that had already been made and the strategic planning processes made to guide these efforts. While the resettlement process had been approved by the Government with considerable support, the remnants of the conflict of the 1990s and subsequent political uncertainty regarding the future political status of Bougainville had resulted in a lack of political will, administrative capacity and the financial means required to take it forward. The communities threatened with displacement had therefore been obliged to take matters into their own hands and organise the resettlement process themselves. Thus, Tulele Peisa was born.

4. Demand for resettlement had increased since the process had started due to the fact that the impact of climate change on the Carterets was becoming ever more apparent. Rising sea levels and loss of land have, among other things, contributed to a decrease in food sources, leading to a situation where the Carterets – an atoll people have called home for more than 200 years – were now unable to sustain the entire population. The situation is deteriorating rapidly; in 2006, only 3 families wished to resettle, while in 2008 some 38 families – more than twelve times the amount of only two years earlier – expressed a desire to relocate to Bougainville. Tulele Peisa estimates that some 300 families (approximately 1750 people) will ultimately need to be relocated if the Carterets continue to disappear into the sea. Families with six or more children will be given priority for resettlement, together with families facing chronic food insecurity. Tulele Peisa hopes to retain a small population on the Carterets, but most of the population will need to resettle on Bougainville, including most of those with income earning potential, in particular 18–45-year-olds. Social mapping and population data collection is already underway.

5. The resettlement process developed by Tulele Peisa first involves a number of steps on the atoll itself. Initially the Council of Elders was mobilised and the plans discussed and approved. The plan was then put before the ABG and endorsed. Once the plan was approved, the group then set out to raise awareness of the issues throughout the islands comprising the atoll and developed a Task Force Committee which became the lead body responsible for elements of the resettlement process. Ceremonial preparations were then carried out, followed by the mobilisation of public and private resources.

6. In terms of activities on Bougainville, the Carterets Integrated Relocation Plan involves a well thought out 14-step process, which when completed, leads ultimately to successful resettlement by those moving to Bougainville. The fourteen steps are:

1. Scoping out available land
2. Identifying traditional land owners

3. Negotiating with land title holders
4. Engaging with landowners
5. Exchange programmes
6. Entering into land negotiations
7. Carrying out social and resource mapping
8. Planting gardens
9. Identify families using objective selection criteria
10. Prepare families for relocation
11. Prepare host families for relocatee arrivals
12. Building homes
13. Moving families to the new resettlement sites
14. Exchanging traditionally valuable items such as shell money.

7. Importantly, the organisation has been very conscious of possible resentment by host communities and has sought to promote inter-marriage between Carteret Islanders and Bougainvillians as one means of developing social cohesion and mutual respect.

8. According to preliminary estimates by Tulele Peisa, some 14 million Kina (US$ 5.3m) will be required between 2009–2019 to resettle all of those who wish to resettle on Bougainville. It was widely agreed by meeting participants that these funds should, to the maximum possible extent, be provided by the PNG national Government in accordance with their legal obligations towards the citizens of the country. Many of the meeting participants expressed strong views that the PNG Government should immediately earmark funds to purchase land on Bougainville for the purposes of resettlement and to provide adequate compensation to those forced to resettle. According to one participant, the fact that the PNG Government's annual budget totalled more than 7 billion Kina (US$ 2.7b) was indicative of the fact that at least a portion of the required funds was available domestically, and that this should be sought and utilised to assist the forced climate migrants from the Carterets and other atolls.

9. At the same time, many present emphasised the need for the international community, in particular the Government of Australia, to seek to identify funds that could be allocated to those involved in the resettlement process, including Tulele Peisa, to assist them in better achieving their objectives on behalf of the resettled population. While Tulele Peisa has already achieved a great deal with very limited financial resources, in order to achieve its longer term objectives of providing housing, land and properly solutions to all of those in need, additional funds will be required. An initial budgetary allocation of 2 million Kina (US$760,000) was earmarked for the purposes of resettling some of the Carteret Islanders, however, to date none of the funds allocated for this purpose were ever used for these purposes according to the official present from the Autonomous Bougainville Government. As a result, the funds were returned to the general budget unspent.

10. The identification of land for the resettlement process was the key challenge facing those needing to resettle. In terms of the resettlement criteria developed by the Carteret community,

some 1,500 hectares (ha) of land would be required to accommodate all 300 families (5ha per family), with an additional 1500 ha of land required for resettlement from the other three affected atolls. Tulele Peisa has developed a laudable land goal for each family which proposes that each resettled family receive land use rights over 5ha of land; 1ha would be allocated for housing and personal gardens, 3ha for livelihood purposes, including the growing of cocoa and copra, and the remaining 1ha set aside for purposes of reforestation. Based on the experience of an earlier resettlement process that had failed, the Islanders felt that it was important that sufficient land was allocated to each family to enable them to earn a livelihood so that the resettlement would be sustainable. This had led to their conclusion that 5ha per family would be required, in order to provide sufficient land for farming cash crops.

11. To date a total of 81ha of Catholic Church land near Tinputz has been identified on Bougainville for use by those to be resettled. This was seen as a good start, but still leaves a dramatic shortfall of more than 1400ha still to be identified and made available for the purposes of resettlement. Because more than 96% of Bougainville is governed by customary land rules and allotted using traditional land arrangements, an emphasis thus far has been placed on securing portions of the remaining 4% of the land which is divided between private owners and State land. The Carteret Islanders did not have the financial resources needed to buy land themselves for the resettlement process, while the Government also lacked the political will to either purchase or expropriate land. Five families have thus far been resettled.

12. There were also a number of obstacles to obtaining clear legal title to land in Bougainville which will make the continued acquisition of new land for resettlement ever more challenging. Most land was subject to competing claims by customary landowners, and establishing clear title was a complex (and often unclear) process. In many instances, there are four levels of land rights holders on Bougainville: traditional owners, the Government, the title holder and the user. Complex layers of rights such as these almost by their nature guarantee that crucial issues such as ownership, control and security of tenure need to be very carefully managed. To date, the Carteret communities have been granted land owned by the Church, but they had also entered into direct negotiations with traditional landowners to maintain good relations and integrate more deeply with local communities upon resettlement. They still do not, however, have clear title to the land on which resettlement would take place. The importance of establishing clear title to ensure security of tenure for all those resettled was emphasised by a wide range of participants.

UN Habitat and UNHCR referred to problems which had occurred in PNG and elsewhere in the world where this had not been resolved, leaving those resettled in an extremely precarious position. The need for ensuring the five indispensable elements of successful resettlement were also emphasised by several participants. These were: land, shelter, infrastructure, environment and livelihood. Ensuring these attributes of acceptable resettlement together with security of tenure and related housing, land and property rights were seen as important ingredients in developing an effective plan to assist the forced climate migrants begin life anew.

13. The representative of the Autonomous Bougainville Government (ABG), Kapeatu Puaria, made a presentation on the constraints faced by the local Government in implementing the

resettlement process. He indicated that "There is a decision and a policy, but there is no Government plan on resettlement". He acknowledged that the ABG had little, if any, capacity for socio-economic planning, and that the ABG had not conducted any planning for the resettlement of those affected by climate change in the Carterets or other outlying island atolls. He confirmed that land expropriation can only be carried out by the PNG National Government. The role of the national PNG government was, therefore, vital in ensuring political and financial support for the process. The uncertainty surrounding the political status of Bougainville in the wake of the ongoing peace process, with a referendum on independence envisaged in several years time, left the ABG with unclear powers and minimal funding from the national budget. Some issues may be resolved following impending elections for a new ABG, allowing it to review and hopefully improve its approach to resettlement. Mr. Puaria indicated that the discussions at the meeting would provide new impetus to the process, as the importance of planning for the human impact of climate change had become clear, and the ABG had learned of many new ideas about how to do this in a sustainable and equitable manner. In response, participants from donor governments indicated that they would be likely to support a new local government plan if it demonstrated that a clear strategy was in place.

14. The ABG official noted that there would also be a need to resettle Islanders threatened with climate change displacement on other outlying island atolls, including the Mortlock, Tasman and Nuguria Islands, but that this would be more difficult than for those from the Carterets. The Carteret Islanders were Melanesians, the same ethnic group as the population of Bougainville, and as a result had strong tribal and cultural ties with Bougainville. Those from the Mortlocks and Tasman Islands and a portion of those from Nuguria, however, were Polynesian which could complicate relationships with host communities and customary landowners. It was emphasised by several participants that carefully managing these cultural differences must form a central theme once resettlement from these atolls commences.

15. A large landowner from Bougainville who had previously offered to sell his sizable landholdings to the Government for the purpose of resettling the entire atoll population which is relocating to the island also made an impassioned presentation during the meeting, covering the political history of Bougainville and his own role in helping to end the violent conflict of the 1990s by involving the United Nations and other mediators. Bougainville's economy and infrastructure had been devastated by the brutal conflict, leaving the large plantations that had been a mainstay of the economy ransacked, and the port and airport destroyed. All of the labourers from mainland PNG had also fled, leaving local businesses without a workforce. While there was now interest in redeveloping land and restarting some of the plantations, vast resources would be required to do so. Due to the fact that he lacked such resources, Mike Forster offered to sell his 2700ha of land at the Raua Plantation – the second largest land allocation on Bougainville, to the Government with a view to using the land to resettle islanders from the four atolls. Unfortunately, he indicated that what could have been a win-win-win deal for all involved was unable to be closed with the local Government. As noted earlier, the national PNG Government had apparently earmarked money for the resettlement

process which could have been used to purchase land, but this money had ultimately gone unspent. Forster was therefore placed in a position where he had to sell the land, which had been in his family since 1923, to other private interests who plan to develop the land for agricultural purposes, and perhaps tourism. Due to his continued interest in ensuring that the Carteret and other Islanders were resettled in a proper way, he had sought assurances from the purchaser that some land would still be made available to them. However, the precise arrangement envisaged by the purchaser was not yet clear. Many of those present applauded the efforts of Mike Forster to offer his land for the purpose of resettling the forced climate migrants, and encouraged the new owners of the Raua Plantation and other private landowners on Bougainville to allocate at least a portion of their lands to resettle Islanders requiring new land sites.

16. The point was made by both representatives of the landowners and UNHCR, based on its experience of resettlement of refugees in the north of PNG, that in order to make resettlement sustainable, there was a need not only to identify and allocate land, but also to develop it together with the necessary infrastructure such as roads. This would also require significant resources and alternative sources of income. Landowner representatives argued that the Carteret Islanders should also consider the possibilities for those resettled to provide labour for some of the plantations that may re-open, or labour in other sectors that may emerge (such as tourism), rather than placing all of their emphasis on those resettled relying on the use of the land as the sole source of income. In this respect, one landowner representative felt that 5ha per family seemed to be somewhat excessive, and queried whether those being resettled could explore other food and income-generating activities such as fishing, the traditional activity of many Islanders. The representative of the Carteret community indicated that the 5ha figure was an ideal and that the critical point was to ensure that those resettled were also given the means to maintain a livelihood so that the resettlement would be sustainable.

17. A leading expert on resettlement, Professor Anthony Oliver-Smith, made a detailed presentation on lessons learned from various resettlement exercises that had been conducted around the world. Some of the key issues necessary for successful resettlement included appropriate site selection, settlement design which was socially and culturally appropriate rather than being driven merely by economic factors, culturally appropriate housing, and community participation in the planning process. Model "resettlement action plans" had, in fact, been developed over the years, which might be adapted for use in the Carteret/ Bougainville process to take it further forward. Other presentations by UN Habitat, the International Commission of Jurists and other organisations highlighted the need to address shelter, infrastructure, environment and livelihoods aspects. The fact that the PNG had recently ratified the International Covenant on Economic, Social and Cultural Rights, which placed obligations on the Government to conduct planning to avoid the homelessness which would arise from the sinking of the Carterets, might also be used to place pressure on the Government to devote more attention to the issue. It was felt that the Covenant could provide a useful framework for developing the best possible resettlement plan.

18. The mobilisation of financial resources was identified as critical, given that land had to be purchased from somewhere in order for it to be allocated to the islanders for the purpose of resettlement, that housing needed to be built and that new livelihood options needed to be developed. The representative of Tulele Peisa noted that under their current plan, it was hoped that 100 families could be moved by 2015. The only land available for this purpose would likely be privately owned and would therefore need somehow to be purchased. There was some discussion about the potential for funding being obtained from private philanthropy organisations, as well as the Global Adaptation Fund established to address climate change, although both were dismissed as unlikely sources.

19. Participants agreed upon a number of concrete steps that should be taken to take the resettlement process forward:

1. A concrete Resettlement Plan was clearly required in order to avoid the mistakes of the past, and also to avoid going back to square one every time more families from the Carterets needed to be moved;

2. A multidisciplinary fact-finding mission should be sent to Bougainville, the Carteret Islands and the PNG capital Port Moresby, comprised of various experts on resettlement, UN and NGOs, to meet with all relevant stakeholders from the community, Government, donors and the private sector, and assist in drawing up such a plan;

3. A key feature of the Resettlement Plan would be to cost the resettlement process, so that it could form the basis of resource mobilisation efforts;

4. During the fact-finding mission, discussions should also be held with the PNG Government about resettlement planning for the other three atolls where the process was expected to be more difficult; and

5. It was noted that the Carterets situation was a vanguard study in terms of being one of the first organised climate change resettlements, so the Resettlement Plan could be used as a model elsewhere.

20. Displacement Solutions Director Scott Leckie closed the meeting by again emphasizing the need for all forced climate migrants to feel a "sense of justice in the universe" and the importance of all participants and others to do whatever possible within their respective fields to ensure this. He again reiterated the commitment of Displacement Solutions to expand its efforts in support of forced climate migrants from the Carteret and other atolls and its willingness to work together with Tulele Peisa and others to ensure that this first large resettlement of forced climate migrants is carried out in a manner that takes both the rights and the needs of those to be resettled seriously. He closed by again stressing the need to treat this first large group of forced climate migrants as rights-holders as this will then set an important benchmark against which all other subsequent movements of forced climate migrants can be judged.

The Bougainville Resettlement Initiative: Meeting Report. Meeting convened by Displacement Solutions in Canberra, Australia on 11 December 2008.

Chapter 48

Tulele Peisa

CARTERETS INTEGRATED RELOCATION
PROGRAM BOUGAINVILLE

[…]

1. Executive Summary

The Carterets Integrated Relocation Program will assist 3,300 people affected by climate change and 10,000 people in three host communities over a 10 year period.

This proposal is a request for PGK 5,972,356 for three years from January 2009–December 2011. A significant portion of these funds is for land purchase in three locations on mainland Bougainville – Tinputz (5 families), Tearouki (20 families) and Mabiri (100 families).

Rising sea levels caused by climate change, have affected the Carterets Islands for over 30 years. Originally there were 6 atolls – but during the past 20 years one of the atolls has been split in half by the sea and there are now 7 tiny atolls. Situated 86kms to the north east of mainland Bougainville in PNG, the atolls are only 1.2 meters above sea level. The population of the Carterets is just over 3,300 people with the majority under 25 years of age.

In 1984 the Provincial Government of Bougainville resettled 10 families from Carterets to mainland Bougainville to relieve the land shortage caused by eroding shorelines. Since that time and despite many promises, very little has been done by the Provincial or PNG Government to assist the Carterets Islands in relocation efforts. Civil war shook the mainland for 10 years from 1989 and the Bougainville mainland still remains unsettled, particularly in the south.

However, the sea relentlessly continues to eat away more and more of the atolls. The Carterets people are now facing severe shortages of food due to saltwater flooding of gardens. Swamp taro, the staple food crop no longer grows on the atolls. Despite the construction of sea walls and planting of mangroves over many years to defend against the rising sea, Carterets Islanders state that more than 50% of their land has now been lost. Tired of empty promises, the Carterets Council of Elders (CoE) formed an association at the end of 2006 in order to organise for voluntary relocation of most of the Carterets population. The association was named Tulele Peisa, which means sailing the waves on our own. The CoE do not want to see Carterets Islanders become dependent on food handouts for their survival. They wish to see Carterets people remain as independent and self-sufficient as possible in a new location.

The Tulele Peisa vision is "To maintain our cultural identity and live sustainably wherever we are." and its mission is to relocate 1,700 Carterets Islanders voluntarily to three safe and secure locations on mainland Bougainville over the next 10 years.

The guiding philosophy is to encourage self-sufficiency and independence through all steps of the relocation process so that Carterets people and host communities do not develop a dependency or cargo mentality but take initiative and action to improve the quality of their personal and community life.

Tulele Peisa is a unique association. In the careful way that it is approaching the relocation process, it represents a community led model for relocation of climate affected communities elsewhere. The Carterets CoE understand that there will be many complexities involved in relocating people from their traditional homelands to integrate into existing communities that are geographically, culturally, politically and socially different. This is why Tulele Peisa will also work with the 10,000 people in three host communities where available land has been negotiated with the Catholic Church and private landholders in Tinputz, Tearouki and Mabiri — two locations in the north and one in Central Bougainville mainland. The Tulele Peisa program will ensure that these host communities will also benefit through upgrading of basic health and education facilities and training programs for income generation. Exchange programs involving chiefs, women and youth from host communities and the Carterets have already begun for establishing relationships and understanding.

Tulele Peisa will provide support, counselling and training to the Carterets people who have voluntarily agreed to relocate as well as with the older generation of Carterets people who are not yet willing to leave their traditional land. Training for adaptation to a different agricultural environment on the mainland to ensure food security and for new income generation opportunities form part of the Tulele Peisa program. Education and health infrastructure upgrades to cope with additional families using these services in the host communities as well community preventive health programs are also an integral part of the holistic approach that Tulele Peisa has developed for the Carterets Integrated Relocation Program (CIRP).

Maintaining links between the relocated Carterets people and their home atolls, sea resources and remaining clan members also form an important part of the program. Establishing a sea transport service for freight and passengers and to enable access to fishing grounds forms part of the CIRP. A marine conservation and management area will also be established to ensure that Carterets people continue to visit and use their marine resources in a sustainable way.

Organisations and programs rely on visionary and practical leadership for them to successfully achieve their goals. Tulele Peisa are lucky to have drawn Ursula Rakova back to her birthplace to be its Executive Director and advocate. Ursula was one of six women in PNG to receive a 'Pride of PNG' award in 2008 for her outstanding contribution to the environment over many years in Non-Government-Organisations in PNG and Bougainville.

Tulele Peisa has been the focus of many international media visits and the story of the Carterets Islanders has now been told many times at international forums to increase awareness of the plight of the Carterets as among the first climate change refugees in the world. The Carterets people now want to see this attention translate to practical assistance for their relocation plans.

[…]

3. History of the Group

In late 2006 the Council of Elders (CoE) of the Carterets Islands held a series of meetings to discuss the worsening effects of sea surges on their islands. The CoE were concerned that progress in establishing a relocation program for the Carterets people was going very slowly, while the erosion of their islands and the destruction of food gardens as a result of sea water surges was increasing at a very fast pace. As early as 2001, the Bougainville government talked about relocation for the Carterets, but as nothing had ever eventuated, the CoE felt the process was not going to happen unless the Carterets people took charge and actively found a solution to their worsening situation.

The CoE contacted Ursula Rakova, the Executive Director of a significant local NGO, (Osi Tanata in Arawa, Bougainville) and requested that she bring her skills and experience 'home' to the Carterets to assist. Ursula responded to the request. She resigned from her paid employment and travelled back to her birthplace in the Carterets. In December 2006, following a series of community meetings, the CoE decided to form a local NGO called Tulele Peisa, which in the local vernacular means "Sailing the Waves on Our Own". It was decided that the Carterets people needed their own indigenous organisation to plan and implement a voluntary relocation program for the next 5–10 years. Tulele Peisa therefore was set up to organise a well planned, staged program of moving Carterets people from their home atolls to the Bougainville mainland, where they could be safe and secure with access to economic opportunities as well as health and education services.

A constitution was drawn up and Tulele Peisa was incorporated (at that time as Carterets Relocation Program) in September 2007 under the Investment Promotion Authority in Papua New Guinea as an indigenous Carterets organisation. The registration number is Association No: 5-3210 and the Document No: is 00009232943. In May this year another form has lodged to change the name from Carterets Relocation Program to Tulele Peisa Incorporated.

[…]

Tulele Peisa goals

- Prepare and work with 3 host communities on the Bougainville mainland to ensure that there is adequate land, infrastructure and economic opportunities to include Carterets people within these communities.
- Prepare and work with 1,700 Carterets people for voluntary relocation to 3 host communities on mainland Bougainville through counselling, agricultural, income generation, education, health and community development training programs.
- Facilitate and establish a Conservation and Marine Management Area for ongoing sustainable use of marine resources for Carterets Islanders.
- Increase awareness at a national, regional and international level about the plight of Carterets people and the complexities surrounding relocation to new communities.
- Build the capacity of Tulele Peisa to carry out its objectives and develop Tulele Peisa as a resource agency and centre for Carterets and host communities on mainland Bougainville over the next 10 years.

Tulele Peisa objectives for three years from 2009–2012

- Tinputz, Tearouki, Raua host communities are welcoming to Carterets people because they have been included in all aspects of planning for additional people moving into their communities.
- Ensure there is sufficient land with secure tenure available for at least 120 Carterets families to establish themselves in each of the three mainland host communities.
- In association with the Catholic Church in Bougainville, provide design and carpentry services and local materials for basic housing for at least 120 families in the three locations.
- In association with the ABG, Catholic Church and Government Health and Education Agencies, ensure there is equal and adequate access to basic health and education services for Carterets people and their host communities.
- Assist Carterets people to overcome fear, anxiety and trauma associated with the need to leave their homeland.
- Facilitate Carterets people to contribute to the purchase of land through making of traditional shell money for sale.
- Facilitate training programs to build the capacity of the Carterets people to adapt to a different lifestyle from the coral atolls to mainland Bougainville – including for small-scale income generation activities.
- Facilitate support, training and resources for new income generation opportunities for the three host communities and for relocated Carterets people.
- Ensure there is a sustainable marine management plan in place for the Carterets.
- Carterets people are able to have ongoing access to their marine resources for local consumption and for sustainable sale of fish and marine resources.
- Carterets people have the skills and knowledge for ongoing management and sustainable harvest of their marine resources.
- Contribute to building an alliance of vulnerable Pacific communities affected by climate change who can lobby and advocate for justice and policies that recognise and support their needs.
- Advocate for a model and process for relocation of climate change an affected cultural group that considers the full range of complexities and needs of affected communities as well as their host communities.
- Build the capacity of Tulele Peisa to carry out its objectives and to continuously learn, change and adapt over time to ensure that it remains a resource agency for Carterets people and their host communities.

3.2 Summary of main activities undertaken 2007–2008

- Support and promote the Voluntary Relocation of the Carterets Islanders affected by rising sea levels.
  - In 2007 a positive partnership was established involving the Catholic Church of Bougainville and Tulele Peisa, which resulted in the church allocating 81 hectares of land to the Carterets Islanders. (The major hindrance for taking up this offer has been the lack of support by the Bougainville Administration's Division of Land to survey the land to enable the process of relocating 5 families to the area.)

- Since January 2007, TP has conducted 8 landholder negotiations in Tinputz for relocation of Carterets families. From these negotiations, there is starting to be a more positive response by communities in these areas to receiving Carterets families into the area.
- In Nov 2007, a social mapping exercise was begun to collect data about the social, cultural, economic, political and environmental lives of the Carterets and Tinputz (mainland Bougainville) peoples in order to find the points of integration for harmonious living between these cultural groups to better inform the relocation program activities.
- Coordinate and facilitate the sustainable well being of Carterets Islanders.
  - TP has engaged 2 male and 2 female facilitators from the Carterets to undertake small business and sustainable livelihoods training that can be shared with their own communities.
  - Continued land negotiations to find suitable land space for the relocation of 1,700 Carterets Islanders.
  - Community mobilisation and awareness in the Carterets Islands about rising sea levels, relocation, sustainable livelihood and economic options, etc.
- Increase awareness and advocacy on the Carterets Relocation Program at local, national and international levels
  - Eight field visits to Tinputz and Wakanai on the Bougainville mainland to raise awareness amongst local landholders about the rising sea levels on the Carterets and the impact on the people there.
  - In October 2007, after continued communication and pressure from TP and the ABG, the PNG National Government allocated K2 million towards the Carterets Relocation Program. However, until now these funds have yet to materialise so they can be used for the purposes of the relocation.
  - Tulele Peisa has hosted many different international media organisations interested in the impacts of climate change on the Carterets. A selection of these include:
    - A short film titled The First Wave made by the acclaimed independent filmmaker Pipp Starr. This movie can be viewed at www.starr.tv or on Tulele Peisa's website at http://www.tulelepeisa.org/.
    - Since this first documentary 14 other media, news network and freelance groups have visited the Carterets Islands to tell the world the Carterets story; these included Australia, New Zealand, Japan, France, England, Germany, United States of America and Denmark.
    - ABC Radio National Program: A Drifting Coconut: The Story of the Carteret Islanders, which can be heard at http://www.abc.net.au/rn/radioeye/stories/2008/2284109.htm.
    - A short film from the series called The Sisters on the Planet available on the Oxfam Australia website from early December 2008 www.oxfam.org.au/climate-change.
  - In September 2007 Bernard Tunim (from Piul Island in the Carterets) and Ursula Rakova toured 5 cities in Australia – Brisbane, Sydney, Newcastle, Canberra and Melbourne – to share their stories and experiences about the impacts of rising sea levels on Carterets. Friends of the Earth Australia and Oxfam Australia sponsored this tour. The other partners were Australian Conservation Foundation, Aid Watch and Green Grants of America. Also view on www.tulelepeisa.org.

- Ursula and two others from the Carterets travelled to Bali to attend the United Nations Conference on climate change in December 2007. The United Nations Forum & Conference on Climate Change (UNFCCC) was hosted by UN, attended by 115 countries with a participation of 15,000 people.
- In 2008 Climate Change awareness workshops were undertaken on five different islands on Carterets to show the local people on what has been done internationally; this includes explanations on climate change and screening of documentaries and other relevant educational materials.

[...]

4.2 Objectives of the Project: what does it aim to achieve?

Tulele Peisa and the Carterets Integrated Relocation Project (CIRP) exist to support and promote the voluntary relocation of the Carterets Islanders who are seriously affected by rising sea levels.

The overarching goal of the CIRP is to coordinate and facilitate a staged relocation of 1,700 Carterets Islanders over a period of 10 years to three disused plantation sites on mainland Bougainville – Tinputz, Tearouki and Mabiri. The Bougainville mainland is 86 kms south west of the Carterets by sea. There are some Carterets people (especially old people) who do not want to move despite the hardship they are experiencing with the inundation of the sea on food gardens and homes. The project will also work with the people who wish to remain on the islands in the next 10 years.

The project aims to achieve this relocation while considering the social, cultural, economic and political implications of moving an entire cultural group to locations with existing populations that have different cultural practices. The CIRP draws on models and learning from international experience in order to develop a local community-led model that will ensure the long-term sustainability of the Carterets relocation. The CIRP aims to enable the Carterets people to maintain links with their atolls through the establishment of a sea link to enable movement of passengers and freight and for returning to traditional fishing grounds.

The CIRP will be a staged relocation over 3 years – 5 families to Tinputz (where temporary housing is already built for families to establish gardens) followed by 20 families to Tearouki and then 100 families to Mabiri.

Key goals for Phase 1 from January 2009–December 2011

- Prepare and work with 3 host communities (Tinputz, Tearouki and Mabiri) on the Bougainville mainland to ensure that there is adequate land, infrastructure and economic opportunities to include Carterets people within these communities.
- Prepare and work with 1,700 Carterets people for voluntary relocation to 3 safe host communities (Tinputz, Tearouki and Mabiri) on mainland Bougainville through counselling, agricultural, income generation, education, health and community development training programs.
- Facilitate and establish a Conservation and Marine Management Area for ongoing sustainable use of marine resources for Carterets Islanders.
- Increase awareness at a national, regional and international level about the plight of Carterets people and the complexities surrounding relocation to new communities.

- Build the capacity of Tulele Peisa to carry out its objectives and develop Tulele Peisa as a resource agency and centre for Carterets and host communities on mainland Bougainville over the next 10 years.

4.3 Background: how were the needs identified?

The need for a relocation project for the Carterets Islanders was identified as early as 2000 with increasing damage to gardens and homes as sea levels continued to rise and inundate the atolls with increasing speed. Tulele Peisa has been formed by the Carterets Islanders themselves because of the need they themselves recognise to relocate most of their population to a safe and secure location on the Bougainville mainland.

4.4 Targeted group: who does the project aim at helping?

- Approximately 3,300 people from the Carterets Islands situated 100kms from the mainland of Bougainville
- Approximately 10,000 people from the Tinputz, Tearouki and Mabiri areas on the mainland of Bougainville.
- Other climate refugees and the international community, who may learn from this integrated, community-led relocation process and model.

4.5 Others working in this field: what other organisations are providing this kind of service?

There are no other organisations in PNG providing this kind of service. However Tulele Peisa will work with a range of local, national and international organisations to provide technical support for aspects of the relocation program. A selection for example:

- Nazareth Rehabilitation Centre will provide food security, organic gardening and home management skills to Carterets Islanders.
- Dumeba will also provide some aspects of food security and land use management.
- Bougainville Centre for Peace & Reconciliation (BCPR) will provide knowledge and skills on community and restorative justice as well as people skills to Carterets Islanders and host communities.
- Bougainville Community Health Program will provide training for community health programs.
- Bougainville Micro-Finance will assist with financial management training. The Nature Conservancy will conduct marine resources survey on the Carterets and provide technical support for establishing a conservation and marine management zone.
- Australian Conservation Foundation will provide ongoing support for reporting, learning and linking with international partners.

[…]

Tulele Peisa, Papua New Guinea, 'Carterets Integrated Relocation Program Bougainville', Project Proposal, 2008.

Chapter 49

Global Humanitarian Forum

KEY POINTS ON CLIMATE JUSTICE

Key Points on Climate Justice

1. Take responsibility for the pollution you cause
2. Act according to capability and capacity
3. Share benefits and burdens equitably
4. Respect and strengthen human rights
5. Reduce risks to vulnerable populations to a minimum
6. Integrate solutions
7. Act in an accountable and transparent manner
8. Act now!

Foreword

The impacts of climate change are being felt today in countries around the world. In some places, environmental changes such as prolonged drought and rising sea levels are threatening entire communities and even nations. If we don't take meaningful and farsighted action now to address climate change, we are not only failing those who suffer today. We are also putting at risk the well-being of our planet and future generations.

Many difficult issues must be faced to chart a more sustainable course and slow the current process of climate change to manageable levels. Government leaders will have to find ways to determine the appropriate distribution of responsibilities for climate change emissions. They will need to forge new agreements on the equitable use of the remaining carbon resources our planet can tolerate and ensure that adequate resources are available to support those forced to adapt to changes in the environment which threaten life and livelihoods.

Finding common ground will require enormous political courage. But making progress clearly requires new commitments and a willingness to move beyond short-sighted and narrow conceptions of national interest. The underlying premises of our efforts to date to address climate change have almost exclusively centred around notions of benevolence, charity or generosity. These remain important and instructive motivations for individual and joint action. But the dangers associated with climate change demonstrate that alone they aren't enough.

A momentum is building around an important idea that could help us find a more promising way forward. A movement around Climate Justice is growing.

Climate Justice suggests that the time has come to think more deeply about our conceptions of obligation and responsibility – not just within nations but also beyond borders. The starting point is to acknowledge the clear injustice of the fact that many decades of carbon emissions in richer parts of the world have led to global warming and caused severe climate impacts in the poorest countries. We must hold governments accountable for putting into practice well-established principles such as the requirement that polluters pay for the environmental damages they cause.

The concept of Climate Justice also recognizes that the world's poorest countries have contributed least to the problem of climate change and acknowledges that although we all have responsibilities to act, because the world's richest economies have contributed and continue to contribute most to the problem, they have a greater obligation to take action and to do so more quickly. That must include providing support to developing countries on a scale that not only ensures they avoid environmentally damaging economic development patterns of the past but also enables them to meet their current and projected energy needs.

Government leaders have acknowledged their responsibility to work together towards social justice and protection of the environment. They have signed treaties and declarations in which they agree to cooperate to protect the climate system and to ensure respect for fundamental human rights. Yet the challenge of climate change makes clear that we must define more precisely what international obligations entail, when they are triggered, and what factors condition our responses.

As a contribution to ongoing effort to shape a shared agenda for Climate Justice, the Global Humanitarian Forum has worked with a range of organizations and individuals to develop these "Key Points on Climate Justice." They reaffirm that the benefits and burdens associated with climate change and its resolution must be allocated in an equitable way. I commend this document as a valuable guide for all who are working to make Climate Justice a reality.

Mary Robinson, President and Founder,
Realizing Rights: The Ethical Globalization Initiative

Introduction

Increasingly unpredictable weather is having a severe impact on people and communities worldwide. We now know that climate change constitutes a serious humanitarian concern and a growing threat to socio-economic development, in particular for the world's poorest communities. These people, already the most vulnerable, contribute least to global emissions of gases that are the principal cause of climate change. They have the fewest mechanisms of protection available and the least means to cope.

However, the poor are not alone in facing the threat of climate change. If emissions of greenhouse gases worldwide continue to intensify, today's dangerous situation will become catastrophic for the entire planet and for all societies. Billions of people are now vulnerable to the indiscriminate impacts of climate change, and significant, immediate and sustained emissions reductions are an urgent priority. Those who already suffer or will suffer from the impacts of climate change, especially the poor, require support in order to adapt to environmental changes.

As the Universal Declaration of Human Rights affirmed more than 60 years ago, every human being is born free and equal in dignity and rights. Yet within the shared biosphere of our planet, the limited resources available to human society have been unequally consumed. The mass consumption of fossil fuels by some has led to a changing climate for the entire planet.

As a global community, we can act together. Efforts to ensure respect for human rights in the 21st century should include protecting the habitat in which human beings coexist. Ultimately, shared concern and commitment to the planet we inhabit and to our neighbours both near and far should be a unifying force.

The following "Key Points on Climate Justice" are intended to serve as principles to guide policy and action on climate change.

Key Points on Climate Justice

1. Take Responsibility for the Pollution You Cause

The polluter should, in principle, bear the costs of pollution."[1] The Rio Declaration on Environment and Development outlined this point nearly two decades ago, and today more than ever, when pollution from greenhouse gases linked to human activity is known to be the principal cause of climate change, polluters should pay for the damage they cause. Pollution should have a price that reflects the full cost of its impact on human society.

Studies by the Global Humanitarian Forum have highlighted how climate change is worsening major health problems, such as malnutrition and malaria, and claiming several hundred thousand lives every year. The Forum estimates the full cost of carbon per year to be in excess of US 100 billion dollars, including large-scale economic losses in the short, medium and long-term for communities around the world.[2]

The so-called polluter-pays principle is not new. It is a well-established concept that seeks to limit the environmental damage of any economic activity by allocating the costs of damage back to the polluter. In economic terms, prices for goods and services should reflect the true and full costs of their production and consumption, including the total costs of their health, environmental, natural resource, social, and cultural harms.[3] As long as these costs remain hidden, markets will react to distorted price signals, leading to inefficient economic choices.

The polluter-pays principle has moved progressively into the legal sphere, where it is now one of the oldest principles of international environmental law.[4] It was developed and has been applied mainly in response to cases of direct environmental hazard caused by toxic pollution.[5] However, the aim of the principle today is to encourage polluters to reduce their environmental impact rather than to sanction polluters after the fact. Effective dissuasive sanctions are necessary in order to trigger fundamental behavioural change among polluters. The cost of the likely sanction must exceed the costs of avoiding pollution.

With respect to climate change, polluters should be encouraged to end or minimize emissions wherever possible. Where this is not achieved, they should still be expected to compensate for any harm or negative effects locally and elsewhere caused by their contribution to overall emissions.

2. Act According to Capability and Capacity

Every actor can and should contribute to resolving climate change according to physical, economic, technical, political and intellectual capabilities and capacities. however, responsibility for the effects of climate change can only be attributed based on an ability to reasonably assume that responsibility. The poor cannot be expected to share the same burden as other groups, since a greater proportion of their existing resources are necessary for survival and a dignified existence.[6]

Many experts agree that in order to avoid catastrophic effects of climate change, global temperature averages must not rise more than 2 degrees Celsius (3.6 degrees Fahrenheit). Indeed, already today, at a temperature rise of less than 0.8 degrees Celsius, the impacts of climate change are substantial. This implies not only contraction (decreasing the amount of emissions in the atmosphere) but, on a medium time-frame, also convergence (people everywhere pollute to a similar, and for some significantly lower, amount). Within globally sustainable levels of greenhouse gas concentration, emerging economies and developing countries must aim at increasing development without reproducing the per capita emission levels of industrialized countries today. Likewise, the emission levels of industrialized nations will have to achieve aggressive reductions.

The world's 50 least developed nations account for less than 2% of greenhouse gas emissions. They lack the same responsibility as developed countries in creating climate change and resolving it. However, it is those very countries who suffer the most serious impacts to their socio-economic well-being and survival.

Ultimately, the aim is not to stop emitting greenhouse gases altogether —which is not, in any case, feasible – but to reduce emissions to globally sustainable levels. However, many countries are emitting below the sustainable level. These countries should not have to reduce emissions at present. The imperative for them should be to increase living standards, even if this means increasing emissions for a while. However, as a global goal, development should proceed as cleanly, sustainably and efficiently as possible. Access to clean and efficient technologies in the poorest countries is crucial to achieving global sustainability of greenhouse gas emissions.[7]

Thus, while overall emissions must contract significantly on a global level, it is clear that for the world's poorest, some amount of emissions growth will need to continue, at least in the medium term, to allow for a transition that does not jeopardize vulnerable populations. Not to do so could have severe impacts on the basic survival of poor communities around the world.

3. Share Benefits and Burdens Equally

As the equality of all human beings is a universally accepted principle, the benefits and burdens associated with climate change and its resolution should be allocated in an equitable way.

Those who have benefited and still benefit from emissions in the form of economic development and wealth, mainly industrialised countries, have the ethical obligation to share benefits and technologies with those who suffer the effects of these emissions, mainly vulnerable people in developing countries.[8]

Human activity is the principal cause of climate change, in particular through the emission of greenhouse gases and deforestation. Greenhouse gas emissions and deforestation are themselves largely by-products of economic and industrial practices that have driven economic growth in different parts of the world. In our carbon-based economy, wealth and high levels of emissions and consumption of resources largely correlate. In the same vein, poverty and low levels of emissions and resource use also correlate.

Economic development and social progress, as humanity has experienced them so far, are ultimately based on the unsustainable consumption of scarce resources. This process has led to the destruction of the natural environment and negative effects on human society, in particular through climate change. The impacts of climate change disproportionately affect the poor.[9]

The consumption of common goods and finite resources has generated positive outcomes, such as technological innovations, including for harnessing renewable energies. Many such benefits today are inaccessible to low-income communities because of intellectual property laws and other contributing factors, which only further entrench global inequities. Granting access to the latest green technologies is an appropriate way to share the benefits of development. Many low and high technology renewable or low-carbon solutions, such as solar-powered cooking appliances, are available. The full introduction of such technologies can greatly promote sustainable development and enhance resilience in the face of climate change.

4. Respect and Strengthen Human Rights

The global human rights framework provides a legal and ethical foundation for the vulnerable to seek support and redress. It also provides governments with a moral standpoint in climate negotiations. Human rights supply legal imperatives but also a set of internationally agreed values and thresholds around which common action can be negotiated and motivated. International cooperation for resolving climate change is vital for the respect and implementation of human rights.

Climate change is already undermining a broad range of human rights, such as rights to health and life; food, water and shelter; rights associated with livelihood, culture, migration and resettlement; gender equality; and personal security. The worst effects of climate change are likely to be felt by individuals and groups whose rights protections are already precarious.[10]

Rights protections are inevitably weakest in resource-poor contexts. But resource shortages also limit the capacity of governments as well as individuals to respond and adapt to climate change. Worse, where governments are poorly resourced, climate change will tend to harm affected populations unevenly, in ways that are de facto discriminatory.

The construction of an international climate change regime has rights implications. At one level, any strategy (or mix of strategies) that is applied globally will determine the long-term access of millions of people to basic public goods. At another level, choices made locally –

such as whether and where to cultivate bio-fuels or preserve forests – will affect food, water and health security and, by extension, the cultures and livelihoods of many of those people.

Adaptation policies raise human rights concerns – for example, if communities or individuals are forcibly removed from disaster or flood-prone areas, or expected to conform to new economic policy imperatives.[11] Adaptive interventions can reduce the likelihood that rights violations might result; interventions after the fact may provide redress where violations have already taken place.

In tackling climate change, states are obliged to attend to human rights risks and consequences. The great majority of states that are party to the UNFCCC are also signatories of the main human rights treaties.[12] In principle, therefore, whatever actions are taken to address climate change should conform to human rights norms as a matter of law.

Beyond this, human rights provide an ethical frame for approaching the problem of climate change and the construction and implementation of policy measures. A human rights lens puts the human person at the centre of analysis, not merely as a potential victim of climate change shocks, but as an active participant in constructing and implementing policy.[13]

5. Reduce Risks to Vulnerable Populations to a Minimum

Climate change involves highly complex and volatile risks because of the increased unpredictability and severity of weather. Given that climate change affects almost the entire spectrum of human society and sustainable development, it is crucial that risk analysis examines all possible implications of climate shocks, in order to minimize the scale of their fallout or escalation.[15]

Today, many indications point to growing risks for populations around the world as a result of climate change. The Global Humanitarian Forum's Human Impact Report noted that some 4 billion people live in areas vulnerable to the impacts of climate change. Some 500 million people over the next decade, however, will live in areas of extreme risk, including mass loss of life and livelihood. Vulnerable populations currently possess only limited adaptation and risk reduction protection. Political actors at the local, national, regional and international levels should make the reduction of these risks a priority.[16]

To tackle the climate crisis, bold decisions will be needed. Some of the decisions might themselves bear risks, such as those connected with structural changes of economies. However, risks should not prevent changing from a carbon-based to a low- or zero-carbon economy. Risk assessment should be done on a global level and on the basis that "business as usual" exposes the world's poorest to intolerable risks. As benefits and burdens should be shared in an equitable manner, risks should also be distributed according to responsibility and capacity to cope with them.

6. Integrate Solutions

An integrated approach is the only viable path to combating global climate change. That means an obligation to pursue multiple tracks in parallel. Emissions should be mitigated in order to stem the root cause of climate change. At the same time, adaptation is imperative for dealing with the unavoidable impacts of climate change. Transfer of technologies, knowledge and experience is necessary to achieve both mitigation and adaptation worldwide. And all these actions can and should be mutually-reinforcing, equitable, sustainable and in respect of human rights.[17]

Adaptation refers to actions that help human beings and natural ecosystems adjust to climate change. Adaptation alone, however, is not always enough. Runaway climate change can trigger impacts so large in scale, such as a massive sea-level rise, that adaptation is simply not possible. Mitigation therefore is crucial. Mitigation means actions that reduce net carbon emissions and limit long-term climate change while continuing development.

There are many cases whereby mitigation and adaptation can be carried out jointly, such as through reforestation projects or mangrove plantations, which may both act as carbon sinks and prevent disaster-exacerbating land or coastal degradation.

Transfer of technologies to poor communities to achieve a low-carbon economic transformation and to provide tools for effective adaptation is crucial. Many possibilities are already at hand and can serve several purposes. Solar cookers, for example, can free people from the burden of collecting firewood. Independence from firewood can contribute to stopping deforestation. An intact forest serves as a natural carbon sink; it prevents landslides and acts as a water reservoir. When looking for solutions, a holistic perspective should be adopted.

In some cases, efforts to tackle climate change can worsen inequities, do greater harm to the natural environment or infringe human rights. Increasing the cost of carbon to a level more reflective of its true cost to human society, for instance, could act as a regressive tax, raising the cost of basic goods and services needed for human survival. The additional financial burden could disproportionately penalize low-income communities due to increases in energy prices. Forestry sink projects may risk infringing the rights of indigenous communities who rely on the access and use of forests for socio-economic and cultural reasons. All such negative effects should be avoided, or where avoidance is not possible, minimized and offset with effective compensating policies.

7. Act in an Accountable and Transparent Manner

To address climate change, we need participation of people everywhere with fair, accountable, transparent, and corruption-free procedures.

Global solutions require a climate of confidence at all levels. On the individual level, each person should do his or her honest best to re-use, reduce and recycle, and should hold governments to policies and actions that do the greatest good for the greatest number. On the government and global levels, promises need to be kept. Procedural justice is necessary for those affected by climate change.

Participation of the public in the decision-making process is a widely accepted principle, both in human rights law[18] and international environmental law.[19] Participation increases accountability and transparency, and mutual accountability is a key value for national and international confidence between actors. States should not only facilitate public awareness and participation, but also encourage such participation. They should ensure that information is accessible to the public, comprehensible, and that people are aware of their right to comment and participate in the decision-making process.

Transparency, accountability, the rule of law and adequate access to information are basic principles for good governance and a democratic society.[20] Participation of the people, and particularly those likely to be affected, will lead to better decisions in the long term. The availability of information facilitates the effectiveness of environmental monitoring by the public, and enhances public understanding of environmental issues and policy goals.[21]

Vulnerable communities have too often been excluded from participating in the negotiation process on issues affecting their communities and lives. But even when included, these states have often suffered under a comparative disadvantage. They can usually afford few delegates to represent them at climate negotiations, where rich countries can provide dozens working on the issue.

Access to justice comprises a final pillar of participatory rights.[22] Bringing claims before judicial courts, for example, may help indigenous communities fighting for the protection of their environment.[23]

Transparent and corruption-free procedures to implement conventions, agreements and received support are an important part of mutual accountability, which builds confidence. Broken promises undermine confidence in political agreements and increase conflicts. Climate justice can only be attained through reliable, accountable and deliverable pledges, which reduce the gap between global aspirations and global actions.

8. Act Now!

Delayed action increases the danger from climate change, leading potentially to more damage and harm for people and communities, as well as for broader socio-economic development. The latest science points to extended delays, estimated at around 20 years, between actions to reduce greenhouse gas emissions in the atmosphere and any corollary impact on climate change. Delay can lead to higher costs because the economic and social damage caused will be higher, and therefore the costs to repair can explode. The science of climate change, as well as the most effective mitigation and adaptation strategies, are sufficiently known and available for implementation. We should take action now.

There is a limited timeframe in which net emissions must begin to fall globally to avoid catastrophic climate change. Current estimates range from around 2015 to 2020. Any later date for peaking of the consumption of fossil fuels would mean negative carbon emissions by 2050 and later. Technology for subtracting carbon out of the atmosphere is not available on an adequate scale at this time, nor can its future availability be considered a reliable basis for decision-making.

Human society is still far from reversing the trend of ever-increasing consumption and emission of greenhouse gases. Since coordinated action worldwide is an imperative for dealing with climate change, an international climate change agreement is a priority. However, any agreement realized at the 2009 United Nations Climate Conference in Copenhagen, Denmark will become effective only after 2012 and will most likely provide for a 2–7 year timeframe for enacting net global emission reductions.

Delayed justice is injustice. Tackling climate change is a highly complex endeavor, and the negotiations seeking international accord in this respect are widely considered among the most complex ever undertaken. Nevertheless, the actions that need to be carried out are straightforward: rigorous cuts in net emissions; promotion of sustainable development to reduce socio-economic vulnerabilities; investing in preventive measures against climate threats, including disaster risk reduction; and ensuring wide dissemination of relevant technologies, particularly those that are renewable.

The knowledge and means to tackle the problem are available today. There should be no further delay.

Global Humanitarian Forum, 'Key Points on Climate Justice', Working Paper, Geneva, 2009.

Notes

1. Principle 16, Rio Declaration on Environment and Development, June 1992.
2. Global Humanitarian Forum, Human Impact Report: Climate Change – The Anatomy of a Silent Crisis, Geneva, 2009. See also Stern, Nicholas, Stern Review: The Economics of Climate Change, Cambridge University Press, Cambridge, 2007; World Health Organisation, The Global Burden of Disease, 2004 Update, Geneva, 2008; World Health Organisation, WHO Malaria Report 2008, Geneva, 2008.
3. Pigou, Arthur Cecil, The Economics of Welfare, 2nd Edition, London, Macmillan, 1924. This principle, that prices should reflect the full cost of production and consumption, relates to the economic theory of externalities. Negative externalities refer to the production or consumption of goods and services that create negative effects – for society as a whole, or for groups or individuals – which carry costs not reflected in the price. For example, excessive use of fertilizers and pesticides, run-off of these into water, and over-abstraction of groundwater produce environmental damages that are economically quantifiable, but are not reflected in the price of pesticides, nor, in most cases, in the price of agricultural produce. In such situations, consumers benefit from market prices that do not reflect the true cost burden of a given good or service on society at large, and producers benefit from the sale of goods or services at an artificially deflated price.
4. At the 1992 Rio Conference, the polluter-pays principle was incorporated into Agenda 21 and Principle 16 of the Rio Declaration on Environment and Development, which states:
   "Internationalization of International Costs – National authorities should endeavour to promote the internationalization of environmental costs and the use of economic instruments, taking into account the approach that the polluter should, in principle, bear the cost of pollution, with due regard to the public interest and without distorting international trade and investment." The principle of responsibility for pollution caused between nation states is expressed in the 1992 United Nations Framework Convention on Climate Change (UNFCCC). Directly drawing from Principle 22 of the 1972 Declaration of the United Nations Conference on the Human Environment (also known as the Stockholm Declaration), the UNFCCC, which enjoys near universal ratification status by 192 countries, recalls in its Preamble: "States have, in accordance with the Charter of the United Nations and the principles of international law [...] the responsibility to ensure that activities within their jurisdiction or control do not cause damage to the environment of other States or of areas beyond the limits of national jurisdiction."
5. See Article 4, Charte de l'environment, La constitutionnelle N°2005, France: "Toute personne doit contribuer à la réparation des dommages qu'elle cause à l'environment, dans les conditions définies par la loi."; Article 225, Constitution of Brazil, 5 October 1988: "Procedures and activities considered as harmful to the environment shall subject the infractors, be they individuals or legal entities, to penal and administrative sanctions, without prejudice to the obligation to repair the damages caused.";

Article L110-1 II 3, Environmental Code, France, Act N° 2002-276, 27 February 2002: "The polluter-pays principle, according to which the costs arising from measures to prevent, reduce or combat pollution must be borne by the polluter;" Article 74, Federal Constitution, Switzerland, 1999: "(1) The Federation adopts rules on the protection of human beings and their natural environment against harmful or irritating effects. (2) The Federation provides for the fact that such effects are avoided. The costs of such avoidance and removal carry the causers […]".

6. The principle of "common but differentiated responsibilities" found in Article 3 of the 1992 UN Framework Convention on Climate Change (UNFCCC) is an expression of climate justice. Coupled with the polluter-pays principle, responsibilities are linked to capabilities and capacities. There is no right to pollute. See also the notion of the "common heritage of mankind" in a number of areas of international law, including the 1970 Declaration of Principles Governing the Seabed and the 1982 UN Convention on the Law of the Sea.

7. Internationally, recognition of the need to differentiate according to socio-economic parameters is reflected in a range of legal and normative documents. The 1972 Stockholm Declaration, Principle 23, emphasised the need to consider "the applicability of standards which are valid for the most advanced countries but which may be inappropriate and of unwarranted social cost for the developing countries." The 1992 Rio Declaration, Principle 6, states that "environmental standards, management objectives and priorities should reflect the environmental and developmental context to which they apply," that "the special situation of developing countries, particularly the least developed and those most environmentally vulnerable, shall be given special priority."

    Assistance should be provided to enable the low-carbon transformation of the poor and leave scope for some unavoidable medium-term expansion of emissions among this group. Efforts should be taken to avoid locking out the poor from enjoying modern forms of energy – a scenario that would have a dramatic negative impact on the socio-economic development of these groups.

8. Treat others as you wish others to treat you. This "Golden Rule" of reciprocity, based on the equality of all human beings and accepted worldwide, is one ethical reason for the sharing of benefits and burdens. Sharing benefits means: The benefits of the earth should be shared by all. If all human beings are equal, they have equal rights to benefit from and use common goods such as air, water and soil. There is no ethical reason why some should have more rights and lesser obligations. Burden sharing means: The burden of the effects of climate change should be shared by all. There is no reason why it should be heavier for vulnerable people who often contribute less to these effects. Climate Justice requires that the burdens should not be carried by those who have contributed least to the problem and are least able to bear the costs.

9. The situation contrasts with Principle 5 of the 1972 Stockholm declaration, whereby "the non-renewable resources of the earth must be employed in such a way as to guard against the danger of their future exhaustion and to ensure that benefits from such employment are shared by all mankind." Furthermore, if "[a]ll human beings are born free and equal in dignity and rights" as per Article 1 of the Universal Declaration of Human Rights, then every human being should also have equal rights to benefit from common goods that are shared among all humanity, such as air and water. So too, Article 29.1 of the same convention stipulates that "everyone has duties to the community in which alone the free and full development of his personality is possible". Thus, while every individual should enjoy the fulfilment of their inalienable rights, they also have a certain obligation to ensure the same is true for others, or not adverse to that.

    In reality, mainly for socio-economic reasons, access to non-renewable resources has been heavily restricted. A number of such resources have neared exhaustion. As a result of over-extraction and climate change, many common goods, such as water, are now severely depleted. Again, it is the world's poorest groups who are disproportionately affected.

    In recognition of the burdens of climate change, which may also include efforts towards resolving the problem, such as through emission reductions or forest preservation, as well as adaptation to climate change, Article 3.2 of the UNFCCC states that: "The specific needs and special circumstances […] of those Parties, especially developing country Parties, that would have to bear a disproportionate or abnormal burden under the Convention, should be given full consideration."

10.   This is partly coincidence. As it happens, the most dramatic impacts of climate change are expected to occur (and are already being experienced) in the world's poorest countries, where rights protections too are often weak. But the effect is also causal and mutually reinforcing. Populations whose rights are poorly protected are likely to be less well-equipped to understand or prepare for climate change effects; less able to lobby effectively for government or international action; and more likely to lack the resources needed to adapt to expected alterations of their environmental and economic situation. A vicious circle links precarious access to natural resources, poor physical infrastructure, weak rights protections, and vulnerability to climate change-related harms.

11.   Adaptation may be reframed as a compensatory or corrective response to potential or actual climate change-related human rights violations. Indeed, discussions of adaptation at international and government level (as opposed to autonomous local measures) already assume a rights basis for policy construction, even if it is rarely articulated in those terms.

12.   See the 1966 International Covenants on Civil and Political, and on Economic, Social and Cultural Rights; the 1989 Convention on the Rights of the Child; the 1965 Convention on the Elimination of All Forms of Racial Discrimination; the 1979 Convention on the Elimination of All Forms of Discrimination against Women; the 1990 Convention on the Protection of the Rights of All Migrant Workers and Members of Their Families; and the 1984 Convention Against Torture, and Other Cruel, Inhuman or Degrading Treatment or Punishment.

13.   Principles of transparency and accountability are vital in the creation of a regime that commands widespread support, is equitable in its outcomes and effective in its reach. The imperative to address the human rights consequences of climate change reinforces the UNFCCC's own imperatives of adaptation and technology transfer. By focusing attention on the protection of the most vulnerable, a human rights lens can further buttress the principles of equity and common but differentiated responsibilities that occur in the UNFCCC.

14.   Africa has a weather monitoring network for climate-related data that is some eight times below the World Meteorological Organization's minimum recommended standard. Therefore basic weather and storm forecasting and early warning services are lacking. Insurance coverage in developing countries, and in particular, least developed countries, is extremely low. For a graphical representation of microinsurance coverage in Africa and South East Asia, see http://knowledge.allianz.com/en/media/graphics/37/pdf (last visited September 2009). Scarce resources and barriers to technology mean that general risk management infrastructure, such as early warning and communication systems, are often rudimentary or totally absent. Further, as climate change alters weather patterns, traditional knowledge on which poor communities typically rely becomes unsuitable for dealing with weather and climate-related concerns.

15.   As is laid out in the Hyogo Framework for Action, which is the key instrument for implementing disaster risk reduction as adopted by the Member States of the United Nations: "The starting point for reducing disaster risk and for promoting a culture of disaster resilience lies in the knowledge of the hazards and the physical, social, economic and environmental vulnerabilities to disasters that most societies face, and of the ways in which hazards and vulnerabilities are changing in the short and long term, followed by action taken on the basis of that knowledge." The link between technology and risk management is outlined in Principle 18 of the 1972 Stockholm Declaration:

"Science and technology, as part of their contribution to economic and social development, must be applied to the identification, avoidance and control of environmental risks and the solution of environmental problems and for the common good of mankind."

16.   In addition to the immediate physical risks, special attention is needed for indirect consequences and secondary effects of risks caused by climate change such as violent struggles for resources, social conflicts of distribution of limited resources, psychic disorientation, despair and depression.

17.   Article 3.3 of the UNFCCC states "The Parties should take precautionary measures to anticipate, prevent or minimize the causes of climate change and mitigate its adverse affects."

18.   See Article 25, International Convention on Civil and Political Rights, December 1966.

19.   Principle 10 of the Rio Declaration states: "Environmental issues are best handled with the participation of all concerned citizens, at the relevant level. At the national level, each individual

shall have appropriate access to information concerning the environment that is held by public authorities, including information on hazardous materials and activities in their communities, and the opportunity to participate in decision-making processes. States shall facilitate and encourage public awareness and participation by making information available. Effective access to judicial and administrative proceedings, including redress and remedy, shall be provided."

20. Participatory rights consist of three pillars: the access to information, public participation and the right to justice. Article 1 of the 1998 Aarhus Convention makes it clear that the right of and access to information is not an end in itself but is a means to the greater goal of assuring "the right of every person of present and future generations to live in an environment adequate to his or her health and well-being." Moreover, the Convention requires States to act proactively to provide periodic reports on environmental risks and systematic updates, as well as making them available to the public. As per Article 7 public participation is equally important as it can influence the decision- making process:

     "Each Party shall make appropriate practical and / or other provisions for the public to participate during the preparation of plans and programmes relating to the environment, within a transparent and fair framework, having provided the necessary information to the public."

21. See Principle 19 of the 1972 Stockholm Convention: "Education in environmental matters, for the younger as well as adults, giving due consideration to the underprivileged, is essential in order to broaden the basis for an enlightened opinion and responsible conduct by individuals, enterprises and communities in protecting and improving the environment in its full human dimension. It is also essential that mass media of communications avoid contributing to the deterioration of the environment, but, on the contrary, disseminates information of an educational nature on the need to project and improve the environment in order to enable mal to develop in every respect."

22. Article 9 of the 1998 Aarhus Convention requires, inter alia, that States permit legal challenges to the substantive or procedural legality of any decision, act, or omission subject to the convention, as well as the enforcement of national environmental law.

23. See Case of the Mayagna (Sumo) Awas Tingni Community v. Nicaragua, Inter-American Court of Human Rights, 2002.

Chapter 50

Robin Bronen

ALASKAN COMMUNITIES' RIGHTS AND RESILIENCE

Forced migration due to climate change will severely challenge the resilience of communities forced to migrate as well as the capacities of local and national governments.

In Alaska, climate change is evident. Temperatures across the state have increased by between 2 and 3.5 degrees Celsius since 1974, arctic sea ice is decreasing in extent and thickness, wildfires are increasing in size and extent, and permafrost is thawing. These ecological phenomena are creating a humanitarian crisis for the indigenous communities that have inhabited the arctic and boreal forest for millennia. Four Alaskan indigenous communities must relocate immediately and dozens of others are at risk; meanwhile, government agencies are struggling to meet the enormous new needs of these communities.

The communities of Shishmaref, Kivalina, Shaktoolik and Newtok on the west coast of Alaska must relocate. The disappearance of sea-ice and sea-level rise are creating stronger storm surges that are eroding the land on which they are situated. These villages have active subsistence lifestyles and have existed on the coast of Alaska for thousands of years. Environmental studies indicate, however, that a catastrophic climatic event could submerge all communities within the next 15 years. There is no sustainable future for these communities in their present locations – and there is no higher ground to which they can move. Their only alternative is migration but, despite the consensus that these communities must relocate, no government funding has been specifically allocated to begin this process.

Each community is involved in an ad hoc process with state and federal government agencies that are struggling to provide protection to the communities while they grapple with the need to work out a relocation process. Government agencies have responded through their traditional methods of erosion control and flooding prevention but these adaptation strategies have proved ineffective in protecting the communities from a rapidly deteriorating environmental habitat.

The 2006 Alaska Village Erosion Technical Assistance Program – established by the US Congress – evaluated the different costs associated with erosion control versus relocation. It also identified a number of critical governance issues that need to be addressed if relocation occurs, noting that there is currently:

- no government agency with authority to relocate communities
- no funding specifically designated for relocation
- no criteria for choosing relocation sites
- no governmental organisation that can address the strategic planning needs of relocation and the logistics of decommissioning the original community location, including hazardous waste clean-up and preservation of cultural sites.

In 2007, the Governor of Alaska created the Alaska Climate Change Sub-Cabinet to implement a climate change strategy for the state. An Immediate Action Workgroup – an advisory group to the Sub-Cabinet – was tasked with identifying the short-term emergency steps that state government needs to take to prevent loss of life and property due to climate change in the communities that must relocate. Both state and federal government representatives co-chair the Workgroup; the multi-level governance structure is unique.

In April 2008, the Workgroup issued its recommendations, in which erosion control and community evacuation plans are central. The Workgroup also recommended that funding be allocated to communities to begin a relocation planning process. In recognition of the complex governance issues identified in the 2006 Alaska Village Erosion Technical Assistance Program report, the Workgroup recommended that one state agency lead the relocation effort and act as the coordinating agency with responsibility of maintaining federal, state and tribal partnerships. The report, however, does not detail the governance structure or jurisdictional authority that will allow the agencies to work together.

> We and our grandfathers have noticed that the water level has been rising, the seasons getting shorter, thinner ice, warmer winters, summers and shorter springs. The loss of land through erosive action and increasing risk to property and lives have caused a dangerous situation for the community of Shishmaref and the culture of its people. The only viable solution is to relocate the community off the island to a nearby mainland location that is accessible to the sea, suitable for the continued subsistence lifestyle of the community, and to preserve the culture and integrity of the community. The constant anxiety caused by the erosion is an excessive burden carried by all members of the community. The 'no action' option for Shishmaref is the annihilation of our community.
>
> (Tony A Weyiouanna Sr, resident, Shishmaref)

Newtok is the most advanced in its relocation efforts, having identified a relocation site and acquired the land through an act of Congress. The state planner facilitating the Workgroup is coordinating the work of the dozens of agencies involved with Newtok's relocation. She has no jurisdiction to require other agencies to join in her relocation efforts but federal and state agencies are working with the Newtok Traditional Council and willingly engaging in the relocation process. However, none of these agencies has a funded mandate to relocate communities endangered by climate change; there is no lead agency to create and coordinate a relocation strategy; and several of the agencies are bound by legal guidelines that throw up serious obstacles. For example, the Alaska Department of Transportation designated with the task of building airstrips and the Alaska Department of Education designated with building schools are unable to move forward with these projects at the relocation sites because regulations require that an existing community with a minimum population be at the site before any infrastructure is built.

The Newtok Traditional Council is a small local tribal government that has only limited capacity to coordinate the relocation work of dozens of federal and state agencies and administer and obtain funding needed for the relocation process.

The humanitarian crisis in Alaska clearly demonstrates the need to create clear principles and guidelines based in human rights doctrine that can serve as a model for other regions. These would help ensure that the social, economic and cultural human rights of individuals and the communities forced to migrate are protected during displacement as well as during resettlement. State and federal governments should be obliged to:

- allow the affected community to be a key player in the relocation process
- ensure culturally and linguistically appropriate mechanisms for participation and consultation
- ensure families and tribes remain together during relocation
- keep socio-cultural institutions intact
- protect subsistence rights and customary communal rights to resources
- safeguard rights to safe and sanitary housing, potable water, education and other basic amenities
- implement sustainable development opportunities as part of the relocation process (and thereby enhance community resilience).

Definition

An accurate definition of this displacement category is essential in order to ensure that the permanent relocation of communities only occurs when there are no other durable solutions. 'Climigration' has been coined as a word to describe this type of displacement. Climigration occurs when a community is no longer sustainable exclusively because of climate-related events and permanent relocation is required to protect people. The critical elements are that climatic events are on-going and repeatedly impact public infrastructure and threaten people's safety so that loss of life is possible. A definition is also critical so that the design and implementation of institutional frameworks of humanitarian response are appropriate. Agencies that have traditionally provided 'disaster relief' and erosion control, for example, will continue to engage in these activities until it is determined that relocation must occur in order to protect the life and well-being of the community. At this point, the community, along with tribal, state and federal governments, will shift their focus to create a relocation process.

Failure to recognise the signals of ecosystem changes will critically impede a community's capacity to adapt and may lead to social and economic collapse. Government agencies will also be hampered if they are unable to identify the early ecological warning signals requiring a community to relocate. Early indicators of community vulnerability may include: repetitive loss of community infrastructure; imminent danger; no ability for community expansion; number of evacuation incidents; number of people evacuated; predicted rates of environmental change; repeated failure of disaster mitigation measures; and viability of access to transportation, potable water, communication systems, power and waste disposal. The sooner a community and governmental agencies recognise that relocation must occur, the sooner all-important funding can be diverted from disaster relief to relocation.

In 2006, the Army Corps of Engineers built a new seawall to protect the community of Kivalina. The day after the dedication ceremony, a storm ruined a critical component of the seawall, leaving the community vulnerable and exposed. In 2007, the community was forced to evacuate when a storm threatened the lives of community members.

Strategies to temporarily evacuate the villages, rebuild public infrastructure and erosion control structures and then return the population to original locations no longer afford adequate protection. Permanent relocation is the only durable solution for Kivalina, as for other Alaskan indigenous communities. The experiences of these communities should be used to guide the creation both of principles that secure their human rights and an institutional response that ensures their safety.

Robin Bronen, 'Alaskan Communities' Rights and Resilience', *Forced Migration Review*, Issue 31, October 2008, pp. 30–31.

Index

9 780415 691345